Advances in Robot Kinematics

Advances in Robot Kinematics

Theory and Applications

Edited by

J. LENARČIČ
Institute Jožef Stefan,
Ljubljana, Slovenia

and

F. THOMAS
Institute of Industrial Robotics (IRI),
Barcelona, Spain

KLUWER ACADEMIC PUBLISHERS
DORDRECHT / BOSTON / LONDON

A C.I.P. Catalogue record for this book is available from the Library of Congress.

ISBN 1-4020-0696-9

Published by Kluwer Academic Publishers,
P.O. Box 17, 3300 AA Dordrecht, The Netherlands.

Sold and distributed in North, Central and South America
by Kluwer Academic Publishers,
101 Philip Drive, Norwell, MA 02061, U.S.A.

In all other countries, sold and distributed
by Kluwer Academic Publishers,
P.O. Box 322, 3300 AH Dordrecht, The Netherlands.

Printed on acid-free paper

Printed in the Netherlands.

The untimely passing of Professor Joseph Duffy, of the University of Florida, is an especially sad event for the community of Advances in Robot Kinematics. Joe Duffy was a good friend and a much respected and admired colleague. When he died on March 12, 2002, the community of Advances in Robot Kinematics lost one of the pioneers in robot kinematics. Joe Duffy distinguished himself as a very capable and product researcher, and as a kind and thoughtful human being. In the name of the Scientific Committee, I extend our most heartfelt condolences to his wonderful wife Anne, and to his coworkers and students at the Centre of Intelligent Machines and Robots.

Bernard Roth, Chair
Scientific Committee

Contents

Performance Evaluation

Design and Control of Special-Type Mechanisms

Redundancy, Singularity, and Self-Motion

Methods in Kinematics

Kinematic Design

Kinematic Analysis

Synthesis of Mechanisms and Motion

Preface

This is the fifth book of the Kluwer's series Advances in Robot Kinematics. The book presents the most recent research advances in the theory, design, control and application of robotic systems, which are intended for a variety of purposes such as manipulation, manufacturing, automation, surgery, locomotion and biomechanics. The issues addressed are fundamentally kinematic in nature, including synthesis, calibration, redundancy, force control, dexterity, inverse and forward kinematics, kinematic singularities, as well as over-constrained systems. Methods used include line geometry, quaternion algebra, screw algebra, and linear algebra. These methods are applied to both parallel and serial multi-degree-of-freedom systems. The results should interest researchers, teachers and students, in fields of engineering and mathematics related to robot theory, design, control and application.

Each contribution in this book had been rigorously reviewed by two or three independent reviewers and 53 articles had been recommended for publication. We are happy to observe that Advances in Robot Kinematics has always attracted the most outstanding authors and has developed a remarkable scientific community in the area. Many important and original scientific results were for the first time reported and discussed in these books. All articles in this book were also reported at the eight international symposium on Advances in Robot Kinematics that was organised in June 2002 in Caldes de Malavella in Spain. The symposium was organised by the Institute of Industrial Robotics (IRI), Barcelona (Spain) under the patronage of the International Federation for the Theory of Machines and Mechanisms.

We are grateful to the authors of the articles for their valuable contributions and for their efficiency in preparing their manuscripts in time, and to the reviewers for their timely reviews. We are also indebted to the personnel at Kluwer for their excellent technical and editorial support.

Jadran Lenarčič and Federico Thomas

Performance Evaluation

A KINEMATIC CALIBRATION OF IN -PARALLEL ACTUATED MECHANISMS USING FOURIER SERIES (EVALUATION INDEX FOR DETERMINATION OF THE SET OF MEASUREMENT PATHS)

Gang Shen and Yukio Takeda
Department of Mechanical Sciences and Engineering
Tokyo Institute of Technology, Japan
glenshen@stu.mech.titech.ac.jp, takeda@mech.titech.ac.jp

Hiroaki Funabashi
Department of Mechanical Engineering
Shibaura Institute of Technology, Japan

Abstract A DBB(Double-Ball-Bar) based kinematic calibration method that uses Fourier series for in-parallel actuated mechanisms is presented. Indices for evaluating the set of the measurement paths used in the calibration were derived, considering the error propagation characteristics of the error equation to be solved for identifying the kinematic parameters of the mechanism. A comparative study of these indices was presented, as well as experimental calibration results based on this study.

Keywords: Robotics, Kinematics, Calibration, Evaluation Index, Condition Number, Parallel Mechanism, Absolute Accuracy, Double-Ball-Bar System

1. Introduction

In-parallel actuated mechanisms (PMs) have advantages in terms of accuracy, operation speed, payload, rigidity and output power compared with serial mechanisms, which make PMs suited to heavy duty applications. In recent years, many researchers have been involved in research and development of high-speed machine tools by using PMs (Proc. PKM2000,2000). For such applications, high absolute accuracy and stiffness are required. We developed a PM-based worktable (Takeda, 1999ICAR, Takeda, 2000) for high-speed machining based on our own kinematic synthesis methods(Takeda, 1995, Takeda, 1996, Takeda,1999JRM, Takeda,1997). Currently our study is focused on improving its absolute accuracy via kinematic calibration.

J. Lenarčič and F. Thomas (eds.), Advances in Robot Kinematics, 3–10.

Various methods of kinematic calibration and improvement of absolute accuracy of PMs have been presented (Vischer, 1998, Ota, 2000, Patel, 2000). However, PMs of sufficient accuracy to be applied to machine tools have not been experimentally realized. For a precise calibration, the error propagation characteristics from the measurement error to the calibration result must be investigated. An appropriate measurement device for calibration must also be chosen. The error propagation characteristics depend on the combination of measurement poses used. Therefore, the index and algorithm for determining the optimal set of measurement poses should be developed. In the literature, some indices and algorithms for such purpose have been proposed (Driels, 1990, Borm, 1991, Zhuang, 1994, Hollerbach, 1996, Nahvi, 1996). However, their precision has not been experimentally confirmed. To minimize the measurement error in a calibration, the use of the measurement devices that directly measure the pose error or the part of the pose errors such as the Double-Ball-Bar(DBB) (Kakino, 1993), is preferable to the use of 3-D position measurement devices.

We have been developing a new method for kinematic calibration of PMs that uses the DBB system. We have already proposed a calibration method by means of Fourier series (Takeda, 2001). In the present paper, we discuss the evaluation index used to determine the optimal set of measurement paths in the calibration.

2. Kinematic Calibration Method by Means of Fourier Series

We use the DBB system shown in Figure 1 as a measuring device of the output displacement(pose) error for the kinematic calibration. In this measuring system, the change in the distance between the two spherical joints P_1 and P_2 can be precisely measured within a limited range when a circular path of the point P_2 on the output link is given as a desired path. The measured value is the error Δr_D of the actual radius r_D from the desired one r_{DO} with respect to the angle ϕ. Once the condition of the measurement path is determined from the desired radius r_{DO}, the center (the coordinate of P_1), the orientation of the output link and the coordinate of P_2, a set of data Δr_D for a circular path measured by the DBB is described as a function of ϕ, that is $\Delta r_D(\phi)$. We describe $\Delta r_D(\phi)$ in terms of an m-th order Fourier series (Takeda, 2001);

$$\Delta r_D(\phi) = a_{D,0}/2 + \sum_{k=1}^{m} \{a_{D,k} \cos(k\phi) + b_{D,k} \sin(k\phi)\}. \qquad (1)$$

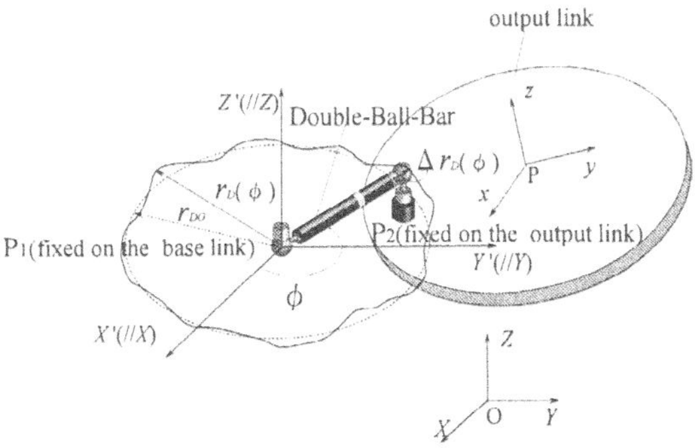

Figure 1. The DBB (Double-Ball-Bar) system

Here, the coefficients $a_{D,k}$ and $b_{D,k}$ are calculated from the measured data at n_m discrete points on the circular measurement path. Using these coefficients, a vector $\boldsymbol{G}_{D,n} = [a_{D,0}, a_{D,1}, \cdots, a_{D,m}, b_{D,1}, \cdots, b_{D,m}]_n^T$ is defined for each measurement path (path number : $n, 1 \leq n \leq n_s$). The relationship between an error e_j of kinematic parameter j and the corresponding error Δr_{ej} in the radius during the whole cycle of $\phi (0 \leq \phi \leq 2\pi)$ is also described as a Fourier series,

$$\Delta r_{ej}/e_j = a_{ej,0}/2 + \sum_{k=1}^{m} \{a_{ej,k} \cos(k\phi) + b_{ej,k} \sin(k\phi)\}. \tag{2}$$

Here, the coefficients $a_{ej,k}$ and $b_{ej,k}$ are theoretically calculated. Using these coefficients, a vector $\boldsymbol{g}_{j,n} = [a_{ej,0}, a_{ej,1}, \cdots, a_{ej,m}, b_{ej,1}, \cdots, b_{ej,m}]_n^T$ is defined for each measurement path.

Using all the data for n_s paths, we can derive the following error equation,

$$\boldsymbol{G}_D = [G]\boldsymbol{E}. \tag{3}$$

Here, $\boldsymbol{G}_D = \left[\boldsymbol{G}_{D,1}^T, \boldsymbol{G}_{D,2}^T, \cdots, \boldsymbol{G}_{D,n_s}^T\right]^T$, $[G] = [\boldsymbol{G}_1, \boldsymbol{G}_2, \cdots, \boldsymbol{G}_{n_e}]$, $\boldsymbol{E} = [e_1, e_2, \cdots, e_{n_e}]^T$, $\boldsymbol{G}_j = \left[\boldsymbol{g}_{j,1}^T, \boldsymbol{g}_{j,2}^T, \cdots, \boldsymbol{g}_{j,n_s}^T\right]^T$ $(j = 1, 2, \cdots, n_e)$. The errors $\boldsymbol{E}$ in the kinematic parameters $\boldsymbol{P}_k = [P_{k1}, P_{k2}, \cdots, P_{kn_e}]^T$ can be obtained by solving the following equation using the measured $\boldsymbol{G}_D$ and calculated $[G]$,

$$\boldsymbol{E} = [G]^+ \boldsymbol{G}_D. \tag{4}$$

Here, $[G]^+$ is the pseudo inverse of the matrix $[G]$, and it is obtained by the following equation,

$$[G]^+ = ([G]^T[G])^{-1}[G]^T = [H]^{-1}[G]^T \tag{5}$$

The values of kinematic parameters $\boldsymbol{P}_k$ are calculated from the following assignment,

$$\boldsymbol{P}_k \leftarrow \boldsymbol{P}_k + \boldsymbol{E}. \tag{6}$$

To eliminate the error caused by the linearization, new $\boldsymbol{P}_k$ is used to calculate the $[G]^+$ of Eq. (4), and calculation of $\boldsymbol{E}$ and $\boldsymbol{P}_k$ is iterated until the values of the kinematic parameters converge.

3. Conditioning Index for Determination of the Set of Measurement Paths

We investigate the error propagation from the measurement error to the identification error. Equation(3) can be rewritten as follows,

$$m_1 \leq \frac{||\boldsymbol{G}_D||}{||\boldsymbol{E}||} \leq M_1, \tag{7}$$

$$\frac{1}{M_2} \leq \frac{||\Delta\boldsymbol{E}||}{||\Delta\boldsymbol{G}_D||} \leq \frac{1}{m_2}. \tag{8}$$

Here, m_1 and M_1 are determined by $[G]$ and m_2 and M_2 are determined by $[G]^{-1}$ if $[G]$ is a square matrix. Bigger values of m_2 and M_2 are preferable. From Eqs. (7) and (8), we obtain the following inequality for a square matrix $[G]$,

$$\frac{||\Delta\boldsymbol{E}||}{||\boldsymbol{E}||} \leq \frac{M_1}{m_2}\frac{||\Delta\boldsymbol{G}_D||}{||\boldsymbol{G}_D||}. \tag{9}$$

The value of M_1/m_2 is called the condition number of $[G]$. A smaller M_1/m_2 value is preferable. Since $[G]$ is not a square matrix in most cases of calibration, the condition number in Eq. (9) cannot be obtained. There are two ways to calculate the condition number for a general case based on a singular value decomposition or the norm.

In the case of the singular value decomposition method, there are two definitions of the condition number. A condition number C_{G1} is defined as the ratio of the maximum singular value $\sigma_{\max[G]}$ and the minimum singular value $\sigma_{\min[G]}$ of $[G]$ (Driels, 1990). That is,

$$C_{G1} = \frac{\sigma_{\max[G]}}{\sigma_{\min[G]}}. \tag{10}$$

The other definition C_{G2} is defined as the product of the maximum singular value $\sigma_{\max[G]}$ of $[G]$ and the maximum singular value $\sigma_{\max[G]^+}$ of $[G]^+$. That is,

$$C_{G2} = \sigma_{\max[G]} \cdot \sigma_{\max[G]^+}. \tag{11}$$

In the case of the norm method, a condition number calculated from M_1 and m_2 is calculated by the following equations when the matrix $[G]$ is square,

$$M_1 = \max \frac{||[G]\boldsymbol{E}||}{||\boldsymbol{E}||} = \max ||\boldsymbol{g}_j||, \tag{12}$$

$$m_2 = \frac{||\Delta \boldsymbol{G}_D||}{||[G]^{-1}\Delta \boldsymbol{G}_D||} = \frac{1}{\max ||\boldsymbol{g}_{i,inv}||}. \tag{13}$$

Here, $\boldsymbol{g}_j$ and $\boldsymbol{g}_{i,inv}$ are the column vectors of $[G]$ and $[G]^{-1}$, respectively. Since $[G]$ is not a square matrix in the general case, replacement of $[G]^{-1}$ by the pseudo inverse matrix $[G]^+$ is expected in the case where $[G]$ is not a square matrix. However, this complicates the calculation. Therefore, we consider a norm based condition number of a symmetric matrix $[H] = [G]^T[G]$ using $[H]$ and $[H]^{-1}$ instead of $[G]$ and $[G]^{-1}$ in Eqs. (12) and (13) because the calculation is simple and the error propagation characteristics of $[G]^+$ in Eqs. (4) and (5) are almost determined by $[H]$. Such a condition number C_{G3} is defined as follows,

$$C_{G3} = ||[H]||\ ||[H]^{-1}||. \tag{14}$$

Here,

$$||[H]|| = \max ||\boldsymbol{h}_j||, \tag{15}$$

$$||\boldsymbol{h}_j|| = \sum |h_{i,j}|, \tag{16}$$

where $h_{i,j}$ denotes the (i, j)-th element of $[H]$.

The following two indices have already been proposed, in addition to the condition number. One is the minimum singular value (Nahvi, 1994) $\sigma_{\min[G]}$ of $[G]$ or the maximum singular value $\sigma_{\max[G]^+}$ of $[G]^+$, and the other is the noise amplification index (NAI)(Nahvi, 1996) defined as,

$$NAI = \frac{\sigma_{\min[G]}}{C_{G1}} = \frac{\sigma^2_{\min[G]}}{\sigma_{\max[G]}}. \tag{17}$$

In order to make it possible to equivalently evaluate all the kinematic parameters that have different dimensions, we use the normalized matrix $[G]'$ instead of $[G]$ for calculating all the indices mentioned above. The normalization is done by using the following equation,

$$\boldsymbol{g}'_j = \frac{\boldsymbol{g}_j}{\max |g_{i,j}|}, \tag{18}$$

where $g_{i,j}$ denotes the (i, j)-th element of $[G]$.

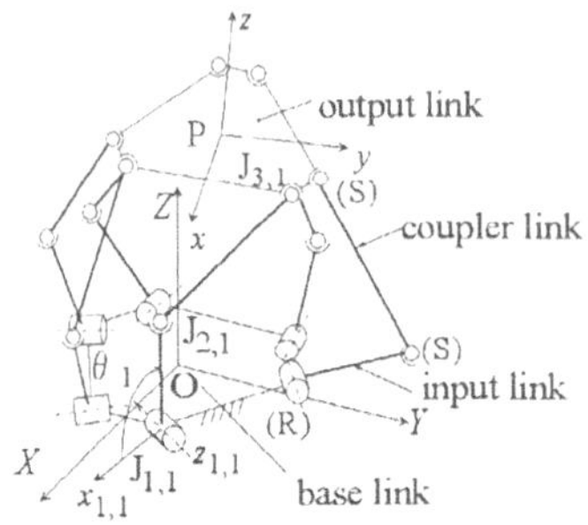

(a) Mechanism configuration

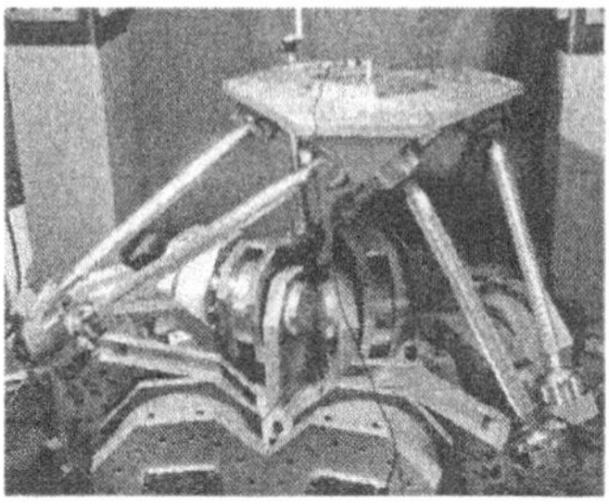

(b) Experimental apparatus

Figure 2. A crank-type spatial PM with six dof

4. Simulations

In this section, the relationships between the indices presented in the previous section and the pose accuracy after calibration are investigated through simulations. The simulations assumed spatial PM with six dof(Takeda, 1999ICAR) shown in Figure 2. Since there exist 11 kinematic parameters in each connecting chain of this PM, total number of parameters would be 66. However, since the DBB system just measures the distance error between P_1 and P_2, some parameters cannot defined independently of the reference coordinate systems on the base and output links. Then, we investigated independence of parameters and clarified 54 independent parameters. And, we considered the position errors of P_1 and P_2 to be identified. These errors were randomly given within the expected range.

We considered the ranges of parameters that determine the measurement paths, and sets of measurement paths were randomly generated. The simulations reflected the accuracy of the DBB systems on the market (offset = 0.5 μm and normal distribution =0.5 μm in 3 σ).

Relationship between indices and pose accuracy after calibration is shown in Figure 3. In order to consider the worst case, we look at the biggest pose error with respect to the index. When the number of paths n_s is the same, a monotone relationship between each index and pose accuracy AP is observed. This means that any index is useful as an evaluation index in the determination of the set of measurement paths when n_s is constant. However, if we compare the cases of different n_s, such monotone relationships still remain only in the indices of C_{G2}, C_{G3} and $\sigma_{\max[G]^+}$. Therefore, we conclude that the indices C_{G2},C_{G3} and $\sigma_{\max[G]^+}$ are appropriate for the determination of the set of measurement paths used in the calibration because the pose accuracy AP can be estimated from their values.

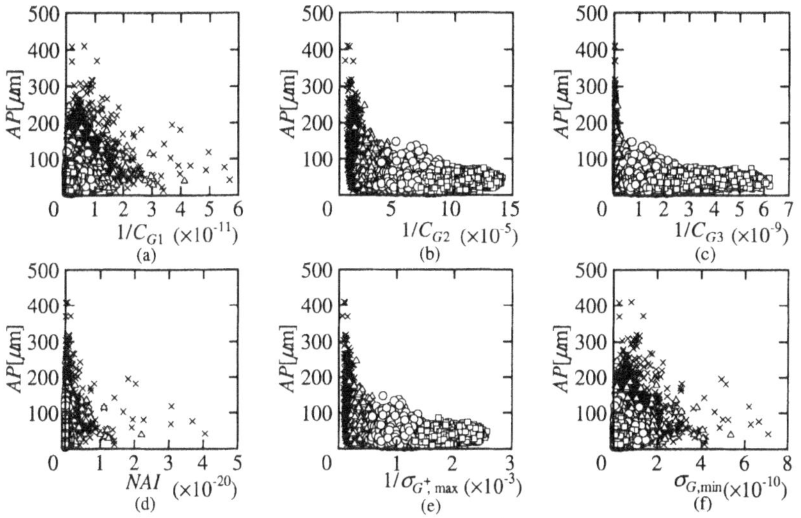

Figure 3. Relationship between indices and AP(pose accuracy after calibration) ($\times : n_s = 10, \triangle : n_s = 12, \circ : n_s = 16, \square : n_s = 24$)

5. Experimental Verification

Based on the results of the simulations, we carried out experiments of kinematic calibration of the PM shown in Figure 2(b). An example of the results we obtained is shown in the case of $n_s = 24$. In this experiment, the optimal set of measurement paths has been determined based on the condition number C_{G3}, and all the kinematic parameters have been successfully identified. To verify the identification result, the accuracy of the PM in the case where all the identified parameters are used was experimentally evaluated. Figure 4 shows the results of a DBB test in which one path that was not used in the calibration is used as a test path. It is evident that the accuracy of the PM was greatly improved by using the proposed calibration method. The circularity was improved from 349 μm to 14 μm in this case.

6. Conclusions

In the present paper, we investigated the evaluation index used for determining the optimal set of measurement paths in the DBB-based kinematic calibration method of in-parallel actuated mechanisms. Appropriate indices were revealed through simulations and, subsequently, were verified in an experiment of kinematic calibration using the experimental mechanism developed by us.

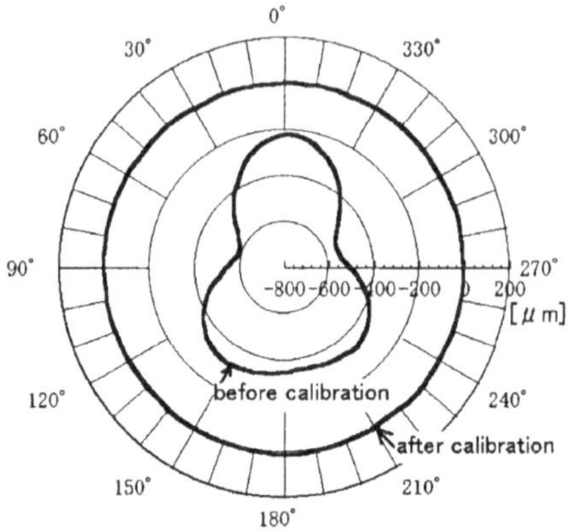

Figure 4. The results of the DBB test before and after calibration (circularity before calibration: 349 μm, circularity after calibration: 14 μm)

References

Borm, J., H. (1991), *Int. J. Rob., Res.*, Vol.10, No.1, pp. 51-63.

Driels, M., R. and Pathre, U., S. (1990), *J. of Robo. Syst.*, Vol.7, No.2, pp. 197-223.

Forsythe, G., E., et al. (1977), *Computer Methods for Mathematical Computations*, Prentice-Hall, Englewood Cliffs, NJ.

Hollerbach, J., et al. (1996), *Int. J. Rob. Res.*, Vol.15, No.6, pp.573-591.

Kakino, Y., et al. (1993), *Accuracy Inspection of NC Machine Tools by Double Ball Bar Method*, Carl Hanser Verlag.

Nahvi, A., et al. (1994), *Proc. IEEE Int. Conf. Robotics and Automation*, pp. 407-412.

Nahvi, A. and Hollerbach, J., M. (1996), *Proc. IEEE Int. Conf. on Rob. and Autom.*, pp. 647-654.

Ota, H., et al. (2000), *Proc. PKM2000*, pp. 57-62.

Patel, A., J., and Ehmann, K., F. (2000), *Int. J. of Machine Tools and Manufacture*, Vol.40, pp. 489-512.

Proc. Year 2000 Parallel Kinematic Machines Int. Conf.(PKM2000) (2000).

Salisbury, J.,K., and Craig, J.,J. (1982), *Int. J. of Robo. Res.*, Vol.1, No.1, pp. 4-17.

Takeda,Y. and Funabashi,H. (1995), *JSME Int. J., Ser. C*, Vol.38, No.4, pp. 749-755.

Takeda, Y. and Funabashi,H. (1996), *JSME Int. J., Ser. C*, Vol.39, No.1, pp. 85-93.

Takeda,Y., et al. (1997), *JSME Int. J., Ser. C*, Vol.40, No.2, pp. 299-308.

Takeda, Y., et al. (1999ICAR), *Proc. 9-th ICAR*, pp. 551-556.

Takeda,Y. and Funabashi,H. (1999JRM), *J. of Robotics and Mechatronics(JRM)*, Vol.11, No.5, pp. 404-410.

Takeda, Y., et al. (2000), *Proc. PKM-2000*, pp. 264-273.

Takeda, Y., et al. (2001), *Proc. 1st Taiwan-Japan Workshop on Mechanical and Aerospace Engineering*, pp. 11-19.

Vischer, P., and Clavel, R. (1998), *Robotica*, Vol.16, pp. 207-218.

Wampler, C., W., et al. (1995), *IEEE Trans. on Rob. and Autom.*, Vol.11, No.5, pp. 710-724.

Zhuang, H., et al. (1994), *Proc. IEEE Int. Conf. on Rob. and Autom.*, pp. 393-398.

DYNAMIC PERFORMANCE INDICES FOR 3-DOF PARALLEL MANIPULATORS

R. Di Gregorio
Department of Engineering, University of Ferrara
Via Saragat, 1; 44100 FERRARA, ITALY
rdigregorio@ing.unife.it

V. Parenti-Castelli
DIEM, University of Bologna
Viale Risorgimento, 2; 40136 BOLOGNA, ITALY
vincenzo.parenticastelli@mail.ing.unibo.it

Abstract This paper presents three new indices for the characterization of the dynamic behavior of 3-dof parallel manipulators. The first index measures the dynamic isotropy of the manipulator, whereas the second and the third indices measure the manipulator swiftness, that is the manipulator capability to produce end-effector acceleration. They allow dynamic behavior of different manipulators to be compared. The indices can be easily expressed as functions of both the parameters locating the end-effector pose and the inertial and the geometric parameters of the links. Moreover, based on their properties, a simple design procedure to size a three-dof manipulator so that it fulfills given dynamic requirements is presented.

Keywords: 3-dof parallel manipulators, dynamic isotropy index, acceleration index

1. Introduction

The designer of a high speed manipulator cannot help but study its dynamic behavior. As a rule of thumb, the dynamic behavior of a manipulator is improved if significant masses, e.g., motors, are moved onto the frame or onto slow moving links. The evaluation of design changes is only possible, however, by means of simulation codes which do not provide, generally, synthetic tools to compare the dynamic performances of different architectures. Synthetic tools would greatly facilitate the design process and are thus welcomed.

Parallel architectures often allow motors to be moved onto or near to the frame. This practice brought about manufacturing parallel manipulators with high dynamic performances (Clavel, 1988; Pierrot et al., 1991). Parallel architectures consist of two rigid bodies, one movable (platform or end-effector) and the other fixed (base or frame),

J. Lenarčič and F. Thomas (eds.), Advances in Robot Kinematics, 11–20.

interconnected by a number of kinematic chains (legs). Most parallel architectures exhibit a number of legs that equals the number of manipulator degrees of freedom (dof) and each leg has only one actuated kinematic pair.

Many dynamic models of parallel architectures were presented in the literature. Some of them are based on Newton-Euler (Do et al., 1988) or on Lagrange (Pang and Shahingpoor, 1994) formulations, and some on others (Baiges and Duffy, 1996; Tsai, 2000; Wang and Gosselin, 1998). All formulations show that the dynamic behavior of manipulators depends on their generalized inertia matrix (GIM) (Angeles, 1997). The GIM of a n-dof manipulator is a n×n symmetric and positive-definite matrix which can be obtained through the analytic expression of the manipulator's kinetic energy. The GIM depends on the manipulator configuration and its eigenvalues are real and positive.

In order to provide a useful tool for the evaluation of manipulator dynamics a geometric interpretation of the manipulator's GIM was presented in (Asada and Cro Granito, 1985; Yoshikawa, 1990; Wiens et al., 1989) that is based on the concept of generalized inertia ellipsoid (GIE). The GIE of an n-dof manipulator is a generalized ellipsoid with n principal axes whose half-lengths are the reciprocals of the square roots of the GIM's eigenvalues. Since the GIM depends on the manipulator configuration, the GIE depends on the configuration too. Configurations where the GIE is a sphere (i.e., the GIM's eigenvalues are all equal) are called dynamically isotropic configurations. When a manipulator assumes a dynamically isotropic configuration, any change of its motion status involves inertia forces whose intensities are nearly independent of the type of change. Therefore, dynamic isotropy is a sought-after quality during design. The GIE can be graphically represented only if the manipulator dof number is less than or equal to three. Nevertheless, even for these manipulators, graphic representations that associate a manipulator configuration with its GIE are complex and difficult to interpret.

Another manipulator dynamic property is the swiftness, i.e. the manipulator capacity of causing high end-effector accelerations for the same variation of the active forces. This property is usually indicated by giving the maximum acceleration the end-effector can reach. Nevertheless, the maximum acceleration does not take into account the manipulator behavior in all its workspace.

Dynamic indices have been proposed in the literature for serial manipulators (Yoshikawa, 1990; Wiens et al., 1989). In particular, with reference to a GIM obtained in joint space, Yoshikawa (1990) presented a dynamic manipulability ellipsoid and an index for the dynamic manipulability measure. A limitation of this approach is that it requires the knowledge of the gravitational and non-linear forces in order to define an actuator maximum virtual torque, which is necessary to evaluate the proposed index. In (Wiens et al., 1989) indices are proposed to characterize the effects of non-linear forces.

This paper analyses the dynamic behavior of 3-dof parallel manipulators. A particular expression of the manipulator's kinetic energy is written, and the GIM reduced to the end-effector (GIMRE) is obtained. Thus the GIMRE is used to define three new indices. The first measures the dynamic isotropy of the manipulator, whereas the second and the third measure the manipulator swiftness and depend on the manipulator configuration. They also allow the dynamic behavior comparison of different manipulators.

The indices can be easily expressed as functions of both the parameters locating the end-effector pose (position and orientation) and the inertial and geometric parameters of the links. Hence, a simple design procedure to size a manipulator that fulfils given dynamic requirements, expressed by assigning the desired values the indices must assume, is proposed.

2. Generalized inertia matrix reduced to the end-effector (GIMRE)

Figure 1 shows a generic 3-dof parallel manipulator. Here, k is the number of legs; r_i, i=1,...,k, is the number of links belonging to the i-th leg; S_p is a reference system fixed to the platform with origin in the centroid, G_p, of the platform, and S_b is a reference system fixed to the base. Hereafter link ij (i=1,...,k; j=1,...,r_i) is the j-th link of the i-th leg.

Figure 2 shows the link ij. S_{ij} is a reference system embedded in it with origin in its centroid, G_{ij}. Moreover, the following notations are used: m_{ij} (m_p) is the mass of link ij (of the platform); $\mathbf{I}_{ij}$ ($\mathbf{I}_p$) is the inertia tensor of link ij (of the platform) referred to point G_{ij} (G_p) projected onto S_{ij} (S_p), and ${}^b\mathbf{I}_{ij}$ (${}^b\mathbf{I}_p$) is the inertia tensor $\mathbf{I}_{ij}$ ($\mathbf{I}_p$) projected onto S_b. $\mathbf{I}_{ij}$ and $\mathbf{I}_p$ are related to ${}^b\mathbf{I}_{ij}$ and ${}^b\mathbf{I}_p$ respectively through the relationships

$$ {}^b\mathbf{I}_{ij} = {}^b\mathbf{R}_{ij}\,\mathbf{I}_{ij}\,{}^b\mathbf{R}_{ij}{}^T, \quad {}^b\mathbf{I}_p = {}^b\mathbf{R}_p\,\mathbf{I}_p\,{}^b\mathbf{R}_p{}^T \tag{1} $$

where ${}^b\mathbf{R}_{ij}$ (${}^b\mathbf{R}_p$) is the rotation matrix transforming vector components measured in S_{ij} (S_p) into vector components measured in S_b, and $(.)^T$ stands for the transpose of (.).

With these notations, the kinetic energy of the link ij, E_{ij}, and of the platform, E_p, can be written as follows

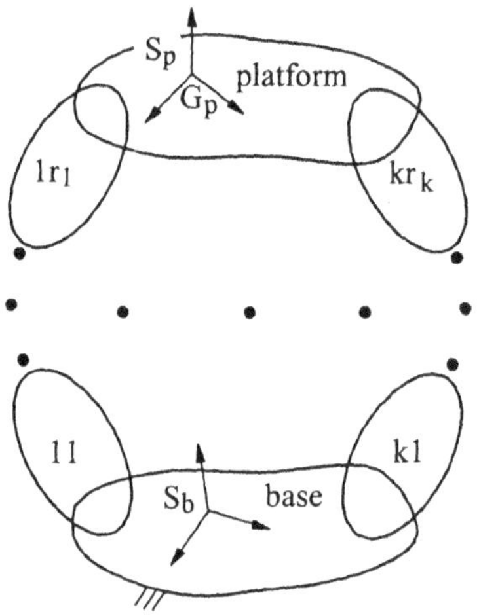

Fig. 1: Generic 3-dof parallel manipulator

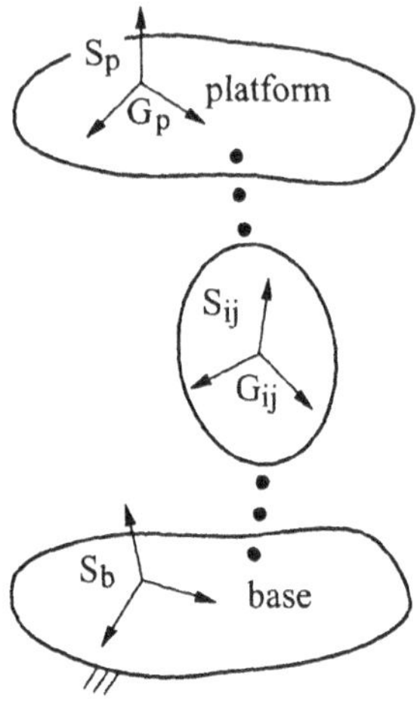

Fig. 2: Generic link ij

$$E_{ij} = \frac{1}{2}[m_{ij}\dot{\mathbf{G}}_{ij}^2 + \omega_{ij}^T\, {}^b\mathbf{I}_{ij}\, \omega_{ij}] \tag{2.1}$$

$$E_p = \frac{1}{2}[m_p\dot{\mathbf{G}}_p^2 + \omega_p^T\, {}^b\mathbf{I}_p\, \omega_p] \tag{2.2}$$

where $\dot{\mathbf{G}}_{ij}$ ($\dot{\mathbf{G}}_p$) and $\boldsymbol{\omega}_{ij}$ ($\boldsymbol{\omega}_p$) are the velocity of the centroid, G_{ij} (G_p), and the angular velocity of link ij (of the platform) measured in S_b.

Moreover, the kinetic energy, E, of the manipulator is

$$E=E_p+\sum_{i=1,k}\ \sum_{j=1,r_i} E_{ij} \tag{3}$$

Since the manipulator has three dof, three out of the six parameters locating the platform pose in S_b can always be used as generalized coordinates of the manipulator.

Henceforth, let these three manipulator generalized coordinates be q_s, s=1,2,3, and $\mathbf{q}$ be the vector $(q_1, q_2, q_3)^T$.

Provided that singular configurations are excluded, the inverse velocity analysis of the manipulator allows the following relationships to be determined in explicit form:

$$\dot{\mathbf{G}}_{ij} = \mathbf{H}_{ij}\,\dot{\mathbf{q}}\,,\quad \omega_{ij} = \mathbf{L}_{ij}\,\dot{\mathbf{q}}\,,\quad \dot{\mathbf{G}}_p = \mathbf{H}_p\,\dot{\mathbf{q}}\,,\quad \omega_p = \mathbf{L}_p\,\dot{\mathbf{q}} \tag{4}$$

where $\dot{\mathbf{q}}$ is the time derivative of $\mathbf{q}$, while $\mathbf{H}_{ij}$, $\mathbf{L}_{ij}$, $\mathbf{H}_p$ and $\mathbf{L}_p$ are 3×3 matrices depending on $\mathbf{q}$ and on the manipulator link geometry.

Taking into account relationships (4), expressions (2) become

$$E_{ij} = \frac{1}{2}\dot{\mathbf{q}}^T[m_{ij}\mathbf{H}_{ij}^T\mathbf{H}_{ij} + \mathbf{L}_{ij}^T\,{}^b\mathbf{I}_{ij}\,\mathbf{L}_{ij}]\dot{\mathbf{q}} \tag{5.1}$$

$$E_p = \frac{1}{2}\dot{\mathbf{q}}^T[m_p\mathbf{H}_p^T\mathbf{H}_p + \mathbf{L}_p^T\,{}^b\mathbf{I}_p\,\mathbf{L}_p]\dot{\mathbf{q}} \tag{5.2}$$

The introduction of expressions (5) into relationship (3) gives the following relationship:

$$E = \frac{1}{2}\dot{\mathbf{q}}^T\,\mathbf{N}\,\dot{\mathbf{q}} \tag{6}$$

where

$$\mathbf{N} = m_p\mathbf{H}_p^T\mathbf{H}_p + \mathbf{L}_p^T\,{}^b\mathbf{I}_p\,\mathbf{L}_p + \sum_{i=1,k}\,\sum_{j=1,r_i}[m_{ij}\mathbf{H}_{ij}^T\mathbf{H}_{ij} + \mathbf{L}_{ij}^T\,{}^b\mathbf{I}_{ij}\,\mathbf{L}_{ij}] \tag{7}$$

Matrix **N** is the generalized inertia matrix referred to the end-effector (GIMRE) of the manipulator. It is a symmetric and positive-definite matrix depending both on **q** and on the inertial and geometric parameters of the links.

3. Dynamic performance indices

The dynamic behavior of the manipulator depends on the values assumed by the three eigenvalues of matrix **N**. Eigenvalues of matrix **N** are the three solutions of the following cubic equation

$$\det(\mathbf{N} - x\,\mathbf{I}_3) = 0 \tag{8}$$

where det(.) is the determinant of (.), x is the unknown and $\mathbf{I}_3$ is the 3×3 identity matrix.

Expanding the left-hand side of Eq. (8) and recalling the relationships between the coefficients of a cubic equation and its solutions, the following relationships between the entries of matrix **N** and its eigenvalues, x_1, x_2 and x_3 (real and positive), can be easily demonstrated:

$$\mathrm{tr}(\mathbf{N}) = x_1 + x_2 + x_3 \tag{9.1}$$

$$\det(\mathbf{N}) = x_1\, x_2\, x_3 \tag{9.2}$$

where tr(**N**) is the trace of matrix **N**.

Explicit forms of tr(**N**) and det(**N**) can be directly obtained from the explicit expressions of the entries of matrix **N**.

In the following paragraphs of this section, three indices will be defined by using only the sum and/or the product of all the eigenvalues of matrix **N**. To reach this goal, without loss of generality, the eigenvalues of matrix **N** will be supposed ordered so that the following inequalities hold

$$0 < x_1 \leq x_2 \leq x_3 \tag{10}$$

3.1 Dynamic isotropy

A three-dof manipulator configuration is dynamically isotropic when the eigenvalues of the matrix **N** in that configuration are all equal, which implies that the following relationship holds:

$$x_1 x_2 x_3 = \frac{(x_1 + x_2 + x_3)^3}{27} \tag{11}$$

Relationship (11) is satisfied only if the three eigenvalues are all equal, because the product of three scalar quantities whose sum is assigned is maximum only if they are all equal. Therefore, the index d_1 defined as follows

$$d_1 = \frac{27 x_1 x_2 x_3}{(x_1 + x_2 + x_3)^3} \tag{12}$$

is equal to one only for dynamically isotropic configurations and is less than one for all the other configurations.

The product and the division of the right-hand side of definition (12) by x_3^3 yield the following alternative definition of d_1

$$d_1 = \frac{27 y_1 y_2}{(y_1 + y_2 + 1)^3} \tag{13}$$

where

$$y_1 = \frac{x_1}{x_3}, \quad y_2 = \frac{x_2}{x_3} \tag{14}$$

Inequalities (10) and definitions (14) give the following bounds on the values of y_1 and y_2

$$0 < y_1 \leq y_2 \leq 1 \tag{15}$$

where equality holds if and only if

$$y_1 = y_2 = 1 \tag{16}$$

Variable y_1 is the ratio between the minimum and the maximum eigenvalues. As a consequence, y_1 directly measures how near to dynamic isotropy a configuration is.

In order to study the pattern of d_1 as a map of y_1 and y_2, the partial derivatives of d_1 with respect to both y_1 and y_2 must be computed and their sign must be analyzed. The differentiation of definition (13) yields the following relationships:

$$\frac{\partial d_1}{\partial y_1} = \frac{27 y_2 (1+y_2-2y_1)}{(y_1+y_2+1)^4}, \qquad \frac{\partial d_1}{\partial y_2} = \frac{27 y_1 (1+y_1-2y_2)}{(y_1+y_2+1)^4} \tag{17}$$

Taking into account inequalities (15), the analysis of the first expression of (17) leads to the conclusion that $\partial d_1/\partial y_1$ is always greater than zero if $y_1 \neq y_2$. Therefore, d_1 has the same pattern as y_1 (i.e., it increases when y_1 increases and decreases when y_1 decreases). Moreover, the two-equation system obtained by equating to zero the right-hand sides of relationships (17) has the only solution $y_1 = y_2 = 1$. Hence, d_1 has only one maximum corresponding to $y_1 = y_2 = 1$.

These two properties of d_1 allow the use of d_1 as an index to measure the dynamic isotropy of a configuration in place of y_1.

Taking into account relationships (9), definition (12) becomes

$$d_1 = 27 \frac{\det(\mathbf{N})}{[\mathrm{tr}(\mathbf{N})]^3} \tag{18}$$

Expression (18) of d_1 contains in explicit form the inertial (m_{ij}, $\mathbf{I}_{ij}$, m_p and $\mathbf{I}_p$) and geometric parameters of the manipulator, and the three generalized coordinates q_s, s=1,2,3, as well.

Since d_1 is equal to one if and only if a dynamically isotropic configuration occurs, the entries of the matrix $\mathbf{N}$ corresponding to a dynamically isotropic configuration must satisfy the following relationship

$$27 \frac{\det(\mathbf{N})}{[\mathrm{tr}(\mathbf{N})]^3} = 1 \tag{19}$$

If the inertial and geometric parameters of the manipulator are assigned, relationship (19) is a scalar equation in three unknowns: the three generalized coordinates q_s, s=1,2,3. In the manipulator configuration space (tridimensional space whose coordinates are the manipulator generalized coordinates), equation (19) is the analytic expression of a surface which is the geometric locus of all points corresponding to dynamically isotropic configurations.

On the other hand, if the inertial and geometric parameters of the manipulator are not assigned, they can in some cases be sized so that equation (19) is matched for any triplet of values of the generalized coordinates q_s, s=1,2,3. A manipulator sized in this way is dynamically isotropic in all its workspace.

3.2 Manipulator swiftness

Manipulator swiftness is the manipulator attitude to cause end-effector accelerations for the same variation of the active forces. Like acceleration, swiftness depends on the manipulator configuration. In particular, the greater the arithmetic average of the eigenvalues, x_s, s=1,2,3, of the matrix **N** corresponding to a manipulator configuration, the lower the accelerations the end-effector can reach starting from that configuration. Therefore, the value of the index d_2 defined as

$$d_2 = \frac{3}{x_1 + x_2 + x_3} \tag{20}$$

is a measure of the swiftness of a manipulator in a given configuration.

Taking into account relationship (9.1), definition (20) becomes

$$d_2 = \frac{3}{tr(\mathbf{N})} \tag{21}$$

Expression (21) of d_2 explicitly contains the same parameters and generalized coordinates as those contained in the expression of d_1.

If the inertial and geometric parameters of the manipulator are given, definition (21) depends only on the three generalized coordinates q_s, s=1,2,3. Moreover, in the manipulator configuration space, the geometric locus of the points corresponding to configurations with the same value, d_{20}, of index d_2 (i.e., with the same swiftness) is a surface whose analytic expression is

$$\frac{3}{tr(\mathbf{N})} = d_{20} \tag{22}$$

A measure of the average swiftness of a manipulator is the average value, d_{2m} (average swiftness index), of index d_2 on the whole useful workspace. Definition (21) allows the average swiftness index d_{2m} to be defined as follows

$$d_{2m} = \frac{\int_W \frac{3\,dw}{tr(\mathbf{N})}}{\int_W dw} \tag{23}$$

where W is the region of the configuration space corresponding to the useful workspace of the manipulator.

The average swiftness index d_{2m} depends only on the inertial and geometric parameters of the manipulator and can be used to compare manipulators with different architectures. Moreover, if the inertia and geometric parameters of a manipulator are not assigned, they can be determined by exploiting definition (23) so that the manipulator has a given value of the average swiftness index d_{2m}.

4. Use of the dynamic performance indices

Indices d_1, d_2, and d_{2m} measure dynamic performances of 3-dof manipulators. Therefore, design requirements regarding manipulator dynamic behavior can be expressed by giving the desired minimum values $d_{1|min}$, $d_{2|min}$, and $d_{2m|min}$ that d_1, d_2, and $d_{2m|min}$, respectively, must have in all the useful workspace W.

Once the desired minimum values are assigned, design requirements can be matched by using the following procedure:

(i) the inertial and geometric parameters of the manipulator are assigned according to the designer's experience;

(ii) using the assigned values of the inertia and geometric parameters, the region V of the configuration space where the following inequality system holds is determined:

$$d_1 \geq d_{1|min}, \quad d_2 \geq d_{2|min} \tag{24}$$

(iii) if the useful workspace W is not included in V, then the inertial and geometric parameters must be changed and the procedure repeated starting from step (ii); otherwise the procedure goes to the next step;

(iv) the average swiftness index d_{2m} is computed;

(v) if the following inequality

$$d_{2m} \geq d_{2m|min} \tag{25}$$

is not satisfied, then the inertial and geometric parameters must be changed and the procedure repeated starting from step (ii); otherwise the procedure stops and the manipulator is sized with the found values of the inertia and geometric parameters.

Variations of the manipulator's geometric parameters involve variations in static and kinematic characteristics of the manipulator and make the repetition of static and kinematic analyses necessary. Therefore, when the above-reported procedure requires changes in the inertial and geometric parameters (steps (iii) and (v)), first only the inertial parameters must be changed and then, only if it is strictly necessary, the geometric parameters must also be changed.

5. Conclusions

Three novel indices for the dynamic performance characterization of 3-dof parallel manipulators have been presented. The first index measures the dynamic isotropy of the manipulator, whereas the second and the third measure the manipulator swiftness, i.e. the manipulator capability to accelerate its end-effector. Since the indices can be expressed as functions of the parameters that define the end-effector pose, the inertia, and the geometry of the links, a simple design procedure is presented for sizing a 3-dof manipulator so that it fulfills given dynamic requirements. The method and the results obtained can be easily extended to any other type of 3-dof manipulator (serial and hybrid).

Acknowledgments

The funds of the Italian MURST and CNR are gratefully acknowledged.

References

Angeles, J. (1997), Fundamentals of robotic mechanical systems, Springer-Verlag, New York (USA), p. 219.

Asada, H. (1983), A geometrical representation of manipulator dynamics and its applications to arm design, *ASME Journal of Dynamic Systems, Measurement, and Control*, 105, No. 3, pp. 131-135.

Baiges, I.J:, and Duffy, J., (1996), "Dynamic modelling of parallel manipulators," *1996 ASME DETC*, Paper 96-DETC/MECH-1136, Irvine, California.

Clavel, R. (1988), DELTA: a fast robot with parallel geometry, *Proc. of the 18th International Symposium on Industrial Robots*, Sydney, Australia, pp. 91-100.

Do, W.Q.D., and Yang, D.C.H. (1988), Inverse dynamic analysis and simulation of a platform type of robot, *Journal of Robotic Systems*, 5, No. 3, pp. 209-227.

Pang, H., and Shahingpoor, M. (1994), Inverse dynamics of a parallel manipulator, *Journal of Robotic Systems*, 11, No. 8, pp. 693-702.

Pierrot, F., Dauchez, P., and Fournier, A. (1991), Hexa a fast six-dof fully parallel robot, *Proceedings of the 5th International Conference on Advanced Robotics*, Pisa (Italy), pp. 1158-1163.

Tsai, L.W. (2000), Solving the inverse dynamics of a Stewart-Gough manipulator by the principle of virtual work, *ASME Journal of Mechanical Design*, 122, No. 1, pp. 3-9.

Wang, J., and Gosselin, C.M. (1998), A new approach for the dynamic analysis of parallel manipulators, *Multibody System Dynamics*, 2, pp. 317-334.

Wiens, G.J., Scott, R.A., and Zarrugh, M.Y. (1989), The role of inertia sensitivity in the evaluation of manipulator performance, *ASME Journal of Mechanical Design*, Vol. 111, pp. 194-199.

Yoshikawa, T. (1990), *Foundations of robotics: analysis and control*, The Mit Press, Cambridge, Massachusetts.

EVALUATION OF A CARTESIAN PARALLEL MANIPULATOR

Han Sung Kim and Lung-Wen Tsai
Department of Mechanical Engineering
University of California, Riverside
hkim@engr.ucr.edu and lwtsai@engr.ucr.edu

Abstract This paper describes a new 3-DOF translational parallel manipulator named a Cartesian Parallel Manipulator (CPM). The machine behaves like a conventional X-Y-Z Cartesian machine due to the orthogonal arrangement of the three supporting limbs. It is shown that there exists a one-to-one correspondence between the input and output of the manipulator. Three possible arrangements of the Z actuator are evaluated by stiffness mapping. A method for compensating actuator misalignment is described.

Keywords: Orthogonal arrangement, decoupled motion, Cartesian manipulator, stiffness mapping

1. Introduction

Parallel manipulators have been studied extensively because they possess high speed, high accuracy, and high stiffness characteristics. However, they have some disadvantages, such as complex forward kinematics, small workspace, and complicated universal and spherical joints.

To overcome these shortcomings, parallel manipulators with fewer than six DOF have been investigated. For example, Lee and Shah (1987) analyzed a 3-DOF parallel manipulator with three *RPS* (revolute-prismatic-spherical) chains. Several other novel 3-DOF parallel manipulators have been developed (Pierrot et al. 1990, and Wenger and Chablat 2000). However, most of these manipulators have coupled motion between the position and orientation of the end effector. Recent research on 3-DOF parallel manipulators has been leaning toward the decoupling of the position and orientation of the end-effector and the elimination of complicated multi-DOF joints. A 3-DOF translational platform that employs only revolute joints and planar parallelograms was designed (Tsai et al. 1996a, and Tsai 1997). Tsai (1996b) presented the design of a spatial 3-*UPU* (universal-prismatic-universal) parallel manipulator. The kinematics of the 3-*UPU* manipulator was studied further by Di Gregori and Parenti-Castelli (1998), and Tsai and Joshi (2000). Zhao and Huang (2000) studied the kinematics of an overconstrained 3-*RRC* (revolute-revolute-cylindrical) translational

J. Lenarčič and F. Thomas (eds.), Advances in Robot Kinematics, 21–28.

manipulator, which was first revealed in a patent by Tsai (1997). Baron and Bernier (2001) studied the design of a 3-DOF Star manipulator conceived by Herve (1992). Carricato and Parenti-Castelli (2001) presented a family of 3-DOF parallel manipulators with pure translational motion.

In this paper, a parallel manipulator that employs only revolute and prismatic joints to achieve translational motion of the moving platform is presented. In what follows, the kinematic structure of the manipulator is described. The kinematics and stiffness of the manipulator are studied. Some design considerations are discussed. Finally, a prototype CPM is demonstrated to show its feasibility.

2. Kinematic Structure

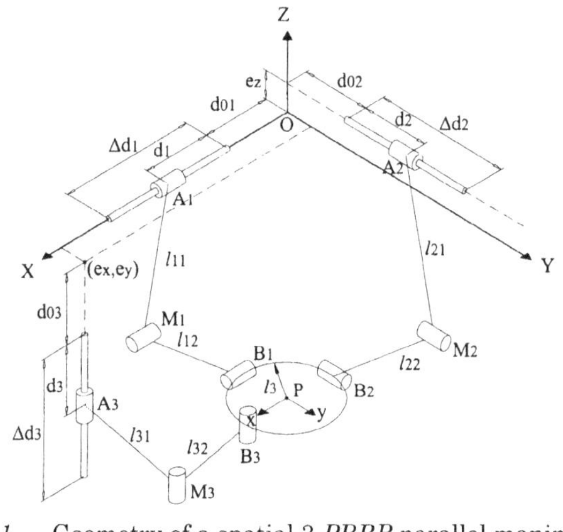

Figure 1. Geometry of a spatial 3-*PRRR* parallel manipulator.

The kinematic structure of the Cartesian parallel manipulator is shown in Fig. 1, where a moving platform is connected to a fixed base by three *PRRR* (prismatic-revolute-revolute-revolute) limbs. The moving platform is symbolically represented by a circle defined by points B_1, B_2, and B_3 and the fixed base is defined by three guide rods passing through A_1, A_2, and A_3, respectively. The three revolute joint axes in each limb are located at points A_i, M_i, and B_i, respectively, and are parallel to the ground-connected prismatic joint axis. Furthermore, the three prismatic joint axes, passing through point A_i for $i = 1, 2, \text{and } 3$, are parallel to the X, Y, and Z axes, respectively. Specifically, the first prismatic joint axis lies on the X axis; the second prismatic joint axis is parallel to the Y axis with an offset e_z in the Z direction; and the third prismatic joint axis is parallel to the Z axis with an offset e_x in the X direction and e_y in the Y direction. Point P represents the center of the moving platform. The link lengths are denoted by ℓ_{i1}, ℓ_{i2}, and ℓ_3, respectively. The starting point of a prismatic joint is defined by d_{0i} and the sliding distance is defined by d_i. Note that each *PRRR* limb is equivalent to a *CRR* limb. In this regard, the 3-*PRRR* parallel manipulator is a kinematic inversion of the 3-*RRC* manipulator.

The general degrees-of-freedom equation (Tsai 2000) predicts that

$$F = \lambda(n - j - 1) + \sum_i f_i = 6(8 - 9 - 1) + 3(2 + 1 + 1) = 0 \tag{1}$$

where F denotes the degrees of freedom of the mechanism, n and j represent the number of links and number of joints, respectively, and f_i denotes degrees of freedom associated with the ith joint. However, due to the special arrangement of the joints, each limb provides two rotational constraints to the moving platform. As a result, there are three redundant constraints on rotation and the mechanism has three translational degrees of freedom.

Two methods of actuation, namely, linear and rotary actuation methods are possible. In the linear actuation method, one actuator drives the prismatic joint in each limb whereas in the rotary actuation method, an actuator drives the first revolute joint of each limb. However, the rotary actuation method is shown to be impractical because of the existence of singularities within the workspace (Kim and Tsai 2002). In what follows, the machine with three linear actuators is presented.

3. Kinematics

The forward and inverse kinematic analyses are trivial since there exists a one-to-one correspondence between the end-effector position and the input joint displacements. Referring to Fig. 1, each limb constrains point P to lie on a plane which passes through point A_i and is perpendicular to the axis of the linear actuator. Consequently, the location of P is determined by the intersection of three planes and a simple kinematic relation can be written as

$$\begin{bmatrix} p_x \\ p_y \\ p_z \end{bmatrix} = \begin{bmatrix} d_{01} + d_1 \\ d_{02} + d_2 \\ d_{03} + d_3 \end{bmatrix} \tag{2}$$

where $\boldsymbol{p} = [p_x, p_y, p_z]^T$ denotes the position vector of the end-effector.

Similarly, the velocities and output forces of the end-effector also have a one-to-one relationship with those of prismatic actuators. Taking the time derivative of Eq. (2) and applying the principle of virtual work yields

$$\begin{bmatrix} \dot{d}_1 \\ \dot{d}_2 \\ \dot{d}_3 \end{bmatrix} = J \begin{bmatrix} \dot{p}_x \\ \dot{p}_y \\ \dot{p}_z \end{bmatrix} \text{ and } \begin{bmatrix} f_x \\ f_y \\ f_z \end{bmatrix} = J^T \begin{bmatrix} f_1 \\ f_2 \\ f_3 \end{bmatrix} \quad (3)$$

where J is a 3×3 identity matrix, $f = [f_x, f_y, f_z]^T$ is the vector of the end-effector output force, and f_i, for $i = 1, 2, \text{and } 3$, denotes the ith linear actuator force. Since J is an identity matrix, the manipulator is isotropic everywhere within its workspace.

4. Stiffness Analysis

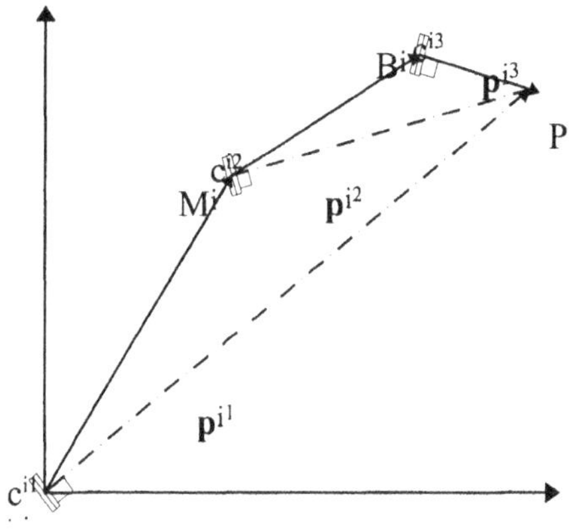

Figure 2. Plan view of a serial manipulator made up of three compliant joints.

Since the revolute joints are free to rotate, external force can only be transmitted in a direction parallel to the common joint axis of each limb and this force will produce a moment about each revolute joint. Assume that all links are rigid, and the major source of compliance comes from the flexibility of the revolute joints. The deflection between two members of a revolute joint can be modeled as an infinitesimal rotation about an axis perpendicular to the axis of revolution. We call such an axis of deflection a *virtual axis* of compliance. In this regard, we may consider each limb together with the moving platform as a serial manipulator having three virtual axes of compliances as shown in Fig. 2. From the result of Kim and Tsai (2002), a *compliance mapping* can be written as

$$\delta \boldsymbol{p} = C\, \boldsymbol{f} \quad (4)$$

where $\delta \boldsymbol{p} = [\delta p_x, \delta p_y, \delta p_z]^T$, and C is a diagonal *compliance matrix* whose diagonal elements are given by

(5)

$$C_{ii} = c_{i1}p_{i1}^2 + c_{i2}p_{i2}^2 + c_{i3}p_{i3}^2$$

where c_{ij} is an angular compliance constant, and $p_{ij} = \|\boldsymbol{p}_{ij}\|$ denotes the distance between a virtual axis of compliance and the end-effector, P. Multiplying Eq. (5) by C^{-1}, we obtain the *stiffness mapping* as

$$f = K\,\delta p \tag{6}$$

where $K = C^{-1}$ is called the *stiffness matrix* of the manipulator.

5. Effect of Actuator Location on Stiffness

The Z actuator can be located anywhere perpendicular to the X-Y plane. In this section, we consider three possible locations of the Z actuator as shown in Fig. 3. The design parameters are selected as

$$\Delta d_i = 1.0,\ d_{0i} = 0.2,\ l_{i1} = l_{i2} = 1.0,\ l_3 = 0.1, \text{ and } e_z = 0\,[\text{m}].$$

Here, the leg lengths are chosen long enough not to restrict the motion of the linear actuators.

To evaluate the three arrangements, the average of stiffness constant in the sense of Euclidean norm is considered,

$$k_{ave} = \sqrt{K_{11}^2 + K_{22}^2 + K_{33}^2} \tag{7}$$

where K_{ii} denotes a diagonal element of the stiffness matrix, K, and the angular compliance constant, $c_{ij} = 1$ [rad/N·m] is assumed. Figure 4 shows the contour maps of stiffness for the three arrangements at three different elevations of the moving platform.

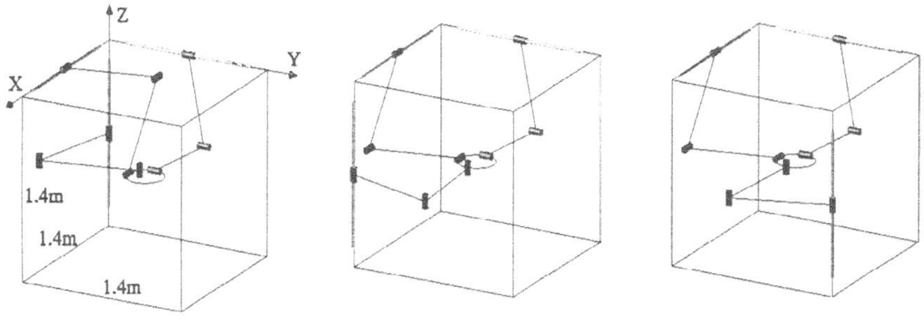

(a) Arrangement I (b) Arrangement II (c) Arrangement III

Figure 3. Three possible arrangements of the Z actuator.

For Arrangement I, the average stiffness is maximum when the end-effector is close to the Z axis. The second arrangement has high average stiffness when the end-effector is close to the X-Z plane. For the last case, the average stiffness is more evenly distributed. For all the cases, the average stiffness becomes higher around the Z actuator. From the distribution point of view, Arrangement III is better than the others.

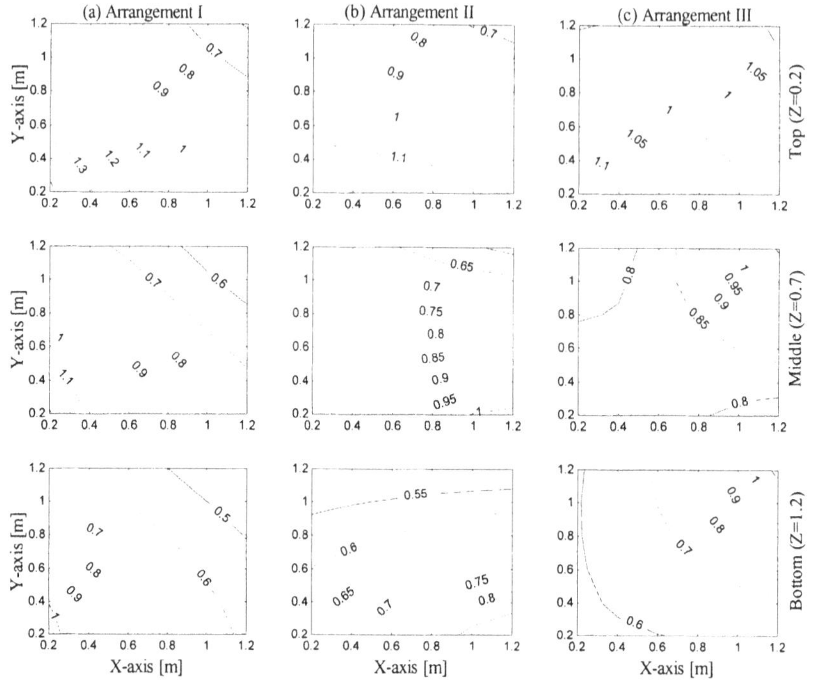

Figure 4. Contour stiffness maps for three arrangements of the Z actuator.

6. Tolerance Consideration

In practice, it may be difficult to fabricate and assemble a perfect orthogonal frame. In this section, we illustrate a method for compensating manufacturing and assembling errors of the linear actuators by one example. We assume that the X actuator is twisted by a small angle $\Delta\theta$ about the Z axis, and the other actuators are mounted perfectly along the Y and Z axes as shown in Fig. 5. When the X and Z actuators are held stationary, the dotted line in Fig. 6 denotes the possible locus of B_2. Hence, if the revolute joint axes of the X limb remain parallel to the prismatic joint as shown in Fig. 6 (a), the x, y, and z motions of the moving platform will be coupled. The coupling problem can be solved by adjusting the revolute joint axes such that they are all parallel to the X axis as shown in Fig. 6 (b). It should be noted that the three revolute joint axes in each limb should be parallel to one another, otherwise, this manipulator may not move or generate pure translational motion.

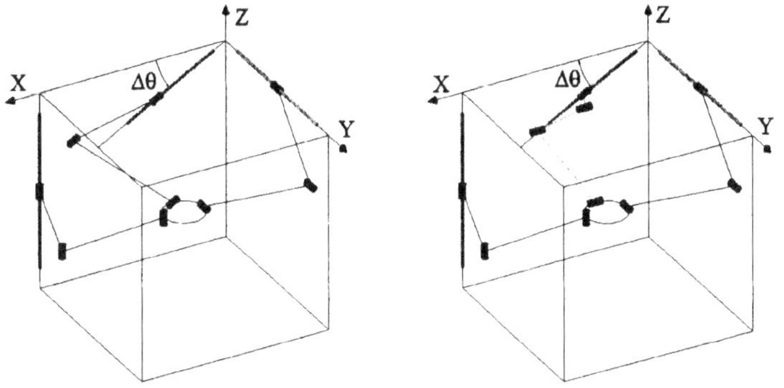

(a) Misaligned X drive (b) Misaligned X drive with compensation

Figure 5. Manipulator with a misaligned X actuator.

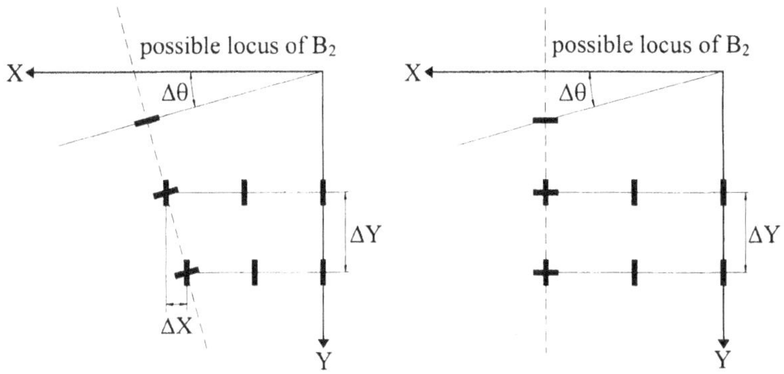

(a) Misaligned X drive (b) Misaligned X drive with compensation

Figure 6. Top view of a misaligned actuator.

7. Conclusions

A new 3-DOF translational parallel manipulator, behaving like a conventional X-Y-Z Cartesian machine, is conceived. It is shown that the x, y, and z motions of the manipulator are decoupled. Some design considerations with regard to the location of the Z actuator and a method for compensating the misalignment of the actuators are discussed. A prototype CPM as shown in Fig. 7 is currently under construction.

References

Baron, L. and Bernier, G., 2001, "The Design of Parallel Manipulators of Star Topology Under Isotropic Constraint," Proc. of the 2001 ASME Design Engineering Technical Conferences, Pittsburgh, PA, DAC-21025.

Carricato, M. and Parenti-Castelli, V., 2001, "A Family of 3-DOF Translational Parallel Manipulators," Proc. of the 2001 ASME Design Engineering Technical Conferences, Pittsburgh, PA, DAC-21035.

Di Gregori, R. and Parenti-Castelli, V., 1998, "A Translational 3-DOF Parallel Manipulator," in *Advances in Robot Kinematics*, Edited by J. Lenarcic and M. L. Husty, Kluwer Academic Publishers, London, pp. 49-58.
Herve, J. M., 1992, "Star, a New Concept in Robotics," 3rd International Workshop on Advances in Robotics Kinematics, Ferrara, Italy, pp. 176-183.
Kim, H. S., and Tsai, L. W., 2002, "Design Optimization of a Cartesian Parallel Manipulator, " Submitted for publication in ASME Journal of Mechanical Design.
Lee, K. and Shah, D. K., 1987, "Kinematics Analysis of a Three Degrees of Freedom In-Parallel Actuated Manipulator," Proc. IEEE International Conf. on Robotics and Automation, Vol. 1, pp. 345-350.
Pierrot, F., Reynaud, C., and Fournier, A., 1990, "DELTA: A Simple and Efficient Parallel Robot," *Robotica*, Vol. 8, pp. 105-109.
Tsai, L. W., Walsh, G. C., and Stamper, R., 1996a, "Kinematics of a Novel Three DOF Translational Platform," Proc. of the 1996 IEEE International Conference on Robotics and Automation, Minneapolis, MN, pp. 3446-3451.
Tsai, L. W., 1996b, "Kinematics of a Three-DOF Platform Manipulator with Three Extensible Limbs," in *Recent Advances in Robot Kinematics*, Edited by J. Lenarcic and V. Parenti-Castelli, Kluwer Academic Publishers, London, pp. 401-410.
Tsai, L. W., 1997, "Multi-degree-of-freedom Mechanisms for Machine Tools and the Like," U.S. Patent, No. 5,656,905.
Tsai, L. W. and Joshi, S., 2000, "Kinematics and Optimization of a Spatial 3-UPU Parallel Manipulator," ASME Journal of Mechanical Design, Vol. 122, No. 4, pp. 439-446.
Tsai, L. W., 2000, *Mechanism Design: Enumeration of Kinematic Structures According to Function*, CRC Press, Boca Raton, FL.
Wenger, P. and Chablat, D., 2000, "Kinematic Analysis of a New Parallel Machine Tool: The Orthoglide," in *Advances in Robot Kinematics*, Edited by J. Lenarcic and M. L. Stanisic, Kluwer Academic Publishers, London, pp 305-314.
Zhao, T. S. and Huang, Z., 2000, "A Novel Three-DOF Translational Platform Mechanism and Its Kinematics," Proc. of the 2000 ASME Design Engineering Technical Conferences, Baltimore, MD, MECH-14101.

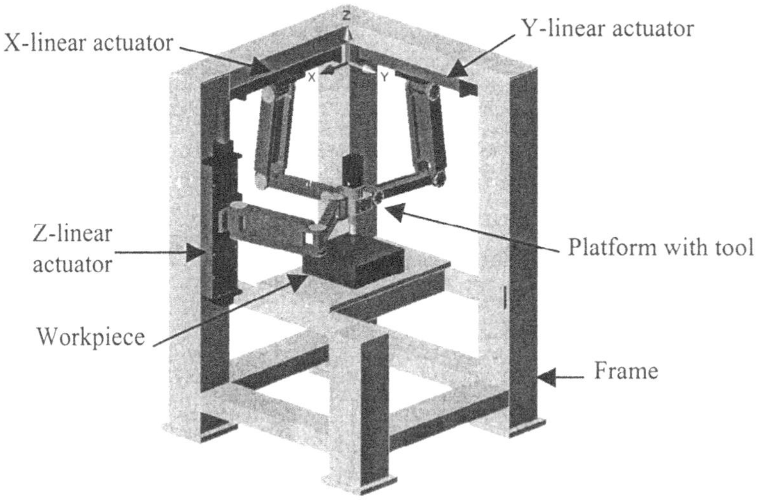

Figure 7. A prototype Cartesian parallel manipulator.

ROTATION SYMMETRY AXES AND THE QUALITY INDEX IN A 3D OCTAHEDRAL PARALLEL ROBOT MANIPULATOR SYSTEM

T. K. Tanev and J. Rooney

Department of Design and Innovation

The Open University, UK

t.k.tanev@open.ac.uk

j.roooney@open.ac.uk

Abstract The geometry of a 3D octahedral parallel robot manipulator system is specified in terms of two rigid octahedral structures (the fixed and moving platforms) and six actuation legs. The symmetry of the system is exploited to determine the behaviour of (a new version of) the quality index for various motions. The main results are presented graphically.

Keywords: parallel robot, quality index, rotation symmetry

1. Introduction

In recent years there has been much interest in the geometry of the Stewart platform (Stewart, 1965) and this work has led to the development of various parallel manipulation systems. Several kinematic perfomance indices have been introduced (Yoshikawa, 1985, Angeles and Lopez-Cajun, 1992, Lee, Duffy and Hunt, 1998, Lee, Duffy and Keler, 1999). In particular the so-called quality index (Lee, Duffy and Hunt, 1998, Lee, Duffy and Keler, 1999) has been used widely. The values of the index are usually plotted for a range of movements of the platform within the workspace. Investigators have usually defined the quality index as the ratio of the *absolute* value of the Jacobian determinant to the *maximal absolute* value of the Jacobian determinant. Hence because of the use of absolute values, the usual quality index lies in the range (0,1). Moreover, symmetry arguments have often been used to restrict examination of the behaviour of the quality index to configurations for which the manipulator is close to its central 'home' position.

In this paper, we investigate a system generalized from the Stewart platform in two ways. Firstly, we generalize the geometry of the manipulator to produce a system consisting of a *3D-solid* moving platform and a *3D-solid* fixed platform (so that neither the six joints on the moving platform nor those on the fixed platform lie in single planes). Secondly, we introduce a new form of the quality index, which does not

J. Lenarčič and F. Thomas (eds.), Advances in Robot Kinematics, 29–38.

involve absolute values. We calculate and plot the actual values (positive, zero and negative) of this new quality index for a *two*-parameter family of movements (either a one-parameter family of rotations about a one-parameter family of rotation axes, or a two-parameter family of screw motions about a single fixed screw axis). The plot yields a surface since two parameters are varied. The paper presents an exploration of the type of behaviour possible in highly symmetrical parallel systems more general than the Stewart platform. Further work is planned to extend the approach to systems with progressively less symmetry. Systems whose geometry differs only slightly from the highly symmetrical case presented here become asymmetrical, yet they retain very similar general characteristics in the behaviour of the quality index.

2. Geometry and Symmetry of 3D Octahedral 6-6 Platforms

The geometry of the Stewart parallel manipulator (Stewart, 1965) is octahedral in the sense that, in a typical position, the moving triangular platform and the fixed triangular base form opposite faces of an octahedron, and the six legs form the edges of the other six triangular faces (Fig. 1) (Lee, Duffy and Hunt, 1998). Often the dimensions are such that the moving and fixed platforms are congruent equilateral triangles and in the 'home' position each of the six legs has the same length, equal to the length of the platform edge. In this position the system forms a regular octahedron, but in most other positions the polyhedral regularity is lost and the system has the geometry of a scalene octahedron. However, since the triangular platforms are rigid they remain regular polygons throughout any motion. Both the moving platform and the fixed platform have an axis of 3-fold rotation symmetry, together with three planes of reflection symmetry perpendicular to the platform surface.

It is possible to generalize the Stewart platform geometry by re-locating the spherical joints in a 2D polygonal arrangement other than an equilateral triangle (see for example Duffy, Rooney, Knight and Crane, 2000, Merlet, 1993, Mayer St-Onge and Gosselin, 2000, Zanganeh and Angeles, 1997). In general this will form a (non-regular) hexagon (Fig.1). The resulting system still consists of an hexagonal platform moving with respect to a fixed hexagonal platform. Similar generalizations lead to quadrilateral, pentagonal and in general planar polygonal platforms [Rooney, Duffy and Lee, 1999].

However, a 3D arrangement of the spherical joints is possible. A particular example of this further generalization is shown in Fig.1 where the joints on the moving 'platform' are located at the vertices of a regular solid octahedron. The joints on the fixed platform may be similarly arranged at the vertices of a different regular solid octahedron and the resulting system then has a more general geometry than previous types of parallel robot manipulator derived from the Stewart platform.

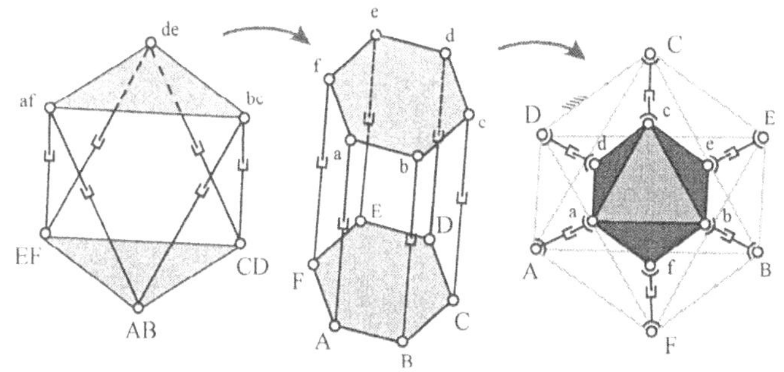

Figure 1. Three types of 6-6 parallel manipulator, progressively more general.

The moving 'platform' now itself has the shape, and hence the symmetry, of a regular octahedron (Fig.2).

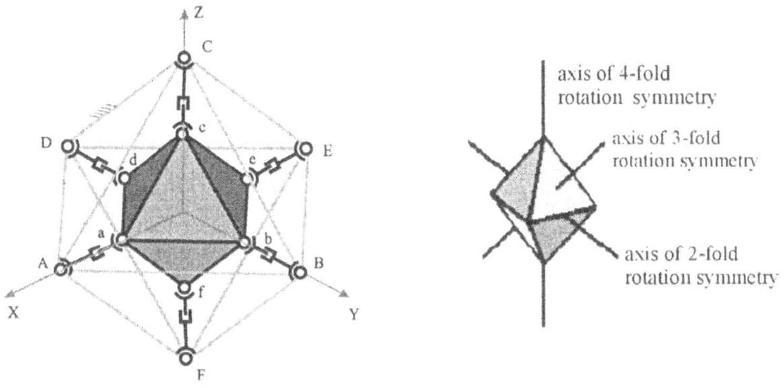

a. The 3D octahedral platform

b. Axes of rotation symmetry

Figure 2. The 3D octahedral platform and some axes of rotation symmetry

The latter has three 4-fold axes of rotation symmetry (each passing through a pair of opposite vertices), four 3-fold axes of rotation symmetry (each passing through the centers of a pair of opposite faces), and six 2-fold axes of rotation symmetry (each passing through the centers of a pair of opposite edges). There are also several types of planes of reflection symmetry, one of which passes through two opposite vertices and bisects the solid octahedron. This generalized parallel manipulator system is referred to as a 3D octahedral platform.

3. The 3D Octahedral Platform and its Quality Index

The octahedral manipulator shown in Fig. 2 is fully parallel and it has a linear actuator on each of its six legs. The six spherical joints of the moving platform are situated at the vertices of a regular octahedron and similarly the six spherical joints of the base are situated at the vertices of another regular octahedron. These two octahedra are each dual to a cube. It is well known that the cube and the octahedron are dual to each other. The first octahedron is formed by joining the centers of adjacent faces of the cube. The size of the second one is chosen in a way that the edge of the octahedron is perpendicular to its parent (the edge of the cube) and cuts through the parent at its midpoint (Fig. 3).

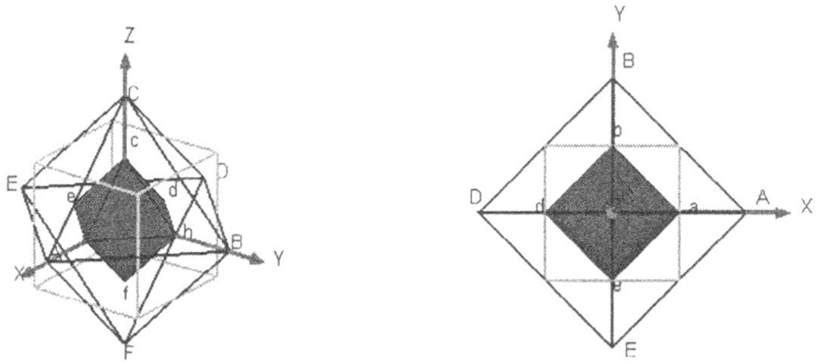

Figure 3. 3D representation and a plan view of the two octahedra and the dual cube.

Let σ denote the edge length of the outer octahedron, then the coordinates of the edges of the octahedrons are [Lee, Duffy and Rooney, 2000]:

$$\begin{aligned} &A:\begin{bmatrix}\frac{\sigma}{\sqrt{2}} & 0 & 0\end{bmatrix},\quad B:\begin{bmatrix}0 & \frac{\sigma}{\sqrt{2}} & 0\end{bmatrix},\quad C:\begin{bmatrix}0 & 0 & \frac{\sigma}{\sqrt{2}}\end{bmatrix},\\ &D:\begin{bmatrix}-\frac{\sigma}{\sqrt{2}} & 0 & 0\end{bmatrix},\ E:\begin{bmatrix}0 & -\frac{\sigma}{\sqrt{2}} & 0\end{bmatrix},\ F:\begin{bmatrix}0 & 0 & -\frac{\sigma}{\sqrt{2}}\end{bmatrix}\end{aligned} \tag{1}$$

$$\begin{aligned} &a:\begin{bmatrix}\frac{\sigma}{2\sqrt{2}} & 0 & 0\end{bmatrix},\quad b:\begin{bmatrix}0 & \frac{\sigma}{2\sqrt{2}} & 0\end{bmatrix},\quad c:\begin{bmatrix}0 & 0 & \frac{\sigma}{2\sqrt{2}}\end{bmatrix},\\ &d:\begin{bmatrix}-\frac{\sigma}{2\sqrt{2}} & 0 & 0\end{bmatrix},\ e:\begin{bmatrix}0 & -\frac{\sigma}{2\sqrt{2}} & 0\end{bmatrix},\ f:\begin{bmatrix}0 & 0 & -\frac{\sigma}{2\sqrt{2}}\end{bmatrix}\end{aligned} \tag{2}$$

The origin of the fixed coordinate frame is placed at the center of the outer octahedron and the origin of the coordinate frame attached to the moving platform is at the center of the inner octahedron.

The Plücker line coordinates of the six legs can be expressed by Grassmann determinants. Hence, the direction ratios of the line are

$$L=\begin{vmatrix}1 & x_1\\ 1 & x_2\end{vmatrix},\quad M=\begin{vmatrix}1 & y_1\\ 1 & y_2\end{vmatrix},\quad N=\begin{vmatrix}1 & z_1\\ 1 & z_2\end{vmatrix}, \tag{3}$$

and the moments of the line segments about the coordinate axes are

$$P=\begin{vmatrix}y_1 & z_1\\ y_2 & z_2\end{vmatrix},\quad Q=\begin{vmatrix}z_1 & x_1\\ z_2 & x_2\end{vmatrix},\quad R=\begin{vmatrix}x_1 & y_1\\ x_2 & y_2\end{vmatrix}, \tag{4}$$

where (x_1, y_1, z_1) and (x_2, y_2, z_2) are coordinates of two point of the line.

Let the position and orientation of the moving octahedron with respect to the fixed octahedron be given by the transformation matrix **A**, where

$$\mathbf{A}=\begin{bmatrix} l_1 & l_2 & l_3 & s_x\\ m_1 & m_2 & m_3 & s_y\\ n_1 & n_2 & n_3 & s_z\\ 0 & 0 & 0 & 1\end{bmatrix} \tag{5}$$

Then using Eqs.(3) and (4), the Plücker line coordinates of the six legs (Fig.2) can be written as:

$$\$_1=\left[\frac{\sqrt{2}}{2}\sigma\left(\frac{l_1}{2}-1\right)+s_x,\frac{\sqrt{2}}{4}\sigma m_1+s_y,\frac{\sqrt{2}}{4}\sigma n_1+s_z;0,-\frac{\sqrt{2}}{2}\sigma\left(\frac{\sqrt{2}}{4}\sigma n_1+s_z\right),\frac{\sqrt{2}}{2}\sigma\left(\frac{\sqrt{2}}{4}\sigma m_1+s_y\right)\right]$$

$$\$_2=\left[\frac{\sqrt{2}}{2}\sigma\left(1-\frac{l_1}{2}\right)+s_x,-\frac{\sqrt{2}}{4}\sigma m_1+s_y,-\frac{\sqrt{2}}{4}\sigma n_1+s_z;0,\frac{\sqrt{2}}{2}\sigma\left(-\frac{\sqrt{2}}{4}\sigma n_1+s_z\right),-\frac{\sqrt{2}}{2}\sigma\left(-\frac{\sqrt{2}}{4}\sigma m_1+s_y\right)\right]$$

$$\$_3=\left[\frac{\sqrt{2}}{4}\sigma l_2+s_x,\frac{\sqrt{2}}{2}\sigma\left(\frac{m_2}{2}-1\right)+s_y,\frac{\sqrt{2}}{4}\sigma n_2+s_z;\frac{\sqrt{2}}{2}\sigma\left(\frac{\sqrt{2}}{4}\sigma n_2+s_z\right),0,-\frac{\sqrt{2}}{2}\sigma\left(\frac{\sqrt{2}}{4}\sigma l_2+s_x\right)\right] \tag{6}$$

$$\$_4=\left[-\frac{\sqrt{2}}{4}\sigma l_2+s_x,\frac{\sqrt{2}}{2}\sigma\left(1-\frac{m_2}{2}\right)+s_y,-\frac{\sqrt{2}}{4}\sigma n_2+s_z;-\frac{\sqrt{2}}{2}\sigma\left(-\frac{\sqrt{2}}{4}\sigma n_2+s_z\right),0,\frac{\sqrt{2}}{2}\sigma\left(-\frac{\sqrt{2}}{4}\sigma l_2+s_x\right)\right]$$

$$\$_5=\left[\frac{\sqrt{2}}{4}\sigma l_3+s_x,\frac{\sqrt{2}}{4}\sigma m_3+s_y,\frac{\sqrt{2}}{2}\sigma\left(\frac{n_3}{2}-1\right)+s_z;-\frac{\sqrt{2}}{2}\sigma\left(\frac{\sqrt{2}}{4}\sigma m_3+s_y\right),\frac{\sqrt{2}}{2}\sigma\left(\frac{\sqrt{2}}{4}\sigma l_3+s_x\right),0\right]$$

$$\$_6=\left[-\frac{\sqrt{2}}{4}\sigma l_3+s_x,-\frac{\sqrt{2}}{4}\sigma m_3+s_y,\frac{\sqrt{2}}{2}\sigma\left(1-\frac{n_3}{2}\right)+s_z;\frac{\sqrt{2}}{2}\sigma\left(-\frac{\sqrt{2}}{4}\sigma m_3+s_y\right),-\frac{\sqrt{2}}{2}\sigma\left(-\frac{\sqrt{2}}{4}\sigma l_3+s_x\right),0\right]$$

Here, the six leg are as follows (Fig.): $\$_1=\mathbf{aA}$, $\$_2=\mathbf{dD}$, $\$_3=\mathbf{bB}$, $\$_4=\mathbf{eE}$, $\$_5=\mathbf{cC}$ and $\$_6=\mathbf{fF}$.

In this paper the quality index (Lee, Duffy and Hunt, 1998, Lee, Duffy and Keler, 1999) is defined in a new form, without using absolute values, as:

$$\lambda=\frac{\det\mathbf{J}}{|\det\mathbf{J}|_{\max}}, \tag{7}$$

where $\mathbf{J}$ is the 6x6 Jacobian matrix of the normalized coordinates of the six legs.

The normalized Jacobian matrix of the six lines, which are changed to unit length, can then be written as

$$\mathbf{J} = \begin{bmatrix} \frac{1}{l_1}\$_1 & \frac{1}{l_2}\$_2 & \frac{1}{l_3}\$_3 & \frac{1}{l_4}\$_4 & \frac{1}{l_5}\$_5 & \frac{1}{l_6}\$_6 \end{bmatrix} \tag{8}$$

where $l_i = \sqrt{L_i^2 + M_i^2 + N_i^2}$, $i = 1..6$.

The quality index given by Eq.(7) takes into account the sign of the determinant of the Jacobian matrix $\mathbf{J}$, and that is why it is bounded between -1 and 1, which (depending on the particular configuration) may differ in sign from the quality index defined by Lee, Duffy and Hunt, 1998, and Lee, Duffy and Keler, 1999. A zero value of the quality index indicates singularities and the value 1 (or -1) for the quality index determines the best configuration of the robot. For a particular system the positive orientation on each leg is fixed by convention and hence the value of λ is relative to this convention.

3.1 Extrema of the Jacobian Determinant

In this paper, the quality index is computed with the edge length σ of the outer octahedron equal to 1. Expressing the elements of the matrix $\mathbf{A}$ given by Eq.(5) in terms of screw parameters (α, β, ϕ, d), the determinant of the Jacobian (non-normalized) for screw motions about axes, which pass through the origin of the coordinate system, can be expressed as:

$$\det \mathbf{J}^* = -\frac{\sqrt{2}}{64}\sin(\alpha)\cos(\alpha)\cos(\beta)\sin(\phi)\left(\cos(\beta)^2 - 1\right)\left(5 - 9\cos(\phi) + 4\cos(\phi)^2\right), \tag{9}$$

where $\mathbf{J}^*$ indicates the non-normalized Jacobian; ϕ is the rotation angle, and α and β are the spherical polar coordinates (longitude and co-latitude) determining the positive direction of the screw axis.

The determinant $\det \mathbf{J}^*$ is equal to zero when: $i)\, \phi = n\pi\ (n = 0,1,2...)$; $ii)\, \alpha = n\pi/2\ (n = 0,1,2...)$; $iii)\, \beta = n\pi/2\ (n = 0,1,2...)$. This implies that the robot is in singular configurations when the screw motion is about the axes of 2-fold and 4-fold rotation symmetry, regardless of the angle of rotation about these axes.

Now, we obtain the maximum value of $\det \mathbf{J}$ the determinant of the normalized Jacobian matrix. We consider rotations about axes that pass through the origin of the coordinate system. In this case, $\det \mathbf{J}$ (not detailed here for lack of space) is a function of three variables,

$$\det \mathbf{J} = f(\alpha, \beta, \phi). \tag{10}$$

Taking the partial derivatives $\partial f/\partial\alpha$, $\partial f/\partial\beta$ and $\partial f/\partial\phi$, and equating to zero yield three equations. We use the tan(half-angle) substitution for $\sin(\phi), \cos(\phi), \sin(\alpha), \cos(\alpha), \sin(\beta)$, and $\cos(\beta)$ in the latter equations. Some of the factors of the above equations lead to the following solutions of the angle $\alpha: \pi/4; -\pi/4; 3\pi/4; -3\pi/4$ (only the solutions wich do not yield singular configurations are listed). These angles imply that there is a local maximum of the determinant of the normalized Jacobian matrix for rotations about axes in a plane of reflection symmetry (the axes of rotation of the moving octahedron pass through the origin and lie in a plane of reflection symmetry that passes through two vertices and cuts an edge at its midpoint) (Fig.4a).

Now we consider rotations about axes in a plane of reflection symmetry ($\alpha = \pi/4$). In this case, $\det \mathbf{J}$ is a function of two variables,

$$\det \mathbf{J} = f(\beta, \phi). \tag{11}$$

Again, taking the partial derivatives $\partial f/\partial\beta$ and $\partial f/\partial\phi$, equating to zero and making the tan(half-angle) substitution t_1 for $\sin(\phi), \cos(\phi)$ and t_2 for $\sin(\beta), \cos(\beta)$ yield two equations. Factorising these two equations and solving them together yield the following solutions (only real solutions are listed here): $(t_2 = 0,\ t_1 = \pm 1.16481293)$, $(t_2 = \pm 1.93185165,\ t_1 = \pm 0.74267285)$, $(t_2 = \pm 0.51763809,\ t_1 = \pm 0.74267285)$. Note that two of the factors lead to a polynomial in the single variable (t_2) of order 64. Hence, the resulting angles, that do not yield singular configurations, are: $\beta_{1,2} = \pm 0.95531661$, $\beta_{3,4} = \pm 2.18627603$; $\phi_{1,2} = \pm 1.27759043$. The four solutions for β determine exactly the axes of 3-fold rotation symmetry of the octahedron. This means that the maximum of $\det \mathbf{J}$ occurs for a rotation at $\phi_{1,2} = \pm 1.27759043$ about the axis of 3-fold rotation symmetry.

A similar analysis can be carried out for a combined translation and rotation about and along the axis of 3-fold rotation symmetry, which yields the following real solutions for d (the translation distance) and angle ϕ: $d = 0$; $\phi_{1,2} = \pm 1.27759043$. In this case, two of the factors lead to univariate (t_1) polynomial of order 20. The obtained result implies that the value of the quality index is smaller when translating from the central position.

Hence, using the values of the variables obtained above, the maximum absolute value of $\det \mathbf{J}$ is obtained as: $|\det \mathbf{J}|_{\max} = 0.23455336$.

3.2 Rotation Axes in a Plane of Reflection Symmetry

In this case, the axes of rotation of the moving octahedron pass through the origin and lie in a plane of reflection symmetry that passes through two vertices and cuts an edge at its midpoint (Fig.4a).

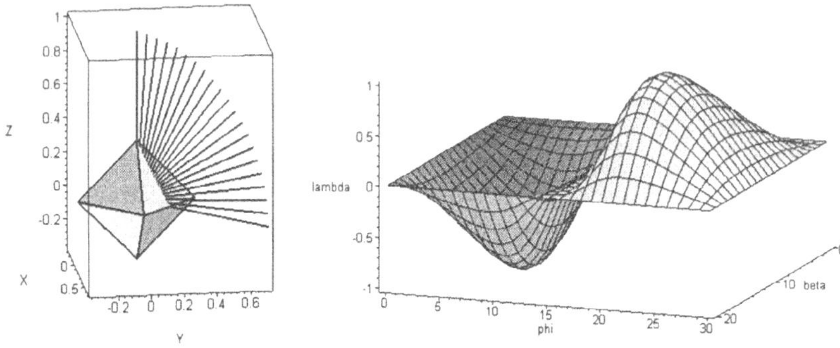

a. Axes of rotation   *b*. The quality index

Figure 4. The axes of rotation and the quality index

Fig.4a shows the axes of rotation for which the quality index is obtained and Fig.4b shows the quality index plotted as a function of ϕ and β. Here, ϕ denotes the angle of rotation about the axis and β is the angle between the axis of rotation and the Z-axis. The convention for measuring a positive sense of rotation is the usual one in which a positive angle advances a right hand screw along the positive direction of the screw axis. For this reason the rotation axes and screw axes are considered to be directed lines. The same notations and conventions will be used further in the paper. In Fig.4 the angle of rotation ϕ varies from $-\pi$ to $+\pi$, and the angle β is within the range $(0, \pi/2)$. The range of ϕ is shown in units of $\pi/15$, and the range of β is shown in units of $\pi/10$. The analysis was carried out for a variation of angle β throughout the full range $(0, 2\pi)$ but only the results for the range $(0, \pi/2)$ are shown in Fig.4. It is clear from Fig.4, that the quality index is zero for rotation of the moving octahedron about the axes of 2-fold and 4-fold rotation symmetry. This means that the robot is in singular configurations regardless of the angle of rotation about these axes. On the other hand, we determined that the robot reaches its best configuration (the quality index is 1 or –1) for a rotation about the axis of 3-fold rotation symmetry.

3.3 Axis of 3-Fold Rotation Symmetry

In this case, the quality index is obtained for translation and rotation along and about the axis of 3-fold rotation symmetry. The obtained quality index and the axis of 3-fold rotation symmetry for the octahedron are shown in Fig. 5.

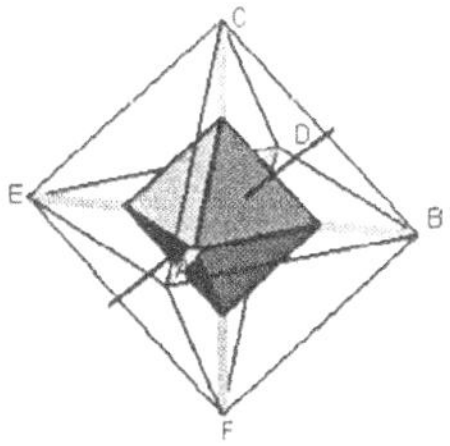

a. The axis of 3-fold rotation symmetry

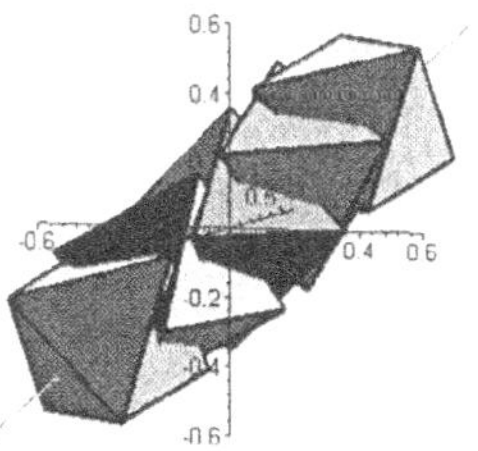

b. Some positions of the moving octahedron during the movement

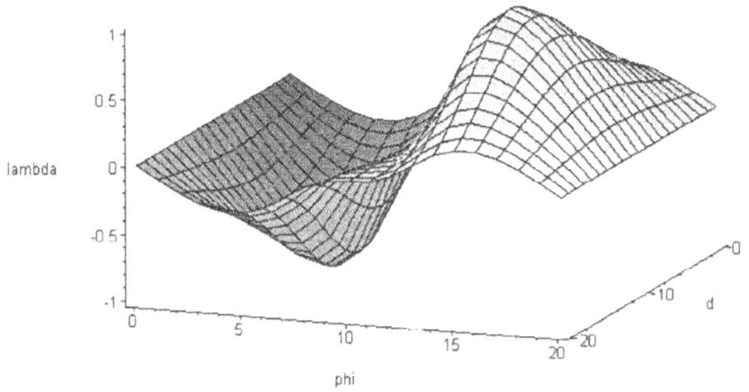

c. The quality index

Figure 5. The axis of rotation, positions of the moving octahedron and the quality index

Here, the angle of rotation ϕ varies from $-\pi$ to $+\pi$, and the translation d varies from -0.5σ to $+0.5\sigma$ (in Fig. 5 the range of ϕ is shown in units of $\pi/10$, and the range of d is shown in units of 0.05σ). Obviously, the quality index is zero when the angle of rotation ϕ is $-\pi$, 0, or $+\pi$, regardless of the translation. On the other hand, the quality index reaches its maximum (1 or –1) when the moving octahedron is in the center of the fixed one ($d = 0$) and rotates about the axis of 3-fold rotation symmetry.

4. Discussion and conclusions

The paper has examined some aspects of the behavior of (a new version of) the quality index for a 3D octahedral parallel manipulator system in terms of symmetry. The quality index has been plotted for two types of motion. The first involved the moving platform rotating about a sequence of axes lying in a plane of reflection symmetry. The second involved the moving platform screwing about an axis of 3-fold symmetry.

Other types of motion have been explored but the details are not reported here.

Three main conclusions have resulted from this work. Firstly, the value of the quality index has been shown to be zero for all rotations about either an axis of 2-fold or an axis of 4-fold symmetry. (It is zero at the 'home' position of course, and it remains zero for any translation away from this position.) Secondly, the maximum value of the quality index occurs at each axis of 3-fold rotation symmetry for a particular angle of rotation. Finally, allowing the quality index to attain negative values leads to two separate types of configuration - one for positive values and the other for negative values. Each of these two configurations has an extremum.

References

Angeles, J., Lopez-Cajun, C. S. (1992), Kinematic isotropy and the conditioning index of serial robotic manipulators, *The International Journal of Robotics Research*, no. 6, vol. 11, pp.560-571.

Duffy, J., Rooney, J., Knight, B. and Crane, C. D. (2000), A review of a family of self-deploying tensegrity structures with elastic ties, *The Shock and Vibration Digest*, no. 2, vol. 32, pp. 100-106.

Lee, J., Duffy, J. and Hunt, K.H. (1998), A practical quality index on the octahedral manipulator, *The International Journal of Robotics Research*, no. 10, vol. 17, pp.1081-1090.

Lee, J. Duffy, J. and Keler, M. (1999), The optimum quality index for the stability of in-parallel planar platform devices, *Journal of Mechanical Design*, vol. 121, pp.15-20.

Lee, J., Duffy, J. and Rooney, J. (2000), An initial investigation into the geometrical meaning of the (pseudo-) inverses of the line matrices of the edges of platonic polyhedra, *Proceedings of a Symposium Commemorating the Legacy, Works, and Life of Sir Robert Stawel Ball Upon the 100th Anniversary of A Treatise on the Theory of Screws*, Trinity College, University of Cambridge, UK., July 9-12.

Mayer St-Onge, B. and Gosselin, C. M. (2000), Singularity analysis and representation of the general Gough-Steward platform, *The International Journal of Robotics Research*, no. 3, vol. 19, pp.271-288.

Merlet, J-P. (1993), Direct kinematics of parallel manipulators, *IEEE Transaction on Robotics and Automation*, no. 6, vol. 9, pp.842-846.

Rooney, J., Duffy, J. and Lee, J. (1999), Tensegrity and compegrity configurations in anti-prism manipulator platforms, *Proceedings of the Tenth World Congress on the Theory of Machines and Mechanisms,* Oulu, Finland, pp. 1278-1287.

Stewart, D. (1965), A platform with six degrees of freedom, *Proceedings of Institution of Mechanical Engineers*, London, UK, no. 15, vol. 180, pp.371-386.

Yoshikawa, T. (1985), Manipulability of robotic mechanism, *The International Journal of Robotics Research*, no. 2, vol. 4, pp.3-9.

Zanganeh, K.E. and Angeles, J. (1997), Kinematic isotropy and optimum design of parallel manipulators, *The International Journal of Robotics Research*, no. 2, vol. 16, pp.185-197.

LINEARIZED KINEMATICS FOR STATE ESTIMATION IN ROBOTICS

K. Parsa, J. Angeles, and A.K. Misra
McGill University,
Department of Mechanical Engineering and
Centre for Intelligent Machines,
3480 University, Montreal, QC H3A 2A7, Canada,
kourosh|angeles|misra@cim.mcgill.ca

Abstract The linerization of the kinematic relations between the twist and the twist-rate of a rigid body of a structurally flexible robotic system and the system generalized coordinates and generalized velocities is reported here. This linearization is required for the state estimators wherein the generalized coordinates and velocities of the underlying mechanical subsystem are to be inferred using the values measured, calculated, or obtained otherwise for the twist and its time derivative, as pertaining to one or more of the links.

Keywords: Kinematics, linearized models, state estimation, twist, twist-rate

1. Introduction

State estimators, a.k.a. observers, use relations between the outputs and the states of a system to estimate the system states from output-variable measurements. A great number of today's well-known state-estimation techniques, e.g., the ones discussed in Misawa and Hedrick, 1989; Canudas de Witt and Slotine, 1991; and Sanchis and Nijmeijer, 1998, among many others, rely on the linearity of the above-mentioned relations. In the case of nonlinear state-output relations, however, it is suggested that, for such techniques, one may linearize the relations and then try the technique using the linearized relations on the system.

Now, let us assume that the acceleration measurements made on the end-effector of a flexible-link robot by a measurement system, such as the one discussed in Parsa et al., 2001, are intended to be used to estimate the *flexural coordinates*[1] of the robot flexible links. In such a case, one has to linearize the state-output relations; to this end, the calculation of the partial derivatives of the twist or the twist-rate of the end-effector, or both, with respect to the states—the generalized coordinates and generalized velocities of the system—is necessary.

The partial derivatives at play may, of course, be obtained by deriving the expressions for the twist and the twist-rate, and then differentiating them sym-

J. Lenarčič and F. Thomas (eds.), Advances in Robot Kinematics, 39–48.

bolically. However, the drawback of such a method is that the differentiations have to be done for each manipulator separately, and that redundant algebraic manipulations and arithmetic operations are unavoidable. In this paper, we derive, in closed form, the required relations in terms of other usually available entities, such as the robot Jacobian, its time derivative, the angular velocity, and the translational velocity. These variables are usually available in the control loop of a robot, so that the number of calculations needed will be the least. In addition to the foregoing advantages, obtaining the partial derivatives of the twist in closed form can ease the dynamics modelling of multibody systems in general.

Although state estimation for structurally flexible robots has been the motive of this work, it should be noted that the applications of linearized models are diverse; for example, they can also be employed in the sensitivity analysis for feedback control and for mechanical design.

Some basic concepts of rigid-body kinematics such as attitude representation through the Euler parameters, the twist vector, and the robot Jacobian are recalled in Section 2. In Section 3, the partial derivatives of the twist and the twist-rate vectors with respect to the generalized coordinates and generalized velocitiess are obtained. Closing remarks are given in Section 4.

All the vectors used in this paper are expressed in the body-frame. In fact, all frames are equivalent insofar as kinematics is concerned. However, since the relations reported herein are intended for real-time control, discriminating between the two types of inertial and local frames is essential. Moreover, throughout this paper, overdots denote *elementwise* differentiation with respect to time.

2. Rigid-Body Kinematics

The main goal of this paper being the linearization of the link twist and its time rate of change in closed form, we start with the linearized relaitons

$$\delta\bar{\mathrm{t}} = \frac{\partial\bar{\mathrm{t}}}{\partial\mathrm{q}}\,\delta\mathrm{q} + \frac{\partial\bar{\mathrm{t}}}{\partial\dot{\mathrm{q}}}\,\delta\dot{\mathrm{q}}, \tag{1a}$$

$$\delta\dot{\bar{\mathrm{t}}} = \frac{\partial\dot{\bar{\mathrm{t}}}}{\partial\mathrm{q}}\,\delta\mathrm{q} + \frac{\partial\dot{\bar{\mathrm{t}}}}{\partial\dot{\mathrm{q}}}\,\delta\dot{\mathrm{q}} + \frac{\partial\bar{\mathrm{t}}}{\partial\dot{\mathrm{q}}}\,\delta\ddot{\mathrm{q}}. \tag{1b}$$

The *Cartesian twist* $\bar{\mathrm{t}}$ of a body and its time rate $\dot{\bar{\mathrm{t}}}$, defined in Subection 2.2, are the output variables. To derive the matrix coefficients of the above equations, a review of some basic concepts is in order.

2.1 Attitude Representation

The attitude of a rigid body can be expressed using many different three- or four-parameter representations; for more information, see Wertz, 1978, and

Goldstein, 1980. Among these different representations, the Euler parameters have proven to be the best regarding the ease of algebraic manipulation, algebraic robustness, and ease of integration into the dynamics models of multi-body sytems.

The Euler parameters constitute a set of unit quaternions and can be expressed as a four-dimensional array $\hat{\mathbf{u}}$ given by

$$\hat{\mathbf{u}} = \begin{bmatrix} \mathbf{u}^T & u_0 \end{bmatrix}^T \triangleq \begin{bmatrix} \mathbf{e}^T \sin\frac{\phi}{2} & \cos\frac{\phi}{2} \end{bmatrix}^T. \tag{2}$$

It can be shown that $\dot{\hat{\mathbf{u}}}$, the time-rate of change of $\hat{\mathbf{u}}$ in the body-frame, is related to the angular velocity $\boldsymbol{\omega}$ of the body, expressed in the same frame, by

$$\dot{\hat{\mathbf{u}}} = \mathbf{H}(\hat{\mathbf{u}})\,\boldsymbol{\omega}, \tag{3}$$

$$\mathbf{H} \equiv \mathbf{H}(\hat{\mathbf{u}}) \triangleq \frac{1}{2}\begin{bmatrix} u_0\mathbf{1} - \mathbf{U} & -\mathbf{u} \end{bmatrix}^T \tag{4}$$

in which $\mathbf{U}$ is the cross-product matrix[2] (CPM) of $\mathbf{u}$, the vector part of the quaternion. A similar relation, given in Angeles, 1997, holds when all the vectors are expressed in an inertial frame.

Furthermore, it can be shown that the inverse relation of Eq. 3 is given by

$$\boldsymbol{\omega} = \overline{\mathbf{H}}(\hat{\mathbf{u}})\,\dot{\hat{\mathbf{u}}}, \tag{5}$$

$$\overline{\mathbf{H}} \equiv \overline{\mathbf{H}}(\hat{\mathbf{u}}) \triangleq 4\mathbf{H}^T \equiv 2\begin{bmatrix} u_0\mathbf{1} - \mathbf{U} & -\mathbf{u} \end{bmatrix}, \tag{6}$$

and that the right-hand sides of Eqs. 3 & 5 are thus the linear transformations of the Euler parameters into angular velocity and vice versa. We now have

Lemma 1.1. *The relations between the time derivative of the Euler-para-meter array and the angular-velocity vector of a rigid body, Eqs. 3 & 5, can equivalently be expressed as*

$$\dot{\hat{\mathbf{u}}} = \mathbf{K}(\boldsymbol{\omega})\,\hat{\mathbf{u}}, \tag{7}$$

$$\boldsymbol{\omega} = \overline{\mathbf{K}}(\dot{\hat{\mathbf{u}}})\,\hat{\mathbf{u}}, \tag{8}$$

respectively, where

$$\mathbf{K} \equiv \mathbf{K}(\boldsymbol{\omega}) \triangleq \frac{1}{2}\begin{bmatrix} -\boldsymbol{\Omega} & \boldsymbol{\omega} \\ -\boldsymbol{\omega}^T & 0 \end{bmatrix}, \tag{9}$$

$$\overline{\mathbf{K}} \equiv \overline{\mathbf{K}}(\dot{\hat{\mathbf{u}}}) \triangleq -\dot{\overline{\mathbf{H}}} \tag{10}$$

in which $\boldsymbol{\Omega}$ is the cross-product matrix of $\boldsymbol{\omega}$.

Another result needed for the derivations reported in Section 3 is recalled below:

Lemma 1.2. (Nikravesh et al., 1985) *Matrices* $\mathbf{H}$ *and* $\overline{\mathbf{H}}$, *defined in Eqs. 4 & 6, obey*

$$\overline{\mathbf{H}} \cdot \dot{\mathbf{H}} \equiv \frac{1}{2}\mathbf{\Omega} \tag{11}$$

where $\mathbf{\Omega}$ *represents the CPM of the angular-velocity vector* $\boldsymbol{\omega}$.

2.2 The Twist Vector and Its Time-Rate

The twist vector of a body is a set of scalar variables that comprise the necessary and sufficient amount of information to determine the velocity field in the body. For a rigid body, these variables can be the three components of the velocity vector of a landmark point and those of the angular-velocity vector of the body. Both vectors are expressed in the same reference frame, whether inertial or not. If the body is flexible, the twist is augmented by a set of generalized velocities representing the body deformation rate. In this paper we use a local frame, i.e., a frame attached to the body, with its origin located at the above-mentioned landmark point.

The twist is usually defined as an array comprising the components of the absolute velocities, both translational and angular, expressed in the local frame. An alternative twist vector that includes the elementwise time derivative of the position vector in the local frame facilitates the partial differentiations needed in this paper. Therefore, two different—but kinematically equivalent—twist vectors are defined herein, accordingly: The one corresponding to the latter is denoted by $\mathbf{t}$ and called briefly the *twist*, while the twist vector corresponding to the former is represented by $\bar{\mathbf{t}}$ and termed the *Cartesian twist*, for its correspondence with the *Cartesian wrench*, i.e.,

$$\mathbf{t} \triangleq \begin{bmatrix} \dot{\mathbf{p}} \\ \boldsymbol{\omega} \end{bmatrix}, \quad \bar{\mathbf{t}} \triangleq \begin{bmatrix} \mathbf{v} \\ \boldsymbol{\omega} \end{bmatrix} \tag{12}$$

where $\mathbf{p}$ and $\mathbf{v}$ are the position vector of the local-frame origin and its absolute velocity, respectively, both expressed in the local frame.

From rigid-body kinematics, one has

$$\bar{\mathbf{t}} = \begin{bmatrix} \dot{\mathbf{p}} + \boldsymbol{\omega} \times \mathbf{p} \\ \boldsymbol{\omega} \end{bmatrix} = \mathbf{R}\,\mathbf{t}, \tag{13}$$

$$\mathbf{R} \triangleq \begin{bmatrix} \mathbf{1} & -\mathbf{P} \\ \mathbf{O} & \mathbf{1} \end{bmatrix} \tag{14}$$

in which $\mathbf{P} \triangleq \mathrm{CPM}(\mathbf{p})$, and $\mathbf{1}$ and $\mathbf{O}$ are the 3×3 identity and zero matrices, respectively.

The translational and angular accelerations $\mathbf{a}$ and $\boldsymbol{\alpha}$ can readily be seen to be related to the time-derivative of the twist vector, called the *twist-rate*, as given below:

$$\mathbf{s} \triangleq \begin{bmatrix} \mathbf{a} \\ \boldsymbol{\alpha} \end{bmatrix} = \dot{\bar{\mathbf{t}}} + \mathbf{W}\bar{\mathbf{t}}, \tag{15}$$

$$\mathbf{W} \triangleq \begin{bmatrix} \boldsymbol{\Omega} & \mathbf{O} \\ \mathbf{O} & \mathbf{O} \end{bmatrix} \tag{16}$$

with $\boldsymbol{\Omega}$ defined already as the cross-product matrix of $\boldsymbol{\omega}$.

2.3 The Jacobian Matrix

In robot kinematics, the Jacobian $\mathbf{J}$ can be defined as the partial derivative of the twist vector with respect to the vector of generalized velocities, i.e.,

$$\mathbf{J} \triangleq \frac{\partial \mathbf{t}}{\partial \dot{\mathbf{q}}} \tag{17}$$

where $\mathbf{q}$ represents the vector of generalized coordinates, $\dot{\mathbf{q}}$ being the vector of generalized velocities. Since the relation between $\mathbf{t}$ and $\dot{\mathbf{q}}$ is linear, the Jacobian is independent of the generalized velocities, and the relation between $\mathbf{t}$ and $\dot{\mathbf{q}}$ can thus be written as

$$\mathbf{t} = \mathbf{J}\dot{\mathbf{q}} + \mathbf{t}_t, \tag{18}$$

with $\mathbf{t}_t$ accounting for the the motion of the base as a prescribed function of time. In general, both $\mathbf{J}$ and $\mathbf{t}_t$ are functions of $\mathbf{q}$ and t, i.e.,

$$\mathbf{J} \equiv \mathbf{J}(\mathbf{q}, t), \quad \text{and} \quad \mathbf{t}_t \equiv \mathbf{t}_t(\mathbf{q}, t).$$

Since all the above arguments are valid for the twist vector of any link, rigid or flexbile, in a kinematic chain, henceforth, the word *twist* can refer to the twist vector of any link.

Given the way in which the twist vector has been defined, one can partition the Jacobian matrix into two blocks, corresponding to the two parts of the twist vector, as

$$\mathbf{J} = \begin{bmatrix} \mathbf{J}_{\dot{p}} \\ \mathbf{J}_{\omega} \end{bmatrix}$$

where

$$\mathbf{J}_{\dot{p}} \triangleq \frac{\partial \dot{\mathbf{p}}}{\partial \dot{\mathbf{q}}}, \quad \text{and} \quad \mathbf{J}_{\omega} \triangleq \frac{\partial \boldsymbol{\omega}}{\partial \dot{\mathbf{q}}}. \tag{19}$$

3. Calculation of the Partial Derivatives

3.1 Partial Derivative of the Angular Velocity

A result needed for deriving closed-form expressions for the partial derivatives of the twist and the twist-rate is given below:

Theorem. *The partial derivative of the angular velocity* $\boldsymbol{\omega}$ *of a rigid body in a kinematic chain with respect to the chain generalized-coordinate vector* $\mathbf{q}$ *can be expressed in terms of the Jacobian* $\mathbf{J}_{\omega}$*, its time rate, and the cross-product matrix* $\boldsymbol{\Omega}$ *of the angular velocity as*

$$\frac{\partial \boldsymbol{\omega}}{\partial \mathbf{q}} = \dot{\mathbf{J}}_{\omega} + \boldsymbol{\Omega}\,\mathbf{J}_{\omega} \tag{20}$$

Proof. We first recall Eq. 5, whereby the angular velocity of a body is expressed as a function of the Euler parameters and their time derivatives, i.e., as $\boldsymbol{\omega} \equiv \boldsymbol{\omega}(\hat{\mathbf{u}}, \dot{\hat{\mathbf{u}}})$. Having this functional relation, one can resort to the chain rule to write the partial derivative of $\boldsymbol{\omega}$ with respect to $\mathbf{q}$ as

$$\frac{\partial \boldsymbol{\omega}}{\partial \mathbf{q}} = \frac{\partial \boldsymbol{\omega}}{\partial \hat{\mathbf{u}}}\frac{\partial \hat{\mathbf{u}}}{\partial \mathbf{q}} + \frac{\partial \boldsymbol{\omega}}{\partial \dot{\hat{\mathbf{u}}}}\frac{\partial \dot{\hat{\mathbf{u}}}}{\partial \mathbf{q}} \tag{21}$$

Then, given that the orientation of a body is, in general, a function of the generalized coordinates as well as time, one can write the first matrix product on the right-hand side of the above relation as

$$\mathbf{T}_1 \triangleq \frac{\partial \boldsymbol{\omega}}{\partial \hat{\mathbf{u}}}\frac{\partial \hat{\mathbf{u}}}{\partial \mathbf{q}} = \frac{\partial \boldsymbol{\omega}}{\partial \hat{\mathbf{u}}}\frac{\partial \dot{\hat{\mathbf{u}}}}{\partial \dot{\mathbf{q}}} = \frac{\partial \boldsymbol{\omega}}{\partial \hat{\mathbf{u}}}\frac{\partial \dot{\hat{\mathbf{u}}}}{\partial \boldsymbol{\omega}}\frac{\partial \boldsymbol{\omega}}{\partial \dot{\mathbf{q}}}$$

which, in light of Eqs. 3, 8, 10, & 19, becomes

$$\mathbf{T}_1 = \overline{\mathbf{K}}\,\mathbf{H}\,\mathbf{J}_{\omega} = -\dot{\overline{\mathbf{H}}}\,\mathbf{H}\,\mathbf{J}_{\omega}.$$

However, since $\overline{\mathbf{H}}\,\mathbf{H} = \mathbf{1}$, upon differentiation with respect to time, we obtain $-\dot{\overline{\mathbf{H}}}\,\mathbf{H} = \overline{\mathbf{H}}\,\dot{\mathbf{H}}$. Thus, according to Lemma 1.2,

$$\mathbf{T}_1 = \frac{1}{2}\boldsymbol{\Omega}\,\mathbf{J}_{\omega}. \tag{22}$$

The second matrix product on the right-hand side of Eq. 21, in turn, can be computed as follows:

$$\mathbf{T}_2 \triangleq \frac{\partial \boldsymbol{\omega}}{\partial \dot{\hat{\mathbf{u}}}}\frac{\partial \dot{\hat{\mathbf{u}}}}{\partial \mathbf{q}} = \frac{\partial \boldsymbol{\omega}}{\partial \dot{\hat{\mathbf{u}}}}\frac{\partial}{\partial \mathbf{q}}\left(\frac{\partial \hat{\mathbf{u}}}{\partial \mathbf{q}}\dot{\mathbf{q}} + \frac{\partial \hat{\mathbf{u}}}{\partial t}\right) = \frac{\partial \boldsymbol{\omega}}{\partial \dot{\hat{\mathbf{u}}}}\left[\frac{\partial}{\partial \mathbf{q}}\left(\frac{\partial \hat{\mathbf{u}}}{\partial \mathbf{q}}\dot{\mathbf{q}}\right) + \frac{\partial}{\partial t}\left(\frac{\partial \hat{\mathbf{u}}}{\partial \mathbf{q}}\right)\right]$$
$$= \frac{\partial \boldsymbol{\omega}}{\partial \dot{\hat{\mathbf{u}}}}\frac{d}{dt}\left(\frac{\partial \hat{\mathbf{u}}}{\partial \mathbf{q}}\right) = \frac{\partial \boldsymbol{\omega}}{\partial \dot{\hat{\mathbf{u}}}}\frac{d}{dt}\left(\frac{\partial \dot{\hat{\mathbf{u}}}}{\partial \dot{\mathbf{q}}}\right)$$

Upon noticing that $\dot{\hat{\mathbf{u}}} \equiv \dot{\hat{\mathbf{u}}}(\hat{\mathbf{u}}, \boldsymbol{\omega})$ and applying the chain rule again the above expression becomes

$$\mathbf{T}_2 = \frac{\partial \boldsymbol{\omega}}{\partial \dot{\hat{\mathbf{u}}}}\frac{d}{dt}\left(\frac{\partial \dot{\hat{\mathbf{u}}}}{\partial \boldsymbol{\omega}}\frac{\partial \boldsymbol{\omega}}{\partial \dot{\mathbf{q}}}\right) = \frac{d}{dt}\left(\frac{\partial \boldsymbol{\omega}}{\partial \dot{\mathbf{q}}}\right) + \frac{\partial \boldsymbol{\omega}}{\partial \dot{\hat{\mathbf{u}}}}\frac{d}{dt}\left(\frac{\partial \dot{\hat{\mathbf{u}}}}{\partial \boldsymbol{\omega}}\right)\frac{\partial \boldsymbol{\omega}}{\partial \dot{\mathbf{q}}}$$

Using Eq. 3, Lemma 1.2, and Eqs. 5 & 19, one can rewrite this relation in the simpler form given below:

$$\mathbf{T}_2 = \dot{\mathbf{J}}_\omega + \frac{1}{2}\boldsymbol{\Omega}\,\mathbf{J}_\omega. \tag{23}$$

Finally, adding up Eqs. 22 & 23 yields the relation sought. □

Example. To verify the validity of the foregoing theorem, assume that for a certain body, its Euler angles (ϕ, θ, ψ) have been taken as the generalized coordinates. We will demonstrate that Eq. 20 holds. To this end, notice that the angular velocity of a rigid body in terms of the Euler angles and their time rates are given by

$$\boldsymbol{\omega} = \begin{bmatrix}\dot{\phi}\sin\theta\sin\psi + \dot{\theta}\cos\psi & \dot{\phi}\sin\theta\cos\psi - \dot{\theta}\sin\psi & \dot{\phi}\cos\theta + \dot{\psi}\end{bmatrix}^T$$

Setting $\mathbf{q} \triangleq \begin{bmatrix}\phi & \theta & \psi\end{bmatrix}^T$, $\mathbf{J}_\omega$ and its time rate can then be found as

$$\mathbf{J}_\omega \equiv \frac{\partial\boldsymbol{\omega}}{\partial\dot{\mathbf{q}}} = \begin{bmatrix}\sin\theta\sin\psi & \cos\psi & 0\\ \sin\theta\cos\psi & -\sin\psi & 0\\ \cos\theta & 0 & 1\end{bmatrix},$$

$$\dot{\mathbf{J}}_\omega = \begin{bmatrix}\dot{\theta}\cos\theta\sin\psi + \dot{\psi}\sin\theta\cos\psi & -\dot{\psi}\sin\psi & 0\\ \dot{\theta}\cos\theta\cos\psi - \dot{\psi}\sin\theta\sin\psi & -\dot{\psi}\cos\psi & 0\\ -\dot{\theta}\sin\theta & 0 & 0\end{bmatrix}.$$

Substituting $\boldsymbol{\omega}$, $\mathbf{J}_\omega$, and $\dot{\mathbf{J}}_\omega$ in the right-hand side of Eq. 20, after some algebraic manipulations, one obtains

$$\dot{\mathbf{J}}_\omega + \boldsymbol{\Omega}\mathbf{J}_\omega = \begin{bmatrix}0 & \dot{\phi}\cos\theta\sin\psi & \dot{\phi}\sin\theta\cos\psi - \dot{\theta}\sin\psi\\ 0 & \dot{\phi}\cos\theta\cos\psi & -\dot{\phi}\sin\theta\sin\psi - \dot{\theta}\cos\psi\\ 0 & -\dot{\phi}\sin\theta & 0\end{bmatrix}$$

which apparently is the partial derivative of $\boldsymbol{\omega}$ with respect to $\mathbf{q}$, thereby confirming the validity of Eq. 20. □

Using the above theorem, one can readily show that, in the inertial frame, a similar relation holds, the only difference being the sign between the two right-hand-side terms of Eq. 20.

3.2 Partial Derivative of the Twist

To derive closed-form expressions for the twist and the twist-rate partial derivatives, yet another result is needed, namely,

Lemma 1.3. *Let* $\mathbf{f} \equiv \mathbf{f}(\mathbf{q}, \dot{\mathbf{q}}, t)$ *be any vector function which is at least twice continuously differentiable, with* $\mathbf{q} \equiv \mathbf{q}(t)$ *a vector function of time, and* $\dot{\mathbf{q}}$ *its elementwise time derivative. The partial derivatives of the elementwise time-derivative of* $\mathbf{f}$ *with respect to* $\mathbf{q}$ *and* $\dot{\mathbf{q}}$ *are given by*

$$\frac{\partial \dot{\mathbf{f}}}{\partial \mathbf{q}} = \frac{d}{dt}\left(\frac{\partial \mathbf{f}}{\partial \mathbf{q}}\right), \tag{24a}$$

$$\frac{\partial \dot{\mathbf{f}}}{\partial \dot{\mathbf{q}}} = \frac{d}{dt}\left(\frac{\partial \mathbf{f}}{\partial \dot{\mathbf{q}}}\right) + \frac{\partial \mathbf{f}}{\partial \mathbf{q}}. \tag{24b}$$

Proof. This lemma can be proven by applying the chain rule. □

Relations similar to these appear quite frequently in the derivation of the Lagrange equations of robotic systems with rigid and flexible links. The main difference with analytical dynamics is that, in the latter, one deals mostly with scalar functions.

By means of the theorem of Subsection 3.1, recalling the definition of the twist $\mathbf{t}$, and applying the first result of Lemma 1.3 on $\dot{\mathbf{p}}$, one can conclude that

$$\frac{\partial \mathbf{t}}{\partial \mathbf{q}} = \dot{\mathbf{J}} + \overline{\mathbf{W}}\,\mathbf{J}, \tag{25}$$

$$\overline{\mathbf{W}} \triangleq \begin{bmatrix} \mathbf{O} & \mathbf{O} \\ \mathbf{O} & \boldsymbol{\Omega} \end{bmatrix}. \tag{26}$$

Then, from the definition of the Cartesian twist, one can show that the derivatives of $\mathbf{t}$ and $\bar{\mathbf{t}}$ with respect to $\mathbf{q}$ are related by

$$\frac{\partial \bar{\mathbf{t}}}{\partial \mathbf{q}} = \mathbf{R}\frac{\partial \mathbf{t}}{\partial \mathbf{q}} + \mathbf{W}\,\mathbf{J} \tag{27}$$

which, by virtue of the above theorem, can be written as

$$\frac{\partial \bar{\mathbf{t}}}{\partial \mathbf{q}} = \mathbf{R}\,\dot{\mathbf{J}} + \overline{\mathbf{R}}\,\mathbf{J}, \tag{28}$$

$$\overline{\mathbf{R}} \triangleq \mathbf{R}\,\overline{\mathbf{W}} + \mathbf{W} = \begin{bmatrix} \boldsymbol{\Omega} & -\mathbf{P}\,\boldsymbol{\Omega} \\ \mathbf{O} & \boldsymbol{\Omega} \end{bmatrix}, \tag{29}$$

the partial derivatives sought thus having been calculated. We now turn our attention to the twist-rate.

3.3 Partial Derivatives of the Twist-Rate

If Eq. 24a is utilized on the twist-rate, we arrive at

$$\frac{\partial \dot{\mathbf{t}}}{\partial \mathbf{q}} = \ddot{\mathbf{J}} + \overline{\mathbf{W}}\,\dot{\mathbf{J}} + \dot{\overline{\mathbf{W}}}\,\mathbf{J}. \tag{30}$$

As pertaining to the time-rate of the Cartesian twist, $\bar{t}$, the same relation leads to

$$\frac{\partial \dot{\bar{t}}}{\partial q} = R\ddot{J} + (\dot{R} + \overline{R})\dot{J} + \dot{\overline{R}}J \tag{31}$$

On the other hand, exploiting Eq. 24b will provide the second set as

$$\frac{\partial \dot{t}}{\partial \dot{q}} = 2\dot{J} + \Omega J \tag{32}$$

$$\frac{\partial \dot{\bar{t}}}{\partial \dot{q}} = 2R\dot{J} + (\dot{R} + \overline{R})J. \tag{33}$$

The last relations reported here are those of the partial derivatives of the vector s, defined in Eq. 15, with respect to q and $\dot{q}$. It can be shown that

$$\frac{\partial s}{\partial q} = \frac{\partial \dot{\bar{t}}}{\partial q} + (W - Y)\frac{\partial \bar{t}}{\partial q}, \tag{34}$$

$$\frac{\partial s}{\partial \dot{q}} = \frac{\partial \bar{t}}{\partial q} + R\dot{J} + \dot{R}J + (W - Y)RJ \tag{35}$$

where $\mathbf{Y}$ is given by

$$Y \triangleq \begin{bmatrix} O & V \\ O & O \end{bmatrix} \tag{36}$$

in which $\mathbf{V}$ is the cross-product matrix of the velocity v.

4. Conclusions

In the realm of robotics, state estimation requires the calculation of the partial derivatives of the twist and the twist-rate with respect to the generalized coordinates and generalized velocities. To obtain these derivatives in closed form, the Euler parameters were used to represent the orientation of a rigid body, and their relation to the angular velocity of a body was recalled. Then, rather general definitions for the twist vector and the Jacobian matrix—in the context of kinematic chains—were discussed. Next, compact, closed-form expressions for the sought partial derivatives were derived in terms of the Jacobian, its time rate, and other variables available. These relations make it possible to evaluate the partial derivatives without resorting to extra symbolic manipulations. The results reported here should be useful in the dynamics of multybody systems at large.

5. Acknowledgements

This research work was supported by the Natural Sciences and Engineering Research Council of Canada, under Research Grants OGP0004532 and OGP0000967.

Notes

1. The flexural coordinates are understood here as a finite set of the generalized coordinates describing the deformation of a flexible body upon discretization. The deformation of a flexible body can be discretized in a number of different manners, e.g., the Finite Elements Method, the Assumed-Modes Method, and the Method of Cubic Splines. In any case, however, the body deformation is in general expressed as

$$\mathbf{u}(\mathbf{r},t) = \boldsymbol{\phi}(\mathbf{r})\mathbf{q}(t)$$

where $\mathbf{r}$ is the local position vector of a point in the body, and t represents time. $\boldsymbol{\phi}(\mathbf{r})$ and $\mathbf{q}(t)$ are called the shape-function matrix and the vector of flexural coordinates, respectively.

2. The cross-product matrix $\mathbf{V}$ of a vector $\mathbf{v}$ is the skew-symmetric matrix given by

$$\mathbf{V} \equiv \mathrm{CPM}(\mathbf{v}) \triangleq \frac{\partial(\mathbf{v}\times\mathbf{x})}{\partial\mathbf{x}}, \quad \forall\mathbf{x}\in\mathbb{R}^3.$$

References

Angeles, J., (1997), *Fundamentals of Mechanical Robotic Systems, Theory, Methods, and Algorithms*, Springer Verlag, New York, NY.

Canudas de Witt, C., and Slotine, J.-J. E., (1991), Sliding Observers for Robot Manipulators, *Automatica*, vol. 27, no. 5, pp. 859–864.

Goldstein, H., (1980), *Classical Mechanics*, Addison-Wesley, Mass.

Misawa, E.A., and Hedrick, J.K., (1989), Nonlinear Observers—A State-of-the-Art Survey, *J. Dynamic Systems, Measurement, and Control*, vol. 111, pp. 344–352.

Nikravesh, P.E., Wehage, R.A., and Kwon, O.K., (1985), Euler Parameters in Computational Kinematics and Dynamics, Part 1, *J. of Mech. Trans. and Auto. in Design*, vol. 107, pp. 366–369.

Parsa, K., Angeles, J., and Misra, A.K., (2001), Pose-and-Twist Estimation of a Rigid Body Using Accelerometers, in *Proc. IEEE Int. Conf. Robotics and Automation*, Seoul, Korea, pp. 2873–2878.

Sanchis, R., and Nijmeijer, H., (1998), Sliding Controller-Sliding Observer Design for Nonlinear Systems, *European J. of Control*, no. 4, pp. 208–234.

Wertz, J.R., (1978), *Spacecraft Attitude Determination and Control*, Astrophysics and Space Science Library, vol. 73, D. Reidel Publishing Company, Dordrecht, The Netherlands.

PERFORMANCE EVALUATION OF THE GRASP OF TWO COOPERATING ROBOTS USING A TYPE MAP

Narcís Gascons
Dept. of Mechanical Engineering
Universitat de Girona, Girona, Spain
narcis.gascons@udg.es

Haijun Su
Robotics Laboratory
University of California, Irvine, USA
suh@eng.uci.edu

J.M. McCarthy
Robotics Laboratory
University of California, Irvine, USA
jmmccart@uci.edu

Abstract For two cooperating robots holding a workpiece, the type map that relates the motion of the workpiece to the extensions of the arms can be used as a tool to characterize the performance of the system. These type maps are built using the discriminants of the closure equations. Several examples are presented to discuss the performance of the system, which is mainly dependant on the geometry of the grasp.

Keywords: Type map, cooperating robots, fingers, grasp

1. Introduction

Systems like two cooperating robots or two fingers holding a workpiece can be studied as spatial parallel kinematic chains. One difficulty of this study is how to represent a workspace or differential workspace that has six dimensions and is not smooth. All representations are partial and focus on some characteristics of the workspace (Merlet, 1997; Gascons, 2001).

The type map, also a partial representation, offers a different perspective of the capacities of the system: first it shows the zones where closure is possible (Cervantes et al., 2001); second, it shows the motion limits for some joints of interest (McCarthy, 2000; Su and McCarthy,

J. Lenarčič and F. Thomas (eds.), Advances in Robot Kinematics, 49–56.

2001). This way, it gives an approach to the limits on the mobility of the system.

2. System description

Fig. 1 shows two cooperating robots or fingers holding a workpiece with their end effectors. Each robot is a SRR open chain. Once both robots grasp the workpiece, the system becomes a closed chain that is described with nine parameters and four generalized coordinates. The parameters are: link lengths l_1, l_2, l_3, l_4; distance g between the S-joints on the ground; and the parameters a (minimal distance), α (twist angle), d_1 and d_2 (distances that locate the joints R_{21} and R_{22} along their axes with respect to the points where these axes intersect with the common normal). These last four parameters characterize the grasp. The generalized coordinates are: the elbow angles ϕ_1 and ϕ_2 for each arm, and the angles θ_1 and θ_2 along the axes of the extreme R-joints.

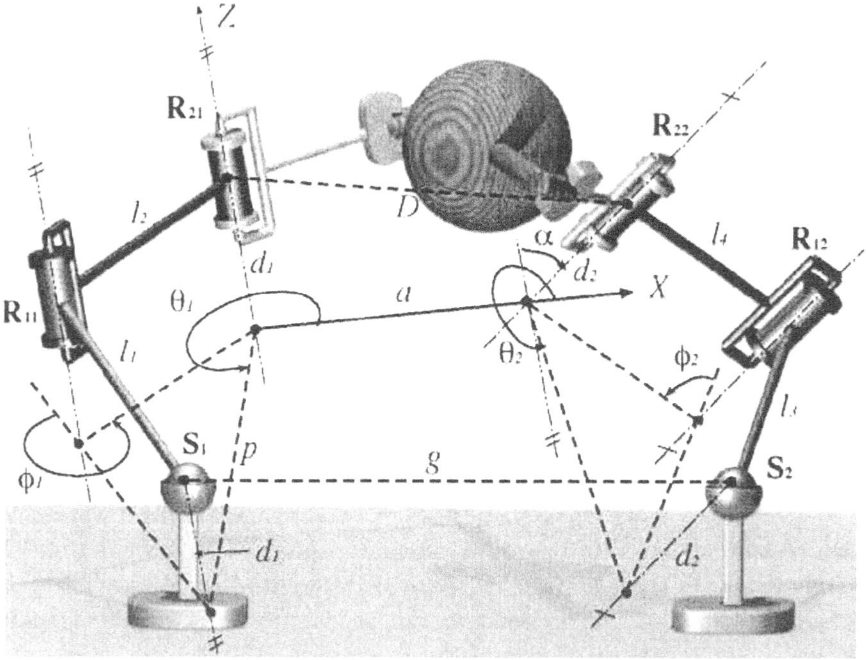

Figure 1. The two robots holding the workpiece, with parameters and coordinates written on them.

In order to fully describe the system's configuration, the set of coordinates should include the rotation of the system around the axis that passes through the two S-joints. However, as this axis is an axis of symmetry for the system's workspace, it can be set apart without altering the validity of the study.

The examples in this paper have the following proportions: $l_1 = l_2 = l_3 = l_4 = 1$ and $g = 2$.

2.1 Grasp

To assign a magnitude to the workpiece, parameter D is made equal to 1. From Fig. 1,

$$D = \sqrt{(d_1 - d_2 \cdot \cos\alpha)^2 + a^2 + (d_2 \cdot sin\alpha)^2} \ . \tag{1}$$

Which means that

$$a = \sqrt{1 + 2 \cdot d_1 d_2 \cdot \cos\alpha - d_1^2 - d_2^2} \ . \tag{2}$$

As a has values between 0 and 1, distances d_1 and d_2 are bounded by Eq. 2 depending on angle α. Fig. 2 shows three examples of valid domains for d_1 and d_2, to be considered in this paper. These three values of α are fully representative of any other different $\{\alpha, d_1, d_2\}$ sets, because $\{\alpha + \pi, d_1, -d_2\}$, $\{-\alpha, d_1, d_2\}$, $\{\pi - \alpha, -d_1, d_2\}$, $\{\alpha, -d_1, -d_2\}$, $\{\pi + \alpha, -d_1, d_2\}$, $\{-\alpha, -d_1, -d_2\}$ and $\{\pi - \alpha, d_1, -d_2\}$ all have the same type map.

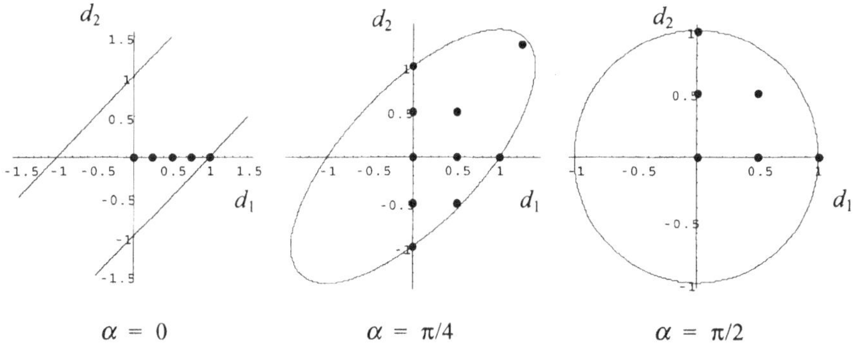

Figure 2. The ellipses that enclose the valid d_1 and d_2 for each value of α.

3. Type maps

A type map is built for any set of values $\{\alpha, d_1, d_2\}$ by checking the motion limits of joints R_{21} and R_{22} (angles θ_1 and θ_2 respectively). The coordinates of the type map could be the angles ϕ_1 and ϕ_2 that give the extension of the arms (Fig. 3). At any point on the type map, these angles are taken as fixed so that the system becomes a SRRS linkage. However, for the sake of generality, instead of ϕ_1 and ϕ_2, the type maps use p and q;

which allows for larger type maps, simplifies formulas and also the curves on the map.

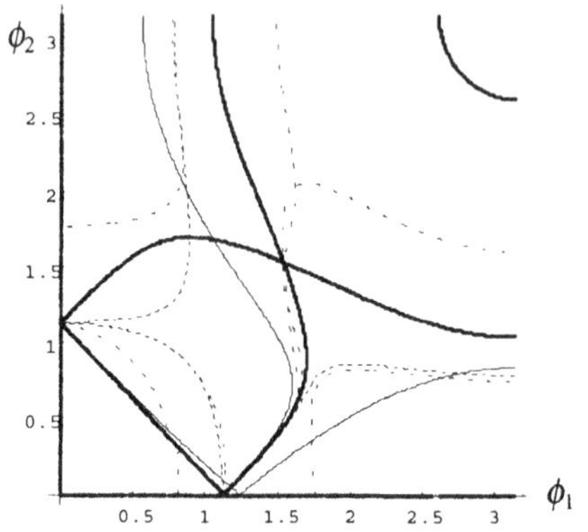

Figure 3. Curves on the type map for $\alpha = \pi/2$, $d_1 = 0.5$, $d_2 = 0.5$ using coordinates ϕ_1 and ϕ_2.

3.1 Classification

The area of the type map is potentially divided in nine zones. The zone where both joints R_{21} and R_{22} behave as cranks (CC) is the most convenient because, then, the workpiece can fully rotate with respect to the arms. Another convenient type is when joint R_{21} is a π non Grashof rocker and joint R_{22} a 0 non Grashof rocker (π0); if the motion intervals are big enough, they simply loose the portions that are more likely to produce interference with the workpiece.

Other types that appear are: CR and RC, when one of the arms is a Grashof rocker with limited motion; RR, when both arms are Grashof rockers; $\pi\pi$, 00 and 0π when both arms behave as non Grashof rockers. For non Grashof rockers, if R_{21} is of 0 type and/or R_{22} is of π type, the angular interval includes the location of the workpiece and so interference is more likely. So that, type 0π can be considered the worst case. A discussion on the distribution of these zones on the type maps follows this paragraf.

3.2 Case $\alpha = 0$

In this case Eq. 2 gives two lines as limits (Fig. 2). Inside them, only the difference $|d_1 - d_2|$ matters; so, any point parallel to those indicated on the figure has the same type map. These type maps are drawn on Fig. 4. When $\alpha = \pi$, the type maps are obviously the same; but the diagram in

Fig. 2 just must be inverted by symmetry around one of the axis of coordinates.

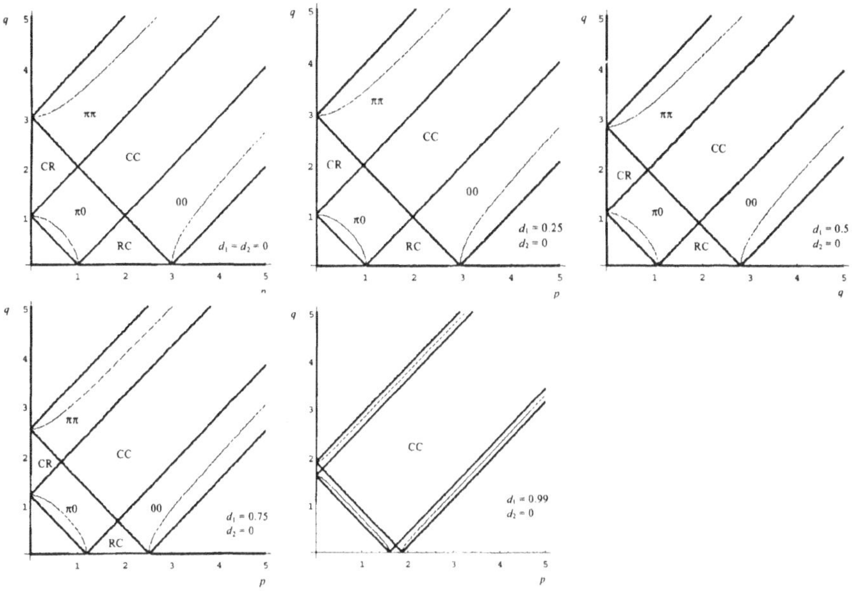

Figure 4. Type maps for the cases with $\alpha = 0$ indicated on Fig. 2.

For any $\{d_1,d_2\}$, the π0 and CC areas are contiguous. The first one near the shorter values of p and q, and the second one for larger values without apparent limit. Other less convenient types appear on the map.

When d_1 increases, the area of the map decreases; while larger portions of it become of the CC type. This can seem advantageous, but it shall be observed that the reason for it is that distance a decreases till zero when $d_1 = 1$. In this limit case, joints R_{21} and R_{22} become redundant and control on workpiece orientation is lost.

3.2 Case $\alpha = \pi/4$

Because of symmetry, only two quadrants of the ellipse in Fig. 2 need to be studied. Moreover, the ellipse for $\alpha = -\pi/4$ is identical; while those for $\alpha = \pm\, 3\pi/4$ are the same, just symmetric around the abscissae axis.

Several representative type maps are shown in Fig. 5. Start with $d_1 = d_2 = 0$, and compare it with the same case in Fig. 4. There is also a fringe alongside $p = q$ with a π0 and a CC area. But the CC area shrinks and is replaced by a RR area (hence, there is less mobility). This fact can be seen on all the type maps; however, as they take other values than zero for d_1 and d_2, their structure becomes more complex.

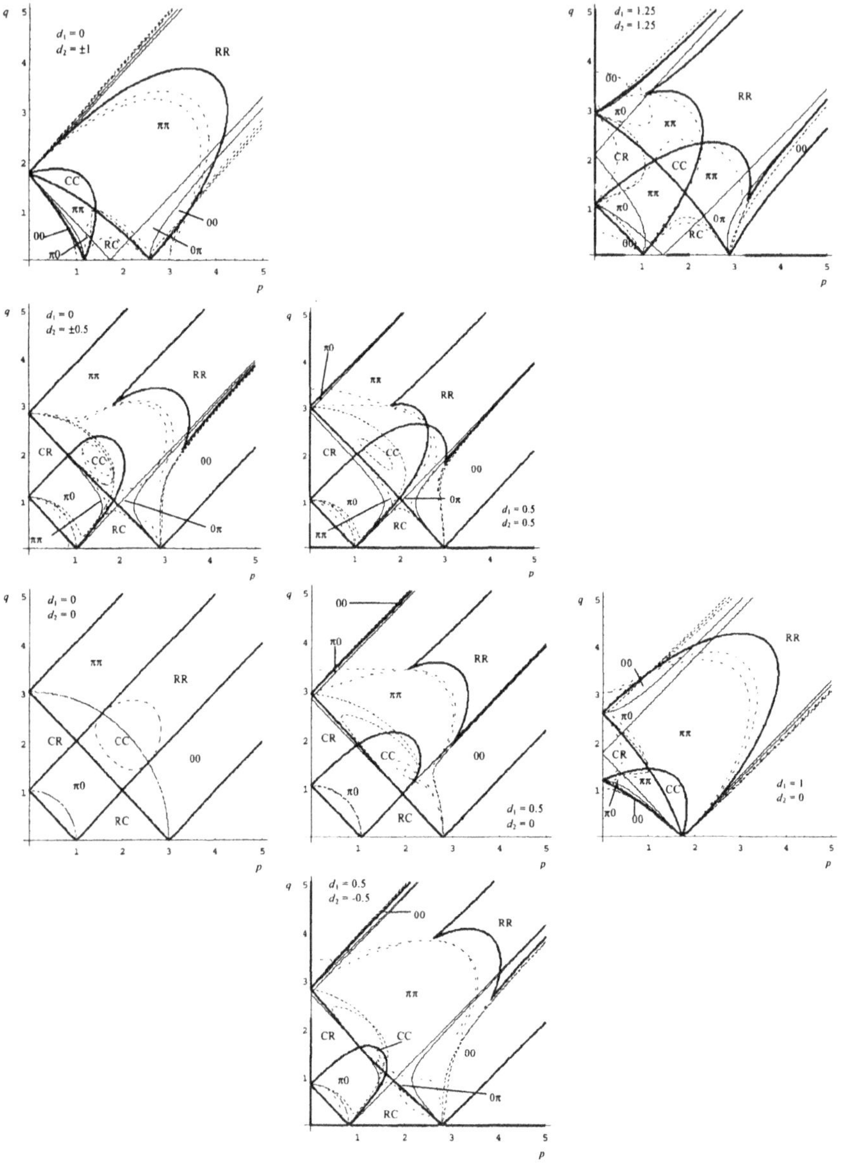

Figure 5. Type maps for the cases with $\alpha = \pi/4$ indicated on Fig. 2.

As d_1 and d_2 approach the limits of Fig. 2, the fringe of RR widens and, also, the space between it and the CC area is occupied by a $\pi\pi$ area. Simultaneously, the optimal CC and $\pi 0$ zones become smaller and also separated by the $\pi\pi$ type. In conclusion, the performance of the mechanism is worse because there is less CC and $\pi0$ areas and, in addition, they are not connected.

Other values of α between the extremes $\alpha = 0$ and $\alpha = \pi/2$ follow similar patterns to that of $\alpha = \pi/4$. Those with $0 < \alpha < \pi/4$ conserve more CC area than those with $\pi/4 < \alpha < \pi/2$, which tend to the type maps of case $\alpha = \pi/2$.

3.3 Case $\alpha = \pi/2$

When $\alpha = \pm\pi/2$ it is only necesary to study a quadrant of the circle in Fig. 2. Some representative cases are shown in Fig. 6.

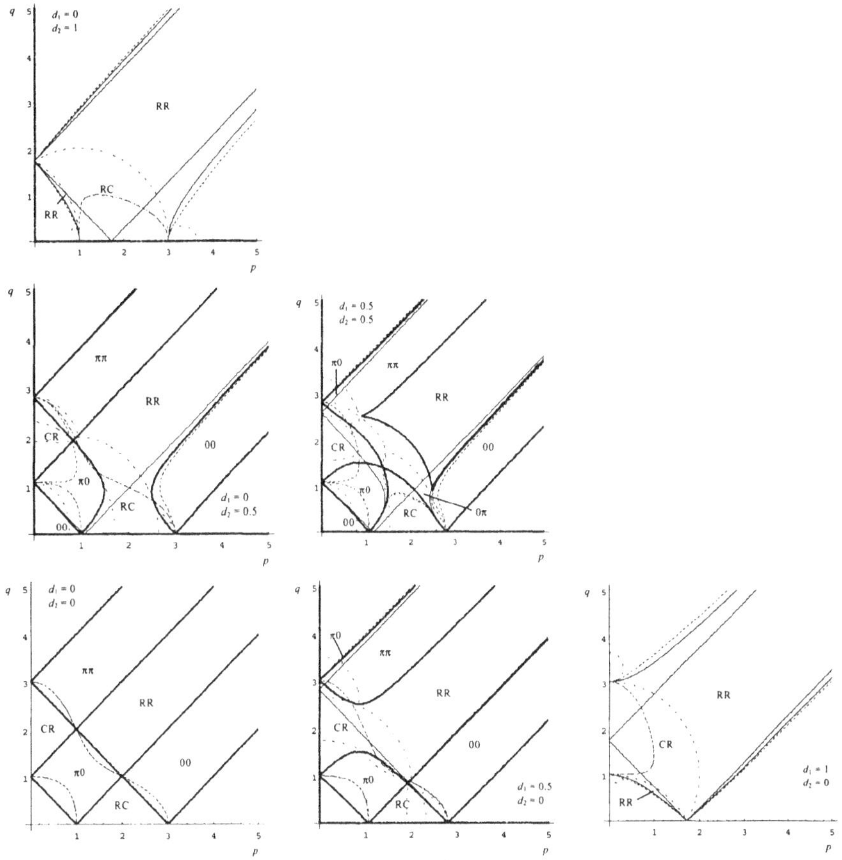

Figure 6. Type maps for the cases with $\alpha = \pi/2$ indicated on Fig. 2.

The case $d_1 = d_2 = 0$ has no CC area (in constrast with the previous α values). So, this is the twist α with more limited motion capacity. As in the previous discussions, $\pi0$ decreases when d_1 and/or d_2 increase. In comparison with the $\pi/4$ cases, $\pi\pi$ does not grow very much; instead of it, RC or RC (Grashof types) are the zones that increase its area and separate the RR from the $\pi 0$ types. That is, Grashof conditions are predominant.

4. Conclusions

The type map is a different tool that gives a representation of the motion capacities of some spatial chains. When applied to two cooperation robots or fingers, it allows a discussion on the performance attending mainly to the parameters that describe the grasp of the workpiece.

For the SRRS system considered in this paper, the planar case (when $\alpha = 0$ and $d_1 = d_2 = 0$) gives the best performance. By changing the twist angle α, a significant reduction on the CC area takes place till it disappears for $\alpha = \pi/2$. Changing the parameters d_1 and d_2 (which are related to the distance between the points of the grasp and the minimal distance between the lines of the extreme R-joints) produces more complicated patterns on the type maps which also reduce the area of the desirable tipologies. It is advisable to choose grasps with zero or low d_1 and d_2 values so that the parameter and the segment a is equal or almost equal to D.

References

Cervantes, J.J., Hernández, J.C. and Angeles, J. (2001), On the kinematic design of the 5R planar, symmetric manipulator, *Mechanimsm and Machine Theory*, no. 36, pp.1301-1313.

Gascons, N. (2001), *Evaluation, along the workspace, of the capacity of an articulated gripper to manipulate a solid*, Ph.D. thesis, Barcelona, UPC.

McCarthy, J.M. (2000), *Geometric design of linkages*, Springer.

Merlet, J-P. (1997), *Les robots parallèles*, Paris, Hermes.

Su, H., and McCarthy, J.M. (2001), Classification of RRSS linkages, *Proceedings of the 2001 ASME Design Engineering Technical Conferences*, Pittsburgh, USA.

ON THE INVARIANCE OF MANIPULABILITY INDICES

E. Staffetti, H. Bruyninckx and J. De Schutter
Katholieke Universiteit Leuven
Department of Mechanical Engineering
Celestijnenlaan 300B, B3001 Heverlee Belgium
Ernesto.Staffetti@mech.kuleuven.ac.be

Abstract This work discusses the invariance of local manipulability indices which are used to convert into a scalar the capability of a robotic device at a given configuration to perform a manipulation task. It is shown that they suffer from non–invariance in the sense that arbitrariness in the choice of the metric functions employed in their definition is unavoidable and they are not invariant under changes of the arbitrary choices introduced. Since there do not exist natural norms in the spaces of generalized forces and velocities, kinematic manipulability is neither an invariant concept nor a natural notion.

Keywords: Manipulability, singularities, invariance.

1. Introduction

The motivation to study manipulability of robotic devices comes from the fact that, in order to perform an end–effector twist or to withstand a wrench acting on the end-effector, the velocities and the efforts at the actuators are, in general, greater at configurations close to singularities.

At singular configurations some twists cannot be executed and, as a consequence, some wrenches can only be passively resisted by the manipulator or the end–effector gains one or more degrees of freedom and, hence, one or more wrenches cannot be resisted even with the actuators locked. Therefore, maintaining a manipulator away from singularities is convenient to execute a task not known a priori.

Although, in general, an exhaustive characterization of the singularities of a manipulator cannot be performed taking into account only the the relation

$$\mathbf{t} = \mathbf{J}(\mathbf{q})\dot{\mathbf{q}} \tag{1}$$

between the joint velocities $\dot{\mathbf{q}}$ and the end–effector twist $\mathbf{t} = (\boldsymbol{\omega}, \mathbf{v})^T$ of the manipulator (Zlatanov et al., 1994), the Jacobian matrix, $\mathbf{J}$, or the matrices $\mathbf{J}\mathbf{J}^T$ and $\mathbf{J}^T\mathbf{J}$, when $\mathbf{J}$ is not square, have been the object of study to characterize manipulability, implicitly regarded as proximity of a configuration to a singularity.

J. Lenarčič and F. Thomas (eds.), Advances in Robot Kinematics, 57–66.

Although several criteria have been proposed to characterize the distance of a configuration from a singularity, many of them, such as condition number, minimum singular value, minimum eigenvalue of the Jacobian are not invariant with respect to change of reference frames, scale or physical units and those that are invariant but do not require any arbitrary choices are inadequate to characterize manipulability. This is the case of the determinant of the Jacobian which is invariant with respect to changes of reference frame and, by normalization, it can be made invariant with respect to changes of scale and physical units. It is one of the most widely used manipulability indices since its introduction in Yoshikawa, 1985, and recent works based on this index are Lee et al., 1998, and Zhang et al., 2000. It is herein shown in Section 2, using elementary geometric considerations, that, in general, it does not represent a measure of distance from a singularity.

Other classes of manipulability indices can be defined that, rather than the proximity to a singularity, study other properties of the forward velocity kinematic map between joint and Cartesian spaces. Park and Kim, 1998 interpreted manipulability as how closely the forward kinematic map of a manipulator approximates an isometry whereas Bicchi and Prattichizzo, 2000 regarded manipulability as the efficiency of the velocity and force transmission between joint space and Cartesian space at a certain configuration of the manipulator.

These indices, described in Section 3 and Section 4, respectively, permit to take into account the task the robotic device has to perform through the choice of a norm in both the tangent space to the end-effector configuration manifold and the tangent space to the joint configuration manifold, and the fact that manipulability indices vary under changes of arbitrary choices can be regarded as a kind of non-invariance.

Since there do not exist natural norms in the space of generalized velocities and forces (see, for example, Park, 1995) it is possible to affirm that manipulability is neither an invariant notion nor a natural concept.

The peculiarity of these alternative notions of manipulability, in which the task to execute is taken into account in their definition, is that the singular configurations are not always penalized.

2. Manipulability as Proximity to a Singular Configuration

In this section the indices that have been used in the robotics literature to characterize manipulability regarded as proximity of a certain configuration of the manipulator to a singularity are reviewed and their adequateness to characterize manipulability is discussed.

Although, as pointed out by Zlatanov et al., 1994, in general, an exhaustive characterization of the singularities of a manipulator cannot be carried out taking into account only its forward velocity kinematic map (1), $\det(\mathbf{J})$, for square Jacobian matrices, and $\det(\mathbf{J}\mathbf{J}^T)$ or $\det(\mathbf{J}^T\mathbf{J})$, for non–square Jacobian matrices, have been proposed as manipulability indices by Yoshikawa, 1985. In this section it is shown, using simple geometric considerations, that they cannot be considered as indices of closeness to a singularity. Other indices that have been used to characterize the distance from a singularity, such as condition number, minimum eigenvalue or minimum singular value of the Jacobian matrix are also reviewed and their adequateness to characterize manipulability of a robotic device is discussed.

The Jacobian for serial manipulators can be expressed in the following form

$$\mathbf{J} = \begin{pmatrix} \mathbf{S}_1 & \cdots & \mathbf{S}_n \\ \mathbf{r}_1 \times \mathbf{S}_1 & \cdots & \mathbf{r}_n \times \mathbf{S}_n \end{pmatrix} \tag{2}$$

where $\mathbf{S}_i$ is a vector along the joint axis i and $\mathbf{r}_i$ is the position vector of any point on the joint axis i, for $i = 1, \ldots, n$, where n is the number of joints of the manipulator. Therefore, $\mathbf{r}_i \times \mathbf{S}_i$ represents the moment of the joint axis with respect to the origin. Its transpose is a mapping from end–effector wrenches $\mathbf{w} = (\mathbf{m}, \mathbf{f})^T$ to joint torques $\boldsymbol{\tau}$

$$\boldsymbol{\tau} = \mathbf{J}^T(\mathbf{q})\mathbf{w}$$

For parallel manipulators in which the contribution of each leg to the wrench space is a pure force, the matrix $\mathbf{G}$, whose column vectors are the basis vector of the wrench space, has the same structure as the Jacobian matrix (2). Hence, the considerations that will be made for the matrix $\mathbf{J}$ of serial manipulators can be extended by duality to the matrix $\mathbf{G}$ of parallel structures.

Since the transformation induced in the twist space by a Euclidean coordinate transformation in the workspace of the manipulator has determinant equal to 1, it follows that $\det(\mathbf{J})$, $\det(\mathbf{J}\mathbf{J}^T)$, and $\det(\mathbf{J}^T\mathbf{J})$ are invariant under Euclidean transformations. By normalizing them with respect to a certain value they can be made insensitive to changes of scale and physical units. In this section it is shown that, besides these interesting properties, these indices do not characterize the proximity of a certain configuration of the manipulator to a singularity.

First of all it is useful to recall that $\det(\mathbf{J})$ is the local scale factor of the n-dimensional volumes of the tangent space to the joint configuration manifold induced by the forward velocity kinematic map of the manipulator. However, this scaling factor at a certain configuration does not give information about the easiness to accomplish manipulation tasks, that is, about manipulability. In the following this statement will be proved using elementary considerations based on a particular expansion of $\det(\mathbf{J})$.

If $\mathbf{A}$ is a square matrix, its determinant can be computed using the Laplace expansion

$$\det(\mathbf{A}) = \sum (-1)^{(i_1+i_2+\cdots+i_k+j_1+j_2+\cdots+j_k)} a_k \, b_{n-k}$$

where

- a_k is the minor of order k formed by the rows $i_1, i_2, \ldots, i_k$ and columns $j_1, j_2, \ldots, j_k$ of the matrix $\mathbf{A}$, and
- b_{n-k} is the determinant of the matrix remaining from the matrix $\mathbf{A}$ by deleting the rows $i_1, i_2, \ldots, i_k$ and columns $j_1, j_2, \ldots, j_k$.

If we expand a 6×6 Jacobian matrix $\mathbf{J}$ having the structure of the matrix (2) with respect to the elements of the first three rows we obtain

$$\begin{aligned}
\det(\mathbf{J}) = &+[\mathbf{S}_1 \ \mathbf{S}_2 \ \mathbf{S}_3] \, [\mathbf{r}_4 \times \mathbf{S}_4 \ \ \mathbf{r}_5 \times \mathbf{S}_5 \ \ \mathbf{r}_6 \times \mathbf{S}_6] \\
&-[\mathbf{S}_1 \ \mathbf{S}_2 \ \mathbf{S}_4] \, [\mathbf{r}_3 \times \mathbf{S}_3 \ \ \mathbf{r}_5 \times \mathbf{S}_5 \ \ \mathbf{r}_6 \times \mathbf{S}_6] \\
&+[\mathbf{S}_1 \ \mathbf{S}_2 \ \mathbf{S}_5] \, [\mathbf{r}_3 \times \mathbf{S}_3 \ \ \mathbf{r}_4 \times \mathbf{S}_4 \ \ \mathbf{r}_6 \times \mathbf{S}_6] \\
&-[\mathbf{S}_1 \ \mathbf{S}_2 \ \mathbf{S}_6] \, [\mathbf{r}_3 \times \mathbf{S}_3 \ \ \mathbf{r}_4 \times \mathbf{S}_4 \ \ \mathbf{r}_5 \times \mathbf{S}_5] \\
&-[\mathbf{S}_2 \ \mathbf{S}_3 \ \mathbf{S}_4] \, [\mathbf{r}_1 \times \mathbf{S}_1 \ \ \mathbf{r}_5 \times \mathbf{S}_5 \ \ \mathbf{r}_6 \times \mathbf{S}_6] \\
&+[\mathbf{S}_2 \ \mathbf{S}_3 \ \mathbf{S}_5] \, [\mathbf{r}_1 \times \mathbf{S}_1 \ \ \mathbf{r}_4 \times \mathbf{S}_4 \ \ \mathbf{r}_6 \times \mathbf{S}_6] \\
&-[\mathbf{S}_2 \ \mathbf{S}_3 \ \mathbf{S}_6] \, [\mathbf{r}_1 \times \mathbf{S}_1 \ \ \mathbf{r}_4 \times \mathbf{S}_4 \ \ \mathbf{r}_5 \times \mathbf{S}_5] \\
&+[\mathbf{S}_3 \ \mathbf{S}_4 \ \mathbf{S}_5] \, [\mathbf{r}_1 \times \mathbf{S}_1 \ \ \mathbf{r}_2 \times \mathbf{S}_2 \ \ \mathbf{r}_6 \times \mathbf{S}_6] \\
&-[\mathbf{S}_3 \ \mathbf{S}_4 \ \mathbf{S}_6] \, [\mathbf{r}_1 \times \mathbf{S}_1 \ \ \mathbf{r}_2 \times \mathbf{S}_2 \ \ \mathbf{r}_5 \times \mathbf{S}_5] \\
&-[\mathbf{S}_4 \ \mathbf{S}_5 \ \mathbf{S}_6] \, [\mathbf{r}_1 \times \mathbf{S}_1 \ \ \mathbf{r}_2 \times \mathbf{S}_2 \ \ \mathbf{r}_3 \times \mathbf{S}_3],
\end{aligned} \tag{3}$$

where the brackets stand for determinants.

The term corresponding to the permutation $(k_1, k_2, k_3, k_4, k_5, k_6)$ of $(1, 2, 3, 4, 5, 6)$ is a product whose first factor is the determinant of the matrix whose column vectors are $\mathbf{S}_{k_1}, \mathbf{S}_{k_2}$, and $\mathbf{S}_{k_3}$, and the second is the determinant of the matrix formed by the moments of the vectors $\mathbf{S}_{k_4}, \mathbf{S}_{k_5}$, and $\mathbf{S}_{k_6}$ with respect to the origin.

The value of $[\mathbf{S}_{k_1} \ \mathbf{S}_{k_2} \ \mathbf{S}_{k_3}]$ is proportional to the oriented volume of the tetrahedron formed by $\mathbf{S}_{k_1}, \mathbf{S}_{k_2}, \mathbf{S}_{k_3}$, and the origin of the reference frame whereas the value of $[\mathbf{r}_{k_4} \times \mathbf{S}_{k_4} \ \ \mathbf{r}_{k_5} \times \mathbf{S}_{k_5} \ \ \mathbf{r}_{k_6} \times \mathbf{S}_{k_6}]$ is proportional to the oriented volume of the tetrahedron formed by $\mathbf{r}_{k_4} \times \mathbf{S}_{k_4}, \mathbf{r}_{k_5} \times \mathbf{S}_{k_5}, \mathbf{r}_{k_6} \times \mathbf{S}_{k_6}$, and the origin of the reference frame.

Since $\mathbf{S}_{k_1}, \mathbf{S}_{k_2}$, and $\mathbf{S}_{k_3}$, are non-zero vectors, $[\mathbf{S}_{k_1}\ \mathbf{S}_{k_2}\ \mathbf{S}_{k_3}]$ is zero when at least two of them are linearly dependent, that is, when at least two of them are parallel or all of them are coplanar. Also in the case in which $\mathbf{S}_{k_1}, \mathbf{S}_{k_2}$, and $\mathbf{S}_{k_3}$ are normalized vectors, the value of this determinant is not a measure of the closeness to a singularity of the subchain composed by the joints k_1, k_2, and k_3. Actually, the oriented areas of the triangles formed by couples of vectors of the set $(\mathbf{S}_{k_1}, \mathbf{S}_{k_2}, \mathbf{S}_{k_3})$ and the origin carry this information.

Similar considerations can be repeated on the determinant formed by the moments of the vectors $\mathbf{S}_{k_4}, \mathbf{S}_{k_5}$, and $\mathbf{S}_{k_6}$ with respect to the origin. Moreover, its value does not depend only on the relative orientation of $\mathbf{S}_{k_4}, \mathbf{S}_{k_5}$, and $\mathbf{S}_{k_6}$ but also on the spatial position of the lines that support them.

Even if we suppose that separately the above two factors characterize the closeness to singularities the product of their values would degrade the information given by them separately. It is clear that in those cases in which it is not possible to locate the reference frame in such a way that the expansion (3) of $\det(\mathbf{J})$ has only one term the situation is even worse.

For parallel manipulators when the reference frame is attached to the upper platform the module of the vectors $\mathbf{r}_{k_4}, \mathbf{r}_{k_5}$, and $\mathbf{r}_{k_6}$ do not change with the configuration of the manipulator. Therefore, for parallel manipulators it is easier (a fortiori if there are symmetries in the parallel device) to find a region of the configuration space such that, limiting the motion of the robotic device to that region, the determinant of the Jacobian does not change sign, its absolute value has one maximum, $|\det(\mathbf{J})|_{\mathrm{m}}$, at a certain configuration and is zero at some singular configurations that bound the region. This is what has been implicitly assumed in Lee et al., 1998 where the manipulability index

$$\frac{|\det(\mathbf{J})|}{|\det(\mathbf{J})|_{\mathrm{m}}} \tag{4}$$

is used for optimal design of parallel manipulators.

When the Jacobian matrix is not square the determinant of the matrices $\det(\mathbf{J}\mathbf{J}^T)$ or $\det(\mathbf{J}^T\mathbf{J})$, which always are square matrices, have been studied instead of $\det(\mathbf{J})$. To study the adequateness of these indices to characterize manipulability the Binet-Cauchy theorem can be employed. It states that if $\mathbf{A}$ is a $m \times n$ matrix, $\mathbf{B}$ is a $n \times m$ matrix, then

$$\det(\mathbf{AB}) = \begin{cases} 0 & \text{if } m > n \\ \sum_{k_1\, k_2\, \dots\, k_m} a_{1\,2\,\dots\,m,\, k_1\, k_2\, \dots\, k_m}\, b_{k_1\, k_2\, \dots\, k_m,\, 1\,2\,\dots\, m} & \text{if } m \le n \end{cases}$$

where

- $k_1 < k_2 < \cdots < k_m$ is one of the $\binom{n}{m}$ class m combinations of the numbers $1, 2, \ldots, n$,
- $a_{1\,2\ldots m,\,k_1\,k_2\ldots k_m}$ is the minor of order m of $\mathbf{A}$ formed by all its rows and by the columns $k_1\,k_2\,\ldots\,k_m$, and
- $b_{k_1\,k_2\ldots k_m,\,1\,2\ldots m}$ is the minor of order m of $\mathbf{B}$ composed by all its columns and by the rows $k_1\,k_2\,\ldots\,k_m$.

If the number of rows of a Jacobian is greater than the number of columns we can apply the Binet-Cauchy theorem to $\det(\mathbf{J}^T\mathbf{J})$. If, for example, we have $n = 6$ and $m < 6$ since

$$a_{1\,2\ldots m,\,k_1\,k_2\ldots k_m} = b_{k_1\,k_2\ldots k_m,\,1\,2\ldots m} = P_{k_1\,k_2\ldots k_m}$$

where $P_{k_1\,k_2\ldots k_m}$ is the element of the Plücker coordinate vector of the twist space $\mathbf{T}$ of the manipulator. After assigning progressive indices to the the Plücker coordinates of $\mathbf{T}$, we obtain that

$$\det(\mathbf{J}^T\mathbf{J}) = P_1^2 + P_2^2 + \cdots + P_{\binom{6}{k}}^2.$$

By duality, adding the squares of the elements of the vector of dual Plücker coordinates of $\mathbf{T}$, $\tilde{P}_i, i = 1, 2, \ldots, \binom{6}{k}$, we obtain that

$$\tilde{P}_1^2 + \tilde{P}_2^2 + \cdots + \tilde{P}_{\binom{6}{k}}^2 = \det(\mathbf{G}^T\mathbf{G}),$$

where $\mathbf{G}$ is the matrix whose column vectors span the wrench space $\mathbf{W}$ of the manipulator whose twist space is spanned by the columns of $\mathbf{J}$. However, besides the coincidence, it is impossible to find a correspondence between this expression for $\det(\mathbf{J}^T\mathbf{J})$ and manipulability.

If the number of rows of a Jacobian is lower of than the number of columns we can apply the Binet-Cauchy theorem to $\det(\mathbf{J}\mathbf{J}^T)$. If $m = 6$, $\det(\mathbf{J}\mathbf{J}^T)$ can be expressed as a sum of the square of all the 6×6 determinants that can be formed with the columns of $\mathbf{J}$. Then, we can repeat the same considerations made for square Jacobian matrices for each term of this expansion and conclude saying that $\det(\mathbf{J}\mathbf{J}^T)$, in general, is not a index of closeness of a manipulator to a singularity. A recent paper in which an index based on $\det(\mathbf{J}\mathbf{J}^T)$ is used is Zhang et al., 2000 where it has been referred to its maximum value $\det(\mathbf{J}_{\mathrm{m}}\,\mathbf{J}_{\mathrm{m}}^T)$. The quality index proposed in this paper is the following

$$\sqrt{\frac{\det(\mathbf{J}\mathbf{J}^T)}{\det(\mathbf{J}_{\mathrm{m}}\,\mathbf{J}_{\mathrm{m}}^T)}}$$

These results are coherent with those obtained for square Jacobian matrices and with the fact that for a linear map from a m-dimensional space into a n-dimensional one with $m \leq n$, as the forward velocity kinematic map of a non–redundant manipulator at a certain configuration, the volume scaling factor is given by $\sqrt{\det(\mathbf{J}^T\mathbf{J})}$. For redundant devices, for which $m > n$, the roles of $\mathbf{J}$ and its adjoint $\mathbf{J}^T$ are interchanged and the volume scaling factor is $\sqrt{\det(\mathbf{J}\mathbf{J}^T)}$.

The singular values of $\mathbf{J}$ are defined as the non–negative square roots of the non–zero eigenvalues of the square matrices $\mathbf{J}\mathbf{J}^T$ and $\mathbf{J}^T\mathbf{J}$. The condition number of the Jacobian, defined as the ratio between its maximum and minimum singular values in Salisbury and Craig, 1982, has often been employed as a manipulability index. The configurations in which the condition number of the Jacobian reaches its optimal value, 1, are called isotropic configurations. In Klein and Blaho, 1987 the minimum singular value of the Jacobian matrix has been proposed as performance index, because it vanishes at the singularities.

The performance indices that rely on eigenvalues overlook the fact that eigenvalues are not meaningful for a Jacobian matrix due to the different physical units of its elements. As a consequence eigenvalues of Jacobian matrices are non–invariant with respect to changes of reference frame, scale and physical units and hence not suitable as performance indices. To overcome this difficulties the concept of dimensionally homogeneous Jacobian has been introduced by Gosselin, 1992. The dimensional homogeneity of the elements of the Jacobian matrix is obtained using the linear velocity of three points on the end–effector instead of the linear and angular velocities of a reference frame attached to it. Nevertheless, the condition number of this Jacobian matrix is not invariant with respect to changes of the three points used to express the velocity of the end–effector.

The correct way to solve this problem is to extend the eigenvalue problem into a generalized eigenvalue problem $\mathbf{J}\,\dot{\mathbf{q}} = \lambda\,\mathbf{W}_q\,\dot{\mathbf{q}}$, where $\mathbf{W}_q$ is a mapping from the tangent space to the joint configuration manifold into the tangent space to the end–effector manifold. However, the choice of $\mathbf{W}_q$ is arbitrary, and non–natural.

The inconsistency of the indices based on the matrices $\mathbf{J}\mathbf{J}^T$ and $\mathbf{J}^T\mathbf{J}$ shows up clearly by noting that $\mathbf{J}$ and $\mathbf{J}^T$ cannot be applied sequentially because the range of the first one does not coincide with the domain of the second one. The solution is to generalize the $\mathbf{J}^T\mathbf{J}$ and $\mathbf{J}\mathbf{J}^T$ operators by inserting a third mapping, from the range of the leftmost to the domain of the rightmost one. The physical meaning of such a bridging operator is clearly of impedance type to map a twist into a wrench, while admittance type operators map in the other direction. Hence, an operator $\mathbf{J}^T\mathbf{W}_C\mathbf{J}$ maps joint velocities into torques at the joints, and permits to define a norm in the space of joint velocities defined by the matrix $\mathbf{W}_q = \mathbf{J}^T\mathbf{W}_C\mathbf{J}$. However, the introduction of a bridging metric function is an arbitrary choice, most often made tacitly by assuming that $\mathbf{W}_C = \mathbf{I}$. There do not exist natural norms in the spaces used to represent generalized velocities and forces (Park, 1995) and, therefore, it does not exist any physically motivated reason to prefer one of them. For this reason manipulability is neither an invariant nor a natural concept.

In the following paragraph some consistent definition of manipulability are outlined.

3. Manipulability as Proximity to an Isometry

Another possible interpretation of manipulability can be based on how far is the forward velocity kinematic map f from an isometry at a certain configuration of the manipulator. Unfortunately, as pointed out in Park and Kim, 1998, a theorem of Gauss states that no global isometry exists between two manifolds of different Gaussian curvatures as the joint configuration space and the end–effector configuration space. If $\mathbf{W}_q$ and $\mathbf{W}_C$ are the matrices that define the Riemannian metrics on the joint space and the end–effector spaces, respectively, the condition for f to be an isometry is that $\mathbf{J}^T\mathbf{W}_C\mathbf{J} = \mathbf{W}_q$ or equivalently $\mathbf{J}^T\mathbf{W}_C\mathbf{J}\mathbf{W}_q^{-1} = \mathbf{I}$. The choice of $\mathbf{W}_C$ and $\mathbf{W}_q$ is again arbitrary, and non–natural and can be based on the task specification.

Since the notion of isometry makes sense if the dimension of the range space of f is at least as large as the dimension of its domain for redundant manipulators that do not fulfill this requirement, the adjoint map, $\mathbf{J}^T$, which maps end–effector wrenches into torques at the joints of the manipulator has to be considered. In this case, since the induced metrics on the cotangent spaces to the end–effector configuration spaces and joint configuration spaces are $\mathbf{W}_C^{-1}$ and $\mathbf{W}_q^{-1}$, respectively, the condition for the inverse velocity kinematic map to be an isometry is $\mathbf{J}\mathbf{W}_C^{-1}\mathbf{J}^T = \mathbf{W}_q^{-1}$ or equivalently $\mathbf{J}\mathbf{W}_C^{-1}\mathbf{J}^T\mathbf{W}_q = \mathbf{I}$.

4. Manipulability as Efficiency of Velocity and Force Transmission

Manipulability can also be interpreted as the efficiency of the velocity transmission, that is as the capability of a robot manipulator to transform (a subset of) the joint velocities of into (a subset of) the twist at the end–effector. Because the ratio must consist of two scalars, a norm in the end–effector velocity space and a norm in the space of joint velocities, represented by the matrices $\mathbf{W}_t$ and $\mathbf{W}_q$, respectively, are needed (Bicchi and Prattichizzo, 2000)

$$\frac{\mathbf{t}^T \mathbf{W}_t \, \mathbf{t}}{\dot{\mathbf{q}}^T \mathbf{W}_q \, \dot{\mathbf{q}}}.$$

To find the configurations that maximize this index, also called manipulator velocity ratio in Dubey and Luh, 1988, with respect to a given twist, it is necessary to find a relationship between the numerator and denominator variables $\mathbf{t}$ and $\dot{\mathbf{q}}$.

A completely similar reasoning applies to the case of the efficiency of force transmission from joint torques $\boldsymbol{\tau}$ to end–effector wrenches $\mathbf{w}$

$$\frac{\mathbf{w}^T \mathbf{W}_w \, \mathbf{w}}{\boldsymbol{\tau}^T \mathbf{W}_\tau \, \boldsymbol{\tau}}.$$

For this index, also called manipulator mechanical advantage in Dubey and Luh, 1988, it is necessary to distinguish between active and passive wrenches at the reference element. The motivation for such a distinction comes from the fact that the wrenches that a manipulator is able to apply most efficiently may differ from wrenches that are most efficiently resisted by the mechanical structure of the manipulator. The passive wrench space is defined as the space of end–effector wrenches that can be resisted without applying any force or torque at the joints. Any element of this space is reciprocal to any element of the twist space. On the contrary, the active wrench space is defined as the space of end–effector wrenches that can be generated by applying joint actuation. Unlike twist space and passive wrench space which are uniquely defined by the robot kinematics, the active wrench space is found by solving the complete dynamics, that is, the dynamics of robot and environment. For example, the active wrench space depends on the velocity and the acceleration of all robot joints. If we are interested in the static case, that is, in the steady state solution of the dynamic equations, stiffness is the only dynamic characteristic of the environment to be considered.

5. Conclusions

In this paper it has been shown that arbitrariness in the choice of the metric functions in unavoidable for a correct definition of manipulability. Since there do not exist natural norms in the spaces used to represent twists and wrenches, and manipulability indices are not invariant under changes of the norms used, manipulability is neither an invariant nor a natural concept. It has been shown that those indices that do not imply this kind of arbitrariness do not represent a consistent characterization of manipulability. Then, some valid definitions of manipulability are reported.

Acknowledgements

The authors would like to acknowledge the support of the K. U. Leuven's Concerted Research Action GOA/99/04 Project.

References

Zlatanov, D., Benhabib, B., and Fenton, R.G. (1994), Singularity Analysis of Mechanisms Via a Velocity-Equation Model of the Instantaneous Kinematics, *1994 IEEE International Conference on Robotics and Automation*, 1994.

Yoshikawa, T., (1985), Manipulability of Robotic Mechanisms, *The International Journal of Robotics Research*, vol. 4, no. 2, pp. 3–9, 1985.

Lee, J., Duffy, J., and Hunt, K. H., (1998), A Practical Quality Index Based on the Octahedral Manipulator, *The International Journal of Robotics Research*, vol. 17, no. 10, pp. 1081–1090, 1998.

Zhang, Y., Duffy, J., and Crane, C., (2000), The Optimum Quality Index for a Spatial Redundant 4-8 In-Parallel Manipulator in J. Lenarčič and M. M. Stanišić (eds.) *Advances in Robot Kinematics*, Kluwer Acedemic Publisher, 2000.

Park, F. C., and Kim, J. W., (1998), Manipulability and Singularity Analysis of Multiple Robot Systems: A Geometric Approach, *Proceedings of the 1998 IEEE International Conference on Robotics and Automation*, 1998.

Bicchi, A., and Prattichizzo, D., (2000), Manipulability of Cooperating Robots with Unactuated Joints and Closed-Chain Mechanisms, *IEEE Transactions on Robotics and Automation*, vol. 16, no. 4, pp. 336–345, 2000.

Park, F. C., (1995), Distance Metrics on the Rigid Body Motions with Applications to Mechanism Design, *ASME Journal of Mechanical Design*, vol. 177, no. 1, pp. 48–54, 1995.

Salisbury, J. K., and Craig, J. J., (1982), Articulated Hands: Force Control and Kinematic Issues, *The International Journal of Robotics Research*, vol. 1, no. 1, pp 4–17, 1982.

Klein, C. A., and Blaho, B. E., (1987), Dexterity Measures for the Design and Control of Kinematically Redundant Manipulators, *The International Journal of Robotics Research*, vol. 6, no. 2, pp. 72–83, 1987.

Gosselin, C. M., (1992), The Optimum Design of Robotic Manipulators Using Dexterity Indices, *Robotics and Autonomous System*, vol. 9, pp. 213–226, 1992.

Dubey, R., and Luh, J. S., (1988), Redundant Robot Control Using Task Based Performance Measures, *Journal of Robotic Systems*, vol. 5, no. 5, pp. 409–432, 1988.

INFLUENCE OF A MANIPULABILITY INDEX ON TRAJECTORY PLANNING FOR ROBOTS IN A WORKSPACE WITH OBSTACLES

Francisco Valero, Vicente Mata, Francisco Rubio and Josep-Lluís Suñer
Department of Mechanical Engineering
Universidad Politécnica de Valencia, Spain
fvalero, vmata, frubio, josuner@mcm.upv.es

Abstract In this article a collision-free robot path planning procedure is presented in which a manipulability index has been included as one of the optimality criteria when generating the configuration space. From the set of collision-free configurations, a standard algorithm for obtaining optimal trajectories has been considered, taking into account the dynamical robot characteristics. It has been analyzed the effect of the manipulability index in parameters such as the total time that is necessary to perform the trajectory, while the characteristics of the actuators are taken into account.

Keywords: Path planning, trajectory planning, manipulability index, obstacle avoidance.

1. Introduction

There is a clear difference between the algorithms devoted to path planning and those devoted to obtaining the optimal trajectory. The first ones try, fundamentally, to obtain a sequence of robot configurations that fulfils some conditions, mainly collision avoidance, whereas the second ones try to obtain a temporal history of the evolution for generalised co-ordinates, by considering aspects to be minimised such as, the required time or the energy consumption. The path planning problem is a very active research field and there are a vaste literature about it, (Latombe, 1999, Ahuactzin et al., 1998, Barraquand and Latombe, 1991, Li and Trabia 1996, Kavraki et al. 1996). In the field of optimal trajectory planning, Lin et al., 1983 and Shin and McKay, 1986 should be pointed out, because of their similarity with the approach used in the trajectory optimisation.

The algorithm proposed in this paper is able to solve a path planning problem of the characteristics previously mentioned, but it shows some distinguishing characteristics against those proposed by other authors.

The main characteristic is that it uses Cartesian co-ordinates to model both the robot and its work environment, which yield some advantages

J. Lenarčič and F. Thomas (eds.), Advances in Robot Kinematics, 67–76.

when it comes to define the problem and the feasible configurations for the robot.

The algorithm starts from an initial feasible configuration and the final one we want to achieve, and it takes into account the static obstacles modelled by using spheres, cylinders and planes.

The algorithm operates in two stages. In the first one, a discrete space of feasible configurations is generated by means of solving successive optimisation problems, by using the robot model and its environment. In the second stage the configuration space is transformed into a weighted graph, which allows us to obtain the sequence of configurations that define the path. Earlier versions of this algorithm can be found in Valero, F. et al., 1997 and Valero, F. et al. 2000. It must be emphasised the great improvement in the quality of the obtained path that has been achieved by using a manipulability index (Yoshikawa, 1990) in the formulation.

From these results, minimum time trajectories will be obtained associating cubic functions to the temporal evolution of the joint variables. A Sequential Quadratic Programming (SQP) algorithm will be used to minimised the time required by the movement. Constraints including prescribed configurations, smoothness of the resulting movement and torque constraints associated to the actuators will be imposed.

A dynamic model of an ideal robot is considered. One of the results obtained for an industrial robot is shown, having planned the trajectory with and without considering the manipulability index and obtaining the minimum time trajectories in both cases.

2. Obtaining Adjacent Configurations

The generation of a configuration space is based on the idea of generating adjacent configurations from the initial to the final feasible configuration by using an optimisation problem (Valero, F. et al. 2000). In particular, from each configuration C^j another six C^1, C^2, C^3, C^4, C^5 and C^6 can be generated in branching with positive and negative increments along parallel axes of the Cartesian Reference System. Next, the branching from the C^1 configuration is repeated to all possible directions, where previous adjacent configuration does not exist yet, and so on until the configuration space is completed.

This procedure has as main advantage that it does not establish a strict order to locate the end-effector for the generation of the configurations, and allows the access to difficult areas from more adequate configurations.

It is necessary to specify how the robot and the work environment have been modelled to understand the procedure used for obtaining the adjacent configurations.

A wire frame model is carried out for the robotic system, so that a configuration C^j ($\alpha^j{}_i$, $p^j{}_k$) is defined without ambiguity by means of the Cartesian co-ordinates of a successive points called significant points $\alpha^j{}_i$, and any other interesting robot point $p^j{}_k$ could be calculated by using the $\alpha^j{}_i$. The significant points are selected on the basis of the robot degrees of freedom, so that this number should be the minimum necessary to define the robot configuration, since they are characteristic of each system. The set of co-ordinates which constitutes the model, will be submitted to the characteristic constraints of the geometry and mobility in the robot joints.

The obstacles are modelled by using spheres, cylinders and plane surfaces.

As mentioned previously, a feasible robot configuration in the workspace is defined by means of the Cartesian co-ordinates of the significant points and the constraints to be satisfied,(particularly adequate for industrial robot geometry), which can be sorted into one of the following types:

1. Robot geometry.

   Type (1) constraints consider information about robot geometry as number and type of kinematic joints, length of links, limits of joint movements, etc.

2. Collisions avoidance.

   Type (2) constraints avoid collisions between links of the robot and obstacles. The obstacles will be modelled by using spheres, cylinders and planes. For every sphere, cylinder and plane used to modelling the robotic system, the characteristic formulation for this type of constraints can be established in terms of some minimum distances.

3. No obstacles are located between adjacent configurations.

   Type (3) constraints avoid the location of obstacles between feasible adjacent robot configurations. Due to the fact that a discrete space of feasible configurations is considered, it is necessary to guarantee that the movement between adjacent configurations C^p and C^k is possible, therefore it must be verified that there are not obstacles between them.

   To achieve that, the procedure used verifies that the distance between the significant points α_i of the configurations mentioned before, satisfies the following condition:

$$\left|\overrightarrow{\alpha_i^p \alpha_i^k}\right| \le 2 \cdot min(r_j) \tag{1}$$

   where r_j is the smallest characteristic dimension in the modelled workspace.

For any robot, a configuration C^j will be determinated by K variables corresponding to the Cartesian co-ordinates of its significant points, where the following expression have to be satisfied:

$$N_{dof} = K - N_{ra}$$

being N_{dof} the number of degrees of freedom and N_{ra} number of equations obtained from constraints of type (1).

A definition for adjacent configuration, applied to the robotic system, is needed in order to obtain the configuration space. Given a C^k configuration, a new one C^p is said to be adjacent to it, if the robot end point position undergoes an unit increment in the discreticed workspace (as a rectangular prism with axes parallel to a predetermined Cartesian reference system), see Valero et al. 2000.

C^p will be obtained by using C^k as an initial estimation for solving an optimisation problem with the objective function

$$\left\| C^p - C^k \right\| = \sum_{i=1}^{K-3} \left(x_i^p - x_i^f \right)^2 + \frac{1}{\upsilon^{p^2}} \qquad (2)$$

where x_i^f belong to the significant point of the given final configuration and $\upsilon^p = \sqrt{\det \mathbf{J}(C^p)\mathbf{J}^{\mathrm{T}}(C^p)}$, is the Yoshikawa's manipulability index associated to the C^p configuration.

The constraints for the optimisation problem are sorted into one of the four types that have been illustrated in the previous section.

An additional advantage associated to the use of significant points is the easiness to obtain the Jacobian matrix of configurations $\mathbf{J}(C^p)$, since both the unit vectors and the position ones can be obtained efficiently with few operations.

3. Obtaining the Sequence of Configurations.

The sequence of configurations that constitutes the path is obtained by associating a weighted graph to the configuration space and looking for the minimum weighted path.

In the graph, the nodes correspond to the robot configurations and the arcs are related to joint displacements between adjacent configurations and the term corresponding to the manipulability index.

The weight corresponding to the arc that goes from node k (C^k robot configuration) to node p (C^p robot configuration), can be given as:

$$a(k,p) = \sum_{i=1}^{K-3} \left(x_i^p - x_i^k \right)^2 + \frac{1}{\upsilon^{p^2}} \qquad (3)$$

when that C^k and C^p are adjacent. In addition C^k and C^p must satisfy both type (3) constraints that avoid the obstacles between configurations and the angle increased from C^k to C^p must be smaller than the magnitude of the forbidden zone for that joint, so that large displacements are avoided for movement between adjacent configurations.

In case the points above mentioned are not satisfied, then we consider that $a(k,p)=\infty$. Finally, the searching is started in the weighed graph with the path that joins the node corresponding to the initial configuration to the node corresponding to the final configuration. Since the arcs satisfy that $a(k,p)\geq 0$, the Dijkstra's algorithm is used to obtain the path that minimises the distance between the initial and final configurations. If this path exists, it is easy to obtain a sequence of *m* robot configurations *S*.

4. Obtaining the Minimum Time Trajectory

The procedure used to obtain trayectories in the joint space according to an optimality criteria is based on the work of Lin et al., 1983.

Given a sequence of *m* configurations *S*, it is desirable to obtain a trajectory that contain them, so that the time required to perform it should be minimum.

Let it be $S_j(q_{i\,j})$ and $S_{j+1}(q_{i\,j+1})$ belonging to *S* and expressed in terms of joint variables. The trayectory beetwen these two configurations can be fit to a polinomial function.

$$q_{ij} = a_{ij} + b_{ij}t + c_{ij}t^2 + d_{ij}t^3 \tag{4}$$

for $i=1,..,N_{dof}$ and $j=1,..,m-1$.

In order to guarantee the smoothness of the trajectory, the following conditions are considered that will be imposed in those points in which the trajectory is discreticed.

- Position

  In each interval *j* the initial and final position must agree with the corresponding to S_j and S_{j+1}, giving $(2N_{dof}(m-1))$ equations.

$$q_{ij} = a_{ij} + b_{ij}\sum_{k=1}^{j-1} t_k + c_{ij}\left(\sum_{k=1}^{j-1} t_k\right)^2 + d_{ij}\left(\sum_{k=1}^{j-1} t_k\right)^3 \tag{5}$$

$$q_{i\,j+1} = a_{i\,j+1} + b_{i\,j+1}\sum_{k=1}^{j} t_k + c_{i\,j+1}\left(\sum_{k=1}^{j} t_k\right)^2 + d_{i\,j+1}\left(\sum_{k=1}^{j} t_k\right)^3 \tag{6}$$

- Velocity

  The initial and final velocity must be null. It will give ($2N_{dof}$) equations.

$$\dot{q}_{i1}(t=0)=0 \tag{7}$$

$$\dot{q}_{im}\left(t=\sum_{k=1}^{m-1} t_k\right)=0 \tag{8}$$

In the passage by the intermediate configurations S_j ($1<j<m$), the final velocity of an interval must agree with the initial of the following one, ($N_{dof}(m-2)$) equations

- Aceleration

  In the passage by the intermediate configurations S_j ($1<j<m$), the final aceleration of an interval must agree with the initial of the following one, ($N_{dof}\,(m-2)$) equations

$$\ddot{q}_{ij}\left(t=\sum_{k=1}^{j} t_k\right)=\ddot{q}_{ij+1}\left(t=\sum_{k=1}^{j} t_k\right) \tag{9}$$

The previous conditions define a system of ($4N_{dof}(m-1)$) independent linear equations when the duration of the intervals defined by the configurations of S is known. Since the complete trajectory has ($4N_{dof}(m-1)$) unknowns corresponding to the coefficients of the polynomials, the linear system can be solved, obtaining the complete trajectory.

Table 1.

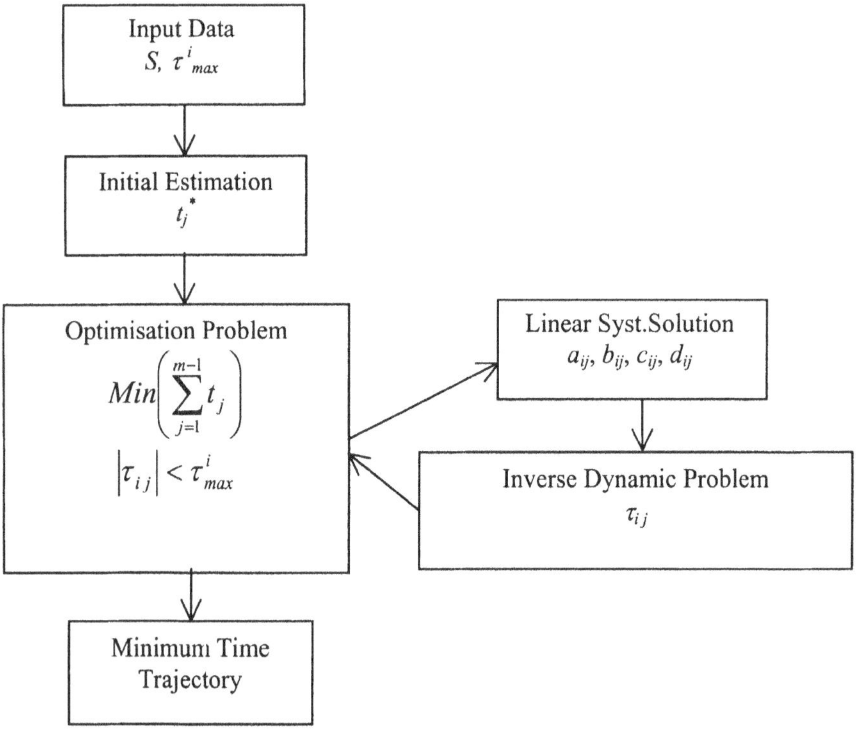

It is necessary to know the time spent on the movements to obtain the solution that is achieved by solving an optimization problem whose objective function is the trajectory total time and the constrains are the maximum torques in the robot actuators. The solution of the optimization problem is approached by means of a SQP algorithm of Fortran mathematical library NAG. In each iterative step is necessary to obtain the coefficients of the previously mentioned polynomials from an estimation of the variables (t_j) and to calculate the maximum torque for each one of the actuators by means of solving the inverse dynamic problem in each interval. An efficient algorithm is used to solve the inverse dynamic problem that can be found in Mata et al., 1999. The process is shown in the block diagram in Tb. 1.

5. Influence of the manipulability index

To evaluate the influence of the manipulability index, an example is presented in which it is applied to a robotic system that consist of a Robot PUMA 560.

The path-planning problem is solved for the following cases:

Case 1 By using the procedure above mentioned.

Case 2 By removing the manipulability index (υ) in obtaining the Configuration Space and the Graph, and obtaining the solution of the problem.

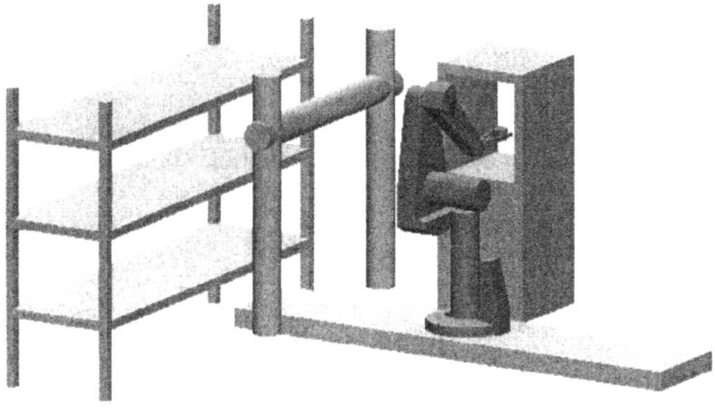

Figure 1. Working Position.

In both cases, we have taken N_x=2, N_y=2 y N_z=6 to calculate the increments associated to the end-effector movement. In the first case we have obtained a space with 776 configurations, a trajectory with 15 configurations and 2.09 seconds. Whereas in the second case 355, 19 configurations and 2.31 seconds are obtained.

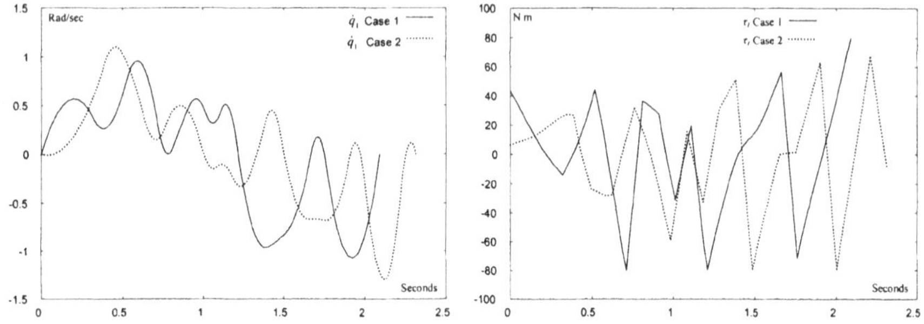

Figure 2. Velocity and Torque in Joint 1.

In order to make a qualitative comparison of the trajectories obtained, special attention is given to the three first joints, since they are responsible for the major movements of the robot, and we represent the velocity and the torque in each one of them corresponding to the trajectories obtained. In Fig. 2 it can be noticed that the trajectory corresponding to Case 2 requires higher velocities, in spite of being made in a greater time. In Fig. 3 it can be noticed that for the trajectory corresponding to Case 2, both the required velocity and the torque have maximum values that exceed those of Case 1. It turns out that this joint is specially relevant since it is the more demanding one for this type of robot.

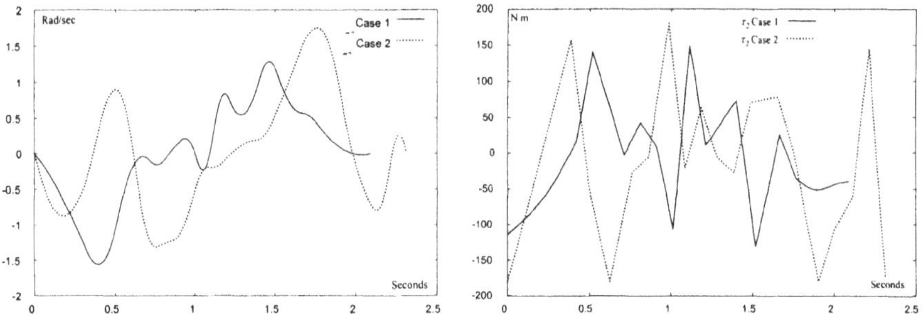

Figure 3. Velocity and Torque in Joint 2

As for joint 3, as it can be noticed in Fig. 4, the maximum values of velocity are next in both trajectories, although keep going higher in those corresponding to the second case.

On calculating the consumed theoretical energy in the trajectories by the actuators of the three first joints, it turns out the value 52.8 Joules for Case 1 while in Case 2 it is 75.2 Joules (29.8% higher). The joint 2 stands out, where it is required 25.8 Joules for case 1 and 51.1 Joules (49.5% higher) for Case 2.

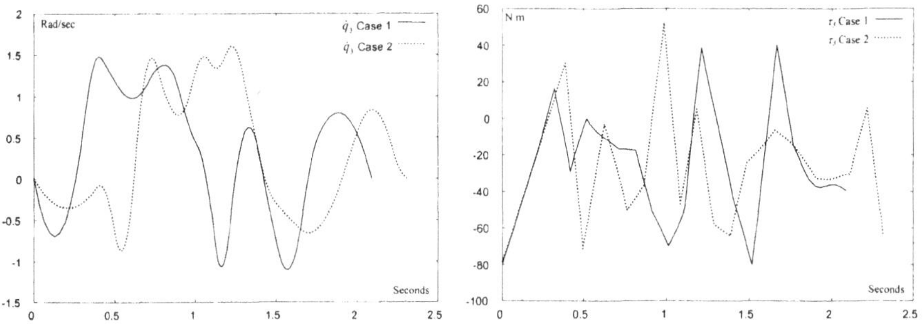

Figure 4. Velocity and Torque in Joint 3

6. Conclusions

The inclusion of the manipulability index in the collision-free path planning process not only contributes advantages as the quality improvement of configurations from the point of view of the manipulability, but also facilitates the later obtaining of optimal trajectories, considering the dynamical robot characteristics. In the examples verified, the invested total time to perform the trajectory obtained from the configurations with active manipulability index, is smaller, without penalising the dynamical response. With respect to the algorithm, it should be emphasised both his flexibility and the reasonable costs of computation in its application to real problems. The flexibility of the proposed algorithm is evident in the easiness to include different criteria when it came to generate the collisions-free configuration space

References

Ahuactzin, J. M., Gupta, K. and Mazer, E. (1998). Manipulation Planning for Redundant Robots: A Practical Approach. *International Journal of Robotics Research* 17(7): 731-747.

Barraquand, J., and Latombe, J-C. (1991). Robot Motion Planning: A Distributed Representation Approach. *International Journal of Robotics Research* 10(6): 628-649.

Kavraki, L.E., Svestka, P., Latombe, J-C., and Overmans, M. (1996). Probabilistic Roadmaps for Planning in High-Dimensional Configuration Spaces. *IEEE Transactions on Robotics and Automation* 12(4): 566-580.

Latombe, J-C. (1999). Motion Planning: A Journey of Robots, Molecules, Digital Actors, and Other Artefacts. *International Journal of Robotics Research* 18(11): 1119-1128.

Li, Z.J., and Trabia, M. B. (1996). Adaptative Path Planning and Obstacle Avoidance for a Robot with a Large Degree of Redundancy. Journal of Robotic Systems 13(3): 163-176.

Lin, C., Chang, P. and Luh, J. (1983). Formulation and Optimization of Cubic Polynomial Joint Trajectories for Industrial Robots. IEEE Transaction on Automatic Control Vol. AC-28, No.12. pp.1066-1073.

Mata, V., Provenzano, S., Cuadrado, J. I. and Valero, F. (1999). Algorithm for Solving the Inverse Dynamic Problem in Robots by Using The Gibbs-Appell Formulation. Proc.of Tenth World Congress on Theory of Machines and *Mechanisms,* Vol. 3, pp.1208-1215. Oulu, Finland.

Shin, K. and McKay, N. (1986). A Dynamic Programming Approach to Trajectory Planning of Robotic Manipulators. *IEEE Transactions on Automatic Control* AC-31 (6): 491-500.

Valero, F., Mata, V., Cuadrado, J.I., and Ceccarelli, M. (1997). A formulation for path planning of manipulators in complex environments by using adjacent configurations. *Advanced Robotics* 11 (1): 33-56.

Valero, F., Mata, V., and Ceccarelli, M. (2000). Path planning in complex environments for industrial robots with additional degrees of freedom. *Theory and Practice of Robots and Manipulators. Proceedings of the Thirteenth CISM-IFToMM Symposium.* Springer Wien New York.

Yoshikawa, T. (1990). Foundations of Robotics: Analysis and Control. The MIT Press.

Design and Control of Special-Type Mechanisms

HUMANOID HUMERAL POINTING KINEMATICS

Jadran Lenarčič
J Stefan Institute
University of Ljubljana, Slovenia
jadran.lenarcic@ijs.si

Michael Stanišić and Eric Schearer
University of Notre Dame
Notre Dame, IN. USA
stanisic@nd.edu

Abstract Humanoid humeral pointing kinematics is implemented in a humanoid robotic shoulder complex consisting of a parallel mechanism as the innermost shoulder girdle and a serial mechanism as the outermost spherical glenohumeral joint. These subsystems operate cooperatively as an offset double pointing system. The humanoid humeral pointing kinematics are also implemented within a passive humanoid shoulder complex.

Keywords: Humanoid Shoulder, Shoulder Girdle, Parallel Mechanism

1. Primary function of the human shoulder complex in human arm motions

The human shoulder complex is divided into two subsystems. The innermost portion, the *shoulder girdle*, consists of the sternum, thorax, torso, clavicle and scapula. The outermost portion, the *glenohumeral joint*, connects the humerus to the scapula Zatsiorsky, 1997. The primary function of the human shoulder complex is pointing the humerus. With a roll about the pointed humeral axis the elbow orientation is controlled. Human shoulder girdle motion is primarily a consequence of this humeral pointing.

In the human shoulder complex (Fig. 1) the girdle is a closed-loop chain, the glenohumeral joint is a serial chain. The girdle bears full load and inertial forces of the arm and manipulated object, while developing a small motion. Without this small motion the human's humeral workspace would be much less than a hemisphere.

On the right side of Fig. 1, the human shoulder girdle is overlayed by a simple kinematic equivalent, consisting of a universal joint centered at

J. Lenarčič and F. Thomas (eds.), Advances in Robot Kinematics, 79–88.

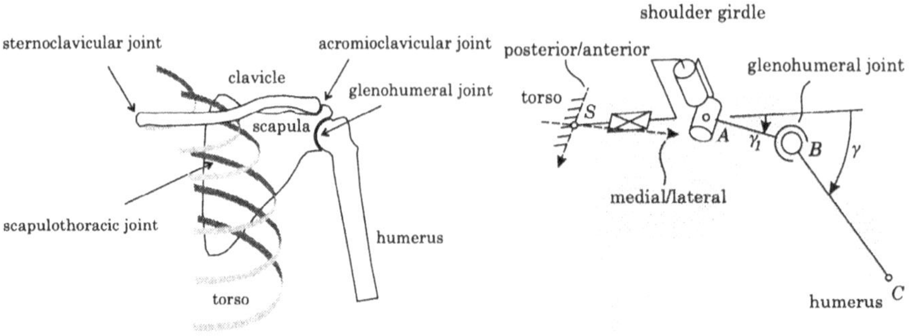

Figure 1. The human shoulder complex (frontal view) and a kinematic equivalent of the human shoulder complex.

point A. The center of the glenohumeral joint is point B. Point S represents a fixed reference placed at the intersection of the posterior/anterior axis passing through the center of the sternoclavicular joint and the medial/lateral axis through the middle of the torso. To mimic the distance between points S and B a translation is introduced, which is a dependent coordinate treated as a function of the rotation angles at A.

In Kohne et al., 2001, the relative position between points S and B was measured in humans performing different manipulation tasks. The maximum working cone of axis $\overline{SB}$ is produced when the humerus is stretched downward. An average maximum cone angle is about 50°. The central cone axis inclines 10° upward and 5° forward with respect to the medial/lateral axis. The distance between points S and B changes about ±10% of the average value as a function of the inclination angle between axis $\overline{SB}$ and the central axis of its working cone. $\overline{SB}$ shortens when the girdle moves upward and backward and lengthens when the girdle is moving downward and forward.

During humerus elevation the glenoid fossa (on which the humerus moves) tilts upward. This conjoint movement is the "shoulder rhythm" Dvir and Berme, 1978, illustrated in two images on the right side of Fig. 3. Measurement of the shoulder rhythm in the frontal plane Inman et al., 1944, Dvir and Berme, 1978, Umek-Venturini and Lenarčič, 1996, show that to a given angle γ corresponds an angle γ_1 (Fig. 3) which is associated with the lower reach of axis $\overline{AB}$. The upper reach is constant and is delimited by the working cone of this axis. Increase in the humeral elevation causes a decrease in the girdle's range, as well as in the kinematic redundancy of the shoulder complex. The ratio between γ_1 and γ is nearly constant for all γ and differs among subjects. In a first approximation, $\gamma_1 : \gamma = 1 : 3$.

The humerus and girdle move conjointly in the horizontal plane and any other plane intersecting axis $\overline{SA}$. *Humanoid pointing is defined as a configuration in which the displacement of the shoulder girdle and the humerus are co-planar and the relationship between angles* γ_1 *and* γ *achieves a value consistent with a measured human ratio.*

2. A humanoid shoulder complex

Fig. 2 shows a serial humanoid shoulder complex developing the shoulder girdle and glenohumeral joint movements shown in the kinematic model on the right side of Fig. 1. Point A in Fig. (2) is the center

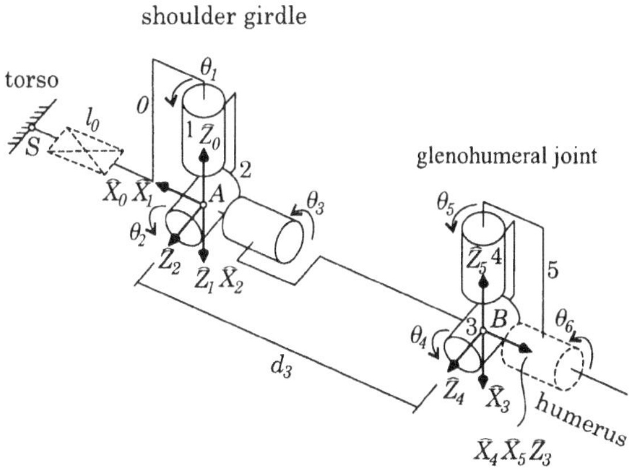

Figure 2. A serial humanoid shoulder complex.

for the non-synovial gliding motion of the scapula over the back of the thorax. Of three rotations associated with center A, θ_1 and θ_2 form the innermost pointing system. θ_3 is introduced to optimize performance of a parallel mechanism introduced later. Linear displacement l_0 develops the bulk of what is a moving center for the motion of the scapula over the thorax.

Point B in Fig. (2) is the center for pointing motion in the glenohumeral joint. This center is not fixed in the human but its movement is small and it is not necessary to cause it to displace. Two rotations, θ_4, θ_5 are necessary for pointing. θ_6, develops the humeral roll and does not contribute to humeral pointing.

The serial model in Fig. 2 incorporates a parallel mechanism shrouding center A which allows for controlling the three rotations at center A with leg lengths l_1, l_2 and l_3, shown on the left side of Fig. 3. The parallel mechanism increases load carrying capacity and rigidity of cen-

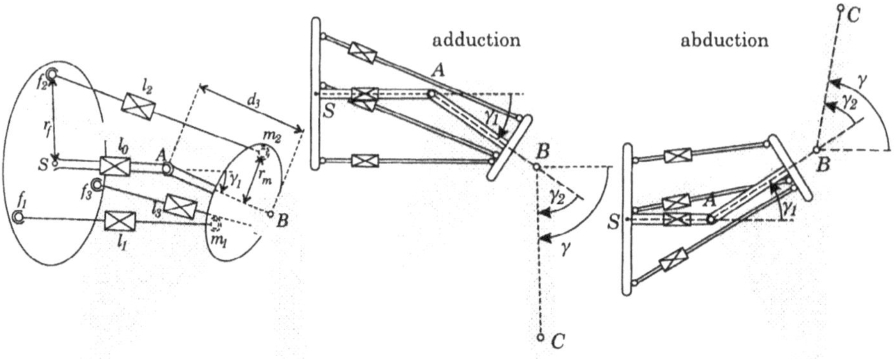

Figure 3. The parallel mechanism humanoid shoulder girdle and the complex in abduction-adduction.

ter A. Legs l_1, l_2 and l_3 represent the combined action of the human clavicle and ligaments which constrain the gliding of the human scapula over the thorax. The two images on the right side of Fig. 3 show the frontal view of the left arm in its minimum and maximum elevation, overlaid by the corresponding configurations of the humanoid shoulder complex. The lateral size of the humanoid shoulder girdle is represented by d_3, the distance between centers A and B in Fig. 2. The maximum inclination angle γ_1 at center A is 32° and the extension/contraction of the central leg varies within $l_0 = d_3 \pm 10\%$. Details on the design of this parallel mechanism are discussed in Lenarčič et al.(a), 2000, Lenarčič et al.(b), 2000.

This shoulder complex has seven degrees of freedom, four in the girdle (l_0,l_1,l_2,l_3) and three in the glenohumeral joint (θ_4,θ_5,θ_6). Neither l_0 or θ_6 contribute to humeral pointing and are not considered further. l_0 depends on γ_1 and θ_6 determines the final elbow orientation. Analogous to human humeral pointing, *humanoid humeral pointing occurs when line segments $\overline{SA}$, $\overline{AB}$, and $\overline{BC}$ are co-planar and the ratio between angles γ_1 and γ_2 achieves an anatomically consistent value.*

3. Humanoid humeral pointing kinematics of the serial shoulder complex

Values of γ_1 and γ_2 vary as functions of the pose of the girdle and have an anatomical background obtained by measuring a human performing the task of pointing. We incorporate this anatomical dependency within the humanoid shoulder complex.

Frame assignments, DH parameters and notation follow Craig, 1989. Table 1 gives the DH parameters for the serial chain in Fig. 2. The $\hat{X}_5$ axis is the axis of the humanoid humerus, pointing from center B to the elbow axis (which is not shown). This serial chain forms a double pointing system, with rotations θ_1, θ_2 and θ_3 at center A and rotations θ_4, θ_5 and θ_6 at center B. d_3 represents the lateral width of the humanoid shoulder girdle.

The forward humeral pointing kinematic equations are developed using the homogeneeous transformations ${}^{i-1}_{i}T$ as in Craig, 1989, with c_i and s_i representing $\cos\theta_i$ and $\sin\theta_i$ respectively. Writing ${}^{i-1}_{i}T$ for each joint i and link $i-1$ using the information in each row of Table 1 and

Table 1. The Denavit-Hartenberg Parameters.

joint i, body $(i-1)$	α_{i-1}	a_{i-1}	d_i	θ_i
$i=1$	180°	0	0	θ_1
$i=2$	90°	0	0	θ_2
$i=3$	−90°	0	d_3	θ_3
$i=4$	90°	0	0	θ_4
$i=5$	−90°	0	0	θ_5

multiplying together yields eqns. (1) and (2). Given θ_1-θ_5, homogeneous transformations (1) and (2) may be computed then multiplied to yield ${}^{0}_{5}T = {}^{0}_{3}T\, {}^{3}_{5}T$. The first column of ${}^{0}_{5}T$ is the homogeneous direction ${}^{0}\hat{X}_5$, the pointing direction of the humanoid humerus. Our interest is in the inverse pointing problem where the humerus pointing direction is given and the joint angles must be determined. The serial humanoid shoulder complex in Fig. 2 is kinematically redundant for the purpose of pointing the humeral link ($\hat{X}_5$). It contains five rotations and pointing is a two degree-of-freedom (dof) task. The system has three redundant dof. To resolve two of the redundant dof the condition of humanoid humeral pointing defined in section 2.0 is applied. The $\hat{X}_5$, $\hat{Z}_3$, and $\hat{X}_0$ axes remain coplanar for pointing motions of the humeral link, defining *the humeral plane* shown in a direct and lateral view in left and right sides of Fig. 4. This constraint provides that the pointing systems work synergistically.

$$
{}^{0}_{3}T = \begin{bmatrix} c_1c_2c_3 - s_1s_3 & -c_1c_2s_3 - s_1c_3 & -c_1s_2 & -d_3c_1s_2 \\ -s_1c_2c_3 - c_1s_3 & s_1c_2s_3 - c_1c_3 & s_1s_2 & d_3s_1s_2 \\ -s_2c_3 & s_2s_3 & -c_2 & -d_3c_2 \\ 0 & 0 & 0 & 1 \end{bmatrix} \quad (1)
$$

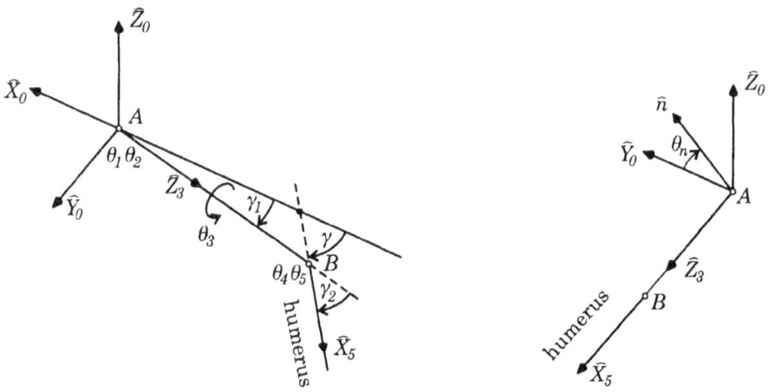

Figure 4. Direct and Lateral views of the humeral plane.

$$ {}^{3}_{5}T = \begin{bmatrix} c_4c_5 & -c_4s_5 & -s_4 & 0 \\ s_5 & c_5 & 0 & 0 \\ s_4c_5 & -s_4s_5 & c_4 & 0 \\ 0 & 0 & 0 & 1 \end{bmatrix} \tag{2} $$

Within the humeral plane three interrelated *inclination angles*, γ, γ_1 and γ_2 exist. γ represents the inclination of the humerus from the torso, γ_1 represents the inclination within the humanoid shoulder girdle (related to the human scapular inclination from the torso) and γ_2 represents the inclination of the humerus relative to the scapula. Inclination angles γ_1 and γ_2 are coupled by the human shoulder rhythm.

$$ \frac{\gamma_2}{\gamma_1} = r \quad . \tag{3} $$

where r is approximately 2 for most individuals Let ${}^{0}\hat{X}_5$, have known desired components x_5, y_5 and z_5,

$$ {}^{0}\hat{X}_5 = \begin{bmatrix} x_5 \\ y_5 \\ z_5 \end{bmatrix} \tag{4} $$

then the normal to the humeral plane, $\bar{n}$, is found from the vector product,

$$ {}^{0}\bar{n} = \frac{{}^{0}\hat{X}_5 \times (-{}^{0}\hat{X}_0)}{|{}^{0}\hat{X}_5 \times (-{}^{0}\hat{X}_0)|} = \begin{bmatrix} 0 \\ -z_5 \\ y_5 \end{bmatrix} \frac{1}{\sqrt{y_5^2 + z_5^2}} \tag{5} $$

and the inclination of the humerus from the torso is found from the scalar product,

$$ \gamma = \cos^{-1}[{}^{0}\hat{X}_5 \cdot (-{}^{0}\hat{X}_0)] = \cos^{-1}[-x_5] \quad . \tag{6} $$

Knowing γ from eqn. (6), γ_1 and γ_2 are found by solving eqn. (3) and knowing that γ_1 and γ_2 sum to equal γ (see Fig. 4). Thus,

$$\gamma_1 = \frac{\gamma}{r+1} , \tag{7}$$

then γ_2 comes from eqn. (3). Knowing inclination angles γ and γ_1 from eqns. (6) and (7) and the normal direction of the humeral plane from eqn. (5), the direction of the ${}^0\hat{Z}_3$ axis (with components x_3 y_3 and z_3), may be computed. With reference to Figs. 4,

$$ {}^0\hat{Z}_3 = \begin{bmatrix} x_3 \\ y_3 \\ z_3 \end{bmatrix} \begin{bmatrix} -\cos\gamma_1 \\ \sin\gamma_1 \sin\theta_n \\ -\sin\gamma_1\cos\theta_n \end{bmatrix} \tag{8}$$

where,
$$\theta_n = \text{Atan2}(y_5, -z_5) \tag{9}$$

With ${}^0\hat{Z}_3$ defined above, the solutions for θ_1 and θ_2 come from equating eqn. (8) to the third column of eqn. (1),

$$\theta_2 = \cos^{-1}[-z_3] \tag{10}$$

and,
$$\theta_1 = \text{Atan2}(s_1, c_1) = \text{Atan2}\left(\frac{y_3}{s_2}, -\frac{x_3}{s_2}\right) \tag{11}$$

θ_3 effects the values of θ_4 and θ_5. Determination of θ_3 considers performance of a parallel mechanism used to actuate the pointing system centered at A. Suppose θ_3 is known from this consideration. The direction cosine matrix ${}^0_3R$, (the upper left (3×3) portion of ${}^0_3T$ in eqn. 1) can be computed. Then,

$$ {}^3\hat{X}_5 = {}^0_3R^T \; {}^0\hat{X}_5 \tag{12}$$

which when equated to ${}^3\hat{X}_5$ in eqn. (2) (the first column of ${}^3_5R$) can be solved for θ_4 and θ_5,

$$\theta_5 = \sin^{-1}[-x_5(c_1c_2s_3+s_1c_3)+y_5(s_1c_2s_3-c_1c_3)+z_5s_2s_3] \tag{13}$$

$$\theta_4 = \text{Atan2}(s_4, c_4) \tag{14}$$

where,
$$s_4 = \frac{-x_5c_1s_2 + y_5s_1s_2 - z_5c_2}{c_5}$$

and,
$$c_4 = \frac{x_5(c_1c_2c_3 - s_1s_3) - y_5(s_1c_2c_3 + c_1s_3) - z_5s_2c_3}{c_5} ,$$

completing the inverse kinematics.

4. The parallel mechanism humanoid shoulder girdle kinematics

Center A of the girdle is shrouded by three extensible legs, l_1, l_2 and l_3 shown in Fig. 3, creating a parallel mechanism for the humanoid shoulder girdle (analogous to the closed-loop human shoulder girdle). The l_i $(i = 1,2,3)$ correspond to the values of θ_i $(i = 1,2,3)$ which come from the inverse kinematics of the serial humanoid shoulder girdle already determined and are found by computing the distances from the fixed centers f_i to their corresponding moving centers m_i, shown in Fig. 3.

Centers of the spherical joints are arranged on the fixed and moving platforms on circles with uniform 120° spacing as shown in Fig. 3. The radius of the circle of centers on the fixed and moving platform is r_f and r_m respectively, reported in Lenarčič et al.(a), 2000, Lenarčič et al.(b), 2000. Positions of the fixed centers f_i in frame $\{0\}$ are (l_0 is assumed known),

$$ {}^0\bar{r}_{f_1} = \begin{bmatrix} l_0 \\ r_f \\ 0 \end{bmatrix}, \; {}^0\bar{r}_{f_2} = \begin{bmatrix} l_0 \\ -\frac{1}{2}r_f \\ \frac{\sqrt{3}}{2}r_f \end{bmatrix}, \; {}^0\bar{r}_{f_3} = \begin{bmatrix} l_0 \\ -\frac{1}{2}r_f \\ -\frac{\sqrt{3}}{2}r_f \end{bmatrix} \tag{15} $$

and positions of the moving centers m_i in frame $\{3\}$ are,

$$ {}^3\bar{r}_{m_1} = \begin{bmatrix} 0 \\ -r_m \\ 0 \end{bmatrix}, \; {}^3\bar{r}_{m_2} = \begin{bmatrix} -\frac{\sqrt{3}}{2}r_m \\ \frac{1}{2}r_m \\ 0 \end{bmatrix}, \; {}^3\bar{r}_{m_3} = \begin{bmatrix} \frac{\sqrt{3}}{2}r_m \\ \frac{1}{2}r_m \\ 0 \end{bmatrix} \tag{16} $$

The ${}^3\bar{r}_{m_i}$ must be transformed into frame $\{0\}$,

$$ \begin{bmatrix} {}^0\bar{r}_{m_i} \\ 1 \end{bmatrix} = {}^0_3T \begin{bmatrix} {}^3\bar{r}_{m_i} \\ 1 \end{bmatrix} \qquad (i = 1,2,3) \tag{17} $$

where ${}^0_3T$ is known from eqn. (1), (θ_1, θ_2 and θ_3 known). Defining ${}^0\bar{l}_i$ as a vector along the extension of leg i $(i = 1,2,3)$,

$$ {}^0\bar{l}_i = {}^0\bar{r}_{m_i} - {}^0\bar{r}_{f_i} \tag{18} $$

then the leg lengths are computed from eqns. , (15)-(18) and (1) by,

$$ l_i = |{}^0\bar{l}_i| \; . \tag{19} $$

which result in the desired values of θ_1, θ_2 and θ_3 satisfying the humeral plane condition.

Finally, the redundancy θ_3 can be resolved by enhancing performance of the humanoid shoulder girdle and/or glenohumeral joint such as,

1 avoiding kinematic singularities of the shoulder girdle,
2 avoiding kinematic singularities of the glenohumeral joint,
3 maximizing overlap within the prismatic pairs of the extensible legs in the shoulder girdle (i.e. maximizing joint stability), or
4 avoiding mechanical interference between the legs of the shoulder girdle.

Control of θ_3 is a balance between these goals. Optimization of one goal does not lead to optimization of another.

5. A passive prototype using elastic elements

The geometric constraint of the humeral plane was implemented into a back-driven model of the shoulder complex. Elastic members attached to the legs of the shoulder girdle and between the moving platform of the shoulder girdle and the humerus constrain the system. The stiffness of the elastic members was "tuned" so that when back-driven, the model would approximate humanoid humeral pointing kinematics. A search was used to tune these stiffnesses as follows. Given a humeral pointing direction the five joint angles which satisfy the condition of humanoid humeral pointing were found by the inverse kinematic equations in section 3.0. The ability to match this optimal performance with the elastically constrained model was measured by the sum of the squared errors, SSE, of the joint angles in the model as described in Eqn. (20).

$$SSE = (\theta_{10} - \theta_1)^2 + (\theta_{20} - \theta_2)^2 + (\theta_{40} - \theta_4)^2 + (\theta_{50} - \theta_5)^2 \qquad (20)$$

where the θ_{i0} ($i = 1, 2, 4, 5$) are joint angles as computed in section 3.0 and the θ_i are the joint angles that result in prototype with the elastic members. The θ_i minimize the potential energy within the elastic members for a given humeral pointing direction. Varying the ratio of the stiffness of the three inner elastic members to the stiffness of the three outer elastic members, R_s, and varying the connection points of the outer elastic members, a sufficiently small value of SSE was found for a wide range of pointing directions. Over a wide range of humerus pointing directions a suitable value of R_s was found to be approximately 3.0, with a maximum SSE of 0.0062 radians^2.

6. Conclusions

This paper presented the direct and the inverse humanoid humeral pointing kinematics of a humanoid robotic shoulder complex. The design of a passive back-driven prototype incorporating elastic members was reported.

References

Tsuji. T., Morasso, P. G., Goto, K., and Ito K. (1995), Human hand impedance characteristics during maintained posture, *Biol. Cybern.*, pp. 475–485.

Lenarčič, J. (1999), Basic kinematic characteristics of humanoid manipulators, *Laboratory Robotics and Automation*, vol. 11, pp. 31–38.

Lenarčič, J., and Umek, A. (1994), Simple model of human arm reachable workspace, *IEEE Trans. on Systems, Man and Cybern.*, vol. 6, no. 4, pp. 1239–1246.

Zatsiorky, V., *Kinematics of Human Motion*, Human Kinetics Pub.

Kohne, A., Klopčar, N., and Lenarčič, J., A robotic model of the shoulder girdle contraction, *Proc. of ERK'01 Electrotecnical and Computer Science Conf. - Part B*, Portorož, Slovenia, pp. 219–222.

Dvir, Z., and Berme, N. The shoulder complex in elevation of the arm: A mechanism approach, *Journal of Biomechanics*, vol. 11, pp. 219–225.

Inman, V. T., Saunders, J. B., and Abbott, L. C., Observation on the function of the shoulder joint, *Journal of Bone and Joint Surgery*, vol. 26, pp. 1–30.

Umek-Venturini, A., and Lenarčič, J., The kinematics of the human arm elevation in the frontal plane, *Proc. of the 11th Int. Symp. on Biomedical Engineering*, Zagreb, Croatia, pp. 105–108.

Lenarčič, J., Stanišić, M. M., and Parenti-Castelli, V.,(a), Kinematic design of a humanoid robotic shoulder complex, *IEEE Intl. Conf. on Robotics and Automation*, San Fransisco, pp. 27-32.

Lenarčič, J., Stanišić, M. M., and Parenti-Castelli, V.,(b), A 4 dof parallel mechanism simulating the movement of the human sternum-clavicle-scapula complex, *Advances in Robot Kinematics*, eds. J. Lenarčic and M.M. Stanišić, pp. 325–333.

Craig, J. J., *Introduction to Robotics: Mechanics and Control*, Addison-Wesley Publishing Co.,

EFFICIENT ALGORITHMS FOR ROBOTS WITH HUMAN-LIKE STRUCTURES AND INTERACTIVE HAPTIC SIMULATION

O. Khatib, O. Brock, K.-S. Chang, D. Ruspini,
L. Sentis, F. Conti, and S. Viji
Robotics Laboratory, Department of Computer Science
Stanford University, Stanford, California 94305
{khatib, oli, kcchang, ruspini, lsentis, conti, sviji}@robotics.stanford.edu

Abstract A new field of robotics is emerging. Robots are today moving towards applications beyond the structured environment of a manufacturing plant. They are making their way into the everyday world that people inhabit. The paper focuses on models, strategies, and algorithms associated with the autonomous behaviors needed for robots to work, assist, and cooperate with humans. In addition to the new capabilities they bring to the physical robot, these models and algorithms and more generally the body of developments in robotics is having a significant impact on the virtual world. Haptic interaction with an accurate dynamic simulation provides unique insights into the real-world behaviors of physical systems. The potential applications of this emerging technology include virtual prototyping, animation, surgery, robotics, cooperative design, and education among many others.

Keywords: Operational Space Control, Dynamic Simulation, Mobile Manipulation, Real-Time Path Modification, Haptics

1. Introduction

The successful introduction of robotics into human environments will rely on the development of competent and practical systems that are dependable, safe, and easy to use. To work, cooperate, assist, and interact with humans, the new generation of robot must have mechanical structures that accommodate the interaction with the human and adequately fit in his unstructured and sizable environment. Human-compatible robotic structures must integrate mobility (legged or wheeled) and manipulation (preferably bi-manual), while providing the needed access to perception and monitoring (head camera) [Hirai et. al., 1998; Takanishi et. al., 1998; Khatib et. al., 1999; Asfour et. al., 1999; Nishiwaki et. al., 2000]. These requirements imply robots with branching structures - tree-like topology involving much larger numbers of degrees of freedom than those usually found in conventional industrial robots. The substantial

J. Lenarčič and F. Thomas (eds.), Advances in Robot Kinematics, 89–98.

increase in the dimensions of the corresponding configuration spaces of these robots renders the set of fundamental problems associated with their modeling, programming, planning, and control much more challenging. In this article we present algorithmic foundations developed in our laboratory to address these issues.

2. Whole-Robot Control: Task and Posture

Human-like structures share many of the characteristics of macro/mini structures [Khatib, 1995]: coarse and slow dynamic responses of the mobility system (the macro mechanism), and the relatively fast responses and higher accuracy of the arms (the mini device). Inspired by these properties of macro/mini structures, we have developed a framework for the coordination and control of robots with human-like structures.

2.1 Task Dynamic Behavior

The joint space dynamics of a manipulator are described by

$$A(\mathbf{q})\ddot{\mathbf{q}} + \mathbf{b}(\mathbf{q}, \dot{\mathbf{q}}) + \mathbf{g}(\mathbf{q}) = \mathbf{\Gamma} \tag{1}$$

where $\mathbf{q}$ is the n joint coordinates, $A(\mathbf{q})$ is the $n \times n$ kinetic energy matrix, $\mathbf{b}(\mathbf{q}, \dot{\mathbf{q}})$ is the vector of centrifugal and Coriolis joint forces, $\mathbf{g}(\mathbf{q})$ is the vector of gravity, and $\mathbf{\Gamma}$ is the vector of generalized joint forces.

The *operational space formulation* [Khatib, 1987] provides an effective framework for dynamic modeling and control of branching mechanisms [Russakow et. al., 1995], with multiple operational points. The generalized torque/force relationship [Khatib, 1987; Khatib, 1995] provides the decomposition of the total torque, $\mathbf{\Gamma}$ (equation 1) into two dynamically decoupled command torque vectors: the torque corresponding to the task behavior command vector and the torque that only affects posture behavior in the null space:

$$\mathbf{\Gamma} = \mathbf{\Gamma}_{task} + \mathbf{\Gamma}_{posture} \tag{2}$$

For a robot with a branching structure of m effectors or operational points, the task is represented by the $6m \times 1$ vector, $\mathbf{x}$, and the $6m \times n$ Jacobian matrix is $J(\mathbf{q})$. This Jacobian matrix is formed by vertically concatenating the m $6 \times n$ Jacobian associated with the m effectors.

The task dynamic behavior is described by the operational space equations of motion [Khatib, 1995]

$$\Lambda(\mathbf{x})\ddot{\mathbf{x}} + \mu(\mathbf{x}, \dot{\mathbf{x}}) + \mathbf{p}(\mathbf{x}) = \mathbf{F} \tag{3}$$

where $\mathbf{x}$, is the vector of the $6m$ operational coordinates describing the position and orientation of the m effectors, $\Lambda(\mathbf{x})$ is the $6m \times 6m$ kinetic energy matrix associated with the operational space. $\mu(\mathbf{x}, \dot{\mathbf{x}})$, $\mathbf{p}(\mathbf{x})$, and $\mathbf{F}$ are respectively the centrifugal and Coriolis force vector, gravity force vector, and generalized force vector acting in operational space.

The joint torque corresponding to the task command vector $\mathbf{F}$, acting in the operational space is

$$\mathbf{\Gamma}_{\text{task}} = J^T(\mathbf{q})\mathbf{F} \tag{4}$$

The task dynamic decoupling and control is achieved using the control structure

$$\mathbf{F}_{\text{task}} = \widehat{\Lambda}(\mathbf{x})\mathbf{F}^{\star}_{\text{motion}} + \widehat{\mu}(\mathbf{x}, \dot{\mathbf{x}}) + \widehat{\mathbf{p}}(\mathbf{x}) \tag{5}$$

where, $\mathbf{F}^{\star}_{\text{task}}$ represents the inputs to the decoupled system, and $\widehat{.}$ represents estimates of the model parameters.

2.2 Posture Behavior

An important consideration in the development of posture behaviors is the interactions between the posture and the task. It is critical for the task to maintain its responsiveness and to be dynamically decoupled from the posture behavior. The posture can then be treated separately from the task, allowing intuitive task and posture specifications and effective whole-robot control. The overall control structure for task and posture is given by equation 2, where

$$\mathbf{\Gamma}_{\text{posture}} = N^T(\mathbf{q})\mathbf{\Gamma}_{\text{desired-posture}} \quad \text{with} \quad N^T(\mathbf{q}) = \left[I - \overline{J}(\mathbf{q})J(\mathbf{q})\right] \tag{6}$$

where $\overline{J}(\mathbf{q})$ is the *dynamically consistent generalized inverse* [Khatib, 1995], which minimizes the robot kinetic energy.

This relationship provides a decomposition of joint forces into two control vectors: joint forces corresponding to forces acting at the task, $J^T\mathbf{F}$, and joint forces that only affect the robot posture, $N^T\mathbf{\Gamma}_{\text{posture}}$ (see Figure 1). For a given task this control structure produces joint motions that minimize the robot's instantaneous kinetic energy.

Dynamic consistency is the essential property for the task behavior to maintain its responsiveness and to be dynamically decoupled from the posture behavior since it guarantees not to produce any coupling acceleration in the operational space given any τ_{null}. In addition, dynamic consistency enables task behavior and posture behavior to be specified independently of each other, providing an intuitive control of complex systems.

For instance, the robot posture can be controlled to maintain the robot total center-of-mass aligned along the $\mathbf{z}_0$ axis of the reference frame. This

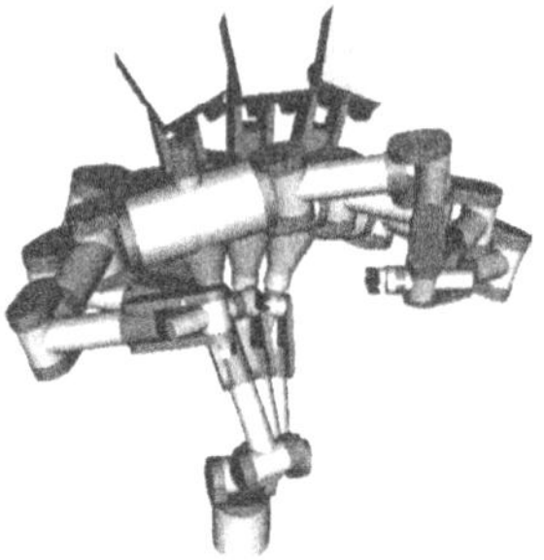

Figure 1. Dynamic Consistency: a sequence of snapshots from the dynamic simulation of a 24 degree-of-freedom humanoid system. The task consists of maintaining the position of the hands. The torso of the robot performs a motion in the posture space without affecting task execution.

posture can be simply implemented with a posture energy function

$$V_{\text{posture-energy}}(\mathbf{q}) = \frac{1}{2}k(x_{\text{CoM}}^2 + y_{\text{CoM}}^2) \tag{7}$$

where k is a constant gain, x_{CoM}, and y_{CoM} are the x and y coordinates of the center of mass. The gradient of this function provides the required attraction to the z axis of the robot center of mass. This is illustrated in the simulation shown in Figure 2.

Figure 2. Posture Behavior: a sequence of snapshots from the dynamic simulation of a 21-degree-of-freedom humanoid system. The hands are haptically teleoperated, while the remaining degrees of freedom are controlled automatically in the posture space. The posture controls the center of mass to balance the robot, while maintaining a natural arm configuration.

More complex posture behaviors can be obtained by combining various posture energies. We are currently exploring the generation of human-like natural motion from motion capture of human and the extraction of motion characteristics using human biomechanical models.

2.3 Efficient Operational Space Algorithms

Early work on efficient operational space dynamic algorithms has focused on open-chain robotic mechanisms. An efficient $O(n)$ recursive algorithm was developed using the spatial operator algebra [Kreutz-Delgado et. al., 1991] and the articulated-body inertias [Featherstone, 1987]. A different approach that avoided the extra computation of articulated inertias also resulted in an $O(n)$ recursive algorithm for the operational space dynamics [Lilly, 1992] Building on these early developments, our effort was aimed at algorithms for robotic mechanisms with branching structures that also address the issue of redundancy and dynamics in the null space.

The most computationally expensive element in the operational space whole-body control structure (equation 2) is the posture control, which involves the explicit inversion operation of the $n \times n$ joint space inertia matrix A,which requires $O(n^3)$. We have developed a computationally more efficient operational space control structure that eliminates the explicit computation of the joint space inertia matrix and its inverse. This elimination was achieved by combining the dynamically consistent null space control and the operational space control in a computationally more efficient dynamic control structure.

Using this control structure, we have developed a recursive algorithm for computing the operational space dynamics of an n-joint branching redundant articulated robotic mechanism with m operational points [Chang and Khatib, 2000]. The computational complexity of this algorithm is $O(nm + m^3)$, while existing symbolic methods require $O(n^3 + m^3)$. Since m can be considered as a small constant in practice, this algorithm attains a linear time $O(n)$ as the number of links increases. This work was extended for the dynamics of closed-chain branching mechanisms with an efficient $O(nm + m^3)$ algorithm [Holmberg and Khatib, 2000].

3. Interactive Haptic Simulation

Beyond their immediate application to physical robots, these efficient dynamic algorithms are making a significant impact on the simulation and interaction with the virtual world. The computational requirements associated with the haptic interaction with complex dynamic environments are quite challenging. In addition to the need for real-time free-motion simulation of multi-body systems, contact and impact resolution and constrained motion simulation are also needed.

Building on the operational space formulation, we developed a general framework [Ruspini and Khatib, 1999] for the resolution of multi-contact between articulated multi-body systems. A contact point is treated as an operational point and a contact space is defined. Similarly to the operational space inertia matrix, a contact space inertia matrix Λ is introduced to provide the effective masses seen at all the contact points and to characterize the dynamic relationships between them. Computing the contact space inertia matrices Λ for a number of m contact point on a branching mechanism is achieved with an efficient $O(nm + m^3)$ recursive algorithm.

The contact space representation allows the interaction between groups of dynamic systems to be described easily without having to examine the complex equations of motion of each individual system. As such, a collision model can be developed with the same ease as if one was considering interaction only between simple bodies. Impact and contact forces between interacting bodies can then be efficiently solved to prevent penetration between all the objects in the environment. Such dynamic interactions are shown in Figure 3.

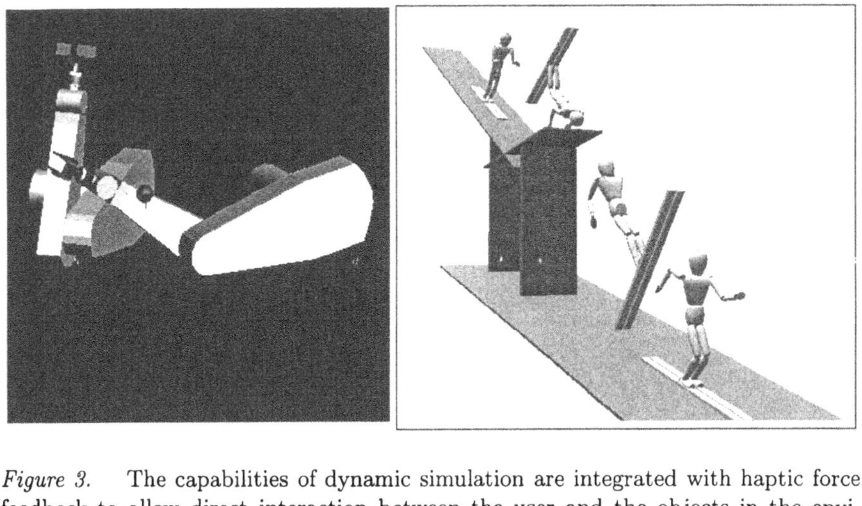

Figure 3. The capabilities of dynamic simulation are integrated with haptic force feedbeck to allow direct interaction between the user and the objects in the environment (left). The user can attach to any object in the scene and feel contact forces resulting from dynamic interactions. Dynamic simulation is capable of generating complicated, physically consistent behavior: The sliding motion and subsequent jump with the landing are simulated dynamically (right).

4. Task-Consistent Elastic Plans

The control methods presented in Section 2 allow the consistent control of task and posture for robots with complex mechanical structures. To perform or assist in the execution of complex actions, however, these control structures have to be linked with motion generated by a planner. Furthermore, since unstructured environments can be highly dynamic, such an integration has to accommodate unforeseen obstacle motion in real time, while conforming to constraints imposed by the task. We have developed algorithms that perform task-consistent, real-time path modification to address this issue.

Motion planners generally perform a global search in configuration space to determine a collision-free motion accomplishing a given task. Due to the high dimensionality of the configuration space of the class of robots we are concerned with in this paper, planning operations are too computationally complex to be performed in real time. As a consequence, motion in dynamic environments cannot adequately be generated by those planners. The elastic band framework [Quinlan and Khatib, 1993] was developed to allow real-time modification of a previously planned path, effectively avoiding a costly planning operation in reaction to changes in the environment. More recently, this framework was complemented by the elastic strip framework [Brock and Khatib, 1998], which allows reactive path modifcation for robots with many degrees of freedom.

In dynamic environments it is desirable to integrate path modification with task behavior. To accomplish this we extend the overall control structure for task and posture behavior (equation 2 by adding torques $\mathbf{\Gamma}_{obstacle}$ representing desired obstacle avoidance behavior:

$$\mathbf{\Gamma} = \mathbf{\Gamma}_{task} + \mathbf{\Gamma}_{posture} + \mathbf{\Gamma}_{obstacle}$$

Both, $\mathbf{\Gamma}_{posture}$ and $\mathbf{\Gamma}_{obstacle}$ have been mapped into the null space of the task, as shown in equation 6. Using this control structure, the task-consistent obstacle avoidance behavior shown in the second image of Figure 4 is achieved. Note how without task-consistency the end effector deviates significantly from the required straight-line trajectory. Using task-consistent obstacle avoidance, the end effector only deviates minimally from the task, as can be seen in the graphs shown in Figure 4.

This approach of integrating task and obstacle avoidance behavior can fail, however, when the torques resulting from mapping $\mathbf{\Gamma}_{obstacle}$ into the null space yield insufficient motion to ensure obstacle avoidance. In such a situation it would be desirable to suspend task execution and to realize obstacle avoidance with all degrees of freedom of the robot.

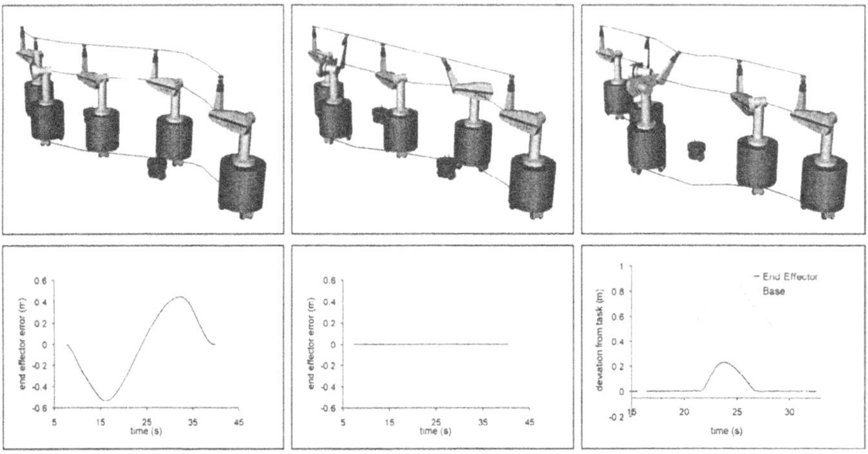

Figure 4. Top images show from left to right: real-time obstacle avoidance without task consistency, with task consistency, and transitioning between task-consistent and non task-consistent behavior. Lines indicate trajectories of the base, elbow, and end effector. The graphs show end effector error and base deviation from the task for the respective experiment.

The coefficient c corresponds to the ratio of the magnitude of the torque vector $\mathbf{\Gamma}_{obstacle}$ mapped into the null space of the task and its unmapped magnitude. This coefficient is an indication of how well the behavior represented by $\mathbf{\Gamma}_{obstacle}$ can be performed inside the null space of the task. We experimentally determine a value c_s at which it is desirable to suspend task execution in favor of the behavior previously mapped into the null space. Once the coefficient c assumes a value $c < c_s$, a transition is initiated. During this transition task behavior is gradually suspended and previous null space behavior is performed using all degrees of freedom of the manipulator. The motion of the manipulator is now generated using the equations

$$\begin{aligned} \mathbf{\Gamma}_{\text{task-consistent}} &= J^T(\mathbf{q})\,\mathbf{F} \;+\; N(\mathbf{q})^T \mathbf{\Gamma}_{\text{obstacle}} \\ \mathbf{\Gamma} &= \alpha\, \mathbf{\Gamma}_{\text{task-consistent}} \;+\; \overline{\alpha}\, \mathbf{\Gamma}_{\text{obstacle}} \end{aligned}$$

where α is a time-based transition variable, transitioning between 1 and 0 during task suspension and between 0 and 1 during resumption of the task, and $\overline{\alpha} = (1 - \alpha)$ is defined as the complement of α.

The experimental results, performed on the Stanford Assistant Manipulator, for such transistioning behavior can be seen in Figure 4. The image on the top right shows how despite task-consistent obstacle avoidance the task has to be suspended to ensure obstacle avoidance. Below, the graph shows how the base deviates significantly from the straight line in response to the obstacle. The end effector, however, maintains the task until it has to be suspended. The graph also shows that the task is resumed in a smooth manner, after the base has passed the obstacle

5. Conclusion

Advances toward the challenge of robotics in human environments depend on the development of the basic capabilities needed for both autonomous operations and human/robot interaction. In this article, we have presented methodologies for whole-robot coordination and control, cooperation between multiple robots, interactive haptic simulation with contact, and the real-time modification of collision-free path to accommodate changes in the environment.

For the whole-robot coordination and control, we presented a framework which provides the user with two basic task-oriented control primitives: task control and posture control. The major characteristic of this control structure is the dynamic consistency it provides in implementing these two primitives: the robot posture behavior has no impact on the end-effector dynamic behavior.

Addressing the computational challenges of human-like robotic structures, we presented efficient $O(nm + m^3)$ recursive algorithms for the operational space dynamics of mechanisms involving branching structures and closed chains. Building on the operational space formulation, we also developed a framework for the resolution of multi-contact between articulated multi-body systems.

The elastic strip framework allows the seamless integration of reactive, real-time obstacle avoidance and the task-oriented control structure. It provides for real-time motion generation that combines obstacle avoidance and task execution. When kinematic or external constraints imposed by obstacles make it impossible to maintain the task, task-consistent obstacle avoidance is suspended and all degrees As the constraints are relaxed, the task is resumed in a smooth manner.

Acknowledgments

The financial support of Honda Motor Company and NSF (grants IRI-9320017) is gratefully acknowledged. Many thanks to Alan Bowling, Arancha Casal, Robert Holmberg, Jaeheung Park, Costas Stratelos, James Warren, and Kazuhito Yokoi for their valuable contributions to the work reported here.

References

Asfour, T., K. Berns, J. Schelling, and R. Dillmann (1999), Programming of Manipulation Tasks of the Humanoid Robot ARMAR, *Proc. of the Intl. Conf. on Advanced Robotics.*

Brock, O. and O. Khatib (1998), Elastic Strips: Real-Time Path Modification for Mobile Manipulation, *Robotics Research*, Springer Verlag.

Chang, K.-S. and O. Khatib (2000), Operational Space Dynamics: Efficient Algorithms for Modelling and Control of Branching Mechanisms, *Proc. of the IEEE Intl. Conf. on Robotics and Automation*, pp. 850–856.

Featherstone, R. (1987), *Robot Dynamics Algorithms*, Kluwer Academic Publishers.

Holmberg, B. and O. Khatib (2000), Development of a Holonomic Mobile Robot for Mobile Manipulation Tasks, *Intl. J. of Robotics Research*, vol. 19, no. 11, pp. 1066-1074.

Hirai, K., M. Hirose, Y. Haikawa, and T. Takenaka (1998), The Development of Honda Humanoid Robot, *Proc. of the IEEE Intl. Conf. on Robotics and Automation*, vol. 2, pp. 1321–1326, Leuven, Belgium.

Khatib, O. (1987), A Unified Approach to Motion and Force Control of Robot Manipulators: The Operational Space Formulation, *Intl. J. of Robotics and Automation*, vol. 3, no. 1, pp. 43–53.

Khatib, O. (1995), Inertial Properties in Robotics Manipulation: An Object-Level Framework, *Intl. J. of Robotics Research*, vol. 14, no. 1, pp. 19-36.

Khatib, O., K. Yokoi, K.-S. Chang, D. Ruspini, R. Holmberg, and A. Casal (1996), Coordination and Decentralized Cooperation of Multiple Mobile Manipulators, *Intl. J. of Robotic Systems*, vol. 13, no. 11, pp. 755–764.

Khatib, O., K. Yokoi, O. Brock, K.-S. Chang, and A. Casal (1999), Robots in Human Environments: Basic Autonomous Capabilities, *Intl. J. of Robotics Research*, vol. 18, no. 7, pp. 175–183.

Kreutz-Delgado, K., A. Jain, and G. Rodriguez (1991), Recursive Formulation of Operational Space Control, *Proc. of the IEEE Intl. Conf. on Robotics and Automation*, pp. 1750–1753.

Lilly, K. W. (1992), *Efficient Dynamic Simulation of Robotic Mechanisms*, Kluwer Academic Publishers.

Nishiwaki, K., T. Sugihara, S. Kagami, F. Kanehiro, M. Inaba, and H. Inoue (2000), Design and Development of Research Platform for Perception-Action Integration in Humanoid Robot: H6, *Proc. of the IEEE/RSJ Intl. Conf. on Intelligent Robots and Systems*, vol. 3, pp. 1559–1564.

Quinlan, S. and O. Khatib (1993), Twowards Real-Time Execution of Motion Tasks, *Experimental Robotics 2*, Springer Verlag.

Russakow, J., O. Khatib, and S. M. Rock, (1995), Extended Operational Space Formation for Serial-To-Parallel Chain (Branching) Manipulators, *Proc. of the IEEE Intl. Conf. on Robotics and Automation*, vol. 1, pp. 1056-1061.

Ruspini, D. C. and O. Khatib (1999), Collision/Contact Models for Dynamic Simulation and Haptic Interaction, *Proc. Intl. Symp. of Robotics Research*, pp. 185–195.

Takanishi, A., S. Hirano, and K. Sato (1998), Development of an anthropomorphic head-eye system for a humanoid robot-realization of human-like head-eye motion using eyelids adjusting to brightness, *Proc. of the IEEE Intl. Conf. on Robotics and Automation*, vol. 2, pp. 1308–1314, Leuven, Belgium.

PARTICULAR ASPECTS IN DESIGNING ANTHROPOMORPHIC MECHANISMS

Cornel Brisan
Department of Mechanisms, Fine Mechanics and Mechatronics
Technical University of Cluj-Napoca, Romania
cornelbrisan@yahoo.com

Manfred Hiller
Institute of Mechatronics and System Dynamics
Gerhard Mercator University of Duisburg, Germany
hiller@mechatronik.uni-duisburg.de

Abstract The paper presents some aspects of anthropomorphic mechanisms modelling. Thus, considering biomechanical criteria regarding the human superior limb, a structural synthesis for two different types of anthropomorphic mechanisms (serial and parallel respectively) was made. Also, for a parallel anthropomorphic mechanism, the direct and inverse geometrical model is presented.

Keywords: human biomechanics, anthropomorphic mechanisms, kinematics, parallel mechanisms

1. Introduction

Anthropomorphic robots are directly related with the human wish to self outrun. Anthropomorphic mechanisms constitute the mechanical architecture of those robots. Most of the anthropomorphic robots prototypes are basically models of the human arm bone system. That means a serial topology, with rotational joints as actuators. Usually, this type of robots have industrial tasks such as mechanical operations (fitting, drilling, punching, screw fastening, boring, reaming). Also, they can be used in human prosthetics or for other biomedical tasks (Gosselin, C.M., et al., 1996, Merlet, J., 1997). There are also few prototypes of anthropomorphic robots which have as a mechanical structure parallel mechanisms (Charoenseang, S. Et al., 1998, Northrup, S., et al, 2001).

2. Biomechanical considerations

Anthropomorphic robots are defined as functional and structural models of the human superior limb (HSL). Biomechanical studies

J. Lenarčič and F. Thomas (eds.), Advances in Robot Kinematics, 99–106.

regarding HSL are praised that there are three main joints in its structure: shoulder, elbow and wrist joint respectively. There are also considered three elementary movements (flexion-extension, adduction-abduction, pronation-supination) which can describe, from kinematical point of view, the behaviour of an arbitrary human joint. The shoulder joint is usually modelled as a spherical joint.

There are two different interpretation regarding the movements which can be performed on the level at the elbow and wrist joint respectively. First of these interpretations considers that at the level of the elbow joint can be performed only the flexion- extension movement. With this approach it results that at the level of the wrist joint can be made three movements: flexion-extension (F-E), abduction – adduction (Ab-Ad) and pronation – supination (P-S). Second interpretation considers that at the level of the elbow joint two movements (F-E and P-S) may be performed. Of course, in this case only two movements can be performed at the level of wrist joint. Based on these biomechanical considerations two models of the HSL, with different distribution of d.o.f., can be considered:

a) first model (Fig.1a) corresponds to the first interpretation and will be called Serial Anthropomorphic Mechanism 313 (SAM313);

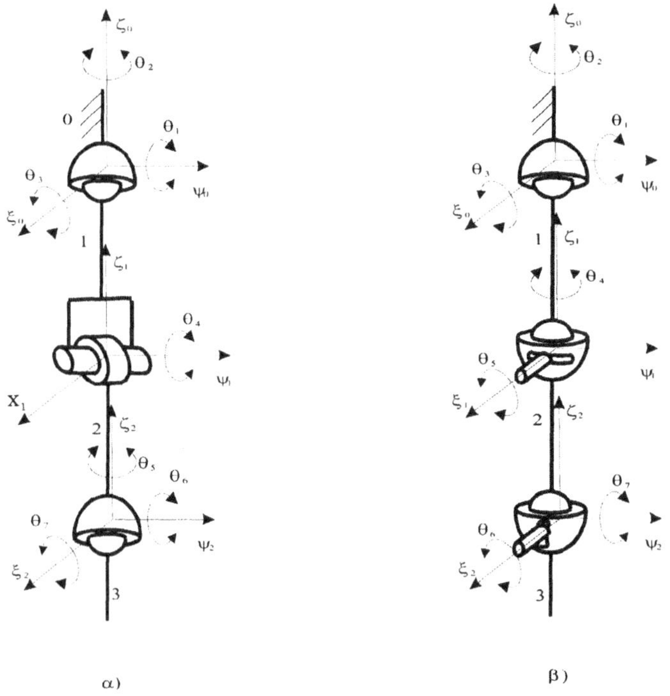

Figure 1. Serial Anthropomorphic Mechanisms

b) second model (Fig.1b) corresponds to the second interpretation and will be called Serial Anthropomorphic Mechanism 322 (SAM322);

We can observe that in fact both of them (SAM313 and SAM322) are models of the human bone system. In order to increase the anthropomorphic coefficient (Morecki, 1978) of these models, the mechanisms from figure 2 (models of the muscle system) can be considered (Brisan, 1995).

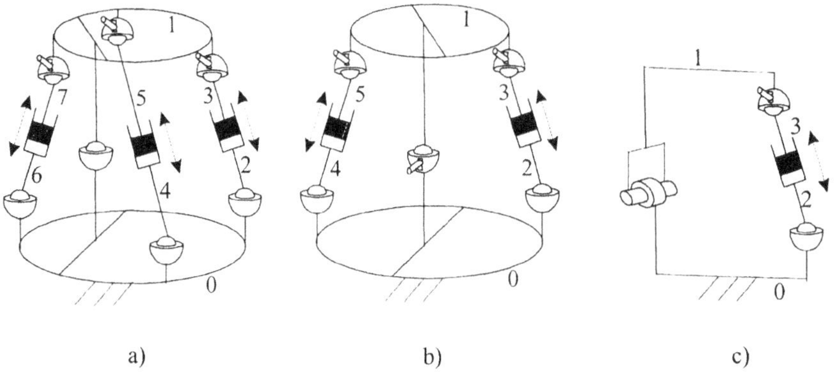

Figure 2. Models of the muscle system

By serial connection of these mechanisms two parallel anthropomorphic mechanisms (PAM313 and PAM322 respectively) are obtained. The graphs of those mechanisms (Fig.3) are praised that each of them has 7 independent loops.

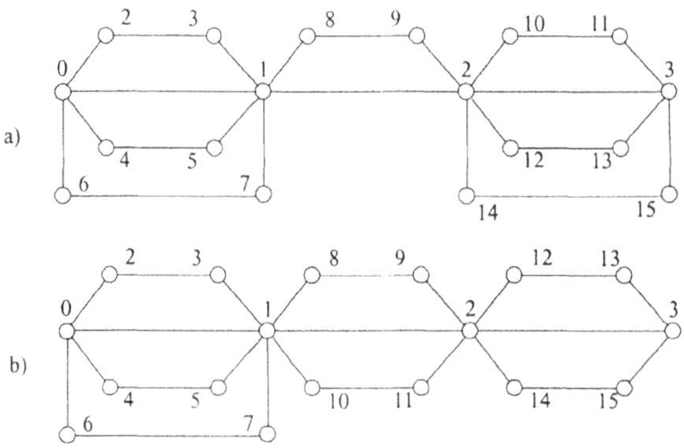

Figure 3. The graphs of the PAM313 and PAM322

3. Geometrical model

3.1 Direct geometrical model

Let (Fig. 4) the PAM322 be the manipulator and $\{T_i\}$, $i=\{0,...3\}$ the coordinate systems, solidary with the platforms of the mechanism.

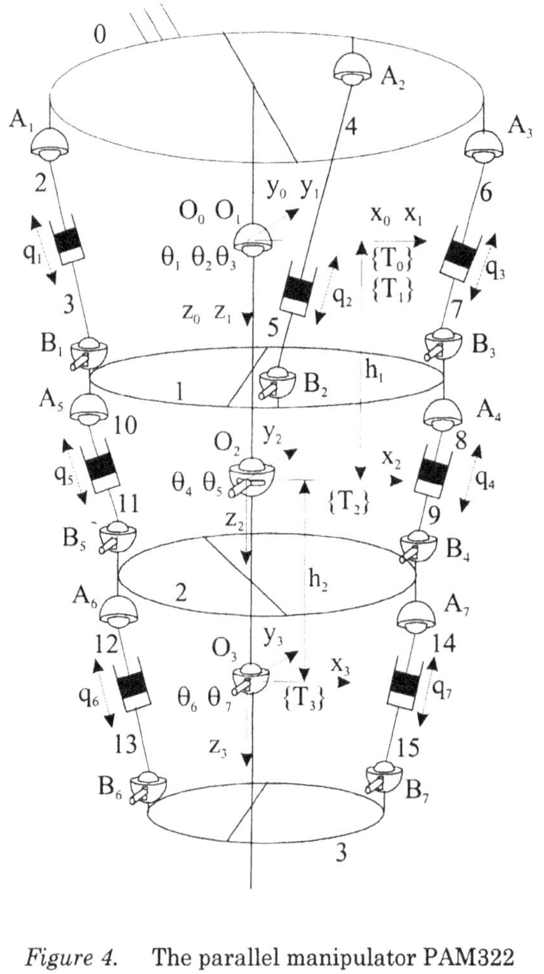

Figure 4. The parallel manipulator PAM322

Also, let θ_i, $i=1,...,7$ be the angular displacements at the level of spherical and universal joints respectively.

We note:

$$\mathbf{q}_i = \mathbf{A}_i\mathbf{B}_i \text{ , } \mathbf{a}_i = \mathbf{O}_j\mathbf{A}_i \text{ , } \mathbf{b}_i = \mathbf{O}_j\mathbf{B}_i \text{ ,} i = \{1,...,7\} \text{, } j = \{1,...,3\}. \tag{1}$$

The positional scalar equations are given by:

$$\left\|\mathbf{a}_i - \mathbf{b}_i\right\| = \left\|\mathbf{q}_i\right\|, \; i = 1,\ldots,7, \tag{2}$$

where $\mathbf{a}_i$ are known constant vectors, $\mathbf{q}_i$ are the generalized coordinates (known in the case of direct geometrical model - DGM) and $\mathbf{b}_i$ are vectors which contain the unknowns:

$$\mathbf{b}_i = \mathbf{b}_i(\theta_i)\;,\; i = 1,\ldots,7. \tag{3}$$

From equations (2), using numerical methods, the angular displacements $\boldsymbol{\theta}_i$ can be found.

It is possible to observe that (DGM) can be solved separately for each of the mobile platforms.

Thus, from the first three equations (2) it results $\theta_1, \boldsymbol{\theta}_2, \boldsymbol{\theta}_3$. The coordinates of the points B_1, B_2, B_3 are given by:

$$\mathbf{X}_{B_i}{}^{0} = \mathbf{H}_{01}(\theta_1, \theta_2, \theta_3) \cdot \mathbf{X}_{B_i}{}^{1}, \tag{4}$$

where $\mathbf{X}_{B_i}{}^{0}$ is the matrix of coordinates of the points $B_i\;, i = 1,\ldots,3$ relating to the $\{T_0\}$ system, $\mathbf{X}_{B_i}{}^{1}$ is the matrix of coordinates of the points $B_i\;, i = 1,\ldots,3$ relating to the $\{T_1\}$ system, and $\mathbf{H}_{01}(\theta_1,\theta_2,\theta_3)$ is the transformation matrix, which describe the relative movements between the coordinates systems $\{T_1\}$ and $\{T_0\}$ respectively.

The matrix H_{01} is given by the relation:

$$\mathbf{H}_{01} = \mathbf{A}_1(\theta_1) \cdot \mathbf{A}_2(\theta_2) \cdot \mathbf{A}_3(\theta_3), \tag{5}$$

where $\mathbf{A}_i(\theta_i), i = \{1,\ldots,3\}$ are the transformation matrices which describe the relative movements at the level of spherical joint.

For the last platform it results:

$$\mathbf{X}_{B_j}{}^{0} = \mathbf{H}_{03} \cdot \mathbf{X}_{B_j}{}^{3}, \tag{6}$$

where $\mathbf{X}_{B_j}{}^{0}$ is the matrix of coordinates of the points $B_j\;, j = \{6,7\}$ relating to the $\{T_0\}$ system, $\mathbf{X}_{B_j}{}^{3}$ is the matrix of coordinates of the points $B_j\;, j = \{6,7\}$ relating to the $\{T_3\}$ system, and $\mathbf{H}_{03}$ is the transformation matrix, which describe the absolute position and orientation of the $\{T_3\}$ coordinate system.

The matrix H_{03} is given by the relation:

$$\mathbf{H}_{03} = \prod_{i=1}^{7} \mathbf{A}_i(\theta_i)\,, \tag{7}$$

where $\mathbf{A}_i = \mathbf{A}_i(\theta_i)$, $i = 1,\dots,7$, are the relative transformation matrices.

It results that the relations (6) and (7) gives the solution of the DGM of the manipulator from figure 4.

This mathematical model was implemented in MOBILE software (Kecskemethy, A., 1999) and in figure 5 a simulation is presented.

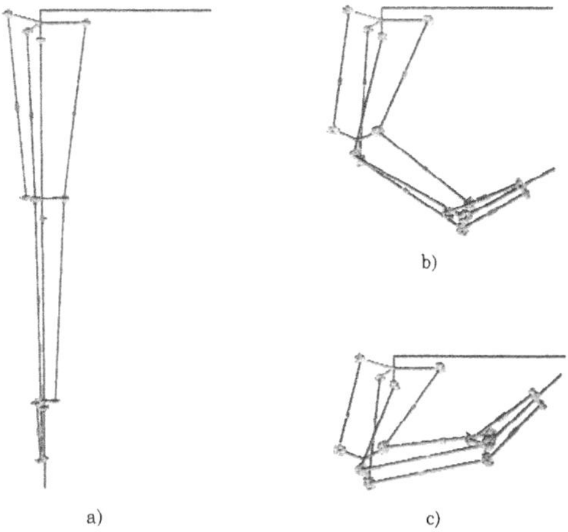

Figure 5. Movement simulation of PAM322

Thus, in figure 5a the initial position of the mechanism corresponds to the vertical position of a HSL and the last position (Fig. 5c) corresponds to the position of the HSL when the hand is near the body. Figure 5b presents an intermediate position.

3.2 Inverse geometrical model

In the case of inverse geometrical model (IGM) the absolute position and orientation of the $\{T_3\}$ is given. The aim is to find the variation laws of the generalized coordinates q_i i=1,...,7.

We observe that the PAM322 is a redundant manipulator. This property can be used for an optimization of the movement (Hiller, 1995).

Using the relations:

$$\prod_{i=j}^{1} \mathbf{A}^{-1}{}_i \cdot \mathbf{H}_{03} = \prod_{i=j+1}^{7} \mathbf{A}_i,\ j = 1,\dots,6, \tag{8}$$

it is possible to find the functions:

$$\theta_i = f_i(\theta_1)\,,\, i = 2,\dots,7. \tag{9}$$

It results that the coordinates of the points A_i ,i=4,...,7 and B_j ,j=1,...,7 could be expressed as functions of θ_1:

$$\begin{aligned} \mathbf{X}_{A_i} &= \mathbf{X}_{A_i}(\theta_1), \\ \mathbf{X}_{B_i} &= \mathbf{X}_{B_i}(\theta_1). \end{aligned} \tag{10}$$

The generalised coordinates q_i are given by relations:

$$q_i = \left[(x_{B_i} - x_{A_i})^2 + (y_{B_i} - y_{A_i})^2 + (y_{B_i} - y_{A_i})^2\right]^{1/2} \tag{11}$$

and based on relations (10) it results that they are also functions of θ_1:

$$q_i = q_i(\theta_1). \tag{12}$$

In the case of human arm a criterion for optimisation could be the minimization of the mechanical work performed by muscular system.

For PAM322 the mechanical work of the actuated joints is given by:

$$L = \sum_{i=1}^{7} F_i \cdot \Delta q_i, \tag{13}$$

where F_i are the forces performed at the level of prismatic joints and Δq_i are the displacements.

Also, based on relation (12) it results:

$$L = L(\theta_1). \tag{14}$$

From relation (14) it is possible to find the value of the θ_1 ($\theta^*{}_1$) which corresponds to ($L_{\min}$):

$$L_{\min} = L(\theta^*{}_1). \tag{15}$$

4. Conclusions

The human superior arm can be modelled by anthropomorphic mechanisms. Using biomechanical criteria it is possible to find two different types of anthropomorphic mechanisms.

First type has serial topology (called SAM) and the other a parallel one (called PAM). For each of them there are also two variants with different distribution of the number of degrees of freedom. The main difference between SAM and PAM is the type of the actuator (rotational and prismatic respectively). From this point of view it is possible to consider that the PAM are models of the human muscle system.

The solutions of direct geometrical model of PAM can be obtained only in numerical form. Because PAM are redundant mechanisms, it is possible to use one of the generalised coordinate in order to optimise the kinematic behaviour. Also, a particularity of inverse geometrical model in the case of PAM is that though there are three mobile platforms it is possible to find an analytical solution. For that the corresponding open loop (SAM) may be used.

5. Acknowledgements

The research work was supported by Alexander von Humboldt Foundation.

6. References

Angeles, J., (1982), Spatial Kinematic Chains: Analysis, Synthesis, Optimisaton, Springer-Verlag, Berlin.

Brisan, C., (1995), Contributions to the Research, Modelling and Simulation of Biped Robots, PhD Thesis, Technical University of Cluj-Napoca.

Carretero, J.A., Podhorodeski, R.P., Nahon, M.A., and Gosselin, C.M. (2000), Kinematic Analysis And Optimization of a New Three-Degree-of-Freedom Spatial Parallel Manipulator," ASME Journal of Mechanical Design, Vol. 122, No. 1, pp. 17-24.

Charoenseang, S. , Srikaew, A. , Wilkes, D.M. and Kawamura, K. ,(1998), 3-D Collision Avoidance for the Dual - Arm Humanoid, IASTED International Conference on Robotics and Manufacturing, Banff, Canada.

Gosselin, C.M., St-Pierre, É., and Gagné, M., (1996), On the Development of the Agile Eye: Mechanical Design, Control Issues And Experimentation, IEEE Robotics and Automation Society Magazine, Vol. 3, No. 4, pp. 29-37.

Hiller, M., (1995). Multiloop Kinematic Chains and Dynamics of Multiloop Systems, in Kinematics and Dynamics of Multi-Body Systems (ed. By J.Angeles and A. Kecskemethy), CISM Courses and Lecture, nr. 360, Springer Verlag.

Kecskemethy, A., (1999). Mobile, User's Guide, Gerhard-Mercator University, Duisburg.

Merlet, J.-P., (1997), Miniature In-Parallel Positionning System Mips for Minimally Invasive Surgery," World Congress on Medical Physics and Biomedical Engineering, Nice, France, September 14-19.

Morecki, A., (1978), Concept of Anthropomorphism in Description of Androidal Robots, 3th World Congress on the Theory of Machines and Mechanisms, Udine, Italy.

Northrup, S., Sarkar, N. and Kawamura, K., (2001), Biologically- Inspired Control Architecture for a Humanoid Robot, 2001 IEEE/RSJ International Conference on Intelligent Robots and Systems, IROS 2001, , Maui, Hawaii.

UNCERTAINTY MODEL AND SINGULARITIES OF 3-2-1 WIRE-BASED TRACKING SYSTEMS

Federico Thomas**, Erika Ottaviano*, Lluís Ros**, Marco Ceccarelli*

*Laboratorio di Robotica e Meccatronica
Università degli Studi di Cassino, Via G. Di Biasio, 43, 03043 Cassino, Italy
{ootaviano, ceccarelli}@unicas.it

** Institut de Robòtica i Informàtica Industrial (CSIC-UPC)
Llorens Artigas 4-6, 08028 Barcelona, Spain
{llros, fthomas}@iri.upc.es

Abstract This paper presents an ellipsoidal set-membership uncertainty model for wire-based tracking systems with wire length uncertainties and joint clearances. Although the proposed model is valid for any number of wires and configurations, including singularities, it has been particularized to a 3-2-1 parallel wire mechanism. The Euler's tetrahedron formula has been used to obtain a numerically stable solution to the direct kinematics of this particular tracking system as well as a compact characterization of its singularities directly expressed in terms of the wire lengths. It is also shown that these singularities, when expressed in the configuration space of the moving element, can be identified as C-surfaces associated with the three basic contacts between polyhedra, which have been widely studied in the context of path planning.

Keywords: Tracking systems, uncertainty manipulation, kinematic singularities. Euler's tetrahedron formula.

1. Introduction

Many systems for measuring position and orientation of moving objects, also known as tracking systems, have been developed. They can be classified according to the measuring principle and used technology (Hightower and Borriello, 2001). Most of the instruments such as cameras, theodolites, laser and wire systems use triangulation or trilateration techniques. Trilateration and triangulation determine the relative position between points by using the geometry of triangles. Triangulation uses measurements of both distances and angles, whereas trilateration uses only distance measurements.

J. Lenarčič and F. Thomas (eds.), Advances in Robot Kinematics, 107–116.

Measurement systems can be also classified according to their characteristics, such as accuracy, resolution, cost, measurement range, portability, and calibration procedure. Laser tracking systems exhibit good accuracy, which can be less than $1\mu m$ if the system is well calibrated. Unfortunately, they are very expensive, their calibration procedure is time consuming, and they are sensitive to the environment. Vision systems have an accuracy of $0.1mm$, they are low cost portable devices but their calibration procedure can be complicate. Wire systems have an accuracy of $0.1mm$, they are also low cost portable devices but capable of measuring large displacements. Moreover, they exhibit a good compromise among accuracy, measurement range, cost and operability.

Wire-based tracking systems consist of a fixed base and a platform connected by six wires whose tension is maintained, while the platform is moved, by pulleys and spiral springs on the base. They can be modelled as six-degree-of-freedom parallel manipulators because wires can be seen as extensible legs connecting the platform and the base by means of spherical and universal joints, respectively. One of these wire mechanisms was used in (Jeong et al., 1999) for robot pose, i.e. position and orientation, measurements. The proposed system was studied as a 2-2-2 Gough-Stewart platform and, as a consequence, its direct kinematics could only be solved by a numerical method. This clearly limited its use in real-time measurements. Alternatively, (Geng, 1994) and (Ceccarelli et al, 1999) used a wire configuration equivalent to a 3-2-1 Gough-Steward platform whose direct kinematics can be solved in closed-form by using trilateration.

Dimension deviations due to fabrication tolerances, wire length uncertainties, and joint clearances may result in an unacceptable performance of wire-based tracking systems. Wire length uncertainties are due, among other factors, to deflections caused by wire self-weights and to discretization errors in the transducers. A way to compensate the self-weight deflections is presented in (Jeong et al., 1999). Joint clearances are due to the uncertainty related to the position of the wires with respect to the corresponding joint holes on the base. They become more evident with time because of the abrasion of the hole due to wire friction. Unlike the effect of the dimension deviations, the effect of wire length uncertainties and joint clearances can not be eliminated by calibration because of their random nature. How these two factors contribute to the measured pose error and how this error evolves when the system approaches a kinematic singularity are the two issues addressed in this paper.

Most error analysis techniques assume that the data are corrupted by random noise whose probability density function is usually assumed to be gaussian, as in (Zhu and Ting, 2000). Real-world uncertainties, however, also include nongaussian, nonwhite noise and systematic errors. These uncertainties can easily be considered in a set theoretic setting which consists in defining bounds for the uncertain variables. The main problem with this kind of uncertainty description is that, although the initial uncertainty sets have simple shapes, the result of principal operations with them have a complicated shape. This is why some canonical sets, that depend on a fixed number of parameters, are introduced for the approximation of uncertainty sets. Among many possibilities, ellipsoids are usually taken as these canonical sets because they: (a) can be concisely described; (b) provide a satisfactory approximation of convex sets in most applications; (c) can be represented using matrices interpretable as weighted covariance matrices; and (d) are invariant, as a class, under affine transformations.

Most ellipsoidal calculus are unable to deal with degenerate cases, that is, with situations in with ellipsoids extend to infinity in some directions. Here this is an important issue because this is what happens when a tracking system approaches a kinematic singularity, as it will be shown later. Fortunately, the calculus presented in (Ros et al., 2002) is immune to degeneracies and can readily applied here.

This paper is organized as follows. Section 2 presents the adopted uncertainty model and Section 3 gives a description of the singularities first in terms of the wire lengths and then in terms of the configuration of the moving platform. Finally, Section 4 gives the conclusions.

2. The uncertainty model

In order to describe the relative position between base and platform, let us introduce an absolute and a relative frame fixed to the base and the platform, respectively, as shown in Figure 2a. The pose of the platform with reference to the base is given by the configuration vector $\mathbf{q} = (\boldsymbol{\Omega}, \mathbf{t})$, where $\mathbf{t}$ stands for the translation vector and $\boldsymbol{\Omega}$ for the orientation vector given in Euler's angles, i.e. $\mathbf{q} \in \Re^3 \times SO(3)$. The corresponding rotation matrix will be noted $\mathbf{R}(\boldsymbol{\Omega})$. The unit vector along wire i will be noted $\mathbf{e}_i$ and l_i the corresponding wire length. The centers of the articulations on the base and the platform for wire i will be denoted $\mathbf{a}_i$ and $\mathbf{b}_i$, respectively, as shown in Figure 2a. According to these definitions $\mathbf{R}(\boldsymbol{\Omega})\mathbf{b}_i + \mathbf{t} = \mathbf{a}_i + l_i\mathbf{e}_i$. Thus, the wire length l_i can be expressed as:

$$l_i = \| \mathbf{R}(\boldsymbol{\Omega})\mathbf{b}_i + \mathbf{t} - \mathbf{a}_i \| \, . \tag{1}$$

Then, its linearization around the nominal values of $\mathbf{a}_i$ and $\mathbf{b}_i$ ($\mathbf{a}_i^0$ and $\mathbf{b}_i^0$), the measured value of l_i (l_i^0), and the calculated value of the configuration $\mathbf{q}$ ($\mathbf{q}^0$) is:

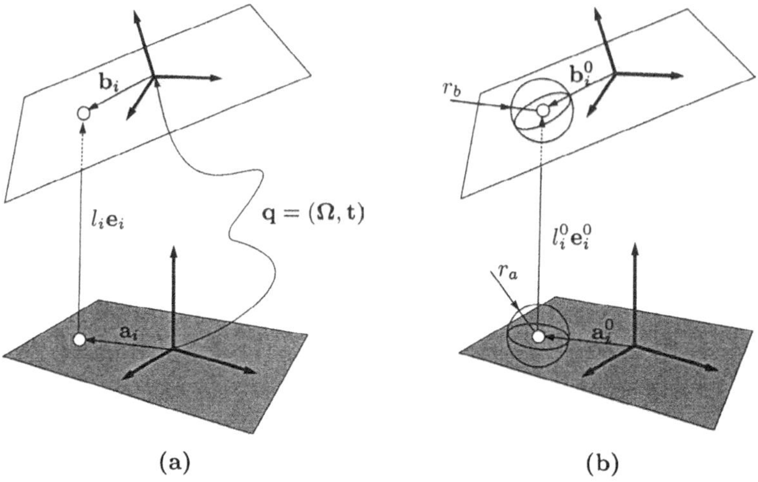

Figure 1. Magnitudes associated with wire i (**a**), and its associated joint clearances (**b**)

$$l_i - l_i^0 = \frac{\partial l_i}{\partial \mathbf{q}}(\mathbf{q} - \mathbf{q}^0) + \frac{\partial l_i}{\partial \mathbf{a}_i}(\mathbf{a}_i - \mathbf{a}_i^0) + \frac{\partial l_i}{\partial \mathbf{b}_i}(\mathbf{b}_i - \mathbf{b}_i^0). \tag{2}$$

It can be proved, by taking derivatives (1), that:

$$\frac{\partial l_i}{\partial \mathbf{a}_i} = -\mathbf{e}_i^t, \quad \frac{\partial l_i}{\partial \mathbf{b}_i} = \mathbf{e}_i^t\mathbf{R}(\boldsymbol{\Omega}), \quad \text{and} \quad \frac{\partial l_i}{\partial \mathbf{q}} = \begin{pmatrix} (\mathbf{t} - \mathbf{a}_i) \times \mathbf{e}_i \\ \mathbf{e}_i \end{pmatrix}^t.$$

Substituting in (2), we get:

$$l_i - l_i^0 = \begin{pmatrix} (\mathbf{t} - \mathbf{a}_i) \times \mathbf{e}_i \\ \mathbf{e}_i \end{pmatrix}^t (\mathbf{q} - \mathbf{q}^0) + v_i - v_i^0, \tag{3}$$

where $v_i = \mathbf{e}_i^t(\mathbf{R}(\boldsymbol{\Omega})\mathbf{b}_i - \mathbf{a}_i)$ and $v_i^0 = \mathbf{e}_i^t\left(\mathbf{R}(\boldsymbol{\Omega})\mathbf{b}_i^0 - \mathbf{a}_i^0\right)$. Finally, collecting (3) for the six wires into a single matrix expression, we get:

$$\mathbf{J}(\mathbf{q} - \mathbf{q}^0) = (\mathbf{l} - \mathbf{v}) - (\mathbf{l}^0 - \mathbf{v}^0) \tag{4}$$

where $\mathbf{J} = \begin{pmatrix} (\mathbf{t} - \mathbf{a}_1) \times \mathbf{e}_1 & (\mathbf{t} - \mathbf{a}_2) \times \mathbf{e}_2 & \cdots & (\mathbf{t} - \mathbf{a}_6) \times \mathbf{e}_6 \\ \mathbf{e}_1 & \mathbf{e}_2 & \cdots & \mathbf{e}_6 \end{pmatrix}^t$,
$\mathbf{l} = (l_1, \ldots, l_6)$, and $\mathbf{v} = (v_1, \ldots, v_6)$.

Note that the above formulation is valid for any wire configuration and can be extended to any number of wires. Note also that the rows of $\mathbf{J}$ are the Plücker coordinates of an arrangement of lines equivalent to that of the wire lines. Actually, $\mathbf{J}$ is the linear mapping between wire extensions and platform velocities, i.e. the inverse Jacobian of the associated parallel mechanism.

Let us assume that the real value of $\mathbf{a}_i$ is inside a sphere of radius r_a centered around $\mathbf{a}_i^0$ as shown in Figure 2b. The same can be done for the platform joint clearances assuming that the radii of the corresponding spheres are r_b. Then, it can be checked that

$$(\mathbf{v}-\mathbf{v}^0)^t(\mathbf{v}-\mathbf{v}^0) \leq 6(r_a+r_b)^2 \tag{5}$$

is a conservative bound for $\mathbf{v}$. We can also assume that all wire lengths have the same error bounds so that

$$(\mathbf{l}-\mathbf{l}^0)^t(\mathbf{l}-\mathbf{l}^0) \leq 6r_l^2 \tag{6}$$

also provides a conservative bound for $\mathbf{l}$. Then, the Minkowski difference between the spheres (6) and (5), that is

$$\left((\mathbf{l}-\mathbf{v})-(\mathbf{l}^0-\mathbf{v}^0)\right)^t\left((\mathbf{l}-\mathbf{v})-(\mathbf{l}^0-\mathbf{v}^0)\right) \leq 6(r_a+r_b+r_l)^2, \tag{7}$$

gives us the uncertainty region for $(\mathbf{l}-\mathbf{v})$ (Ros et al., 2002).

Equation (7), together with (4), leads to:

$$(\mathbf{q}-\mathbf{q}^0)^t\mathbf{J}^t\mathbf{J}(\mathbf{q}-\mathbf{q}^0) \leq 6(r_a+r_b+r_l)^2, \tag{8}$$

which is an ellipsoidal bound for the uncertainty region associated with $\mathbf{q}$, i.e. the configuration of the moving platform. Since $\mathbf{J}$ is a square matrix, it can be shown that the volume of the region defined by (8) is proportional to the inverse of $\det(\mathbf{J})$. In other words, the uncertainty region using this model is unbounded if, and only if, $\mathbf{J}$ is singular. The next section is devoted to the characterization of these singularities.

3. Singular configurations

Since the rows of $\mathbf{J}$ are the Plücker coordinates of lines, it is possible to characterize the singularities in terms of the geometry of linearly dependent sets of lines (Pottmann and Wallner, 2001). The set of conditions is typical for many parallel manipulators (McCarthy, 2000). Nevertheless, herein we will use an ad hoc approach that takes advantage of the 3-2-1 wire configuration.

First of all, let us define the function

$$\Xi(\mathbf{p}_1,\mathbf{p}_2,\mathbf{p}_3,\mathbf{p}_4) = \begin{vmatrix} 0 & r_{12} & r_{13} & r_{14} & 1 \\ r_{21} & 0 & r_{23} & r_{24} & 1 \\ r_{31} & r_{32} & 0 & r_{34} & 1 \\ r_{41} & r_{42} & r_{43} & 0 & 1 \\ 1 & 1 & 1 & 1 & 0 \end{vmatrix}, \tag{9}$$

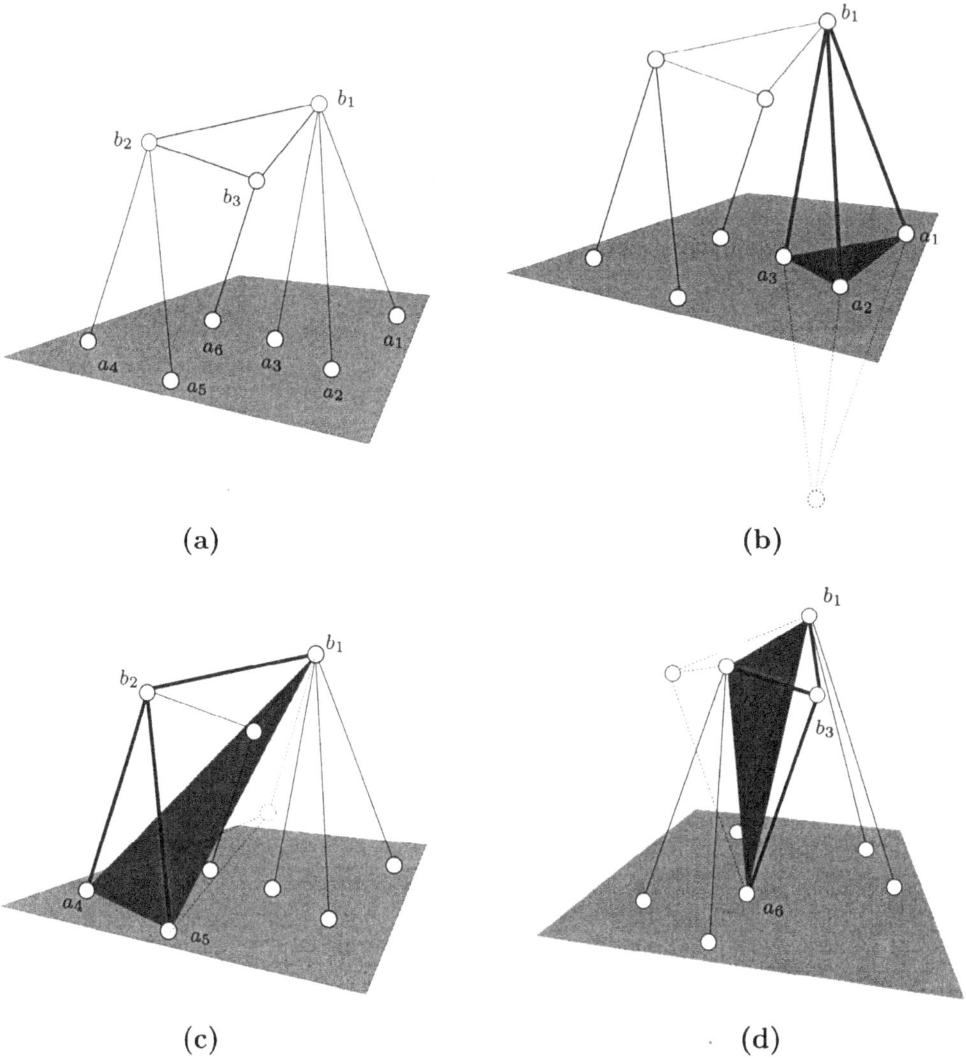

Figure 2. General kinematic model of a 3-2-1 tracking system (**a**). There are up to 8 possible solutions for the configuration of the platform compatible with a set of wire lengths due to the two possible solutions for the location of $\mathbf{b}_1$ (**b**), $\mathbf{b}_2$ (**c**), and $\mathbf{b}_3$ (**d**).

where $\mathbf{p}_1, \ldots, \mathbf{p}_4$ are four points in $\Re^3$ and r_{ij} is the square of the distance between $\mathbf{p}_i$ and $\mathbf{p}_j$. It can be shown that the volume, V, of the tetrahedron defined by these four points is related to (9) through the formula $288V^2 = \Xi(\mathbf{p}_1, \mathbf{p}_2, \mathbf{p}_3, \mathbf{p}_4)$. If determinant (9) vanishes, $\mathbf{p}_1$, $\mathbf{p}_2$, $\mathbf{p}_3$, and $\mathbf{p}_4$ lie on the same plane. If it gives a negative value, the tetrahedron cannot be assembled with the given distances. Although it is generally accepted that this remarkable formula derives from a development given by Euler in 1753 (Dörrie, 1965), the Italian painter Piero della Francesca already obtained an equivalent result around 1490 (Peterson, 1997).

Formula (9) has a 2D counterpart (known as Herron's formula) that allows us to obtain the area of a triangle in terms of the lengths of its edges. As a consequence, any of the heights of a tetrahedron can readily be obtained by dividing its volume by the corresponding base area using these formulas. The important point here is that (9) and its 2D counterpart are numerically stable and coordinate-free. Hence, this permits to solve the direct kinematics of a 3-2-1 tracking mechanism in an alternative way to that presented in (Ceccarelli et al, 1999), as explained bellow.

In a 3-2-1 wire-based tracking system, some of the joints in the platform coincide in the way shown in Figure 2a. This configuration greatly simplifies the analysis of the direct kinematics of the system. Indeed, given l_1, l_2, and l_3, the location of $\mathbf{b}_1$ can be obtained using the volume and area formulas given above. Note that there are two possible solutions which are specular with respect to the plane defined by $\mathbf{a}_1$, $\mathbf{a}_2$, and $\mathbf{a}_3$ (Figure 2b). Once one of these two solutions for $\mathbf{b}_1$ is chosen, $\mathbf{a}_4$, $\mathbf{a}_5$, $\mathbf{b}_1$ and $\mathbf{b}_2$ define another tetrahedron with known edge lengths. Again, there are two possible specular locations for $\mathbf{b}_2$, in this case with respect to the plane defined by $\mathbf{a}_4$, $\mathbf{a}_5$, and $\mathbf{b}_1$ (Figure 2c). Finally, after choosing one of the two solutions, $\mathbf{a}_6$, $\mathbf{b}_1$, $\mathbf{b}_2$, and $\mathbf{b}_3$ define another tetrahedron with known edge lengths. In this case there are two specular possible locations for $\mathbf{b}_3$ with respect to the plane defined by $\mathbf{a}_6$, $\mathbf{b}_1$ and $\mathbf{b}_2$ (Figure 2d). We conclude that, if $\Xi(\mathbf{a}_1, \mathbf{a}_2, \mathbf{a}_3, \mathbf{b}_1)$, $\Xi(\mathbf{a}_4, \mathbf{a}_5, \mathbf{b}_2, \mathbf{b}_3)$, and $\Xi(\mathbf{a}_6, \mathbf{b}_1, \mathbf{b}_2, \mathbf{b}_3)$ are all different from zero, there are 8 possible configurations for the moving platform compatible with a given set of wire lengths. Let us choose one of these solutions and let us infinitesimally modify the wire lengths. This will univocally give a new configuration for the platform. This means that the Jacobian relating the platform and wire velocities is well-defined if, and only if, none of these determinants vanish. In other words, these three determinants fully characterize all the singularities of the tracking system. This characterization is given in terms of the wire lengths, but in most cases we want to give it in terms of the configuration of the platform. The solution to this problem curiously comes from the developments in motion planning for polyhedra.

Note that $\Xi(\mathbf{p}_1, \mathbf{p}_2, \mathbf{p}_3, \mathbf{p}_4)$ is zero if, and only if, $\mathbf{p}_1$, $\mathbf{p}_2$, $\mathbf{p}_3$, and $\mathbf{p}_4$ lie on a plane. In other words, if, and only if, $\begin{vmatrix} \mathbf{p}_1 & 1 \\ \mathbf{p}_2 & 1 \\ \mathbf{p}_3 & 1 \\ \mathbf{p}_4 & 1 \end{vmatrix} = 0$. As a consequence, when at least one of $\begin{vmatrix} \mathbf{a}_1 & 1 \\ \mathbf{a}_2 & 1 \\ \mathbf{a}_3 & 1 \\ \mathbf{b}_1 & 1 \end{vmatrix}$, $\begin{vmatrix} \mathbf{a}_4 & 1 \\ \mathbf{a}_5 & 1 \\ \mathbf{b}_2 & 1 \\ \mathbf{b}_3 & 1 \end{vmatrix}$, or $\begin{vmatrix} \mathbf{a}_6 & 1 \\ \mathbf{b}_1 & 1 \\ \mathbf{b}_2 & 1 \\ \mathbf{b}_3 & 1 \end{vmatrix}$ vanishes, the system is in a singular configuration. Now, the problem is to explicitly express these determinants in function of the configuration of the platform $\mathbf{q} \in \Re^3 \times SO(3)$.

These three determinants, when equated to zero, can be seen as the implicit equations of the C-surfaces in $\Re^3 \times SO(3)$ associated with each of the three basic contacts between polyhedra (see (Thomas, 1995) for details). In particular, with the face-vertex, edge-edge and vertex-face contact, respectively. The algebraic expressions for these surfaces were first given in (Donald, 1984). They are rather long (particularly the one corresponding to the second determinant), thus requiring a computer algebra system for their manipulation.

Each C-surface, a variety of dimension 5, divides $\Re^3 \times SO(3)$ into two half-spaces. Then, the three C-surfaces lead to a partition of the configuration space into 8 regions with congruent signs for the 3 determinants. During normal operation, a 3-2-1 wired-based tracking system should work in one of these regions without getting out of it to avoid ambiguities. If one of the above determinants goes below a certain threshold the system should stop under the risk of trespassing a C-surface. The explicit description of these 8 regions is not easy but, if the moving platform is constrained to move in a plane, the configuration space becomes $\Re^2 \times SO$ and the technique used in (Avnaim et al., 1988) for obtaining such descriptions can be applied. Iterating this for different parallel planes, a discretization can be obtained.

A motion of the platform defines a trajectory in the configuration space, and each point of this trajectory has an associated ellipsoidal uncertainty region. These ellipsoids, as obtained in the previous section, are unbounded if, and only if, the configuration is on a C-surface. In this case, it can be shown that the ellipsoid degenerates into a strip containing the tangent plane to the C-surface at the considered configuration (Figure 3). Thus, even in the case the system is in a singularity, the associated uncertainty remains bounded in the direction orthogonal to the corresponding C-surface.

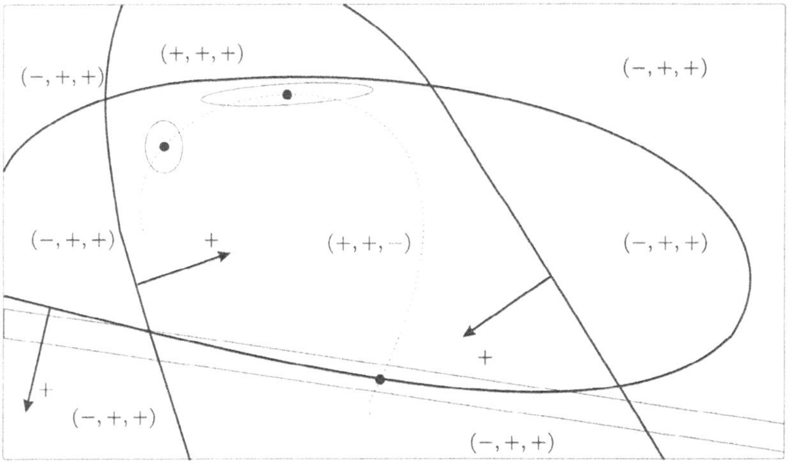

Figure 3. The C-surfaces induce a partition of the configuration space of the moving platform with the same signs for the associated determinants. A motion of the platform corresponds to a trajectory (shown here in dotted line). Each point of this trajectory has an associated uncertainty ellipsoid which degenerates into a strip when trespassing a C-surface.

4. Conclusions

An ellipsoidal uncertainty model for wire-based tracking systems with wire length uncertainties and joint clearances has been presented. Using this model, a one-to-one correspondence between kinematic singularities of the tracking system and configurations with unbounded uncertainty arises. Then, although it is possible to characterize the singularities in terms of the geometry of linearly dependent sets of lines, we have presented an ad hoc approach that takes advantage of the 3-2-1 wire configuration. It has also been shown that these singularities, when expressed in the configuration space of the moving platform, can be identified as C-surfaces associated with the three basic contacts between polyhedra. Since the contacts between wires can be formulated in the same terms, this representation of the problem leads to a unifying formalism where the region for safe tracking –i.e. free from singularities and without wire wrapping problems– is a region of the configuration space of the moving platform bounded by a set of C-surfaces. This point, as well as the experimental validation of the presented results concentrates our current efforts.

Acknowledgements

This work has been partially supported by the Italian-Spanish Bilateral Cooperation Committee and the Spanish CICYT under contract TIC2000-0696.

References

Avnaim, F., Boissonnat, J.D., and Faverjon, B. (1988), A practial motion planning algorithm for polygonal objects amisdst polygonal obstacles, *Proc. of the IEEE Int. Conf. on Robotics and Automation*, pp. 1656-1661.

Ceccarelli, M., Avila-Carrasco, C., and Ottaviano, E. (2000), Error analysis and experimental tests of CATRASYS (Cassino Tracking System), *2000 Int. Conf. on Industrial Electronics, Control and Instrumentation, IECON 2000*, paper SPC11-SP2-4, Nagoya.

Ceccarelli, M., Toti, M.E., and Ottaviano, E. (1999), CATRASYS (Cassino Tracking System): A new measuring system for workspace evaluation of robots, *Proc. of the 8th Int. Workshop on Robotics in Alpe-Adria-Danube Region RAAD'99*, pp. 19-24, Munich.

Donald, B.R. (1984), *Local and Global Techniques for Motion Planning*, Master Science Thesis, MIT, Boston, Massachusetts.

Dörrie, H. (1965), *100 Great Problems of Elementary Mathematics. Their History and Solution*, Dover Publications, New York.

Geng, Z., and Haynes, L.S. (1994), A 3-2-1 kinematic configuration of a Stewart platform and its application to six degree of freedom pose measurements, *Robotics and Computer-Integrated Manufacturing*, Vol. 11, No. 1, pp. 23-34.

Hightower, J., and Borriello, G. (2001), A survey and taxonomy of location systems for ubiquitous computing, *UW CSE-01-08-03*, University of Washington, Department of Computer Science and Engineering.

Jeong, J.W., Kim, S.H., and Kwak, Y.K. (1999), Kinematics and workspace analysis of a parallel wire mechanism for measuring a robot pose, *Mechanism and Machine Theory*, Vol. 34, No. 6, pp. 825-841.

McCarthy, J.M. (2000), *Geometric Design of Linkages*, Springer-Verlag, New York.

Peterson, M.A. (1997), The geometry of Piero della Francesca, *The Mathematical Intelligencer*, Vol. 19, No. 3, pp. 33-40.

Pottmann, H., and Wallner, J. (2001), *Computational Line Geometry*, Springer-Verlag, Berlin.

Ros, L., Sabater, A., and Thomas, F. (2002), An ellipsoidal calculus based on propagation and fusion, *IEEE Trans. on Systems, Man and Cybernetics. Part B*, Vol. 32, No. 4, to appear.

Tadokoro, S. at al. (1996), On the fundamental design of wire driven parallel manipulators with redundancy, In *Japan/USA Symposium on Flexible Automation*, pp. 151-158.

Thomas, F. (1995), An approach to the movers' problem that combines oriented matroid theory and algebraic geometry, *Proc. of the 1995 IEEE Int. Conf. on Robotics and Automation*, Vol. III, pp. 2285-2293.

Zhu, J., and Ting, K-L.(2000), Uncertainty analysis of planar and spatial robots with joint clearances, *Mechanism and Machine Theory*, Vol. 35, pp. 1239-1256.

TENSION DISTRIBUTION IN TENDON-BASED STEWART PLATFORMS

Richard Verhoeven, Manfred Hiller
Institute of Mechatronics and System Dynamics
Gerhard Mercator University of Duisburg, Germany
{verhoev,hiller}@mechatronik.uni-duisburg.de

Abstract When controlling the motion of tendon-based Stewart platforms, it is necessary to maintain the tendon forces inside an acceptable range, between a minimum and a maximum. This leads to a polyhedral set in the force space, where optimal solutions are represented as points of minimum norm. Optimal solutions may fail to be continous along a trajectory, but they can be approximated by almost optimal solutions which are continous except in singular postures.

Keywords: tendon-based parallel manipulator, tension, optimization, polyhedron, continuity

1. Introduction

Tendon-based Stewart platforms represent an innovative technology that can be used for tasks such as manipulation of heavy loads [Bostelman et al., 1994] or for high acceleration of small masses [Ming and Higuchi, 1994b]. In [Ming and Higuchi, 1994a] it has been shown that for n degrees of freedom (DOF's) of the end-effector, at least $n+1$ tendons are necessary; however, redundant systems [Tadokoro et al., 1996] with $m > n+1$ tendons are of particular interest. While the workspaces of classical manipulators with rigid links are restrained due to joint limitations, the main restriction to the workspace of tendon-based systems is the condition that tendon forces must be positive [Verhoeven and Hiller, 2000]. More precisely, they must remain inside an *acceptable* range, i. e. between a positive minimum force $f_{\min}$ and a maximum force $f_{\max}$.

This paper presents a method fo finding such acceptable solutions and discusses the problem of their continuity along a trajectory. We consider the homogeneous problem where the only forces of interest are those in the tendons; i. e. inertia and gravity can be neglected (or gravity does not apply, as in space applications). This case is special because its solutions are scalable: given a solution, for example doubling the force in all tendons still yields a solution, which is not possible in the inhomogeneous case.

J. Lenarčič and F. Thomas (eds.), Advances in Robot Kinematics, 117–124.

2. Acceptable and Minimum Ratio Solutions

Let $\underline{h} \in \mathbb{R}^m$ be a vector whose components $h_1, h_2, \ldots, h_m$ represent the (signed) amount of force in tendon 1, tendon 2, ..., tendon m; $\underline{h}$ is then said to be a *solution* of the homogeneous tension problem if

$$\underline{A}^{\mathrm{T}}\,\underline{h} = \underline{0} \quad \text{and} \quad \underline{h} > \underline{0} \tag{1}$$

where $\underline{A}^{\mathrm{T}} \in \mathbb{R}^{n \times m}$ is a matrix which is continously dependent on the posture of the end-effector [Verhoeven et al., 1998] and the expression $\underline{h} > \underline{0}$ means that each entry of $\underline{h}$ must be positive. Thus, a solution represents a force distribution such that force equilibrium is achieved and all tendons are under tension. Given the technical bounds $0 < f_{\min} \leq f_{\max}$ for the tendon forces, a solution $\underline{h}$ is called *acceptable* if

$$\underline{h} \;\in\; [f_{\min}, f_{\max}]^m \;\; , \tag{2}$$

i.e. each entry h_μ, $\mu = 1, \ldots, m$, belongs to the interval $[f_{\min}, f_{\max}]$.

There is no straightforward way to find acceptable solutions. As the kernel of $\underline{A}^{\mathrm{T}}$ is an $(m-n)$ dimensional vector space, one could try fixing some $m - n$ components of $\underline{h}$ and then compute the others, but this is not likely to supply an acceptable solution. As shown in Fig. 1, choosing $h_1 = f_{10}$ or $h_2 = f_{20}$ does not result in acceptable solutions.

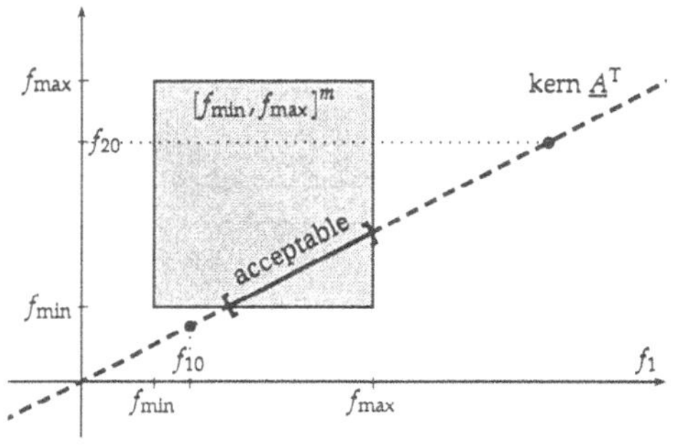

Figure 1. Fixing some forces may not yield acceptable solutions

The case with $m = n + 1$ tendons is simple, then the kernel of $\underline{A}^{\mathrm{T}}$ is one-dimensional. Solutions exist if all components of an arbitrary element $\underline{h}$ have the same sign, and acceptable solutions exist if the ratios $h_\mu / h_{\mu'}$ of any two components of $\underline{h}$ are all less or equal to $f_{\max}/f_{\min}$.

If $m > n+1$, this is more complex (see Fig. 2 a)). Here the acceptable solutions are the intersections of a 2-dimensional vector space with a cube, which is a convex polyhedron. The basic idea is as follows: once

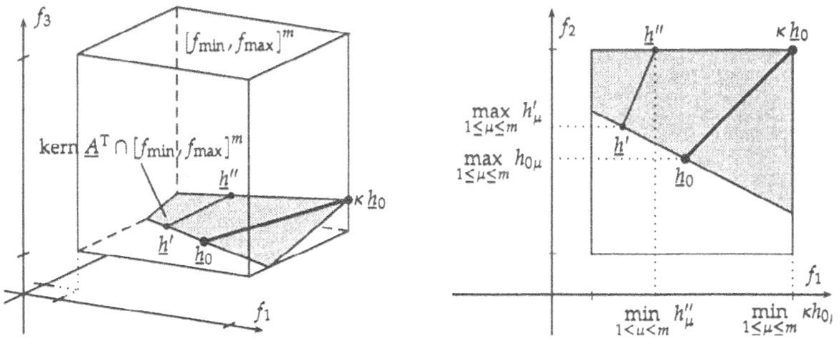

a) 3D view b) $f_1 f_2$ projection

Figure 2. The idea of minimal solution

we have an acceptable solution, we easily obtain a range of acceptable solutions by taking scalar multiples inside the cube $[f_{\min}, f_{\max}]^m$. For example the segment from $\underline{h}'$ to $\underline{h}''$ is a range of scalar multiples of $\underline{h}'$. Such scaling can be used to fix the tensions as low as possible (if we want to save energy) or as high as possible (if we are looking for high eigenfrequencies). Now the solution $\underline{h}_0$ is better then $\underline{h}'$ in the sense that it provides a longer segment of acceptable solutions.

More precisely, $\underline{h}_0$ is an acceptable solution such that the maximum $\max\{h_{0,1}, \ldots, h_{0,m}\}$ is minimal among all acceptable solutions, and $\kappa\,\underline{h}_0$ is an acceptable solution such that the minimum of these tensions is maximal (see Fig. 2 b)). For $\underline{h}'$, the tendon with highest tension is tendon 2, and this tension is higher than the highest one occurring in $\underline{h}_0$. Thus, when seeking low tensions, $\underline{h}_0$ is better than $\underline{h}'$. Conversely, $\kappa\,\underline{h}_0$ is better than $\underline{h}''$ when high tensions are required. Hence in both cases scaled versions of $\underline{h}_0$ supply optimal results.

We call such a solution $\underline{h}_0$ *minimum ratio* if

$$\frac{\max\limits_{1\le\mu\le m} h_{0,\mu}}{\min\limits_{1\le\mu\le m} h_{0,\mu}} = \underbrace{\min\left\{ \frac{\max\limits_{1\le\mu\le m} h_\mu}{\min\limits_{1\le\mu\le m} h_\mu} : \underline{A}^{\mathrm{T}}\underline{h} = 0 \wedge \underline{h} > \underline{0} \right\}}_{=:\,\kappa} . \qquad (3)$$

Then *acceptable solutions exist if and only if there are minimum ratio solutions that are acceptable*, e. g. if there is a minimum ratio $\underline{h}_0$ with

$$\frac{f_{\min}}{\min_{1\le\mu\le m} h_{0\mu}}\,\underline{h}_0 \in [f_{\min}, f_{\max}]^m . \qquad (4)$$

This is basically the same idea as for systems with $m = n + 1$ tendons. The difference is that here it is not sufficient to consider an arbitrary solution. Instead we have to look at an arbitrary *minimum ratio* solution (indeed, for $m = n + 1$, all solutions are minimum ratio.)

3. Finding Minimum Ratio Solutions

This notion of a mimimum-ratio solution is unpleasant because it involves the ratio of two components of a vector, but the components to choose depend on the entire vector. This is not originally a linear problem, but we can eliminate the quotient. As the set of solutions is invariant to positive scaling, we can restrict ourselves to those vectors where all components are greater or equal to 1, i. e. we consider the convex polyhedron

$$\mathcal{P} := \left\{ \underline{h} \in \mathbb{R}^m : \underline{A}^{\mathrm{T}} \underline{h} = \underline{0} \wedge \underline{h} > \underline{1} \right\} \tag{5}$$

which is nonempty if and only if the set of solutions is nonempty. Then we get for the minimum ratio κ defined in Eq. (3)

$$\kappa = \min \{ \|\underline{h}\|_\infty : \underline{h} \in \mathcal{P} \} \tag{6}$$

where $\|\underline{h}\|_\infty$ denotes the maximum of the components of $\underline{h}$. Now we define the set of *minimal* solutions as

$$\mathcal{M} := \{ \underline{h} \in \mathcal{P} : \|\underline{h}\|_\infty = \kappa \} \ . \tag{7}$$

These are precisely those minimum ratio solutions which have at least one component equal to 1. Then all minimum ratio solutions are scaled minimal solutions, and for $\underline{h}_0 \in \mathcal{M}$, the vectors

$$f_{\min} \underline{h}_0 \quad \text{and} \quad \frac{f_{\max}}{\kappa} \underline{h}_0 \tag{8}$$

are optimal acceptable solutions in the sense that there is no acceptable solution with a strictly greater (strictly smaller) tension.

Now we need some convex polyhedra theory. Given a polyhedron $\mathcal{P}$ (possibly unbounded, as in Eq. (5)), the set

$$\text{char.cone}\,\mathcal{P} := \left\{ \underline{y} \in \mathbb{R}^m : \underset{\underline{x} \in \mathcal{P}}{\forall} \ \underline{x} + \underline{y} \in \mathcal{P} \right\} \tag{9}$$

is called the *characteristic cone* of $\mathcal{P}$ [Schrijver, 1999, p. 100]. In our case, the characteristic cone of $\mathcal{P}$ is contained in $[0, \infty[^m$, and therefore it does not contain affine spaces greater than points (i. e. lines, planes etc.). Such polyhedra are called pointed, and there is a representation theorem [Schrijver, 1999, p. 106] that states: If $\mathcal{P}$ is a pointed polyhedron and $\mathcal{Q}$ the convex hull of its vertices, then

$$\mathcal{P} = \mathcal{Q} + \text{char.cone}\,\mathcal{P} \tag{10}$$

in the sense that $\mathcal{P}$ consists in all the sums of an element of $\mathcal{Q}$ and an element of char.cone $\mathcal{P}$. This implies in our case that

$$\begin{aligned} \min\{\|\underline{h}\|_\infty : \underline{h} \in \mathcal{P}\} &= \min\{\|\underline{h}+\underline{y}\|_\infty : \underline{h} \in \mathcal{Q} \wedge \underline{y} \in \text{char.cone}\,\mathcal{P}\} \\ &= \min\{\|\underline{h}\|_\infty : \underline{h} \in \mathcal{Q}\}\ ; \end{aligned} \tag{11}$$

hence if there are solutions, then $\mathcal{M}$ intersects $\mathcal{Q}$, and it is sufficient to search for minimal solutions inside the bounded set $\mathcal{Q}$. Often $\mathcal{M}$ contains a single point $\underline{h}_{\min}$, as shown in Fig. 3.

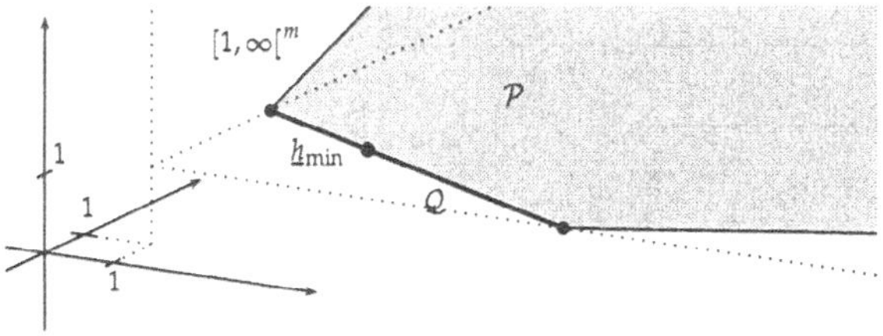

Figure 3. Minimal solution in $\mathcal{Q}$

4. Discontinuity Example

But in some posture, $\mathcal{M}$ may contain several solutions. An example manipulator is shown in Fig. 4. A point $\boldsymbol{p}$ is moved by five tendons (drawn as prismatic links) in the plane. Considering postures along the central vertical axis satisfying $\sin\alpha < \epsilon$ and $\sin\beta < \frac{1}{2} - \epsilon$ with some small $\epsilon > 0$, we demonstrate the algorithm for finding a minimal solution and show that it can fail to be continuous.

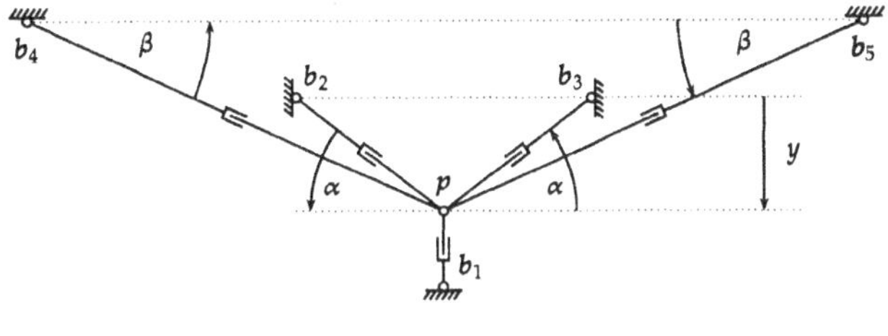

Figure 4. Example manipulator

a) *Identify* $\underline{A}^{\mathrm{T}}$. Here we have translation only, hence $\underline{A}^{\mathrm{T}}$ consists of the unit vectors along the tendons [Verhoeven et al., 1998]:

$$\underline{A}^{\mathrm{T}} = \begin{pmatrix} 0 & -\cos\alpha & \cos\alpha & -\cos\beta & \cos\beta \\ -1 & \sin\alpha & \sin\alpha & \sin\beta & \sin\beta \end{pmatrix}. \tag{12}$$

b) *Compute the kernel of* $\underline{A}^{\mathrm{T}}$. Here the kernel of $\underline{A}^{\mathrm{T}}$ consists of all linear combinations of the columns of

$$\underline{H} = \begin{pmatrix} 2\sin\alpha & 2\sin\beta & 0 \\ 1 & 0 & \cos\beta \\ 1 & 0 & -\cos\beta \\ 0 & 1 & -\cos\alpha \\ 0 & 1 & \cos\alpha \end{pmatrix} . \tag{13}$$

It can be shown that for reasons of symmetry, all minimum ratio solutions are linear combinations of the first two columns. Hence we consider the polyhedron $\mathcal{P}'$ given by

$$\underline{h} = (\ 2\lambda_1 \,\mathrm{s}\,\alpha + 2\lambda_2 \,\mathrm{s}\,\beta \quad \lambda_1 \quad \lambda_1 \quad \lambda_2 \quad \lambda_2\ )^{\mathrm{T}} \quad \text{with} \quad \underline{h} > \underline{0} . \tag{14}$$

c) *Compute the vertices of* $\mathcal{Q}'$. For a 2-dimensional $\mathcal{P}'$, they are characterized by the fact that at least two entries are equal to 1. It can be shown that $h_1 = 1$ for all $\underline{h} \in \mathcal{P}'$. Thus, there are only two possibilities:

λ_1	λ_2	κ	
1	$\frac{1-2\sin\alpha}{2\sin\beta}$	λ_2	is always a vertex of $\mathcal{Q}'$
$\frac{1-2\sin\beta}{2\sin\alpha}$	1	λ_1	is only a vertex of $\mathcal{Q}'$ for $\alpha > 0$

(15)

d) *Find minimal solutions in* $\mathcal{Q}'$. For $\alpha < 0$ there is a single vertex and hence a unique minimal solution:

$$\underline{h}_{\mathrm{min}} = \begin{pmatrix} 1 & 1 & 1 & \frac{1-2\sin\alpha}{2\sin\beta} & \frac{1-2\sin\alpha}{2\sin\beta} \end{pmatrix} . \tag{16}$$

For $\alpha > 0$, candidates for minimal solutions are found in the segment between the two vertices, and the result is unique too:

$$\underline{h}_{\infty} = \begin{pmatrix} 1 & \frac{1}{2(\sin\alpha+\sin\beta)} & \cdots & \frac{1}{2(\sin\alpha+\sin\beta)} \end{pmatrix} . \tag{17}$$

Thus, for $\sin\beta < \frac{1}{2}$ this minimal solution is discontinous at $\alpha = 0$:

$$\underline{h}_{\mathrm{min}} \overset{\alpha\nearrow 0}{\longrightarrow} \begin{pmatrix} 1 & 1 & 1 & \frac{1}{2\sin\beta} & \frac{1}{2\sin\beta} \end{pmatrix} \tag{18}$$

$$\underline{h}_{\infty} \overset{\alpha\searrow 0}{\longrightarrow} \begin{pmatrix} 1 & \frac{1}{2\sin\beta} & \frac{1}{2\sin\beta} & \frac{1}{2\sin\beta} & \frac{1}{2\sin\beta} \end{pmatrix} . \tag{19}$$

This problem can be solved if, instead of the $\|\cdot\|_\infty$ minimal elements of $\mathcal{P}$, we consider minimal elements of $\mathcal{P}$ with respect to some $\|\cdot\|_p$-norm, defined by

$$\|\underline{h}\|_p := \sqrt[p]{\sum_{\mu=1}^{m} h_\mu^p} \quad \text{for} \quad p \in [1, \infty[\ . \tag{20}$$

These norms are known to approximate the $\|\cdot\|_\infty$-norm as $p \to \infty$. Fig. 5 shows the tensions h_3 and h_5 for the solutions to $p = 2, 4, 8, \infty$ in some postures. The base point coordinates in this example are:

	$\boldsymbol{b}_1$	$\boldsymbol{b}_2$	$\boldsymbol{b}_3$	$\boldsymbol{b}_4$	$\boldsymbol{b}_5$
x	0.00	-2.00	2.00	-5.67	5.67
y	-2.50	0.00	0.00	1.00	1.00

(21)

In all the postures shown we have $h_1 = 1$ and $\kappa = h_5$. The $\underline{h}_5$ diagram shows that here it does no harm to use $p = 2$ instead of $p = \infty$ because κ changes very little, while the h_3 curve for $p = 2$ is much smoother than that for $p = \infty$.

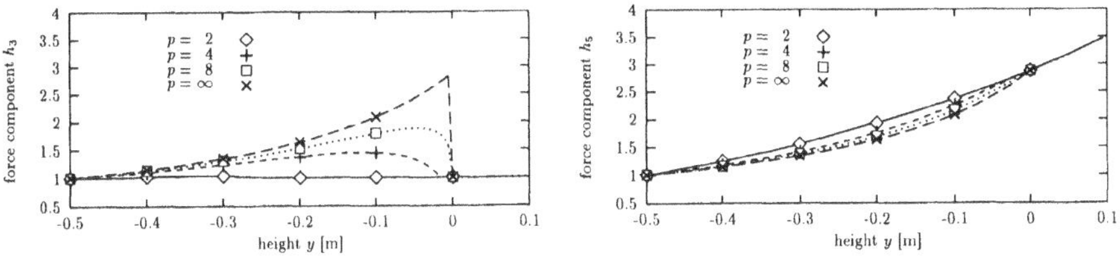

Figure 5. Comparison of solutions

5. Continuity of Solutions

It is possible to show that:

- For each $p \in [1, \infty]$, if the $\|\cdot\|_p$ minimal solution is unique in each posture on a trajectory, then it is continous.
- For $p \in]1, \infty[$, the $\|\cdot\|_p$ minimal solution is always unique, except at singularities.
- At singularities, any such approach leads to discontinous solutions.

The proof will be published in R. Verhoeven's Ph. D. thesis and involves advanced theory of parametric optimization, e. g. [Bank et al., 1983]. The idea is to prove that the mapping from $\underline{A}^{\mathrm{T}}$ to the minimum value of the $\|\cdot\|_p$ norm is lower and upper semicontinous, and thus continous. Then, one can show that the mapping of $\underline{A}^{\mathrm{T}}$ to the set $\mathcal{M}$ has a property called *upper semi continuity according to Berge*, which becomes continuity when the set $\mathcal{M}$ consists of a single point. The assertion on singularities is shown using an example.

Hence it is a good idea to choose $\|\cdot\|_p$ minimal solutions if continuity is important. If $\mathcal{Q}$ has a unique vertex as in the above example, they are also minimum ratio, otherwise they supply approximations for minimum ratio solutions. Near the border of the workspace, this may not lead to acceptable solutions even though they exist.

6. Conclusions

The problem of finding acceptable force distributions can be transformed to a norm minimization problem on a convex polyhedron. Candidates for good solutions are found in the convex hull of the vertices. The minima for the $\|\cdot\|_\infty$ norm provide optimal solutions, but they may fail to be continous along a trajectory. However, they can be approximated with $\|\cdot\|_p$ minima which are always continous except at singularities.

7. Acknowledgements

The research work reported here was made possible by DFG (Deutsche Forschungsgemeinschaft) Grant No. HI 370/18-2.

References

Bank, B., Guddat, J., and Klatte, D. (1983). *Non-linear parametric optimization.* Birkhäuser Verlag, Basel, Switzerland.

Bostelman, R. V., Albus, J. S., Dagalakis, N. G., Jacoff, A., and Gross, J. (1994). Applications of the NIST ROBOCRANE. In *Proc. 5th Int. Symp. on Robotics and Manufacturing: Research, Education, and Applications*, pages 403–410, Maui, HI, USA. ASME.

Ming, A. and Higuchi, T. (1994a). Study on multiple degree-of-freedom positioning mechanism using wires (part 1) – concept, design and control. *Int. Journal of the Jap. Soc. for Precision Engineering*, 28(2):131–138.

Ming, A. and Higuchi, T. (1994b). Study on multiple degree-of-freedom positioning mechanism using wires (part 2) – development of a planar completely restrained positioning mechanism. *Int. Journal of the Jap. Soc. for Precision Engineering*, 28(3):235–242.

Schrijver, A. (1999). *Theory of linear and integer programming.* Wiley Eastern Ltd., New York, USA.

Tadokoro, S., Nishioka, S., Kimura, T., Hattori, M., Takamori, T., and Maeda, K. (1996). On fundamental design of wire configurations of wire-driven parallel manipulators with redundancy. In *Proc. 1996 Japan-U.S.A. Symposium on Flexible Automation*, pages 151–158, Boston, MA, USA.

Verhoeven, R., Hiller, M., and Tadokoro, S. (1998). Workspace, stiffness, singularities and classification of tendon-driven Stewart platforms. In Lenarčič, J. and Husty, M. L., editors, *Advances in Robot Kinematics: Analysis and Control*, pages 105–114, Strobl/Salzburg, Austria. Kluwer Academic Publishers.

Verhoeven, R. and Hiller, M. (2000). Estimating the controllable workspace of tendon-based Stewart platforms. In Lenarčič, J. and Stanišić, M. M., editors, *Advances in Robot Kinematics*, pages 277–284, Portorož, Slovenia. Kluwer Academic Publishers.

TRAJECTORY TRACKING CONTROL FOR A CABLE SUSPENSION MANIPULATOR

T. Heyden, T. Maier and C. Woernle
Institute of Drive Systems and Mechatronics
University of Rostock, Germany
woernle@mbst.uni-rostock.de

Abstract The underconstrained cable suspension manipulator CABLEV moves a payload platform in space by three spatially arranged cables with computer-controlled winches. To make the payload platform track prescribed reference trajectories in space, a two-stage control concept is presented. A nonlinear feedforward control that exploits the flatness property of the system generates control inputs for the undisturbed motion along reference trajectories. Asymptotically stable tracking behaviour is achieved by a superimposed linear feedback of actual state variables.

Keywords: cable suspension manipulator, nonlinear control, flat systems, parallel kinematics

1. Introduction

A cable suspension manipulator supports a payload platform in space by several spatially arranged cables with computer-controlled winches. Compared to conventional cranes, it is possible to control not only the translational motion of the payload but also its orientation. By this, cable suspension manipulators combine the ability of cranes to support heavy payloads in a large workspace with the accuracy of robot manipulators. Various configurations have been proposed (Dagalakis et al., 1991, Arai and Osumi, 1992, Maier and Woernle, 1998, Tadokoro et al., 1999). Applications are, for example, precise handling and assembling large and heavy components on construction sites or on shipyards.

If the position of the payload platform is not completely defined by the cable lengths, the cable suspension manipulator is said to be underconstrained (Ming and Higuchi, 1994). The most simple example is an overhead crane with a load mass. In this paper the kinematically hybrid, underconstrained cable suspension manipulator CABLEV is investigated (Fig. 1). It has been developed at University of Rostock as a prototype system to validate control concepts. Its payload platform is supported by three cables with winches mounted on trolleys that move them on a common gantry. The control task is to make the payload platform track desired trajectories in space. A control principle is obtained by means of

J. Lenarčič and F. Thomas (eds.), Advances in Robot Kinematics, 125–134.

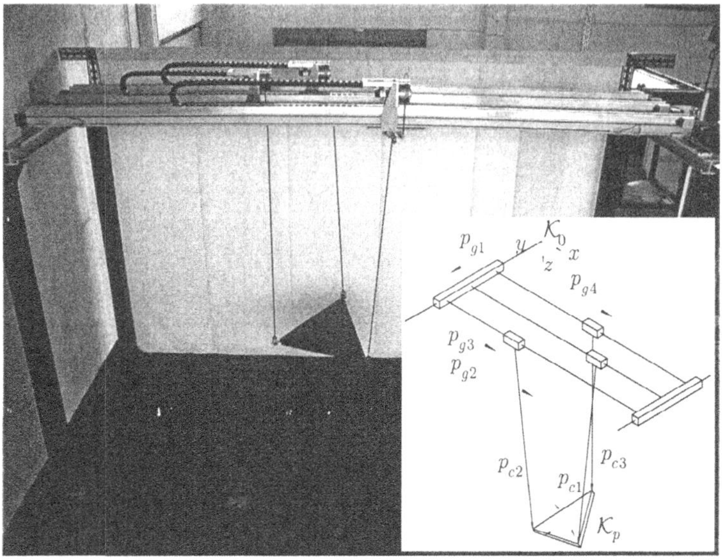

Figure 1. Cable Suspension Manipulator CABLEV

the so-called flatness property of the system dynamics of CABLEV. According to Fliess et al. (1992) a system is called flat if its state variables as well as its control inputs can be algebraically expressed in terms of the control outputs and their time derivatives up to a certain order. This inversion of the system dynamics can be directly used as a feedforward control law. As successfully demonstrated by experiments with CABLEV, flatness-based feedforward control provides exact motions of the platform along desired trajectories as long as there are no unknown disturbances, e.g., wind forces, acting on the platform. To counteract these effects, feedback of actual, measured state-space variables is needed.

For flat systems, trajectory tracking control by exact state-space linearization with nonlinear state feedback can be achieved (Delaleau and Rudolph, 1995). Control of an overhead crane by this approach is described in Maier and Woernle (2000). For control of CABLEV a less complex linear feedback of actual state variables is proposed under the assumption that the feedforward control already ensures motions in the neighborhood of the reference trajectory.

The paper is organized as follows. First the flatness-based feedforward control of CABLEV is briefly reviewed. Based on the linearized equations of motion a linear state-space controller is developed afterwards. Simulations show the effects of combined nonlinear feedforward and linear feedback control.

2. Dynamic Model of CABLEV

Coordinates

The independently controllable robot coordinates $\boldsymbol{p} \in \mathbb{R}^7$ are the gantry displacement p_{g1}, the trolley displacements p_{g2}, p_{g3}, p_{g4} and the cable lengths p_{c1}, p_{c2}, p_{c3}, see Fig. 1,

$$\boldsymbol{p} = \begin{bmatrix} \boldsymbol{p}_g \\ \boldsymbol{p}_c \end{bmatrix} \quad \text{with} \quad \boldsymbol{p}_g = \begin{bmatrix} p_{g1} \\ p_{g2} \\ p_{g3} \\ p_{g4} \end{bmatrix}, \quad \boldsymbol{p}_c = \begin{bmatrix} p_{c1} \\ p_{c2} \\ p_{c3} \end{bmatrix}. \tag{1}$$

The platform coordinates $\boldsymbol{y} \in \mathbb{R}^6$ describe the spatial position and orientation of the platform-fixed coordinate system $\mathcal{K}_p$ relative to the inertial system $\mathcal{K}_0$ by, for example, three cartesian coordinates r_x, r_y, r_z of the origin of $\mathcal{K}_p$ and three Bryant angles φ_1, φ_2, φ_3,

$$\boldsymbol{y} = \begin{bmatrix} \boldsymbol{r} \\ \boldsymbol{\varphi} \end{bmatrix} \quad \text{with} \quad \boldsymbol{r} = \begin{bmatrix} r_x \\ r_y \\ r_z \end{bmatrix}, \quad \boldsymbol{\varphi} = \begin{bmatrix} \varphi_1 \\ \varphi_2 \\ \varphi_3 \end{bmatrix}. \tag{2}$$

The spatial velocity $\dot{\boldsymbol{s}}$ of the platform relative to $\mathcal{K}_0$ includes the translational velocity $\boldsymbol{v}$ of the origin of $\mathcal{K}_p$ and the platform-fixed angular velocity $\boldsymbol{\omega}$. The relation between the time derivatives of $\boldsymbol{y}$ and $\dot{\boldsymbol{s}}$ is then given by

$$\begin{aligned} \begin{bmatrix} \dot{\boldsymbol{r}} \\ \dot{\boldsymbol{\varphi}} \end{bmatrix} &= \begin{bmatrix} \boldsymbol{I} & \boldsymbol{0} \\ \boldsymbol{0} & \boldsymbol{H}_\omega(\boldsymbol{\varphi}) \end{bmatrix} \begin{bmatrix} \boldsymbol{v} \\ \boldsymbol{\omega} \end{bmatrix} \\ \dot{\boldsymbol{y}} &= \quad \boldsymbol{H}(\boldsymbol{y}) \qquad \dot{\boldsymbol{s}}\,, \end{aligned} \tag{3}$$

whereby the second matrix equation is the kinematic differential equation related to the Bryant angles.

With three supporting cables the payload platform can perform sway motions with three degrees of freedom. The sway motions are described, e.g., by the sway coordinates

$$\boldsymbol{\eta} = [\, r_x \quad r_y \quad \varphi_3 \,]^{\mathrm{T}}. \tag{4}$$

Dynamic Equations

With three longitudinally stiff cables there exist three implicit constraints between the robot coordinates $\boldsymbol{p}$ and platform coordinates $\boldsymbol{y}$ (Fig. 2)

$$g_i(\boldsymbol{p}, \boldsymbol{y}) \equiv \boldsymbol{c}_i^2 - p_{ci}^2 = 0, \qquad i = 1, 2, 3, \qquad \text{or} \qquad \boldsymbol{g}(\boldsymbol{p}, \boldsymbol{y}) = \boldsymbol{0}, \tag{5}$$

whereby the cable vectors $\boldsymbol{c}_i$ can be expressed by

$$\boldsymbol{c}_i = \boldsymbol{r} + \boldsymbol{d}_i - \boldsymbol{b}_i, \quad i = 1,2,3 \; . \qquad (6)$$

To derive the feedforward controller model, the equations of motion of CABLEV are formulated as differential-algebraic equations in terms of the coordinates $\boldsymbol{p}$ and $\boldsymbol{y}$. They consist of the kinematic differential equation (3), the constraint equation (5), the seven dynamic equations for the assembly of the gantry, the trolleys and the winches

$$\boldsymbol{M}_p(\boldsymbol{p})\, \ddot{\boldsymbol{p}} = \boldsymbol{K}_p \boldsymbol{w} + \boldsymbol{G}_p^{\mathrm{T}}(\boldsymbol{p}, \boldsymbol{y})\boldsymbol{\lambda} \; , \qquad (7)$$

and the six dynamic equations of the platform

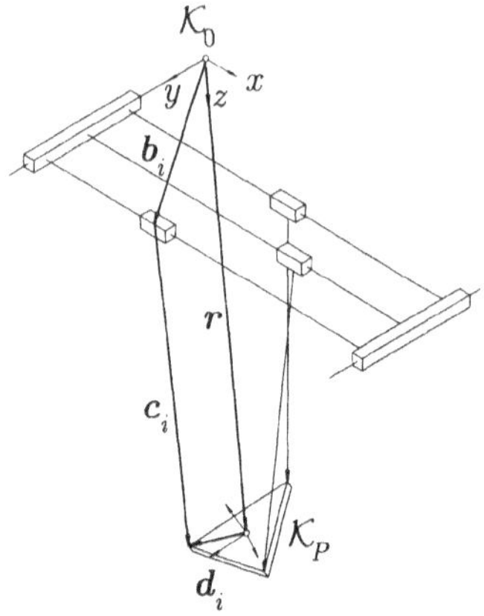

Figure 2. Kinematic constraint between $\boldsymbol{p}$ and $\boldsymbol{y}$

$$\boldsymbol{M}_s(\boldsymbol{y})\, \ddot{\boldsymbol{s}} + \boldsymbol{k}_s(\boldsymbol{y}, \dot{\boldsymbol{s}}) = \boldsymbol{Q}_s^e(\boldsymbol{y}) + \boldsymbol{G}_s^{\mathrm{T}}(\boldsymbol{p}, \boldsymbol{y})\boldsymbol{\lambda} \; . \qquad (8)$$

Here, $\boldsymbol{M}_p \in \mathbb{R}^{7,7}$ and $\boldsymbol{M}_s \in \mathbb{R}^{6,6}$ are mass matrices, $\boldsymbol{k}_s \in \mathbb{R}^6$ contains generalized centrifugal and Coriolis forces, $\boldsymbol{Q}_s^e \in \mathbb{R}^6$ includes all applied generalized forces, and $\boldsymbol{K}_p \in \mathbb{R}^{7,7}$ projects the control forces $\boldsymbol{w} \in \mathbb{R}^7$ (drive forces of the trolleys and winch torques) in direction of the robot coordinates $\boldsymbol{p}$. The second summands on the right-hand sides of (7), (8) result from the cable forces with the cable force coordinates $\boldsymbol{\lambda} \in \mathbb{R}^3$ and the functional matrices

$$\boldsymbol{G}_s = \frac{\partial \boldsymbol{g}}{\partial \boldsymbol{y}} \boldsymbol{H} \quad , \quad \boldsymbol{G}_p = \frac{\partial \boldsymbol{g}}{\partial \boldsymbol{p}} \; . \qquad (9)$$

The equations of motion are obtained by means of the object-oriented multibody C++library MOBILE (Kecskeméthy, 1995).

3. Flatness-Based Feedforward Control

For flat systems an inversion of the system dynamics exist in such a way that the control input as well as the state variables can be algebraically expressed in terms of the output and their time derivatives up to a certain differential order (Fliess et al., 1992). This property leads directly to a feedforward control law that moves the platform along sufficiently smooth trajectories. Control of conventional rigid-link manipulators by inverse dynamic algorithms (computed torque) is a special case of flatness-based feedforward control. For CABLEV the inversion of the system dynamics is obtained in two steps.

Generalized inverse kinematics

First the robot coordinates $\hat{\boldsymbol{p}}$ are expressed in terms of the prescribed outputs $\hat{\boldsymbol{y}}$ and their time derivatives (Maier and Woernle, 2000). With seven robot coordinates $\boldsymbol{p}$ and six platform coordinates $\boldsymbol{y}$ the system is kinematically redundant. This redundancy is here treated by introducing a geometric control constraint between the trolley coordinates p_{g2}, p_{g3}, p_{g4},

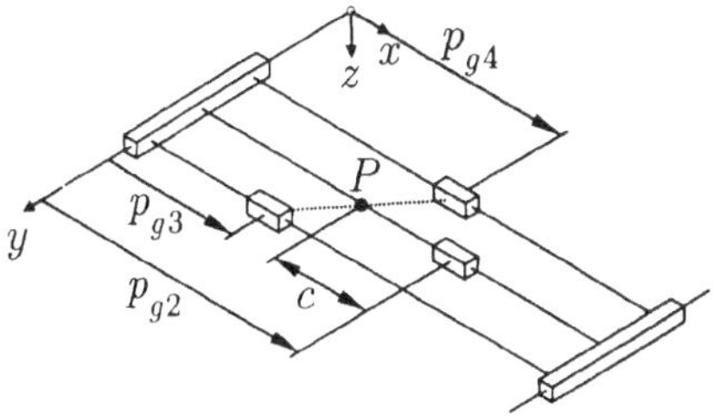

Figure 3. Control constraint

$$g_0(\boldsymbol{p}) = p_{g2} - \tfrac{1}{2}(p_{g3} + p_{g4}) - c = 0, \quad c = \text{const} . \tag{10}$$

According to Fig. 3 this constraint means that the positions of the outer trolleys are always symmetrical to a point P on the intermediate rail with constant distance c to the inner trolley. With (8), (5), (10) a set of ten nonlinear equations is available to calculate the unknowns $\hat{\boldsymbol{p}} \in \mathbb{R}^7$ and $\hat{\boldsymbol{\lambda}} \in \mathbb{R}^3$ for given $\hat{\boldsymbol{y}}$, $\dot{\hat{\boldsymbol{y}}}$, $\ddot{\hat{\boldsymbol{y}}}$:

$$\left.\begin{array}{rcl} \boldsymbol{G}_s^{\mathrm{T}}(\hat{\boldsymbol{p}}, \hat{\boldsymbol{y}})\hat{\boldsymbol{\lambda}} & = & \boldsymbol{M}_s(\hat{\boldsymbol{y}})\,\ddot{\boldsymbol{s}} + \boldsymbol{k}_s(\hat{\boldsymbol{y}}, \dot{\boldsymbol{s}}) - \boldsymbol{Q}_s^e(\hat{\boldsymbol{y}}) \\ \boldsymbol{g}(\hat{\boldsymbol{y}}, \hat{\boldsymbol{p}}) & = & \boldsymbol{0} \\ g_0(\hat{\boldsymbol{p}}) & = & 0 \end{array}\right\} . \tag{11}$$

The velocities $\dot{\boldsymbol{s}}$ and the accelerations $\ddot{\boldsymbol{s}}$ can be expressed in terms of the output derivatives $\dot{\boldsymbol{y}}$, $\ddot{\boldsymbol{y}}$ using (3). System (11) can be solved numerically with respect to $\hat{\boldsymbol{p}}$ and $\hat{\boldsymbol{\lambda}}$. In practice, the numerical algorithm leads to a particular solution

$$\hat{\boldsymbol{p}} = \boldsymbol{\Psi}(\hat{\boldsymbol{y}}, \dot{\hat{\boldsymbol{y}}}, \ddot{\hat{\boldsymbol{y}}}) \tag{12}$$

that is close to an initial guess for $\hat{\boldsymbol{p}}$. Although other solutions exist, this approach is suitable, since a good initial guess is available from the preceding time step. The function (12) is now called generalized inverse kinematics. It can be regarded as a generalization of inverse kinematics for rigid-link manipulators.

Inverse Dynamics

The second step of the inversion of the system dynamics is the calculation of the physical control forces $\boldsymbol{w}$ in terms of $\hat{\boldsymbol{p}}$. They are obtained directly by solving (7) and (8) with respect to $\boldsymbol{w}$, leading to $\boldsymbol{w} = \boldsymbol{\xi}(\hat{\boldsymbol{p}}, \dot{\hat{\boldsymbol{p}}}, \ddot{\hat{\boldsymbol{p}}}, \hat{\boldsymbol{\lambda}})$. Thus, the first and second order time derivatives of the generalized inverse kinematics (12) are needed, too. Altogether a feedforward control law is obtained that calculates $\boldsymbol{w}$ for output trajectories $\hat{\boldsymbol{y}}(t)$ specified up to the fourth-order time derivative (notation $\frac{\mathrm{d}^n y}{\mathrm{d}t^n} \mathrel{\hat{=}} y^{(n)}$)

$$\boldsymbol{w} = \boldsymbol{\Phi}(\hat{\boldsymbol{y}}, \dot{\hat{\boldsymbol{y}}}, \ddot{\hat{\boldsymbol{y}}}, \hat{\boldsymbol{y}}^{(3)}, \hat{\boldsymbol{y}}^{(4)}). \tag{13}$$

In practice, the computational amount for the feedforward control can be considerably reduced. The prototype system CABLEV is actuated by electric motors with reduction gears of high ratio (motor and gear constants $\boldsymbol{K}_c$, $\boldsymbol{K}_g$, generalized drive forces $\boldsymbol{w}_c$ (winches), $\boldsymbol{w}_g$ (trolleys), inputs of the motor current-controllers (current-controller dynamics neglected) $\overline{\boldsymbol{w}}_c$, $\overline{\boldsymbol{w}}_g$) that tend to linearize and to decouple the dynamics of the drive axes. This justifies well the use of individual cascaded axis controllers that make the actual robot coordinates $\boldsymbol{p}$ follow the sequence of values $\hat{\boldsymbol{p}}$ calculated by means of the generalized inverse kinematics (12). To further improve the tracking behaviour, the velocities $\dot{\hat{\boldsymbol{p}}}$ can be calculated by differentiation of (11) and added to the velocity inputs $\boldsymbol{v}_c$, $\boldsymbol{v}_g$ of the axis controllers. The block diagram in Fig. 4 shows this feedforward control scheme, that has been successfully verified by experiments.

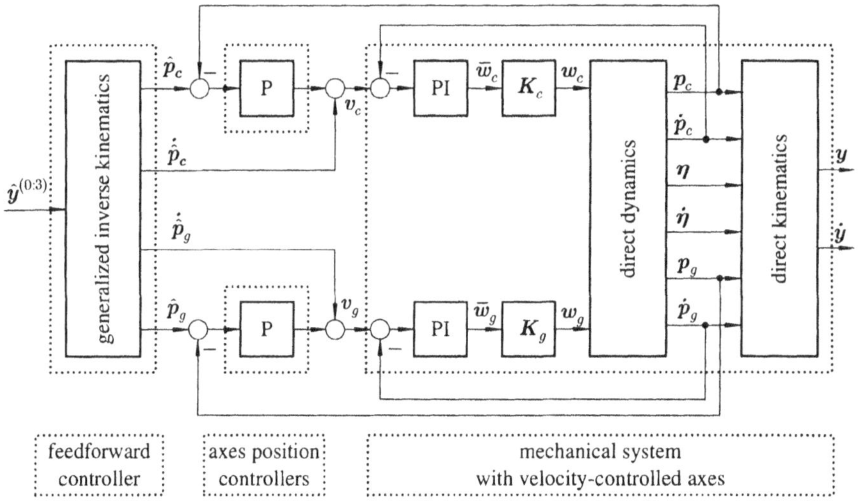

Figure 4. Flatness-Based Feedforward Control for CABLEV

4. Closed Loop Trajectory Tracking Control

For active damping of sway motions of the platform a feedback of actual state variables is provided that is superimposed to the described feedforward control. The idea is to achieve sway control by appropriate motions of the four gantry/trolley coordinates $\boldsymbol{p}_g$ but not by changes of the cables lengths $\boldsymbol{p}_c$. A similar approach was used by Sawodny, Aschemann et al. (1999) for trajectory control of an overhead crane. For controller design a linear state-space model of CABLEV with velocity-controlled axes according to Fig. 4 is derived. It is based on the linearized equations of motion that describe small motions in the vicinity of an equilibrium position of the platform.

Linearized Sway Motion Model

To apply established methods for linear control design the linearized equations of motion are formulated as ordinary differential equations in terms of small deviations $\Delta\hat{\boldsymbol{\eta}}$ of the sway coordinates from an equilibrium position $\hat{\boldsymbol{\eta}}_E$ of the platform. For that purpose the differential-algebraic equations (3), (5), (8) describing the sway motion are linearized and transferred into ordinary differential equations by eliminating the cable force coordinates $\boldsymbol{\lambda}$ and introducing the independent sway coordinates $\Delta\boldsymbol{\eta}$. The linearized sway dynamics model is then obtained as

$$\boldsymbol{M}_\eta \Delta\ddot{\boldsymbol{\eta}} + \boldsymbol{C}_\eta \Delta\boldsymbol{\eta} = \boldsymbol{B}_1 \Delta\boldsymbol{p}_g + \boldsymbol{B}_2 \Delta\ddot{\boldsymbol{p}}_g \tag{14}$$

with constant coefficient matrices $\boldsymbol{M}_\eta \in \mathbb{R}^{3,3}$, $\boldsymbol{C}_\eta \in \mathbb{R}^{3,3}$, $\boldsymbol{B}_1 \in \mathbb{R}^{3,4}$, $\boldsymbol{B}_2 \in \mathbb{R}^{3,4}$. According to the block diagram in Fig. 4 the control inputs for the mechanical system are the reference velocities $\boldsymbol{v}_g$ of the axis velocity controllers. Dynamic transfer behaviour from the reference velocities $\boldsymbol{v}_g$ to the actual velocities $\Delta\dot{\boldsymbol{p}}_g$ of the gantry/trolleys appearing in (14) is approximated by a first order dynamics,

$$\boldsymbol{T} \Delta\ddot{\boldsymbol{p}}_g + \Delta\dot{\boldsymbol{p}}_g = \boldsymbol{v}_g \, , \tag{15}$$

with time constants $\boldsymbol{T} = \mathrm{diag}(T_1, \ldots, T_4)$. Introducing the state vector

$$\boldsymbol{x}_{\mathrm{sway}} = \begin{bmatrix} \Delta\boldsymbol{\eta} \\ \Delta\dot{\boldsymbol{\eta}} \\ \Delta\boldsymbol{p}_g \\ \Delta\dot{\boldsymbol{p}}_g \end{bmatrix} \quad \text{with} \quad \Delta\boldsymbol{\eta} = \begin{bmatrix} \Delta r_x \\ \Delta r_y \\ \Delta\varphi_3 \end{bmatrix}, \; \Delta\boldsymbol{p}_g = \begin{bmatrix} \Delta p_{g1} \\ \Delta p_{g2} \\ \Delta p_{g3} \\ \Delta p_{g4} \end{bmatrix},$$

the reference velocities $\boldsymbol{v}_g$ as inputs, i.e. , $\boldsymbol{u}_{\mathrm{sway}} = \boldsymbol{v}_g$, and the sway coordinates as outputs, i.e. , $\boldsymbol{y}_{\mathrm{sway}} = \Delta\boldsymbol{\eta}$, equations (14), (15) can be brought into standard state-space form

$$\dot{\boldsymbol{x}}_{\mathrm{sway}} = \boldsymbol{A}\,\boldsymbol{x}_{\mathrm{sway}} + \boldsymbol{B}\,\boldsymbol{u}_{\mathrm{sway}} \, , \qquad \boldsymbol{y}_{\mathrm{sway}} = \boldsymbol{C}\,\boldsymbol{x}_{\mathrm{sway}} . \tag{16}$$

Linear State-Space Controller

The sway dynamics (16) is stabilized by a linear state feedback

$$\boldsymbol{u}_{\mathrm{sway}} = -\boldsymbol{R}\,\boldsymbol{x}_{\mathrm{sway}} \, . \tag{17}$$

For trajectory tracking control the small deviations $\Delta\boldsymbol{\eta}$, $\Delta\dot{\boldsymbol{\eta}}$, $\Delta\boldsymbol{p}_g$, $\Delta\dot{\boldsymbol{p}}_g$ in $\boldsymbol{x}_{\text{sway}}$ are referred to the values of the corresponding variables from the reference trajectory. The feedback gain matrix $\boldsymbol{R} \in \mathbb{R}^{4,14}$ is here determined by LQR-design. Since the linear feedback (17) is only designed for the dynamics of CABLEV in the vicinity of an equilibrium position in the workspace, an increasing decline of the control behaviour is expected if the platform moves away from this position. Though simulation results show a good robustness of the controller, it is planned to define regions in the workspace with correspondingly adapted feedback gains, that are, e.g., provided by pre-computed lookup-tables.

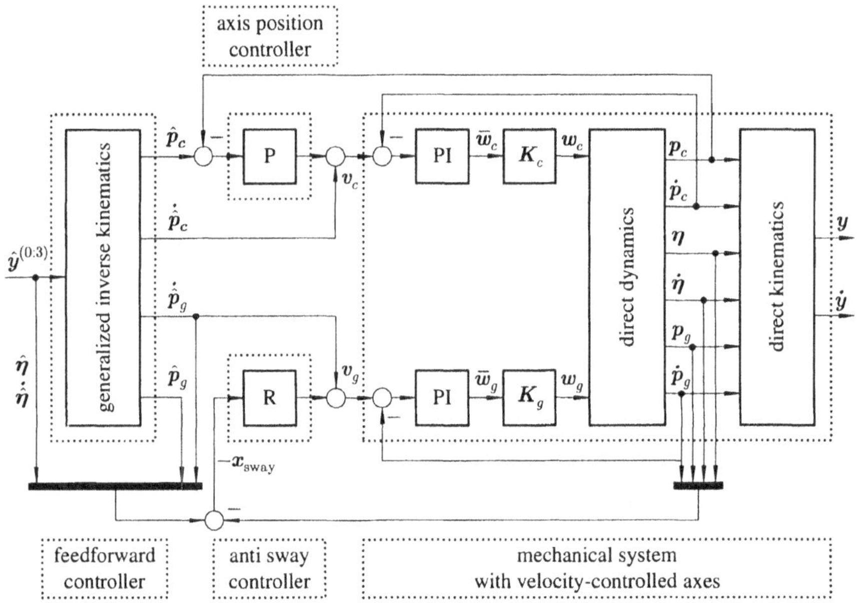

Figure 5. Trajectory Tracking Feedback Control for CABLEV

Simulation Results

A straight-line translational motion and a rotation around a space-fixed vertical axis between rest positions are chosen as reference trajectories. To show the effects of the feedback control, non-vanishing initial conditions, therefore being not consistent with the reference trajectory, are assumed. The large changes of the coordinate $r_x(t)$ for the translational motion in Fig. 6a and $\varphi_3(t)$ for the rotational motion in Fig. 6b are a result of the feedforward control. The superimposed feedback control dampens the initial disturbances. Note that the rotation around the vertical axis is mainly achieved by opposite motions of the outer trolleys.

a) Linear motion of the payload platform

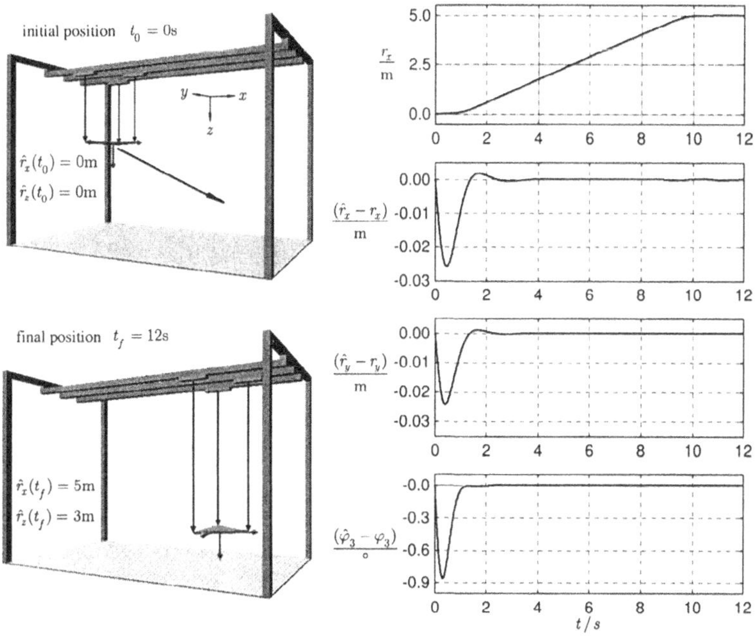

b) Rotation of the payload platform around a fixed vertical axis

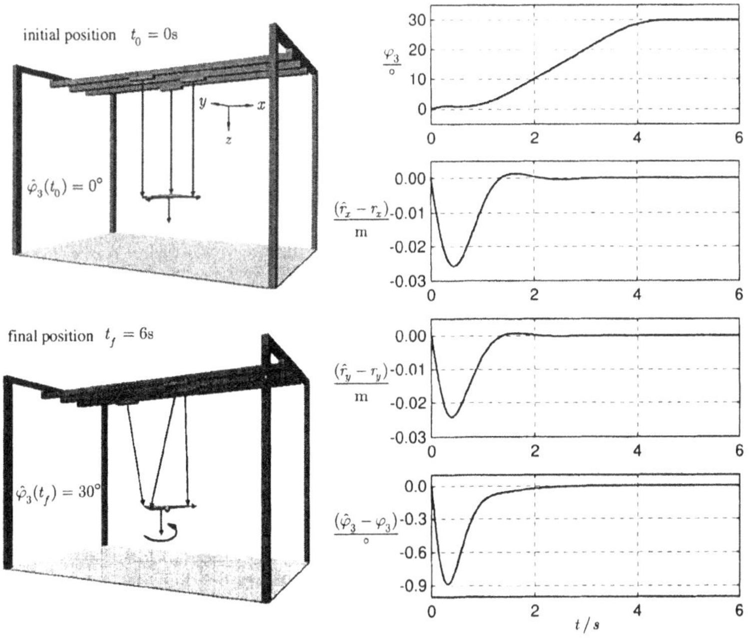

Figure 6. Examples for controlled motions of CABLEV

5. Conclusions

For the kinematically underconstrained cable suspension manipulator CABLEV a trajectory tracking control algorithm is developed that consists of a flatness-based feedforward control law and a linear feedback of measured state variables. Simulation results confirm this approach. Ongoing work is devoted to the experimental validation.

6. Acknowledgements

This research is supported in part by the German Research Foundation (DFG).

References

Arai, T. and Osumi, B., (1992), Three Wire Suspension Robot, *Industrial Robot*, no. 19, pp. 17-22.

Dagalakis, N.G., Albus, J.S., Wang, B.L., Unger, J., and Lee, J.D. (1991), Stiffness Study of a Parallel Link Robot Crane for Shipbuilding Applications, *ASME Journal of Offshore Mechanics and Arctic Engineering*, no. 111, pp. 183-193.

Delaleau, E. and Rudolph, J. (1995), Decoupling and Linearization by Quasi-Static Feedback of Generalized States, *Proc. 3rd European Control Conference ECC'95*, Rome, pp. 1069-1074.

Fliess, M., Lévine, J., Martin, P., and Rouchon, P. (1992), On Differentially Flat Nonlinear Systems, In Fliess, M. (Ed.): *Nonlinear Control System Design*, Pergamon Press, pp. 408-412.

Kecskeméthy, A. (1995), Object-Oriented Modeling of Mechanical Systems, In: Angeles, J. and Kecskeméthy, A. (Eds.): *Kinematics and Dynamics of Multibody Systems*, Berlin, Springer-Verlag, pp. 217-276.

Maier, T. and Woernle, C. (1998), Inverse Kinematics for an Underconstrained Cable Suspension Manipulator, In Lenarčič, J. and Husty, M. (Eds.): *Advances in Robot Kinematics: Analysis and Control*, Kluwer Academic Publishers, Dordrecht, pp. 97-104.

Maier, T. and Woernle, C. (2000), Dynamics and Control of a Cable Suspension Manipulator, *The* 9th *German-Japanese Seminar, Nonlinear Problems in Dynamical Systems -Theory and Applications*, University of Duisburg, Germany.

Maier, T. (2001), *Bahnsteuerung eines seilgefuehrten Handhabungssystems - Modellbildung, Simulation und Experiment*, Dissertation, University of Rostock.

Ming, A. and Higuchi, T. (1994), Study on Multiple Degree-of-Freedom Positioning Mechanism Using Wires, *Int. Journal of the Japanese Society for Precision Engineering*, no. 28, pp. 131-138 and 235-242.

Sawodny, O., Aschemann, H., Lahres, S., and Hofer, E.P. (1999), Tracking Control for Automated Bridge Cranes, In: Tzafestas, S. (Ed.): *Advances in Manufacturing*, Berlin, Springer Verlag, pp. 310-320.

Tadokoro, S., Maeda, K., Takamori, T., Hiller, M., and Verhoeven, R. (1999), Design of Parallel Robot Driven by Redundant Cables: WARP Manipulator. In: Kecskeméthy, A., Schneider, M., and Woernle, C. (Eds.): *Advances in Multibody Systems*, Graz University of Technology, pp. 359-369.

IS DESIGN OF NEW DRUGS A CHALLENGE FOR KINEMATICS?

Kazem Kazerounian
Professor of Mechanical Engineering
University of Connecticut
Storrs, CT 06269-3139
kazem@engr.uconn.edu

Abstract The systematic study of kinematics can be traced to the writings of the ancient Greeks, Egyptians, Romans and Persians as far back as 500 B.C.. For many centuries kinematics (along with geometry) was regarded as one of the basic sciences that explained observed physical phenomena, and was used to engineer machines. Though it may seem unlikely, today a valid claim may be that: kinematics (and in particular robot kinematics) is one of the disciplines that can significantly contribute to the understanding of the function of the biological systems at the microscopic level, and can be utilized to engineer new diagnostic tools, treatments and drugs for various diseases. Given the vast body of knowledge in the theoretical, applied and analytical kinematics and robotics, the obvious absence of the kinematics community and their contribution in the field of protein fold prediction, protein docking, protein engineering and drug design is puzzling. In this paper we will discuss these potential areas and challenges in biotechnology that might be pertinent and of interest to the kinematics and robotics research community.

Keywords: Drug design, Protein conformation, Kinematics, Robotics

1. Background

1.1 Protein Structure

Proteins are large molecules that are made up from 20 different amino acids. Each one of these amino acids is referred to as a residue. Amino acids are the building blocks (monomers) of proteins. Twenty different types of amino acids are used to synthesize proteins. About fifty to hundreds of these residues (amino acid monomers) are connected together in a serial manner to create a long chain. This chain is known as a polypeptide chain. The structure of an amino acid is schematically shown in Figure 1. In this figure, R represents one of the 20 possible side chains the simplest of which is a single hydrogen atom in the *Glycine*. The protein representations in figure 1 is a linear representation that merely describe the sequence of amino acids residues (and hence atoms)

J. Lenarčič and F. Thomas (eds.), Advances in Robot Kinematics, 135–144.

in the chain, and by no means entail any information on the three dimensional structure of the protein. This representation is commonly known as the *"primary"* or *"linear"* structure of the protein and can be determined from the sequence of the gene in the DNA that initiated the construction of the protein.

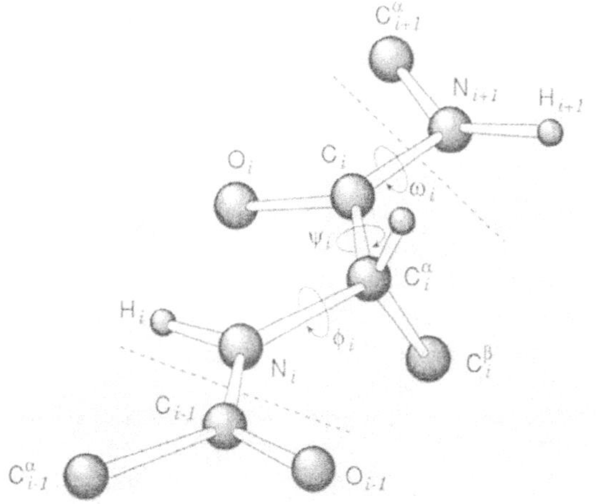

Figure 1. The ith residue in a long peptide chain

There are various ways to represent the three dimensional structure of proteins. Probably one of the most illustrating methods is to represent atoms as "balls" and bonds as "sticks". Please note that when entering the arena of quantum mechanics, these simplifications are meaningless. However at a molecular mechanics level, this representation is justified and facilitates the study and analysis of the molecules. Figure 1 illustrates a ball and stick model of the peptide chain. The "back bone" of the chain is a repeat of the -N-C_α-C- sequence, with side chains resembling branches coming off the tree trunk. The bond lengths (stick lengths) are relatively constant. Also most of the angles that define the relative three dimensional orientation of these bonds (sticks) with respect to each other can also be reasonably assumed to be constant. A comprehensive list of these lengths and angles is included in (Engh, R.A. & Huber R., 1991). The rotation angle around C-N (usually referred to as ω_i) is very close to either 0 degrees or 180 degrees. Therefore the C_α-C-N-C_α chain is always in one plane (known as peptide plane i). There are however two angles that are free to change. These angles are free rotations (spins) around the two bonds N-C_α and C_α-C and are commonly known as ϕ_i and ψ_i for the residue (i) of the protein chain (residues are numbered starting from the amino side of the chain. The value of these angles determines the three dimensional structure of the backbone of the protein molecule (secondary structure). The 3-D structure of the protein is also known as the protein conformation or protein folding. It is the values of these angles (ϕ_i and ψ_i for i=1 to N) and therefore the three dimensional conformation of the protein that makes the protein perform its function. For more details, the reader is referred to many excellent books on protein structures (e.g. Branden C. & Tooze J.1998; Lesk A. M., 2001).

1.2 Protein Conformation (Folding)

A protein molecule can exist in a so called "*denatured*" state in which it looks like an open long string without any particular importance to its shape. However, when placed in its appropriate environment (such as cell environment: appropriate solvent, PH, and temperature) it folds on itself to a complicated and unique shape (*native state*). This process may take anywhere between .001 to a few seconds. A protein in the *native* state can be *denatured* (by changing solvent or its PH or temperature) and when placed back in the appropriate environment it returns again to its *native* state. This repeatable process proves that the final three dimensional structure (*secondary structure*) of the protein is intrinsic to its *primary structure*. Therefore, in principal, it must be a predictable state. The folding occurs under the effect of nuclear forces (among protein atoms as well as between protein atoms and the solvent's atoms). The final conformation is a relatively stable configuration for which the total potential energy is globally minimized (or in other words the system is in stable static equilibrium).
The function of a protein can be loosely compared to a key and lock combination (*protein docking*). The three dimensional structure of the protein acts like a unique key that fits only in a particular lock (usually in the form of a cavity of specific geometry on the surface of another protein). Sometimes a small flexibility and compliance is needed in both the lock and the key (*receptor* and the *ligand*) for the protein *docking* to take place. This is usually known as *induced fit*.

1.3 Prediction of Protein Fold

Being able to accurately predict the three dimensional structure of the protein based on the known sequences of amino acids in the chain is the key to fully understanding the biological functions of the protein and therefore being able to control such functions as part of disease treatments. **Despite more than 50 years of intense work on this subject, the protein prediction problem remains largely**

unsolved. Protein prediction is referred to as the "**grand challenge**" or **the most challenging computational and scientific problem of the century**. To date, the structure of about 6000 proteins are experimentally observed (either through x-ray crystallography or Nuclear Magnetic Resonance (NMR) technologies. There are an estimated 100,000 to 200,000 proteins in the human body. Many of these, due to their complicated structure or denaturing tendencies in crystallization, will not be accurately determined with current technologies. In addition, in order to better understand their function, a reliable prediction methodology is an essential tool.

There are three major categories of methods for predicting protein folding. 1) Comparative modeling, 2) Fold recognition and 3) *ab initio* methods. In comparative modeling there is significant similarity between the primary structure (sequence) of protein of interest and a protein of already known structure. Therefore the predicted structure is assumed to have significant similarities to the know structure. In fold recognition, there is no significant sequence similarity, but the target proteins structure turns out to have similar structural aspects to a known protein. The 3rd method, *ab initio*, does not assume a priori knowledge of the structure, and attempts to predict the final fold based on the principles of physics in conjunction with various optimization and computational techniques. "Critical Assessment of Protein Structure Prediction (CASP)" is an initiative conducted in the form of a competition (four have been conducted so far). In these, the primary structure of the proteins (for which the secondary structure is experimentally determined but not publicly known yet) are announced for prediction by various methodologies. In CASP4 about 100 research team participated. For an overall description of these methods and their performance, the reader is referred to Moult, J., 1999). Complete results for the CASP1-4 are available on the web site: (http://predictioncenter.llnl.gov).

Based on the performance of participating algorithms, there is still a long way to go before the *ab initio* problem is solved. Nevertheless, accurate and efficient ab initio methods are the ultimate goal of protein prediction.

1.4 *ab initio* Prediction Methods

The ab intio prediction methods generally have a model of the force field (potential field) of the protein molecule. These methods, then attempt to change the geometry of the molecule to optimize the potential energy. Local optimization in this process is practically worthless. A rather comprehensive review of the global optimization based *ab initio* techniques is included in reference (Floudas *et al,* 2000). In the

optimization based methods the variables are the Cartesian coordinates of the atoms of the protein molecule.
Althought there are many variations, the potential energy (as the objective function in the *ab initio* optimization) is typically defined as:

$$E_{total} = E_{bnd\text{-}lenght} + E_{bnd\text{-}angle} + E_{torsion} + E_{van\ der\ Waals} + E_{Electrostatic} \quad (1)$$

Where, $E_{bnd\text{-}lenght}$ = Bond energy calculated by Hook's law = $\Sigma\ k_b\ (l\text{-}l_0)$ where k_b is experimentally known and l and l_0 are actual and ideal bond lengths.

$E_{bnd\text{-}angle}$ = Bond Angle (bending) energy calculated by hook's law = $\Sigma\ k_\theta\ (\theta\text{-}\theta_0)$ where k_θ is experimentally known angular stiffness of the bond, and θ and θ_0 are actual and ideal bond angles.

$E_{torsion}$ = $\Sigma\ A\ (1+\cos(n\varphi\text{-}\varphi_0))$ = where A is an experimentally determined constant , and is φ the torsion angle (the rotation angle around the bond between the 2nd and the 3rd atom in any serially connected four atoms. The torsion energy is mainly used to correct the bond and bending energies to make the results agree with experimental results.

$E_{van\ der\ Waals}$ = The energy due to non bonded forces between two atoms i and j = $\Sigma\ \Sigma\ (-a_{ij} / r_{ij}^6 + b_{ij} / r_{ij}^{12})$ where a and b are experimentally known and r is the distance between the two atoms.

$E_{Electrostatic}$ = Electrostatic energy between two atoms = $\Sigma\ \Sigma\ q_i q_j / r_{ij}$ where q,s are known and r is the distance between the two atoms.

For a detailed study of the molecular mechanics the user is referred to Molecular Mechanics (Burkert U. and Allinger N., 1982; Rappe, A. and Casewit, C. 1997). The computation of the potential energy as the cost function in the optimization based *ab initio* methods is rather straight forward. However the computational complexity is mind bugling. To put things in perspective, let's examine the computational requirement of a rough exhaustive search for a protein molecule that contains 100 residues (200 revolute joints). If samples on the joint angles are taken at every 36 degrees (10 samples per joint- so rough that the results are meaningless), and the energy calculations for each state of the protein takes 10^{-20} seconds (using many parallel super computers) then the total computational time required is:
Computation time = $(360 / 36)^{200} * 10^{-20} = 10^{180}$ seconds = $3.16 * 10^{172}$ years

2. Challenge for Kinematicians?

From a kinematics point of view, a protein in denatured state is a serial linkage with N+1 solid links connected by N revolute joints (N might be up to several hundreds).

Several researchers have drawn analogies between protein folding methodologies and robotics methodologies (Parson et al, 1994; Karavi et al 1996, 1998;, Song and Amato 2000). However, the majority of the actual works implemented are in the fields of motion planning and path planning and are undertaken by researchers in the computer science, optimization and artificial intelligence communities.
More recently, work based on theoretical, and analytical kinematics is starting to appear amongst the vast amount of literature on protein folding. Chirikjian, G.S. and co workers (2000 and 2001) have used motion groups (rotation and motion) for the statistical study of the mechanics of the polymer molecules. Manocha D. *et al* (2000) used Ragavan &Roth's (1989) solution to the general 6-R serial manipulator inverse kinematics problem to facilitate solving the receptor-ligand docking problem for the short side chains. In addition there are sporadic applications of some of the kinematics notations and procedures in this field.

2.1 Kinematics notation

The commonly used convention for describing the geometry of the protein is to use a set of the Cartesian coordinates of its atoms and their bonds (connectivity). This author is not aware of any earnest attempt to define a protein chain as a three dimensional kinematics chain with solid links. Potential kinematics notations that could be utilized are: Denavit-Hartenberg notation (1955), vector notation and analysis (Chase, M. 1964), tensor methods (Kislitsin A. P.1954; Osman & Mansour1971), screw coordinates (Yuan and Freudenstein, 1971), Dual numbers (Yang and Freudenstien, 964), quaternion operators (Sandor, 1968), constant distance equation method (Osman and Sergev, 1972), spherical trigonometry method (Duffy, 1980), Zero Position Method (Gupta, 1984) and train components method (Osman *et al*, 1981). An in-depth study of these notations and their applicability to the protein analysis is needed. Some of the factors that are of paramount importance in this process are: (1) Computational efficiency and (2) Geometric stability: small changes in the geometry of the protein should not lead to large changes in the kinematic parameters (Qian and Kazerounian, 1996).

2.2 Hydrogen Bonds and Serial VS. Parallel Robotic Manipulators

In the process of folding, as various portions of the long snake type chain get closer together, a hydrogen bond is created between the hydrogen of an amino group and the oxygen of a carboxyl group. This bond acts as a solid bond that cerates a closed loop within the otherwise serial linkage. Some of these developed loops lack internal mobility, while in other cases, these loops have a mobility of one or higher. These internal mobilities act in a similar manner to parallel manipulators, and are decisive factors in the continuation of the folding, or in the function of the protein.

2.3 Direct Kinematics

As mentioned earlier, the current practice in the protein molecular simulation, analysis or fold prediction is to use the coordinates of each one of the atoms as generalized coordinates (variables), and constrain the model by forcing constant distance constraints between bonded atoms. Alternatively such constraints may be added as penalty functions to the optimization cost functions (as bond energies of terms 1 and 2 in equation 1). By using the torsion angles (ϕ_i and φ_i) as generalized coordinates, the computational complexity of the analysis is expected to reduce tremendously. For this however, we need kinematics based robot methodologies that can produce the position of all atoms (which are actually body points of various links of the chain) based on the values of the torsion angles. The challenge in this process is:

- to accommodate and account for the changing structure of the linkage due to hydrogen bonds,
- to check for and avoid collision of various links (atoms),
- To design the algorithm (by decoupling the direct kinematics of various sections in the long chain) to take maximum advantage of parallel computing.

The reduction in the computational time may be tremendous. If we assume that each amino residue in the chain, on average has 10 atoms (most have more), then the number of optimization variables in the majority of *ab initio* methods reduces from 10*3 to 2N variables. Considering that most optimization methods have a computational cost proportional to the square of the number of variables (global search based methods such as simulated annealing and genetic algorithms excluded), the expected computational time (number of cost evaluations) reduction is 15^2 or 225 times. Additionally, in evaluating the cost function, the bond and angle energies (terms 1 and 2 in equation 1) are not needed anymore.

2.3 Inverse Kinematics and Drug Design

The inverse kinematics problem in protein fold prediction appears in two forms: The first is when simulating the behavior of the long snake type chain under the influence of field forces (atomic attractions and repulsions). This process in principal is similar to that of redundancy resolution of robotic manipulators under energy favored criteria (Kazerounian, 1987, Kazerounian and Nedungadi, 1988 & 89 and Wang and Kazerounian 1989). The difference however is that the end link trajectory is not specified, and therefore the optimization process is not constrained by the velocities and acceleration of the end link. In this process, the manipulator chain (*denature* protein chain) should be allowed to find a local energy favored path (local minimum potential energy path) to a final configuration. This final configuration is shown to be a minimum potential energy configuration (Kazerounian and Wang 1988). The second form of the inverse kinematics problem that arises in the protein structure analysis is the problem of protein docking (ligand-receptor docking). In this process, the geometric characteristics of the docking location on a receptor (usually in the form of a cavity on a protein molecule) are known. These geometric characteristics are defined by positional information of certain atoms of the mating protein chain (ligand). The problem is then finding a protein (by searching in the data bank or designing a new protein) that would result in the specified position for these atoms.

3. Conclusion

This paper presents a discussion on the similarities of the protein folding in biological systems and the manipulation of robotics linkages. There are tremendous opportunities for contribution by kinematics and in particular robot kinematics researchers to the field of protein fold prediction and therefore new drug design. The number of degrees of freedom in a protein chain and the force environment under which it folds and functions create mind bugling analysis and computational complexities. However the vast body of knowledge in theoretical, analytical and applied kinematics and robotics offers numerous valuable methodologies that can potentially make significant advances in this multi disciplinary field.

Acknowledgement

The author wishes to acknowledge many fruitful discussions with Professors Eric Jordan and James Knox and Mr. Alireza Senejani.

References

Branden C. and Tooze J.(1991), Introduction to Protein Structure, 2nd edition, *Garland Publishing*.

Burkert U. and Allinger N. (1982) , Molecular Mechanics, *Oxford Press.*
Cantor, C.R. and Schimmel, P.R. (1997), Biophysical Chemistry. The conformation of biological macromolecules, *W.H. Freeman & Co.*1997.
Chase, M. (1964), Vector analysis of Linkages, *Transactions of the ASME Journal of Engineering for Industry*, Vol. 85, No. 2, June 196, pp 300-308.
Chirikjian, G.S. and Wang, Y.F., (2000), Conformational Statistics of Stiff Macromolecules as Solutions to PDEs on the Rotation and Motion Groups, *Physical Review* , pp. 880-892, Vol. 62, No. 1, July 2000.
Chirikjian, G.S. (2001), Conformational Statistics of Macromolecules Using Generalized Convolution, *Computational and Theoretical Polymer Science*, pp143-153, Vol. 11, February 2001.
Chirikjian, G.S. and Kyatkin, A.B. (2000), An Operational Calculus for the Euclidean Motion Group: Applications in Robotics and Polymer Science, *Journal of Fourier Analysis and Applications*, Vol. 6, No. 6, pp.583-606, 2000.
Denavit, J. and Hartenberg, R. S. (1955), A Kinematic Notation for Lower-Pair Mechanisms Based on Matrices, *ASME Journal of Applied Mechanics*, 1955, pp. 215-221.
Engh R.A. and Huber R. (1991), Accurate bond and Angle parameters for X-ray protein structure refinement, *Acrta Cryst*, A47, 392-400.
Floudas C. A., Klepeis, J. L., and Pardalos, P. M., (2000), Global Optimization Approaches in Protein Folding and Peptide Docking, *DIMACS Series in Discrete Mathematics, and Theoretical Computer Science.*
Gupta, K.C. (1986), Kinematic Analysis of Manipulators Using the Zero Reference Position Description, *The International* J. *of Robotic Research,* Vol. 5, No. 2, pp. 5-13.
Kavraki. L. (1996), Geometry and the discovery of new ligands, *In Proc. Int. Workshop on Algorithmic Foundations of Robotics (WAFR),* pages 435--448, 1996.
Kavraki, L. and Latombe J.C. (1998), Probabilistic roadmaps for robot path planning. *Practical Motion Planning in Robotic*, K. Gupta and A. del Pobil (Eds), Wiley Press, 1998.
Kazerounian, K. and Qian, Z. (1989), Kinematics Calibration of Robotic Manipulators, *ASME J. of Mech. Trans. & Aut. in Des.*, Vol. 111.
Kazerounian, K., Nedungadi, A. (1989), A Local Solution with Global Characteristics for torque optimization in Redundant Manipulators, *Intl. J. of Robotic Systems*, Vol. 6.5.
Kazerounian, K. and Nedungadi, A. (1988), Redundancy Resolution of Serial Manipulators Based on Robot Dynamics, *Mechanism and Machine Theory*, Vol. 22, No. 4.
Kazerounian, K. & Wang, Z. (1988), Global Versus Local Optimization in Redundancy Resolution of Robotic Manipulators, *The International J. of Robotics Research,* Vol. 7, No. 5.
Kazerounian K. (1987), Optimal Manipulation of Redundant Robots, *The International J. of Robotics and Automation,* Vol. 2, No. 2 .
Kislitsin A.P., (1954), Tensor Methods in the Theory of Spatial Mechanisms, *Trudi Seminar Po Teroii Mashin I Mekhanizmov, Akedemia Nauk, USSR*, Vol 14, 1954, pp. 51-57.
Lesk A.M. (2001), Introduction to Protein Architecture, *Oxford University Press,* 2001.

Moult, J. (1999), Predicting Protein Three-dimensional Structure, *Current Opinions in Biotechnology*,10: pp. 583-588, 1999.

Osman, M.O. and Mansour W. M. (1971), The Proximity Pertubation Method for the Kinematic Analysis of Six Link Mechanisms, *Journal of Mechanisms,* Vol. 6, No. 2, pp. 203-212.

Osman, M.O. and Segaev, D.N. (1972), Kinematic Analysis of Spatial Mechanisms by Means of Constant Distance Equations, *Transactions of the Canadian Society of Mechanical Engineers*, Vol 1, No. 3, pp 129-134.

Osman, M.O., Bahgat, B. M. and Dukkipati, R.V. (1981), Kinematic Analysis of Spatial Mechanisms Using Train Components, Transaction of the ASME Journal of Mechanical Design, Vol 103, pp. 823-830.

Parsons, D. and Canny, J. F. (1994), Geometric Problems in Molecular Biology and Robotics, *In Proceedings of the Second International Conference on Intelligent Systems for Molecular Biology, Palo Alto, CA,* August, 1994.

Rappe, A. Casewit, C. (1991), Molecular Mechanics Across Chemistry, *University Science B*ooks.

Sandor, G. N. (1968), Principles of General Quaternion-Operator Method of Spatial Kinematic Synthesis, *ASME paper number 68-APM-1.*

Song, G. and Amato, N. M. (2000), Motion planning approach to folding: From paper craft to protein structure prediction. *Technical Report TR00-001, Department of Computer Science, Texas A&M University.*

Voet, D. & Voet, J. (1995), *Biochemistry*, 2nd Ed., *John Wiley & Sons.*

Yang, A.T. and Freudenstein, F. (1964), Application of Dual-Number Quaternions Algebra to the Analysis of Spatial Mechanisms, *Transactions of the ASME Journal of Applied Mechanics*, Vol 86, No. 2, pp. 300-308.

Yuan, M. S. and Freudenstien, F. (1971), Kinematic Analysis of Spatial Mechanisms by Means of Screw Coordinates, *Transactions of the ASME, Journal of Engineering for Industry*, Vol 93, No. 1, pp. 61-73.

Redundancy, Singularity, and Self-Motion

A REDUNDANT SPATIAL STEWART-GOUGH PLATFORM WITH A MAXIMAL FORWARD KINEMATICS SOLUTION SET

Manfred Husty

Institute for Engineering Mathematics, Geometry and Computer Science
University of Innsbruck, Austria
Manfred.Husty@uibk.ac.at

Sonja Mielczarek, Manfred Hiller

Institute for Mechatronics and System Dynamics
Gerhard Mercator University of Duisburg, Germany
{mielczar,hiller}@mechatronik.uni-duisburg.de

Abstract This paper discusses the possibility of designing a redundant spatial parallel manipulator with a maximal forward kinematics solution set. We show that when the anchor points for the joints of the platform and base are planar, in almost every case an infinite number of legs may be added without changing the kinematics of the mechanism. The possible anchor points lie on a cubic which may be split into a line and a conic.

Keywords: forward kinematics, overconstrained spatial parallel manipulator

1. Introduction

Tendon based Stewart platforms represent a special class of Stewart-Gough platform (SGP) where tendons attached to the base and platform are used instead of rigid legs. By changing the lengths of the tendons a desired platform position and orientation (pose) can be reached. As tendons represent only one-sided constraints, a necessary condition for positioning using tendons is positive tension in each tendon. Otherwise the motion of the platform is no longer controllable. Ming and Higuchi [1994] distinguish therefore between "Incompletely Restrained Positioning Mechanisms" (IRFMs) and "Completely Restrained Positioning Mechanisms" (CRPMs) conditional on whether positive tension in the tendons is realised by the dynamics or kinematics of the mechanism. It follows that CRPMs represent, from a geometrical point of view, overconstrained systems because at least $n + 1$ tendons are required for n degrees of freedom (d.o.f.). Normally an overconstrained parallel mechanism has no solution for the forward kinematics. Therefore coordination

J. Lenarčič and F. Thomas (eds.), Advances in Robot Kinematics, 147–154.

of the lengths of the actuators is necessary for positioning the platform, and to ensure that a forward kinematics solution exists. In this paper a pose of a redundant SGP with m actuators and n d.o.f. ($m > n$) is determined using the lengths of n legs or tendons, and then the lengths of the remaining $m - n$ actuators are computed to retain this pose.

The maximum number of assembly modes for a general SGP is 40 [Wampler, 1996]. In this contribution we investigate whether a redundant SGP whose leg lengths are coordinated in the way described above, can also possess 40 solutions for the forward kinematics. It has been shown that in a plane an overconstrained 4RPR parallel manipulator with six assembly modes can exist and that to almost every 3RPR parallel manipulator (this is the planar analogue to the SGP) we can add a fourth leg in such a way that it has no effect on the direct kinematics [Husty et al., 2001]. The coordinates of the additional joints depend only on the geometry but not on the pose of the manipulator. As a consequence the singular surfaces of both manipulators coincide. This result is therefore important when we wish to design an overconstrained manipulator and to attach an additional leg, both to avoid singularities and to increase the workspace. The anchor points of the additional constraint must not lie near any of the solutions for the fourth leg.

In this paper we discuss the spatial case. We show that an infinite number of additional legs may be attached to almost every manipulator where the anchor points of the base joints lie in one plane and of the platform in a second plane, without changing the kinematics of the manipulator (here these mechanisms are called SGP with planar base and platform (SGPP)). The algorithm we propose is based on kinematic mapping of spatial displacements. A numerical example to illustrate this procedure is given.

2. Spatial Kinematic Mapping

A detailed discussion of spatial kinematic mapping can be found in [Study, 1903] or [Weiss, 1935]. Here we summarize the most important facts. The idea of kinematic mapping is to describe different poses of the moving space with respect to a fixed space as points in a higher dimensional projective image space. Study has shown that each displacement of a three dimensional space E with respect to a three dimensional space Σ can be mapped to a six dimensional hyper-quadric S_6^2 (called Study quadric) in the seven dimensional projective space P^7 with homogeneous coordinates $x_i, y_i,\ i = 0, \dots, 3$, called Study parameters. Husty [1996] has shown that an image of the motion, where a given point $M \in E$ is constrained to lie on a sphere centered at $F \in \Sigma$ with radius r, is represented in the kinematic image space P^7 by an algebraic quadric surface H_{FM} having the form

$$H_{FM}\ :\ \sum_{i,j=0}^{3} h_{x_i x_j}\, x_i\, x_j\ + \sum_{\substack{i,j=0\\ i\neq j}}^{3} h_{x_i y_j}\, x_i\, y_j\ +\ 4\sum_{i=0}^{3} y_i^2 = 0\,, \qquad (1)$$

where coefficients $h_{x_i x_j}$, $h_{x_i y_j}$, $i,\, j = 0, \ldots, 3$, are functions of the coordinates of the points F and M and the radius r. The explicit form of the surface H_{FM} is given in [Husty, 1996].

3. Designing a Redundant Actuated SGP with a Maximal Forward Kinematics Solution Set

Consider a SGP (F_i, M_i, r_i) given by six points $F_i \in \Sigma$ with coordinates (A_i, B_i, C_i) with respect to a fixed Cartesian frame $\mathcal{F}$, six points $M_i \in E$ with coordinates (a_i, b_i, c_i) with respect to a moving Cartesian frame $\mathcal{M}$ and six leg lengths r_i, $i = 1, \ldots, 6$. The solution for the forward kinematics can be determined from the intersection points of the six constraint surfaces $H_{F_i M_i}$ we get for the six legs and the quadric S_6^2 in the kinematics image space P^7 [Husty, 1996]. An additional pair $(F_7 \in \Sigma\,, M_7 \in E)$ such that M_7 lies on a sphere centered at F_7 with radius r_7 for each of the 40 given poses can be found if a seventh constraint surface $H_{F_7 M_7} \in P^7$, $H_{F_7 M_7} \neq H_{F_i M_i}\,, i = 1, \ldots, 6$, exists, is different from the six given and passes through all 40 image points on S_6^2. The sufficient condition therefore is, that the seventh constraint surface $H_{F_7 M_7}$ vanishes on the one-dimensional intersection manifold of the six given constraint surfaces $H_{F_i M_i}$, $i = 1, \ldots, 6$.

In algebraic geometry [Cox et al., 1992], the set of all solutions of the system of equations $Q_1 = 0, \ldots, Q_s = 0$ is called the affine variety defined by $Q_1, \ldots, Q_s$. Every other polynomial S that also vanishes on this variety must be an element of the ideal generated by polynomials $Q_1, \ldots, Q_s$. This means S must be a linear combination of Q_i, $S = \sum_{i=1}^{s} f_i\, Q_i$ with polynomial coefficients f_i. Thus the seventh surface $H_{F_7 M_7}$ we are looking for must be an element of the ideal generated by polynomials $H_{F_1 M_1}, \ldots, H_{F_6 M_6}$. As $H_{F_7 M_7}$ also has to be a constraint surface, $H_{F_7 M_7}$ must have the form of eq. (1) and be a quadric. Therefore we can write its equation as a linear combination of $H_{F_i M_i}$ with scalar coefficients λ_i

$$H_{F_7 M_7} = \sum_{i=1}^{6} \lambda_i\, H_{F_i M_i}\,, \quad \lambda_i \in \mathbb{R},\ i = 1, 2, \ldots, 6\,. \qquad (2)$$

Substituting the design parameters into eq. (2) and comparing the coefficients of the image space coordinates $x_i x_j$, $x_i y_j$, y_j^2, $i,\, j = 0, \ldots, 3$, from eq. (1) and eq. (2) we get a following system of 17 equations

$$p_{y_0^2}: \sum_{i=1}^{6} \lambda_i - 1 = 0\,,$$

$$p_{x_0^2}: \sum_{i=1}^{6} \lambda_i\,(R_i - 2\,a_i\,A_i - 2\,b_i\,B_i - 2\,c_i\,C_i) - R_7 + 2\,a_7\,A_7 + 2\,b_7\,B_7 + 2\,c_7\,C_7 = 0\,,$$

$$p_{x_1^2}: \sum_{i=1}^{6} \lambda_i\,(R_i - 2\,a_i\,A_i + 2\,b_i\,B_i + 2\,c_i\,C_i) - R_7 + 2\,a_7\,A_7 - 2\,b_7\,B_7 - 2\,c_7\,C_7 = 0\,,$$

$$p_{x_2^2}: \sum_{i=1}^{6} \lambda_i\,(R_i + 2\,a_i\,A_i - 2\,b_i\,B_i + 2\,c_i\,C_i) - R_7 - 2\,a_7\,A_7 + 2\,b_7\,B_7 - 2\,c_7\,C_7 = 0\,,$$

$$p_{x_3^2}: \sum_{i=1}^{6} \lambda_i\,(R_i + 2\,a_i\,A_i + 2\,b_i\,B_i - 2\,c_i\,C_i) - R_7 - 2\,a_7\,A_7 - 2\,b_7\,B_7 + 2\,c_7\,C_7 = 0\,,$$

$$p_{x_0x_2}: \sum_{i=1}^{6} \lambda_i\,(a_i\,C_i - A_i\,c_i) - a_7\,C_7 + A_7\,c_7 = 0\,,$$

$$p_{x_1x_3}: \sum_{i=1}^{6} \lambda_i\,(a_i\,C_i + A_i\,c_i) - a_7\,C_7 - A_7\,c_7 = 0\,,$$

$$p_{x_0x_1}: \sum_{i=1}^{6} \lambda_i\,(B_i\,c_i - b_i\,C_i) + b_7\,C_7 - B_7\,c_7 = 0\,,$$

$$p_{x_2x_3}: \sum_{i=1}^{6} \lambda_i\,(B_i\,c_i + b_i\,C_i) - b_7\,C_7 - B_7\,c_7 = 0\,,$$

$$p_{x_1x_2}: \sum_{i=1}^{6} \lambda_i\,(A_i\,b_i + a_i\,B_i) - a_7\,B_7 - A_7\,b_7 = 0\,,$$

$$p_{x_0x_3}: \sum_{i=1}^{6} \lambda_i\,(A_i\,b_i - a_i\,B_i) + a_7\,B_7 - A_7\,b_7 = 0\,,$$

$$p_{x_3y_2}: \sum_{i=1}^{6} \lambda_i\,(A_i + a_i) - a_7 - A_7 = 0\,,$$

$$p_{x_0y_1}: \sum_{i=1}^{6} \lambda_i\,(A_i - a_i) + a_7 - A_7 = 0\,,$$

$$p_{x_3y_1}: \sum_{i=1}^{6} \lambda_i\,(B_i + b_i) - b_7 - B_7 = 0\,,$$

$$p_{x_0y_2}: \sum_{i=1}^{6} \lambda_i\,(B_i - b_i) + b_7 - B_7 = 0\,,$$

$$p_{x_2y_1}: \sum_{i=1}^{6} \lambda_i\,(C_i + c_i) - c_7 - C_7 = 0\,,$$

$$p_{x_3y_0}: \sum_{i=1}^{6} \lambda_i\,(C_i - c_i) + c_7 - C_7 = 0\,,$$

where $R_i = a_i^2 + b_i^2 + c_i^2 + A_i^2 + B_i^2 + C_i^2 - r_i^2$, $i = 1, \ldots, 7$. The coordinates of the anchor points F_7 and M_7, of the seventh actuator, its length r_7 and the linear coefficients λ_i, $i = 1, \ldots, 6$, yield 13 variables. Therefore a solution for this problem can be found only if the geometry of the manipulator satisfies certain conditions. It is not difficult to verify that in case of a SGPP some of these equations vanish, so that only 10 equations in 11 unknowns are left. In most cases the solution of our problem is then a one parameter manifold in $\mathbb{R}^2$. Because of the limited length of this paper we cannot offer the full analysis of all cases, but will present a classification of the cases for symmetrical SGPPs and give an example. We introduce the following denotations

$$\begin{array}{lll} S_1 = p_{x_0x_2} + p_{x_1x_3}\,, & S_2 = p_{x_0x_1} + p_{x_2x_3}\,, & S_3 = p_{x_1x_2} + p_{x_0x_3}\,, \\ D_1 = p_{x_0x_2} - p_{x_1x_3}\,, & D_2 = p_{x_0x_1} - p_{x_2x_3}\,, & D_3 = p_{x_1x_2} - p_{x_0x_3}\,, \\ S_4 = p_{x_3y_2} + p_{x_0y_1}\,, & S_5 = p_{x_3y_1} + p_{x_0y_2}\,, & S_6 = p_{x_2y_1} + p_{x_3y_0}\,, \\ D_4 = p_{x_3y_2} - p_{x_0y_1}\,, & D_5 = p_{x_3y_1} - p_{x_0y_2}\,, & D_6 = p_{x_2y_1} - p_{x_3y_0}\,. \end{array}$$

The unknown r_7^2 appears only in four equations of the system ($p_{x_i^2}$, $i = 0, \ldots, 3$) and we use $p_{x_0^2}$ to compute it. Then 16 equations in 12 unknowns are left and they do not depend on the lengths of the six legs.

4. SGP with Planar Base and Platform

Without loss of generality let $\mathcal{F}$ and $\mathcal{M}$ be attached in such a way that $c_i = C_i = 0$, $i = 1, \ldots, 6$. Then we obtain

$$S_6 = -2\,C_7 = 0 \quad \Leftrightarrow \quad C_7 = 0\,, \quad D_6 = -2\,c_7 = 0 \quad \Leftrightarrow \quad c_7 = 0\,. \quad (3)$$

Therefore the seventh base and platform points must lie in the given planes, and it follows that four other equations also vanish

$$D_1 = D_2 = S_1 = S_2 = 0.. \quad (4)$$

Futhermore we note, that for a SGPP the following condition holds

$$p_{x_3^2} = p_{x_2^2} + p_{x_1^2} - p_{x_0^2}\,, \quad (5)$$

thus only 10 equations in 11 unknowns remain.

Symmetrical Case. Assume that platform and base have symmetry axes which are aligned with the y-axis of platform and base coordinate systems, furthermore that $b_1 = b_2 = B_1 = B_2 = 0$, $b_i \neq 0$, $B_i \neq 0$, for $i \neq 1, 2$, $b_3 \neq b_4$, $B_3 \neq B_4$, $a_i \neq 0$ and $A_i \neq 0$ for $i = 1, \ldots, 6$. If the following condition is satisfied

$$a_3 A_4 b_4 B_3 - a_4 A_3 b_3 B_4 \neq 0\,, \quad (6)$$

then the equations $p_{y_0^2}$, D_3, D_4, D_5 and S_3 can be solved for λ_i, $i = 1, \ldots, 5$. Substituting the solutions for λ_i into the remaining equations $p_{x_1^2}$, $p_{x_2^2}$, S_4 and S_5 we get four equations in five unknowns λ_6, a_7, A_7, b_7

and B_7. Condition (6) guarantees that we can eliminate λ_6, a_7 and b_7 with $p_{x_1^2}$, S_4 and S_5. The platform parameters a_7 and b_7 are given by

$$a_7 = A_7 \frac{(a_4B_4(a_1A_3-a_3A_1)-a_3B_3(a_1A_4-a_4A_1))b_7+a_1(a_3A_4b_4B_3-a_4A_3b_3B_4)}{(A_3b_3(a_4A_1-a_1A_4)+A_4b_4(a_1A_3-a_3A_1))B_7+A_1(a_3A_4b_4B_3-a_4A_3b_3B_4)},$$
$$b_7 = B_7 \frac{b_3b_4(B_4-B_3)}{(b_3B_4-b_4B_3)B_7-B_3B_4(b_3-b_4)}. \quad (7)$$

Substituting these into the remaining equation $p_{x_2^2}$ yields a cubic $k_{A_7B_7}$ in A_7 and B_7. There are two values for B_7 which must be excluded, otherwise a_7 and b_7 cannot be computed and it follows that equations $p_{x_1^2}, p_{x_2^2}$ would not be fulfilled. For every other point $F_7 = (A_7, B_7, 0) \in k_{A_7B_7} \subset \Sigma$ there exists a unique point $M_7 = (a_7, b_7, 0) \in E$ such that M_7 lies on the sphere centered at F_7 for every pose of the platform. The coordinates of M_7 and F_7 do not depend on the lengths of the six legs.

The case that condition (6) is not satisfied must be studied carefully. Normally it is not difficult to compute all linear coefficients λ_i and platform joint coordinates a_7 and b_7. Furthermore we note that the curve $k_{A_7B_7}$ splits into a conic and the line $B_7 = 0$ (but excluding the point $(0,0,0)$). However if the design parameters also fulfill the equation

$$A_3b_3(a_4A_1 - a_1A_4) + A_4b_4(a_1A_3 - a_3A_1) = 0, \quad (8)$$

then one linear coefficient depends on the other coefficients. It follows, that the constraint surfaces $H_{F_iM_i}$, $i = 1, \ldots, 6$ are redundant, so the platform has self-motions. Thus in this particular case we can find all possible pairs of anchor points which admit self-motions. The existence of such platforms has already been shown by [Karger and Husty, 1997].

An Example. The design parameters are listed in the Table 1. We obtain the following equation for the curve $k_{A_7B_7}$

$$9503\,B_7^3 - 47312\,B_7^2 - 1599420\,B_7 + (302400 - 16296\,B_7)\,A_7^2 - 2721600 = 0$$

The set of all possible base anchor points $F_{7_k} \in \Sigma$ and the corresponding set of the platform anchor points $M_{7_k} \in E$, such that the redundant manipulator $((F_i, M_i), (F_{7_k}, M_{7_k}))$, $k = 1, \ldots, 10$, has the same solution for the forward kinematics as the initial one is shown in Fig. 1 and Fig. 2. The three points W_1, W_2 and W_3 which must be excluded are also shown. Some connections like (F_{7_1}, M_{7_1}) and (F_{7_4}, M_{7_4}) are of no practical use, but positioning the additional anchor points near the curves passing through the base and platform points, like (F_{7_2}, M_{7_2}) or (F_{7_3}, M_{7_3}), should be avoided if we intend to build a manipulator without singularities. The algorithm used to calculate the example displayed here can be seen in the internet at `http://www.mechatronik.uni-duisburg.de/mielczar/SGP` .

Table 1. Design parameters

$\mathcal{F}$	F_1	F_2	F_3	F_4	F_5	F_6
(A_i, B_i, C_i)	$(-3, 0, 0)$	$(3, 0, 0)$	$(10, 10, 0)$	$(6, 16, 0)$	$(-6, 16, 0)$	$(-10, 10, 0)$
$\mathcal{M}$	M_1	M_2	M_3	M_4	M_5	M_6
(a_i, b_i, c_i)	$(-5, 0, 0)$	$(5, 0, 0)$	$(7, 3, 0)$	$(2, 10, 0)$	$(-2, 10, 0)$	$(-7, 3, 0)$

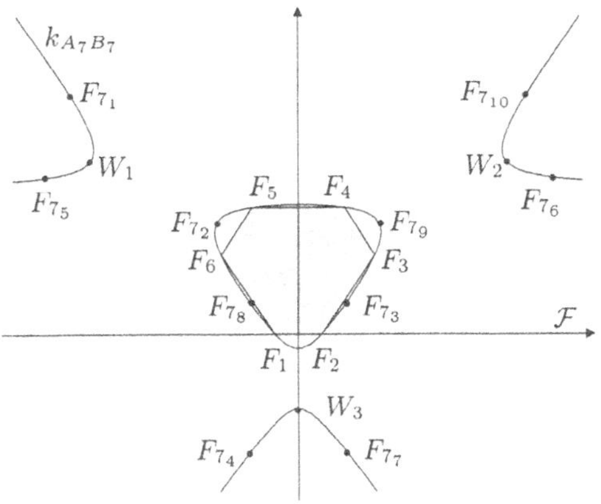

Figure 1. Base, all possible base anchor points F_{7_k} and the points W_1, W_2 and W_3 which must be excluded

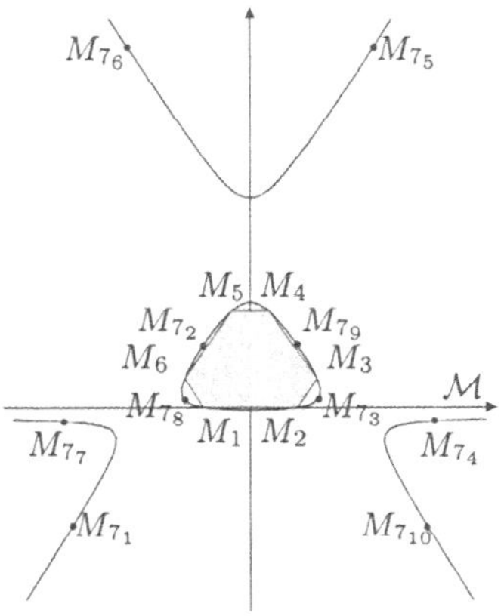

Figure 2. Platform and the corresponding set of the platform anchor points M_{7_k}

Non-Symmetrical Case. The discussion of all special cases exceeds the space available here and is devoted to further publications, but it is worth mentioning, that unlike the symmetrical case, it may occur that that the seventh base and platform points, F_7 and M_7, coincide with those for one of the other six legs so it is not possible to attach an additional leg without changing the kinematics of the manipulator (e.g. Griffis-Duffy type platform [Husty and Karger, 2000] or when the platform can achieve Borel-Bricard motion, [Borel, 1908], [Bricard, 1906]).

5. Conclusions

A new method based on spatial kinematic mapping has been developed that makes it possible to design a redundant actuated SGP with a maximal forward kinematics solution set. We show that for every symmetrical SGPP an infinite number of legs can be added which have no effect on the forward kinematics of the mechanism. The possible anchor points lie in the plane on a cubic which depends only on the design parameters of the manipulator, and not on the lengths of the six legs. The results obtained in this paper should be useful for design of parallel manipulators when we intend to attach an additional leg to avoid singularities.

6. Acknowledgements

The research work reported here was made possible by Grant No. HI 370/18-2 from the DFG (Deutsche Forschungsgemeinschaft) and by Grant No. P14986 of the Austrian FWF (Fonds zur Förderung der wissenschaftlichen Forschung).

References

Borel, E. (1908). Mémoire sur les déplacements à trajectoires sphériques. In *Mém. Acad. Sci.*, pages 1–128, Paris.

Bricard, M. R. (1906). Mémoire sur les déplacements à trajectoires sphériques. In *Journ. École Polyt.*, pages 1–96, Paris.

Cox, D., Little, J., and O'Shea, D., editors (1992). *Ideals, Varieties, and Algorithms.* Springer Verlag, Berlin, Heidelberg, New York.

Husty, M. L. and Karger, A. (2000). Self-Motions of Griffis-Duffy Type Platforms. In *Proc. of IEEE conference on Robotics and Automation (ICRA 2000)*, pages 7–12, San Francisco, USA.

Husty, M. L., Mielczarek, S., and Hiller, M. (2001). Constructing an Overconstrained Planar 4RPR Manipulator with Maximal Forward Kinematics Solution Set. In *Proc. of the RAAD'01 Robotics in the Alpe-Adria-Danube Region*, Vienna, Austria.

Husty, M. L. (1996). An algorithm for solving the direct kinematic of Stewart-Gough-type platforms. *Mechanism and Machine Theory*, 31(4):365–380.

Karger, A. and Husty, M. L. (1997). Singularities and Self-motions of a Stewart-Gough Platforms. In Angeles, J. and Zakhariev, E., editors, *Proc. NATO-ASI Computational methods in Kinematics*, pages 279–288, Varna, Bulgaria. NATO.

Ming, A. and Higuchi, T. (1994). Study on multiple degree-of-freedom positioning mechanism using wires (part 1) – concept, design and control. *Int. Journal of the Jap. Soc. for Precision Engineering*, 28(2):131–138.

Study, E. (1903). *Geometrie der Dynamen.*

Wampler, C. W. (1996). Forward displacement analysis of general six-in-parallel SPS (Stewart) platform manipulators using soma coordinates. *Mechanism and Machine Theory*, 31(3):331–337.

Weiss, E. A. (1935). *Einführung in die Liniengeometrie und Kinematik.* Teubner Verlag, Leipzig, Berlin, Germany.

SINGULARITIES AND SELF-MOTIONS OF A SPECIAL TYPE OF PLATFORMS

Adolf Karger

Faculty of Mathematics and Physics

Charles University Praha, Czech Republic

karger@karlin.mff.cuni.cz

Abstract In this contribution we discuss singular positions and self-motions of a special class of parallel manipulators (Stewart-Gough platforms) with planar base and platform, where the plane of the platform is in a projective correspondence with the plane of the base. As a special case we obtain results for parallel manipulators for which the projective transformation simplifies to affine correspondence or a similarity or a congruence. We describe all singular positions of such manipulators. We show that such a manipulator is in a singular position iff points of the platform lie on a conic section and then the platform is architecturally singular or the singularity of the position does not depend on the points of the platform and depends only on the projective correspondence. The condition for the projective correspondence, which leads to a singular position is geometrically interpreted. The paper is a substantial generalization of results in [Karger, 2001], where only the similarity case was treated. We also prove a surprizing result that manipulators of this type have practically no self-motions in spite of the fact that they have a large variety of singular positions.

Keywords: Stewart-Gough platform, singular position, self motion.

1. Introduction

In this paper we discuss singularities and self-motions of parallel manipulators (Stewart-Gough type platforms) with planar platform and base such that there is a projective correspondence between the platform and base (both planes are supposed different). Let us denote m_i $(M_i), i = 1, ..., 6$ points of the platform (base), respectively. For simplicity we shall suppose that no three points of the platform as well as of the base lie on a straight line. With this assumption we can construct the unique projective transformation P from the plane α of the platform to the plane β of the base such that $P(m_i) = M_i$ for $i = 1, 2, 3, 4$. We say that there is a projective correspondence between the platform and the base iff moreover $P(m_5) = M_5, P(m_6) = M_6$.

J. Lenarčič and F. Thomas (eds.), Advances in Robot Kinematics, 155–164.

The projective transformation is the most general correspondence which maps lines into lines. Let $\{x, y\}$ and $\{X, Y\}$ be Cartesian coordinates in the plane α and β, respectively. The general projective correspondence P between planes α and β is given by equations

$$X = \frac{p_{11}x + p_{12}y + p_{13}}{p_{31}x + p_{32}y + p_{33}}, Y = \frac{p_{21}x + p_{22}y + p_{23}}{p_{31}x + p_{32}y + p_{33}},$$

where $\det(p_{ij}) \neq 0$. Parallel manipulators for which a special case of the projective transformation exists were already treated. Frequently used special cases are:
The affine correspondence given by equations

$$X = p_{11}x + p_{12}y + p_{13}, Y = p_{21}x + p_{22}y + p_{23},$$

the equiform correspondence (similarity)

$$X = p_{11}x + p_{12}y + p_{13}, Y = -p_{12}x + p_{11}y + p_{23},$$

and the congruence

$$X = \cos(\phi)x + \sin(\phi)y + p_{13}, Y = -\sin(\phi)x + \cos(\phi)y + p_{23}.$$

This means that all properties proved for the general case of the projective correspondence remain valid if we consider one of the mentioned special cases. (compare to [Mayer St.- Onge, Gosselin, 2000]).

2. One theorem from projective geometry

The description of singular positions of parallel manipulators with planar platform and base (and we shall consider only such cases) lies on a theorem which is of a projective nature and therefore we shall present it in its projective version at first.

Let us consider the projective space P^3 of dimension three with homogeneous coordinates $[x, y, z, t]$. Let us suppose that we are given two different planes α and β in P^3 and a projective correspondence P from α to β. Without loosing generality we can suppose that the plane α is the plane $x = 0$, the plane β is the plane $y = 0$. A point $m \in \alpha$ ($M \in \beta$) is given by its coordinates $[0, b, c, d]$($[A, 0, C, D]$), respectively. Projective Plücker coordinates of the line $l = mM$ connecting m with M are subdeterminants of order two of the matrix v of homogeneous coordinates of both points,

$$v = \begin{pmatrix} 0 & b & c & d \\ A & 0 & C & D \end{pmatrix}. \tag{1}$$

We write these Plücker coordinates in a row

$$X = [-Ad,\ bD,\ cD - dC,\ bC,\ -Ac,\ -Ab], \tag{2}$$

where the ordering of those subdeterminats is not important, as we will be interested only in a determinant from them. Let us choose six points

$m_1, ..., m_6$ in α and six points $M_1, ..., M_6$, in β, coordinates of which are labeled with corresponding indices. We obtain six rows $X_1, ..., X_6$ which we arrange in a six by six matrix J. If $\det(J) = 0$, we have a singular position of the corresponding parallel manipulator. Let us suppose moreover that there is a projective correspondence P from α to β. P is described by equations

$$A = p_{11}b + p_{12}c + p_{13}d,\ C = p_{21}b + p_{22}c + p_{23}d,\ D = p_{31}b + p_{32}c + p_{33}d. \quad (3)$$

We substitute these equations into the matrix J and compute its determinant. The expanded form of it has 15840 terms, but it factorizes into 3 factors, $\det(J) = Q_1.Q_2.Q_3$, where Q_1 has 720 terms and $Q_1 = 0$ iff points $m_1, ..., m_6$ lie on a conic section, Q_2 is the determinant of the matrix $p_{ij}, i, j = 1, 2, 3$ and therefore it is not equal to zero and

$$Q_3 = p_{13}(p_{12}p_{33} - p_{13}p_{32}) + p_{12}(p_{12}p_{23} - p_{13}p_{22}). \quad (4)$$

This shows that if points $m_1, ..., m_6$ do not lie on a conic, the fact whether the manipulator is in a singular position depends only on the correspondence P and not on the mutual position of points $m_1, ..., m_6$ in the plane α.

Let us now find out what is the meaning of the equation $Q_3 = 0$. We shall look for conditions under which the correspondence P has a point, which corresponds to itself. For such a point $[0, b, c, d]$ and $[A, 0, C, D]$ must be proportional. This yields:

$$A = p_{11}b + p_{12}c + p_{13}d = 0,\ b = 0,$$

$$C = p_{21}b + p_{22}c + p_{23}d = \rho c,\ D = p_{31}b + p_{32}c + p_{33}d = \rho d,$$

where $\rho \neq 0$ is the proportionality factor. We obtain the following system of linear equations:

$$p_{12}c + p_{13}d = 0, (p_{22} - \rho)c + p_{23}d = 0, p_{32}c + (p_{33} - \rho)d = 0. \quad (5)$$

This is a system of homogeneous linear equations for c and d and it must have a non-trivial solution (as $a = b = 0$). This is possible iff all determinants of degree two from (5) are equal to zero. This yields

$$p_{12}p_{23} - p_{13}(p_{22} - \rho) = 0, p_{12}(p_{33} - \rho) - p_{13}p_{32} = 0$$

$$(p_{22} - \rho)(p_{33} - \rho) - p_{23}p_{32} = 0. \quad (6)$$

It is easy to see that the compatibility condition for (6) is $Q_3 = 0$.

We obtain an interesting special case if $p_{12} = p_{13} = 0$. In that case the line $x = 0, y = 0$ corresponds to itself and the equation $Q_3 = 0$ is always satisfied. The invariant points are then the invariant points of the projectivity induced on this line. We have proved the following

Theorem. Let us have two different planes in the projective space with a projective correspondence between them. Choose six points in one of them and join them with corresponding points in the other plane. Then the connecting lines belong to a linear complex iff either choosen points lie on a conic or the correspondence has a fixed point (and the property is independent of the choice of points).

3. Singular positions of parallel manipulators

Let us formulate the previous theorem in the language of parallel manipulators.

Theorem. Let us suppose that we have a manipulator of the considered type, where $m_1, ..., m_6$ are points of the platform and $M_1, ..., M_6$ are points of the base with the property that there is a projective correspondence P from the plane of the platform to the plane of the base such that $P(m_i) = M_i, i = 1, ..., 6$. Then there are two possibilities:

1) Points $m_1, ..., m_6$ lie on a conic. Then the manipulator is architecturally singular and all positions are singular.

2) Points $m_1, ..., m_6$ do not lie on a conic.Then the singularity of the position does not depend on the distribution of points in the plane of the platform, it depends only on the mutual position of the plane of the platform and of the plane of the base and on the correspondence between them. The position is singular iff one of the legs can be replaced by a leg of zero length or two legs can be replaced by colinear legs.

This yields a complete description of all singular positions of this type of parallel manipulators.

Now we shall discuss some special cases in more detail. We use the Study representation of the displacement group and the notation of [Husty, 1996],[Husty, Karger 2000] or [Karger, 2001] which we now briefly explain.

Let g be a matrix of a spatial displacement,

$$g = \begin{pmatrix} 1 & 0 \\ t & \gamma \end{pmatrix}, \tag{7}$$

where t is the translation vector and $\gamma = (a_{ij}), i, j = 1, 2, 3$ is the orthogonal matrix of the orientation. We parametrize g in the usual way (see for instance [Bottema, Roth, 1990]) by Euler and Study parameters.

Let us suppose that the parallel manipulator is given by six points $M_i = (A_i, B_i, 0)$ in the fixed space (the base of the manipulator) and six points $m_i = (a_i, b_i, 0)$ in the moving space (the platform), $i = 1, ..., 6$, where the moving and fixed spaces have their own Cartesian coordinate systems such that $a_1 = b_1 = b_2 = A_1 = B_1 = B_2 = 0$.

The connection between coordinates in the moving and fixed systems is described by a displacement matrix g from (7) in the following way:

Let $q = [1, a, b, c]^T$ be coordinates of a point in the moving space and its coordinates with respect to the fixed space be $P = [1, x, y, z]^T$. Then $P = g.q$, where g is given by (7). Plücker coordinates X_i of the line m_iM_i are

$$X_i = [t_1 + a_{11}a_i + a_{12}b_i - KA_i, t_2 + (a_{21}a_i - K)A_i + a_{22}b_i - KB_i,$$
$$t_3 + a_{31}a_i + a_{32}b_i, B_i(t_3 + a_{31}a_i + a_{32}b_i), -A_i(t_3 + a_{31}a_i + a_{32}b_i),$$
$$A_i(t_2 + a_{21}a_i + a_{22}b_i) - B_i(t_1 + a_{11}a_i + a_{12}b_i)], \tag{8}$$

where $K \equiv x_0^2 + x_1^2 + x_2^2 + x_3^2 = 1$.

Let the correspondence P between α and β be affine. Then it is given by equations

$$X = [A_2b_3x + (A_3a_2 - a_3A_2)y]/(a_2b_3), \quad Y = B_3y/b_3,$$

where X, Y (x, y) are coordinates in the base (platform), respectively. Substitution shows that the factor Q_3 in $\det(J)$ factorizes into the following two factors

$Q_{31} = (x_0^2x_2^2 - x_1^2x_3^2)(a_3A_2 - a_2A_3) + x_1x_2(x_0^2 + x_3^2)(B_3a_2 - b_3A_2) - x_0x_3(x_1^2 + x_2^2)(B_3a_2 + b_3A_2)$,

which is independent of translations and

$Q_{32} = a_2b_3[t_1(a_{31}a_{22} - a_{32}a_{21}) + t_2(a_{32}a_{11} - a_{12}a_{31}) + t_3(a_{21}a_{12} - a_{11}a_{22})] + K[B_3t_3(a_{11}a_2 - KA_2) - B_3t_1a_2a_{31} - t_2A_2(a_{32}b_3 + a_{31}a_3) + a_2A_3(t_2a_{31} - t_3a_{21}) + t_3A_2(a_{22}b_3 + a_{21}a_3)]$,

where we for simplicity give the expression without substitution of Study parameters. Q_{31} and Q_{32} depend only on the affine correspondence (given by first three points of the platform), Q_{31} is independent of translations. Special cases for similarity or congruence are obtained by substitution (see [Karger, 2001]).

4. Factorization of the Jacobian

As we have seen in **2** for manipulators with a correspondence, the Jacobian factorizes into simpler parts. This is not surprizing, there are more ways how to factorize the Jacobian for parallel manipulators with planar platform and base. The Jacobian J of such a manipulator is of the form $J = \sum_{i=1}^{4} K_iL_i$, (compare to [6]) where

$K_1 =$

$$(A_3a_2 - A_2a_3)q_{456} - (A_4a_2 - A_2a_4)q_{356} + (a_2A_5 - A_2a_5)q_{346} - (A_6a_2 - A_2a_6)q_{345} = 0,$$

$$K_2 = b_3q_{456} - b_4q_{356} + b_5q_{346} - b_6q_{345} = 0,$$

$$K_3 = A_3(a_2 - a_3)q_{456} - A_4(a_2 - a_4)q_{356} + A_5(a_2 - a_5)q_{346} - A_6(a_2 - a_6)q_{345} = 0,$$

$$K_4 = A_3b_3q_{456} - A_4b_4q_{356} + A_5b_5q_{346} - A_6b_6q_{345} = 0, \qquad (9),$$

and $q_{ijk} = B_iB_jB_k[a_i(b_j - b_k) + a_j(b_k - b_i) + a_k(b_i - b_j)]$.

$L_i, i = 1, ..., 4$ are (rather long) expressions in coordinates and Study parameters.

Let us remark that K_i depend only on the architecture of the manipulator. In the case of projective correspondence between platform and base substitution shows that $K_2 = K_4 = 0$ and K_1 and K_3 have a common factor Q_1. There are also other possibilities how to factorize the Jacobian. It is enough to impose three linear constrains on $K_1, ..., K_4$, then one constraint remains and it will be a factor in the Jacobian. Let for instance $K_2 \neq 0$. From equations $K_1 - \lambda_1 K_2 = 0, K_1 - \lambda_3 K_2 = 0, K_4 - \lambda_4 K_2 = 0$, where λ_i are arbitrary numbers, we express A_4, A_5, A_6 and after substitution K_2 will factorize out of the Jacobian. All such factorizations are projectively invariant, which means that the constraints depend only on incidence relations and no length or angle is involved. The geometrical meaning of such factorization remains to be found. Conditions (9) are necessary and sufficient conditions for the manipulator to be architecturally singular. Another interesting example of a parallel manipulator with factorizing Jacobian is the Duffy-type platform, which can be verified by direct computation.

5. Self-motions

We shall discuss the problem of self-motions of the considered class of parallel manipulators. In was shown in [Karger, 2001] that in the case of similar correspondence the manipulator has no self-motions. Let now the correspondence between base and platform be an affinity, which is not a similarity.

During a self-motion of a parallel manipulator all six distances $\rho_i = m_iM_i$ must remain constant. Denoting $h_i = (m_iM_i)^2 - \rho_i^2, i = 1, ..., 6$, we obtain six constrains $h_i = 0$ for the motion of the platform (see [Husty, Karger, 2000] for details). Together with the Study condition $U_1 = x_0y_0 + x_1y_1 + x_2y_2 + x_3y_3 = 0$ we have 7 equations for homogeneous coordinates $(x_j, y_j), j = 0, ..., 3$ of the displacement of the platform.

The manipulator has self-motions iff those equations have infinitely many solutions, not counting the homogeneity factor. Let us look for such solutions. Equations

$$U_1 = 0, \; U_i = h_i - h_1 = 0, i = 2, ..., 6, \qquad (10)$$

are treated as a system of 6 linear equations (not homogeneous) in 4 unknowns $y_0, ..., y_3$. The solvability of such a system depends on the rank of the matrix M of (10). In our case M has rank at most 3, let it be 3. From $U_1 = 0, U_2 = 0, U_3 = 0$ we obtain expressions for three of $y_j, j = 0, ..., 3$ and the last one is determined by the equation $h_1 = 0$, which is quadratic with nonzero quadratic term.This means that equations $U_4 = 0, U_5 = 0, U_6 = 0$ must have a solution with non-constant x_j. They are of the form $U_i = f_i x_0^2 + g_i x_3^2 + k_i x_0 x_3 + T_i$, where $i = 4, 5, 6$ and T_i is quadratical in x_1, x_2.

Let us consider these equations as linear in unknowns $x_0^2 = X, x_3^2 = Y, x_0 x_3 = Z$, let the determinant D of this system be nonzero. We express X, Y, Z and write the numerator of $XY - Z^2$, which is a homogeneous polynomial of degree four in x_1, x_2, and it must disappear. The difference of coefficients of x_1^4 and x_2^4 has 22193 terms, but it factorizes into $Q_1 D(A_2 b_3 - a_2 B_3)$, which yields $A_2 = ka_2, B_3 = kb_3, k \neq 0$. Substitution shows that the sum of coefficients of $x_1^3 x_2$ and $x_2^3 x_1$ yields $A_3 = ka_3$, a contradiction.

Let now $D = 0$. Determinants of coefficients of previously considered equations, $H_1 = \det |f_i, g_i, T_i|$, $H_2 = \det |k_i, g_i, T_i|$ must be equal to zero.

The sum of the coefficient of x_1^2 in H_2 and of x_2^2 in H_1 yields $A_2 = ka_2, A_3 = ka_3$, substitution yields $A_3 = ka_3$, a contradiction.The case of congruent platform and base is treated in the same way as the similarity case. This yields a partial result:

If the platform and base are affinely equivalent and the rank of M is equal to three, the platform has no self-motions.

Now we shall treat the remaining case, where rank of M is equal to two. M is the matrix which has in the first row elements (x_0, x_1, x_2, x_3) and in the i-th row ($i = 2, ..., 6$) elements $(a_i - A_i)x_1 + (b_i - B_i)x_2, (A_i - a_i)x_0 + (b_i + B_i)x_3, -(a_i + A_i)x_3 + (B_3 - b_i)x_0, (a_i + A_i)x_2 - (b_i + B_i)x_1)$.

We shall consider only first three rows of the matrix M, as the remaining ones do not bring anything new. The rank of M is equal to two iff all four subdeterminants of order three are equal to zero, let V_i be the subdeterminat obtained by leaving out the i-th column. For instance
$V_4 = x_0(x_0^2 + x_1^2 + x_2^2)(A_2 - a_2)(B_3 - b_3) + x_0 x_3^2(A_2 + a_2)(B_3 + b_3) - 2x_3(x_0^2 + x_1^2)(A_2 a_3 - a_2 A_3) + 2x_1 x_2 x_3(a_2 B_3 - A_2 b_3)$.
Expressions for remaining V_i are similar and they are omitted. To solve these equations we compute resultants of any two of them with respect to x_0 and obtain 6 equations. These equations have two common factors, let us denote them G_1 and G_2 and it is a matter of detailed discussion to show that at least one of them must be zero, we leave this discussion

out as not interesting. The first factor is

$$G_1 = (A_2 - a_2)(B_3 + b_3)x_1^2 + 2px_1x_2(a_2 + A_2)(B_3 - b_3)x_2^2,$$

where $p = A_2a_3 - A_3a_2$, the second factor is more complex, it is

$$G_2 = x_3^4(p^2 + q^2) + x_3^2[x_1^2q^2 + x_2^2(q^2 + 2p^2) + 2x_1x_2pr] + x_2^2(x_1r + x_2p)^2,$$

where $q = B_3a_2 + b_3A_2$, $r = a_2B_3 - A_2b_3$. The equation $G_2 = 0$ represents a curve of degree four in homogeneous coordinates x_1, x_2, x_3. In what follows we show, that all its real points lie on the curve given by equation $G_1 = 0$. We proceed in the following way: Let us write $A_2 = ka_2$, where k is some unknown factor. $G_2 = 0$ is then a quadratic equation in parameter k and its discriminant must be nonnegative to obtain real solutions. Its discriminant is $D = -x_3^2(x_1^2 + x_2^3 + x_3^2)D_1^2$, where

$$D_1 = 2B_3b_3x_1x_2 - (A_3b_3 + a_3B_3)(x_2^2 + x_3^2).$$

This means that we must have $D = 0$, and it is easy to show that $x_3 = 0$ yields $G_1 = 0$. Therefore we have $D_1 = 0$, and similarly we show that $x_2 = 0$ also yields $G_1 = 0$. If $x_2 \neq 0$, we express x_1 from $D_1 = 0$ and substitute the result into G_2, we obtain

$$G_2 = (kb_3 + B_3)^2[(A_3b_3 - a_3B_3)^2x_2^4 + (A_3b_3 + a_3B_3)^2 + 2(A_3^2b_3^2 + B_3a_3^2 + 2b_3^2B_3^3)x_2^2x_3^2],$$

which is always positive unless $B_3 = -kb_3$. Substitution follows $k = -1, A_3 = -a_3$ and $G_1 = 0$.

This shows that we can obtain self-motions only if $G_1 = 0$. This is a quadratical equation for the ratio x_1/x_2, and it is identically satisfied iff the platform and base are congruent or $x_1 = x_2 = 0$. We shall treat this cases later. Let now the platform and base be not congruent. We can suppose that they are also not equiform, because in that case $G_1 = (k^2 - 1)(x_1^2 + x_2^2)$. The discriminant of G_1 is $d = (A_2a_3 - A_3a_2)^2 - (B_3^2 - b_3^2)(A_2^2 - a_2^2)$. If $d \geq 0$, we obtain real translations, which depend on one parameter, as only one equation remains for x_0, x_3. We illustrate the situation by a simple example:

Let us choose $m_2 = [1,0], m_3 = [5,-4], m_4 = [2,-6], m_5 = [-4,-6]$, $m_6 = [-4,-3]$, $M_2 = [2,0], M_3 = [2,2]$. Then $G_1 = (x_1 + 9x_2)(x_1 - x_2) = 0$, we choose $x_1 = -9x_2$. From $V_5 = x_2(x_0 - 3x_3)^2 = 0$ we obtain $x_0 = 3x_3$, as the case $x_2 = 0$ will be discussed later.The norming condition is $82x_2^2 + 10x_3^2 = 1$, which shows that we have a one-parametric system of orientations we can choose to obtain a translatory self-motion. We choose $x_2 = 1/10$, then $x_3 = 3\sqrt{5}/50$. The matrix of the orientation is

$$\gamma = \frac{1}{125}\begin{pmatrix} 118 & -36 & -9\sqrt{5} \\ -9 & -82 & 42\sqrt{5} \\ -18\sqrt{5} & -39\sqrt{5} & -80 \end{pmatrix}.$$

Further we choose $R_1 = R_2 = -6$, we obtain $R_3 = 15146/375$. The translation factor of the motion can be parametrized as

$$t_1 = -10384/6375 + \kappa(225\cos(\phi) - 81\sqrt{17}\sin(\phi)),$$

$$t_2 = -236/2125 + \kappa(-1250\cos(\phi) - 122\sqrt{17}\sin(\phi))$$

$$t_3 = -472\sqrt{5}/2125 + \kappa\sqrt{5}(-205\cos(\phi) + 131\sqrt{17}\sin(\phi)),$$

where $\kappa = \sqrt{12667083}/2734875$, see figures below, where we show one position of the manipulator in four different views.

It remains to treat the case $x_1 = x_2 = 0$. In that case we obtain one equation for x_0, x_3, namely $G_3 = (B_3 - b_3)(A_2 - a_2)x_0^2 + 2px_0x_3 + (B_3 + b_3)(A_2 + a_2)x_3^2 = 0$, which is similar to $G_1 = 0$ and it has the same discriminant. (In the similarity case we have specially $G_3 = (k-1)^2x_0^2+(k+1)^2x_3^2$.) This shows that we obtain two more translatorial self-motions where the platform is rotated around the z-axis - the expected trivial case.

For congruent platform and base we obtain a special case of the general result. The rank of M is equal to two iff $x_3 = 0$. If that is true we obtain a translatory self-motion for any position of the platform obtained by a rotation around a horizontal axis. If $x_0 = 1, x_1 = x_2 = x_3 = 0$, we obtain a two-parametric translation, as known. This covers all possible cases.

6. Conclusions

We have explicitly described all singular positions of parallel manipulators with planar platform and base with a projective corespondence between platform and base. We have shown that there are only two possibilities:

1. Points of the platform lie on a conic section and the manipulator is architecturally singular and always movable.

2. Points of the platform do not lie on a conic section. In that case the singularity of a given position depends only on the projective correspondence between platform and base and not on points of the platform. The manipulator is at singular position iff the correspondence has an invariant point or an invariant line (one leg can be replaced by a leg of a zero length or two legs can be replaced by colinear legs).

We have found all self-motions of parallel manipulators with an affine correspondence between platform and base. Such manipulators can have only translatory self-motions, in general we have one-parametric systems of one-parametric self-motions, in the congruent case we have additionally a two-parametric system of translations.

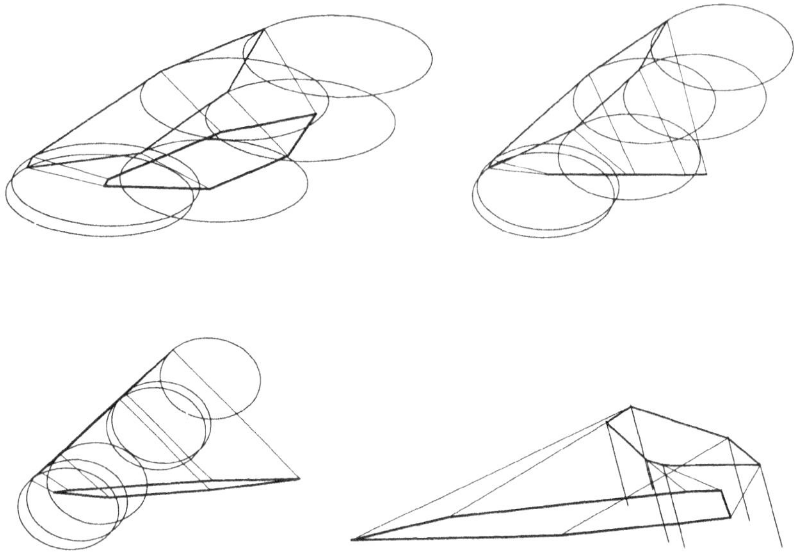

7. Acknowledgements

The research of this contribution was supported by the Grant nr. 201/02/0616 of the Czech Grant Agency and partially by grant MSM113200007 of the Ministery of Education of the Czech Republic

References

Bottema, O., Roth, B: Theoretical Kinematics, Dover Publishing, 1990.

Husty, M. L.: An Algorithm for Solving the Direct Kinematics of General Stewart-Gough Platforms. Mech. Mach. Theory 31, 4(1996), 365-380.

Husty M. L., Karger A.:Architecture Singular Parallel Manipulators and their Self-Motions. In J. Lernarčič and M. M. Stanišič, Advances in Robot Kinematics, Kluwer Acad. Publ. 2000, ISBN 0-7923-6426-0, pp.355-364.

Karger A.: Singularities and self-motions of equiform platforms. MMT **36**, 801-815, 2001.

Ma O.,Angeles J.: Architecture Singularities of Parallel Manipulators. Int. Journal of Robotics and Automation 7, 23-29, 1992.

Mayer St-Onge B., Gosselin C. M.: Singularity Analysis and Representation of the General Gough-Stewart Platform. The International Journal of Robotics Research, **19** , No. 3, 2000, pp. 271-288.

Ma O., Angeles J.: Architecture Singularities of Parallel Manipulators. Int. Journal of Robotics and Automation 7, 23-29, 1992.

Merlet J.P.: Singular Configurations of Parallel Manipulators and Grasssmann Geometry. Int. Journal of Robotics Research, 8, 45-56, 1992.

INVESTIGATION OF SINGULARITIES AND SELF-MOTIONS OF THE 3-UPU ROBOT

A. Wolf and M. Shoham
Robotics Laboratory
Department of Mechanical Engineering
Technion - Israel Institute of Technology
http://robotics.technion.ac.il
[alonw,shoham]@tx.technion.ac.il

F.C. Park
School of Mechanical and Aerospace Engineering
Seoul National University
http://robotics.snu.ac.kr
fcp@plaza.snu.ac.kr

Abstract We investigate the singular configurations and self-motion phenomena of the 3-UPU robot. Drawing principally upon tools from line geometry and screw theory, we first derive, as a set of six lines, the 6X6 transformation matrix mapping external wrenches acting on the moving platform to the internal forces/moments of the moving platform's joints. The closest linear complex to these six governing lines is then obtained. The linear complex's axis and pitch not only provide additional information and understanding on the type and location of the singularities, but also on the nature of any instantaneous motions arising from manufacturing tolerances and low rigidity. It is found that at the home position of certain 3-UPU architectures, the corresponding lines are contained in two zero-pitch linear complexes, causing an instantaneous two-parameter rotational motion about a line pencil at the intersection point of the 3-UPU's limbs. In the vicinity of the home position the lines are contained in two close linear complexes, offering a theoretical explanation of the observed sensitivity of this mechanism to manufacturing tolerances.

Keywords: Parallel robot, 3-UPU, singularity, self motion, line geometry.

1. Introduction

The three DOF 3-UPU parallel mechanism, first introduced in the literature by Tsai (1996), has recently been the subject of kinematic analysis by several researchers (Di Gregorio and Parenti-Castelli 1998,

J. Lenarčič and F. Thomas (eds.), Advances in Robot Kinematics, 165–174.

1999, Tsai and Joshi 2000). Yet, it was only at the 2001 Computational Kinematics Workshop in Seoul that F.C. Park demonstrated with a working prototype what looked like a singular or a self motion behavior of the 3-UPU robot. Di Gregorio and Parenti-Castelli (1998) had predicted a limited form of this behavior of the mechanism. The authors pointed out that while in some singular configurations the 3-UPU mechanism has, at least locally, more than three DOF (Degrees of Freedom), meaning that though designed to have three pure translational DOF, the mechanism gains extra uncontrollable rotational DOFs. Yet, it was apparent from the hardware demonstrations that the self-motions were in fact global in nature, occurring over a continuous and arbitrary range of leg lengths. The velocity equation-based analysis presented in Di Gregorio and Parenti-Castelli (1999) predicted the existence of two groups of singularities, termed translational and rotational, for the general 3-UPU mechanism, but as before this was limited to a finite set of configurations. Zlatanov, Bonev, and Gosselin (2001) also verify that the demonstrated 3-UPU architecture is in a singular configuration at the home position.

The present investigation suggests a different approach to the singularity analysis of parallel robots, which not only determines the singular configurations, but also predicts the mechanism behavior near these configurations. Our approach utilizes line geometry to obtain a 6X6 matrix of the robot's governing lines, and then applies an algorithm presented by Pottmann et al. (1999) to find the closest linear complex to this set of six lines, in a given position of the robot. When the six lines are contained in a linear complex, the robot is in a singular configuration and the possible self motion is determined by the linear complex's direction and pitch. The lines may be contained in two or three different linear complexes; in this case the self motions are respectively defined by a two or three parameter motion. Also, sensitivity to manufacturing tolerances can be evaluated by the closeness of the linear complex to the set of lines. This method is demonstrated for the 3-UPU robot.

2. Linear Complex Approximation Algorithm

Given k lines $L_i = (l_i, \bar{l}_i)$, $i = 1, \ldots, k$ $(k \geq 6)$ by their Plücker coordinates, the algorithm finds among all linear complexes the closest one, denoted by C, to this given set of lines L_i. These linear complexes will be denoted by $X = (x, \bar{x}) \in \Re^6$. According to Klein (1921), the moment of a line L with respect to a linear complex X is given by

$$m(L, X) = \frac{|\bar{x} \cdot l + x \cdot \bar{l}|}{\|x\|} \tag{1}$$

Hence, one can compute C as a minimum of

$$\sum_{i=1}^{k} m(L_i, X)^2 \tag{2}$$

among all linear complexes X, given by $X = (x, \bar{x}) \in \Re^6$. This is equivalent to minimizing the positive semidefinite quadratic form

$$F(X) = \sum_{i=1}^{k} (\bar{x} \cdot l_i + x \cdot \bar{l}_i)^2 = X^T M X \tag{3}$$

under the normalization condition $1 = \|x\|^2 = X^T D X$, where $D = \mathrm{diag}(1,1,1,0,0,0) = \Gamma$, and M is the Gramian matrix. Defining L in axis coordinates as $L_{A,i} = (\bar{l}_{a,i}, l_{a,i})$, M is then the Gramian matrix defined as

$$M = \sum_{i=1}^{k} L_{Aj} \cdot L_{Aj}^T \tag{4}$$

The solution of (3) is a general eigenvalue problem. Using Lagrange multipliers λ, one obtains (where λ are the eigenvalues and X the eigenvectors)

$$(M - \lambda D)X = 0, \quad X^T D X = 1 \tag{5}$$

Hence, λ is the root of the equation

$$\det(M - \lambda D) = 0 \tag{6}$$

For any root λ and corresponding eigenvector $X = (x, \bar{x})$ we have

$$F(X) = X^T M X = \lambda X^T D X = \lambda \tag{7}$$

where all roots are nonnegative and the solution for the linear complex, C, is an eigenvector corresponding to the smallest eigenvalue λ. Given the lines $L_{A,i} = (\bar{l}_{a,i}, l_{a,i})$ and the linear complex C, the standard deviation of the lines from C is given by $\sigma = \sqrt{\lambda/(k-5)}$. Moreover, given the closest linear complex C, its axis A and pitch p are given by $(a, \bar{a}) = (c, \bar{c} - p \cdot c)$ and $p = c \cdot \bar{c}/c^2$, respectively.

When solving Equation 6 two small eigenvalues may appear, meaning that there are two nearly equally good solutions for C, (C_1, C_2). Hence the lines $L_{A,i}$ can be well approximated by the lines of the intersection of the two linear complexes $C_1 \cap C_2$ (this is a two-parameter family of lines—a linear congruence). Analogously, three small eigenvalues $\lambda_1, \lambda_2, \lambda_3$ define three linear complexes (a bundle of complexes). The intersection forms a one-parameter family of lines such as a regulus, a pair of lines, a union of lines or a whole plane.

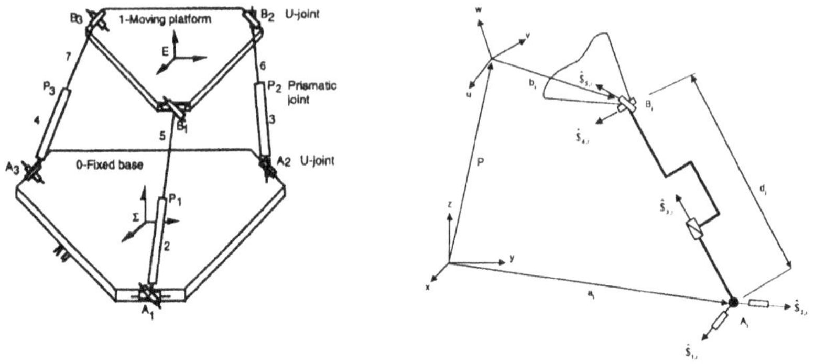

(a) 3-UPU robot (Tsai 1996). (b) Equivalent limb structure.

Figure 1. The 3-UPU robot and its equivalent kinematic structure.

3. Application to the 3-UPU Robot

It has been shown by Hunt (1978), Merlet (1992), and Husty and Karger (1997), that for a parallel robot with $k \geq 6$ (k is the number of lines describing the limbs of the robot), the robot position is singular if and only if the axes of the limbs lie in a linear complex. For the Stewart-Gough platform, the Jacobian matrix of the manipulator is composed of the lines along its limbs (Ciblak and Lipkin 1999, Tsai *et al* 2000). When a wrench is applied along L_i, $i = 1, \ldots, 6$, then the smallest λ_i produces the least amount of power for a given instantaneous motion. When the reciprocal product is zero, then there is no power generated by the wrenches on the respective twist axis. We use these facts to investigate the 3-UPU parallel robot's structure to gain a better understanding of the robot's singularity and self motion.

Consider the 3-UPU robot in Figure 1-a. The Jacobian matrix of the manipulator is composed of the three screws along its limbs. These screws are the reciprocal screws to all the unactuated joint screws in each limb (Tsai 1998, 2000). In order to define the Jacobian matrix, it is necessary to describe all five joint screws of the manipulator. Note that the only actuated joint in each limb is the third one. Let $s_{i,j}$ be a unit vector along the j-th joint axis of the ith limb. Using this, one can denote the five unit screws of each limb as (see Figure 1-b)

$$\hat{\$}_{1,i} = \begin{bmatrix} \hat{s}_{1,i} \\ (b_i - d_i) \times \hat{s}_{1,i} \end{bmatrix}, \quad \hat{\$}_{2,i} = \begin{bmatrix} \hat{s}_{2,i} \\ (b_i - d_i) \times \hat{s}_{2,i} \end{bmatrix}, \quad \hat{\$}_{3,i} = \begin{bmatrix} 0 \\ \hat{s}_{3,i} \end{bmatrix},$$

$$\hat{\$}_{4,i} = \begin{bmatrix} \hat{s}_{4,i} \\ b_i \times \hat{s}_{4,i} \end{bmatrix}, \ \hat{\$}_{5,i} = \begin{bmatrix} \hat{s}_{5,i} \\ b_i \times \hat{s}_{5,i} \end{bmatrix} \tag{8}$$

where $b_i = PB_i$, $d_i = A_iB_i = d_i s_{3,i}$, and $\hat{\$}_{3,i}$ is a prismatic joint with infinite pitch. When expressing the instantaneous twist of the moving platform in terms ofthe motion screws and regarding each limb and the platform as an open-loop chain, one obtains (Tsai 1998)

$$\$_p = \dot{\theta}_{1,j}\hat{\$}_{1,j} + \dot{\theta}_{2,j}\hat{\$}_{2,j} + \dot{\theta}_{3,j}\hat{\$}_{3,j} + \dot{\theta}_{4,j}\hat{\$}_{4,j} + \dot{\theta}_{5,j}\hat{\$}_{5,j} \tag{9}$$

Since the axis of all passive joints in each limb intersects the line passing through points A_i, B_i, a wrench that is reciprocal to all the motion screws is given by

$$\$_{r3,i} = \begin{bmatrix} s_{3,i} \\ b_i \times s_{3,i} \end{bmatrix} \tag{10}$$

Taking the reciprocal product $\Omega(\cdot,\cdot)$ of both sides of $\$_p$ with $\$_{r3,i}$ one obtains

$$\Omega(\$_{r3,i}\$_p) = \dot{d}_i, \quad i = 1, 2, 3 \tag{11}$$

Writing this for each limb one gets $J_x\dot{x} = J_q\dot{q}$, where

$$J_x = \begin{bmatrix} (b_1 \times s_{3,1})^T & s_{3,1}^T \\ (b_2 \times s_{3,2})^T & s_{3,2}^T \\ (b_3 \times s_{3,3})^T & s_{3,3}^T \end{bmatrix} \tag{12}$$

with J_q the 3×3 identity matrix, $\dot{x} = (\omega_x, \omega_y, \omega_z, v_{px}, v_{py}, v_{pz})$, and $\dot{q} = (\dot{d}_1, \dot{d}_2, \dot{d}_3)$. Taking $\dot{x}$ to be the velocity of a point p on the moving platform, and $\dot{q}$ as the vector of actuator joint rates, one can define the relation using J_x.

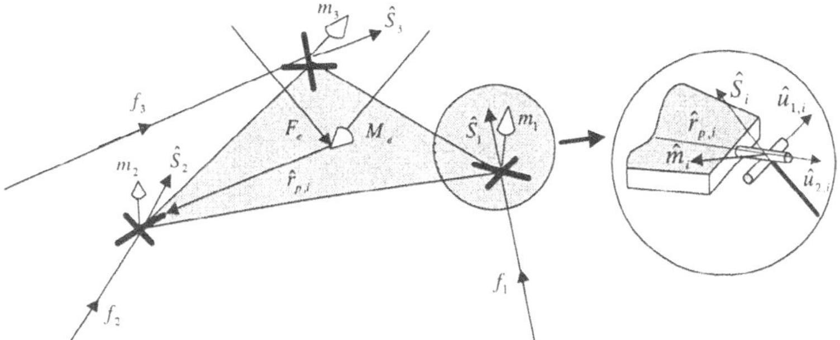

Figure 2. Force and moment transmitted to the moving platform.

The above scheme provides a 3×6 matrix. However, in order to obtain the full 6×6 matrix that maps the external wrench to internal joints' reactions, the static equilibrium of forces and moments about the center of the moving platform has to be derived. These static equilibrium equations are given by (see Figure 2 for definitions)

$$\sum_{i=1}^{3} f_i \hat{s}_i - F_e = 0 \tag{13}$$

$$\sum_{i=1}^{3} m_i \hat{u}_i + \sum_{i=1}^{3} {}^{w}R_P \hat{r}_{p,i} \times f_i \cdot \hat{s}_i - M_e = 0 \tag{14}$$

where $\hat{u}_i$ is a unit vector normal to the two axes of the upper U joint of the limb i, and ${}^{w}R_p$ is a rotation matrix from the platform coordinate system to the reference coordinate system.

Observing Figure 2, $\hat{u}_{2,i}$ is a unit vector along $\hat{r}_p$ (in platform coordinates), $\hat{u}_{1,i}$ is a unit vector along the second pair of the U joint (connected to limb i), and $\hat{s}_i$ is a unit vector along limb i. Due to the U joint structure, $\hat{u}_{1,i}$ is perpendicular to both $\hat{u}_{2,i}$ and $\hat{s}_i$, hence

$$\hat{u}_{1,i} = \frac{\hat{u}_{2,i} \times \hat{s}_i}{|\hat{u}_{2,i} \times \hat{s}_i|} \tag{15}$$

Substituting $\hat{u}_{2,i} =^{w} R_p \hat{r}_{p,i}$ yields

$$\hat{u}_{1,i} =^{w} R_p \hat{r}_{p,i} \times \hat{s}_i \tag{16}$$

Defining

$$\hat{u}_i = \hat{u}_{2,i} \times \hat{u}_{1,i} \tag{17}$$

and substituting (15) and (16) into (17) yields

$$\hat{u}_i =^{w} R_p \hat{r}_{p,i} \times \left(\frac{{}^{w}R_p \hat{r}_{p,i} \times \hat{s}_i}{|{}^{w}R_p \hat{r}_{p,i} \times \hat{s}_i|} \right) \tag{18}$$

The matrix representation of Equations (13) and 14) is given by $GY = B$, where

$$G = \begin{bmatrix} \hat{s}_1 & \hat{s}_2 & \hat{s}_3 & 0 & 0 & 0 \\ {}^{w}R_p \hat{r}_{p,1} \times \hat{s}_1 & {}^{w}R_p \hat{r}_{p,2} \times \hat{s}_2 & {}^{w}R_p \hat{r}_{p,3} \times \hat{s}_3 & \hat{u}_1 & \hat{u}_2 & \hat{u}_3 \end{bmatrix} \tag{19}$$

with $Y = (F_e, M_e) \in \Re^6$ and $B = (f_1, f_2, f_3, m_1, m_2, m_3) \in \Re^6$. Defining $J = G^{-1}$, the forces at the robot's joints are then given by $JY = B$; this result is corroborated by the work of Ciblak and Lipkin (1999), where they developed a model of a rigid body connected to the ground by springs. For the 3-UPU robot, the model consists of three linera springs and three torsional springs in parallel, which resembles J. Observing J, one can see that its columns correspond to lines lying on the *Klein quadric* M_2^4, as they satisfy Klein's equation (Klein 1871, Hunt 1978)

$$p_{01}p_{23} + p_{02}p_{31} + p_{03}p_{12} = 0 \tag{20}$$

3.1 Simulation Results

We now simulate the 3-UPU mechanism using the linear complex approximation algorithm (LCAA) presented above. In the given simulation, the radii of the base and moving platform are $r_b = 25.9806cm$ and $r_p = 20.2072cm$, respectively (as in the real model of Park). The limbs of the robot are equally divided every 120°, both in the base and in the moving platform. The location of the mid moving platform while in the home position is ${}^wP_p = (0, 0, 50)$.

The three axes of the linear complexes as found by the algorithm are:

$$\begin{aligned} A_1 &= (-0, 9382, 0, 346, 0, -34.5987, -93.8239, 0) \\ A_2 &= (0.346, 0.9382, 0, -93.8239, 34.5987, 0) \\ A_3 &= (0, 0, 1, 0, 0, 0) \end{aligned}$$

where $\lambda_1 = \lambda_2 = \lambda_3 = 0$. Using (refeqn:Sr3i) one obtains $\sigma_1 = \sigma_2 = 0$ and $\sigma_3 = 1.7321$, and the corresponding pitches are all zero.

3.2 Discussion

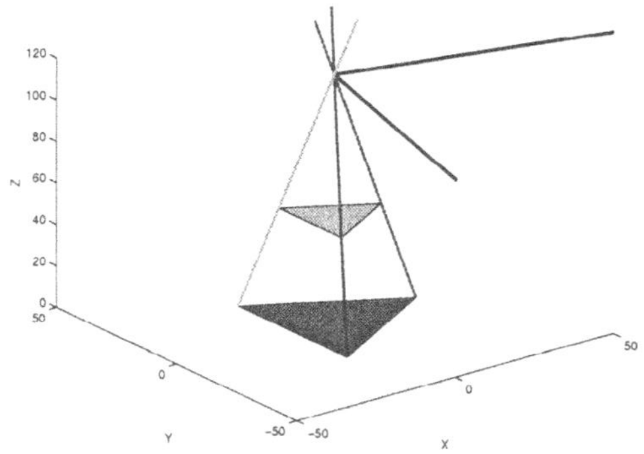

Figure 3. 3-UPU: two zero-pitch linear complexes.

Applying the method of the closest linear complex on the 3-UPU robot in its zero position $P = (0, 0, p_z)$, the lines of J are contained in two zero-pitch linear complexes C_1 and C_2, passing at the intersection point of the extension of the three limbs of the actuator (see the black lines in Figure 3). The intersection of $C_1 \cap C_2$ defines a two-parameter family of lines: a congruence. This result implies that the robot gains an instantaneous two-parameter rotational motion about any horizontal axes passing through the intersection point of the robot's limbs. These axes are the linear combination of the two zero-pitch screws as they define a plane pencil whose vertex is the intersection point of the limbs.

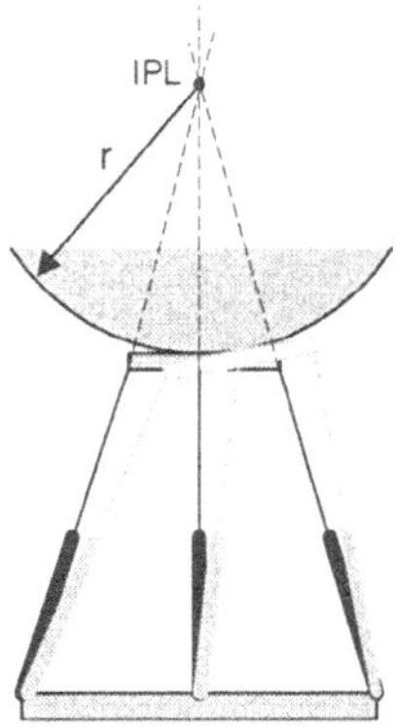

Figure 4. Platform at zero position and at points on a sphere centered at the IPL.

When the robot moves in the vicinity of its base configuration (Figure 4), then it is no longer in a singular configuration. Observing equations (7), one can see that λ has the meaning of the sum of squares of the mutual moment of the lines L_i with respect to C. Hence, σ can be interpreted as the sum of the squares of errors from L_i to C. Moreover, observing the definition of the pitch of the closest linear complex, one can observe that p has a meaning of the projection of the moment of C on a unit screw along C. Hence, the moment m, the STD σ, and the pitch p are distances in Euclidean geometry. Therefore, all results ofthe LCAA must be analyzed relative to the object dimension and the error tolerances utilized during manufacturing. This means that even if for nonsingular configuration, for low values of σ (smaller than error tolerances), the robot may still rotate around the intersection point of the limbs (IPL) uncontrollably.

The following simulation presents motion of the robot on a sphere centered at the original intersection of the limbs, and at each point the LCAA has been applied. It can be seen that the robot is no longer in a singular position. However, there are still two linear complexes with low values of σ. According to measurements taken on Park's prototype, the robot had a self-motion within a radius of approximately 14 cm from the

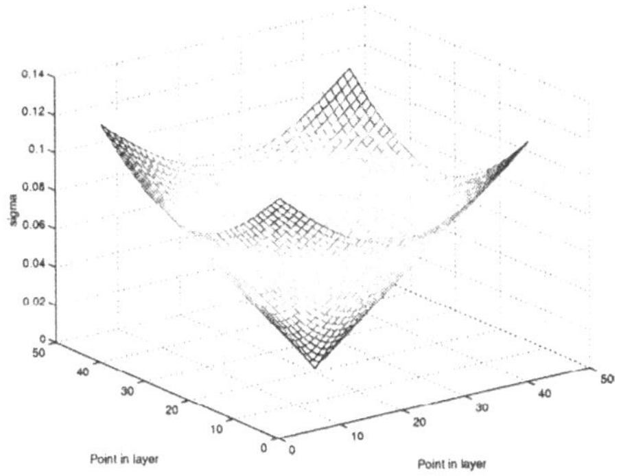

Figure 5. The two minimum σ (rigidity) values corresponding to the two LC axes of Figures 3, 4.

home position. According to the given simulation (Figure 5), σ value within this region attain a maximum value of 0.05 and a minimum of 0 (in centimeters). These results, together with the distance meanings of λ, σ, and p, point to a possibility of an uncontrolled motion of the platform as was observed.

4. Conclusions

In this investigation, the self-motions of the 3-UPU manipulator as was presented by F.C. Park at Computational Kinematics 2001 in Seoul is analyzed. Investigation of the 6×6 Jacobian matrix J of the 3-UPU manipulator in its home position reveals that the matrix is singular. Moreover, by using the linear complex approximation method it has been shown that the lines of J are contained in two axes of two zero-pitch linear complexes (linear congruence). This result implies that the robot gains an instantaneous mobility of two-parameter motion. This motion is a pure rotation about any screw axis, which belongs to the flat pencil, defined by the two axes of the linear complexes. Moreover, when the robot moves along a sphere centered at the initial limb intersection point, it is no longer at a singular point, yet J is still close to two linear complexes with low values of σ. This might still allow for two uncontrolled motions of the moving platform due to manufacturing tolerances or low rigidity, which points to the sensitivity of this mechanism design. In a related paper (Han *et al* 2002), the contribution of the universal joints' torsional clearance to self-motions is analyzed. We believe a comprehensive analytical description of the self-motion phenomenon still remains a topic of further investigation.

References

Han, C.H., Kim, Jinwook, Kim, Jongwon, and Park, F.C. (2002), Kinematic sensitivity analysis of the 3-UPU parallel mechanism, *Mechanism and Machine Theory* (to appear).

Ciblak N., and Lipkin, H. (1999), Synthesis of cartesian stiffness for robotic applications, *Proc. 1999 IEEE Int. Conf. Robotics and Automation*, Detroit, Michigan., pp. 2147-2152.

Di Gregorio, R. and Parenti-Castelli, V. (1998), A translational 3-DOF parallel manipulator, *Advances in Robot Kinematics: Analysis and Control*, J. Lenarcic and M. L. Husty (eds), Dordrecht, Kluwer, pp. 49-58.

Di Gregorio, R. and Parenti-Castelli, V. (1999), Mobility analysis of the 3-UPU parallel mechanism assembled for a pure translational motion, *Proc. IEEE/ASME Int. Conf. Advanced Intelligent Mechatronics*, Atlanta, GA, Sept. 19-23, pp. 520-525.

Hunt, K.H (1983), Structural kinematics of in-parallel actuated robot arms, *ASME J. Mechanisms, Transmissions and Automation in Design*, no. 4, vol. 105, pp. 705-712.

Karger, A., and Husty, M. (1996), On self-motions of a class of parallel manipulators, *Advances in Robot Kinematics*, J. Lenarcic and V. Parenti-Castelli (eds), Dordrecht, Kluwer, pp. 339-348.

Klein F. (1921), *Die Allgemeine Lineare Transformation der Linienkoordinaten*, Gesammelte math. Abhanlungen I, Berlin, Springer.

Merlet, J.P. (1992), Singular configurations of parallel manipulators and grassmann geometry, *Int. J. Robotics Research*, no. 5, vol. 8, pp. 45-56.

Pottmann, H., Peternell, M., and Ravani, B. (1999), An introduction to line geometry with applications, *Computer Aided Design*, vol. 31, pp. 3-16.

Pottmann H., Wallner, J. (2000), *Computational Line Geometry*, Springer, Berlin.

Tsai, L-W. (1996), Kinematics of a three-DOF platform with three extensible limbs, *Recent Advances in Robot Kinematics,* J. Lenarcic and V. Parenti-Castelli (eds.), Dordrecht, Kluwer, pp. 401-410.

Tsai L-W. (1998), The Jacobian analysis of a parallel manipulator using reciprocal screws, *Advances in Robot Kinematics: Analysis and Control,* J. Lenarcic, and M.L. Husty, (eds), Dordrecht, Kluwer, pp. 327-336.

Tsai, L-W., and Joshi, S. (2000), Kinematics and optimization of a spatial 3-UPU parallel manipulator, *ASME J. Mechanical Design*, pp. 439-446.

Zlatanov D., Bonev, L., and Gosselin, C. (2001), Constraint Singularities, *NewsLetter on: http://www.parallemic.org*, July.

ON CLOSURE MODES AND SINGULAR CONFIGURATIONS OF KINEMATIC CHAINS

J.E. Baker
School of Mechanical and Manufacturing Engineering
The University of New South Wales, NSW 2052, Australia
jebaker@it.usyd.edu.au

Abstract Certain singular configurations of linkages characterised by inactive freedoms or increased mobility are well documented. Less widely known is the fact that the closure modes of a kinematic chain can give rise to various degrees of mobility. These issues are sometimes interwoven as a consequence of simultaneities of special positions and accompanying coincidences of different closure modes. The phenomena are here discussed and placed in perspective, and a set of recommendations is put forward in the interest of obviating misconceptions in dealing with these important exceptional features of kinematic chains.

Keywords: closure modes, singular configuration, kinematic chain

1. Background

The notions of stationary and "uncertainty" configurations of linkages are formally introduced by (Hunt, 1978), who explains the concepts in detail and generalises them by defining the incidence of a multiple state of each. Stationarity is to be generally anticipated in a linkage of no special proportions and is typically (although not necessarily) accompanied by a reversal of relative motion at some articulation following an extremum there of a joint variable. The essence of the phenomenon is that portion of the chain is instantaneously still while the remainder continues to move. As presented by Hunt, an uncertainty configuration arises from special dimensional properties of the linkage whereby its mobility overall is instantaneously heightened, a state capable of being succeeded by one of a selection of closure modes of the chain under some circumstances. The classic example of such an instance is that which applies to the parallelogram/antiparallelogram planar loop at an in-line position.

Sometimes these two types of special configuration appear to merge, such as for that in-line position of a kite-shaped planar loop which can be followed by a degenerate and permanent folding of its members about two coincident revolutes. In his discussion Hunt pays particular atten-

J. Lenarčič and F. Thomas (eds.), Advances in Robot Kinematics, 175–182.

tion to the matter of joint actuation for such special configurations, rather than to any issue of closure modes which might arise. He acknowledges the possible incompleteness of his deliberations but proffers the conviction that screw system theory could be deployed to investigate other kinds of special configuration which might exist. General analytical techniques for determining stationary positions are put forward by (Duffy and Gilmartin, 1969), (Gilmartin and Duffy, 1969), (Baker and Waldron, 1974) and (Baker, 1978, 1979, 1980); some special-purpose methods have also been developed. (Baker, 1980) includes a treatment of uncertainty positions.

Now, although Hunt's definition of an uncertainty configuration is such as to be applicable to ordinary constrained spatial linkages, the essential aspect of special parametric conditions suggests immediately that these positions are of greater relevance to overconstrained chains. Planar linkages are, of course, overconstrained. It is shown in (Baker, 1993b) that Altmann's linkage, a particular line-symmetric loop, has stationary positions, as expected, but it also passes through collapsed configurations which are neither stationary nor uncertain, as understood above. They are certainly singular, however, because they are characterised by the linkage degenerating instantaneously there into a pair of superimposed three-bars. For a special case of the linkage at such a position it is demonstrated, further, that two of the joints become stationary and may be locked permanently by a degenerate folding of the loop through coincident vertices.

(Wohlhart, 1996) refers to "bifurcation" positions at which a linkage might adopt one of a pair of possible closures of mobility 1, following instantaneously augmented mobility. He uses this term, rather than Hunt's, in order to introduce and differentiate "kinematotropy", whereby a closure of permanent, greater mobility may ensue from the singular configuration. His principal example is his own "queer square" three-loop linkage which can manifest closures with mobilities 1 and 2, the singular position exhibiting mobility 5. He also cites a chain due to Wunderlich and another from the time of Christopher Wren.

The special positions outlined above and their associated phenomena are obviously of potential significance to linkages which might be employed in manipulators. They indicate regions which should mostly be avoided, or taken into account when joint actuation is being considered. Certainly, paraphrasing (Hunt, 1978), designers of manipulators should be aware of the possible hazards which they pose. In what appears below we attempt to clarify some issues in relation to special configurations and the various closures which might accompany them. To a large extent these matters can be illustrated by means of four-revolute planar loops.

2. Closure Modes

For the four-hinge planar linkage (Fig. 1), we can write

$$2b\,d\cos(\theta-\phi)+2d\,a\cos\theta-2a\,b\cos\phi+c^2-d^2-a^2-b^2=0.$$

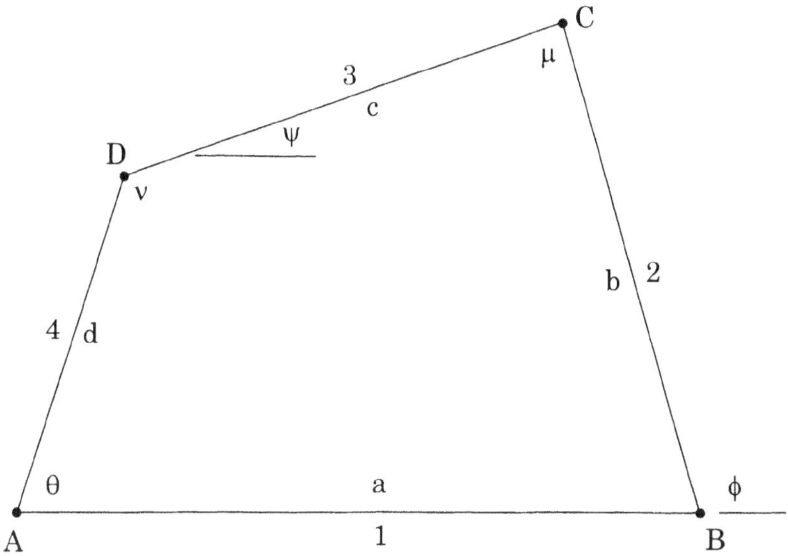

Figure 1.

Putting

$$2b\,d=A,\quad 2d\,a=B,\quad -2a\,b=C,\quad c^2-(d^2+a^2+b^2)=D,$$

we can re-express the result as

$$t_\phi=\frac{-2A\,t_\theta\pm\sqrt{E\,t_\theta^4+F\,t_\theta^2+G}}{[(A-C)-(B-D)]\,t_\theta^2-[(A+C)-(B+D)]}\,, \tag{1}$$

where

$$t_\phi=\tan\frac{\phi}{2}\qquad t_\theta=\tan\frac{\theta}{2}$$

$$\begin{aligned}E&=(A-C)^2-(B-D)^2\\F&=2[A^2+B^2+C^2-D^2]\\G&=(A+C)^2-(B+D)^2.\end{aligned}$$

When a real solution exists the two closure modes are thereby indicated; for a given value of θ, these two modes are commonly manifested by crossed and uncrossed configurations. For a cycle of motion the two series of closure, each taken as a whole, are commonly mirror-images of each other. A generic four-revolute, planar linkage has a limiting closure mode, the (trivial) structural form in which, say,

$$a = b + c + d.$$

When

$$a > b + c + d$$

the chain cannot be closed.

In (Baker, 1986) the cyclohexane molecule is modelled by a special octahedral, line-symmetric loop characterised by a particular, single angle of skew. It is pointed out there that such a loop has two limiting angles of skew beyond which it cannot be closed. For each of these angles the linkage degenerates to a plane structure. The linkage has a mobile closure mode known as the "boat" form among chemical conformational analysts in relation to the molecular ring of cyclohexane. The loop can also be closed in a so-called "chair conformer", a stable state of the molecule, which is rigid. Now, although the latter mode is an alternative closure of the loop, it is not line-symmetric, and so it cannot validly be deemed a closure mode of the linkage as such. That is, closure modes are delineated primarily by an ordered set of parameters and may not all be representative of the linkage which gives rise to them. It should be kept in mind, however, that the categorisation of linkages is somewhat arbitrary, and so it is plausible that a grouping might be broadened to embrace a closure which falls outside an initial schema.

Let us consider another example in order to illustrate this point more forcefully. (Baker, 1993a) includes a description of a variant six-revolute chain formed, in the manner due to (Goldberg, 1943), from a "series" of Bennett linkages. It is depicted schematically here in Fig. 2, and its closure equations may be given as follows.

$$\theta_6 + \theta_1 + \theta_3 + \theta_4 = even\ \pi$$

$$\theta_2 + \theta_5 = 2\pi$$

$$\tan\frac{\theta_2}{2}\tan\frac{\theta_3}{2} = \sin\frac{\alpha_1+\beta_1}{2} / \sin\frac{\alpha_1-\beta_1}{2}$$

$$\tan\frac{\theta_5}{2}\tan\frac{\theta_6}{2} = \sin\frac{\alpha_3+\beta_1}{2} / \sin\frac{\alpha_3-\beta_1}{2}$$

$$\tan\frac{\theta_1+\theta_3}{2}\tan\frac{\theta_2}{2} = \sin\frac{\alpha_2-\beta_1}{2} / \sin\frac{\alpha_2+\beta_1}{2}.$$

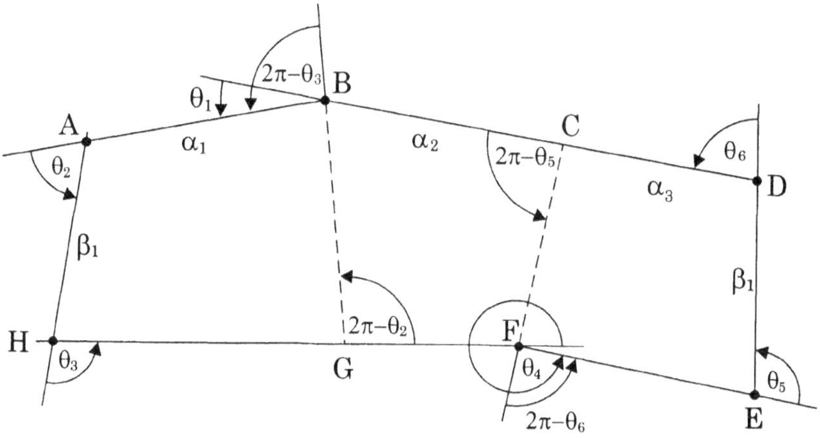

Figure 2.

We now make the loop line-symmetric in character by putting

$$\alpha_3 = \alpha_1,$$

whence the closure equations are reduced to those below:

$$\begin{gathered}\theta_1 + \theta_4 = \theta_2 + \theta_5 = \theta_3 + \theta_6 = 2\pi \\ \tan\tfrac{\theta_2}{2} = \sin\tfrac{\alpha_1+\beta_1}{2} / \sin\tfrac{\alpha_1-\beta_1}{2} \tan\tfrac{\theta_3}{2} \\ \tan\tfrac{\theta_1+\theta_3}{2} = \tan\tfrac{\theta_3}{2} \sin\tfrac{\alpha_1-\beta_1}{2} \sin\tfrac{\alpha_2-\beta_1}{2} / \sin\tfrac{\alpha_1+\beta_1}{2} \sin\tfrac{\alpha_2+\beta_1}{2}.\end{gathered}$$

Inspection of these relationships shows that they have a single solution for a selected input variable, and we note the anti-symmetric form of the first three equations. If we select an arbitrary set of parameters satisfying the dimensional constraints required of linkages based upon Bennett chains and test it numerically, we find that there are four closures. One of them matches our displacement equations, two are strictly line-symmetric six-bars and one manifests degeneracy to a Bennett loop, a pair of opposite joints of the putative six-bar having been locked. Hence, three closures of the numerical loop are spurious modes of the linkage which engendered it. We observe, too, that we could render our closure equations strictly line-symmetric in form by reversing the senses of alternate axes of the loop and so adding π to all skew angles. We can, as well, point to places where pairs of closure modes become identical to each other and are therefore indicative of uncertainty positions under, perhaps, a new definition.

3. In-line Configurations

With reference to Fig. 1 we find

$$2b\,c\cos\mu + d^2 + a^2 = 2d\,a\cos\theta + b^2 + c^2,$$

whence, for small changes,

$$b\,c\,\delta\mu\sin\mu = d\,a\,\delta\theta\sin\theta.$$

A four-revolute planar linkage, of Grashof type or not, commonly changes from a crossed to an uncrossed configuration and vice-versa, once each, in its cycle of motion. The transformation can be represented here by θ becoming zero as bars 1 and 4 are aligned, and $\delta\mu$ concomitantly becoming zero as μ takes a minimum value. So it is a stationary position for joint C. If, however, the loop is transitional between the Grashof and non-Grashof classifications, so that

$$a + b = c + d$$

for this representation, we see that θ and μ reach zero simultaneously, whereby the latter need not experience an extremum. Consequently, the character of the loop of parallelogram form is to remain uncrossed through the in-line configuration, a fact verified by substitution of its special dimensional conditions into eqn. (1). Although it is possible to impose the crossed closure at this juncture, it is necessary at the same time to reverse the sense of $\delta\mu$, that is, to enforce a stationarity there. The kite-shaped linkage displays no crossed configuration, having two uncrossed closures for a given joint angle where solutions exist. The alternative closure is not available at an in-line configuration but, as already mentioned, an effective two-bar degeneracy is a possibility, although it would be again necessary to enforce stationarity on two joints.

Collapsed or in-line configurations offer the potential for simultaneity of the phenomena referred to as stationarity, uncertainty, bifurcation and kinematotropy. Permanent part-chain mobility following such a configuration can arise from degeneracy, as for the kite-shaped planar loop and the Altmann linkage. Permanently greater mobility can be due to disengagement from a single linkage into independent chains, as for Wunderlich's twelve-bar planar loop and Wohlhart's queer square. The latter functions as a four-slider loop in one format and a pair of incorporated two-slider chains in the other.

4. Recommendations

It is evident that the terminology applied to special configurations should be tightened so as to avoid confusion over the phenomena observed at or about these significant positions. I therefore suggest adoption of the following items as guidelines in this matter.

(i) The term stationarity is accurate and unequivocal in reference to a position where some joint freedoms are instantaneously inactive while the remainder of the linkage continues to be mobile, and Hunt's technical definition is satisfactory for the purpose.

(ii) Although Hunt's formal definition of his uncertainty configuration is again easily adequate, I find that the term itself is unsuitable, in view of the foregoing discussion. He recognises the position in question as one of heightened mobility, where all joint screws belong to a screw system of order lower than that which applies to the gross motion, but his word implies, rather, just a choice of subsequent configurations. In order to introduce precision in considering the sometimes coincident phenomena, the designation *supramobility* might be employed.

(iii) Having now isolated the issue of diverse closure modes from those of other singular positions, I emphasise that there is no compelling reason to provide additional distinctive appellations in this context for closures which might be due to degeneracy, enforced stationarity or disengagement. Hence, I submit that the simple expression *forking* could be used in such instances.

(iv) Closure modes should be perceived as different realisations of the same ordered set of parameters. They may or may not belong to the same category of linkage, the category itself possibly subject to variation.

References

J.E. Baker and K.J. Waldron. Limit positions of spatial linkages via screw system theory. *ASME Design Engineering Technical Conference*, New York, N.Y., U.S.A., October 5-9, Paper no. 74-DET-107 (1974).

J.E. Baker. On the investigation of extrema in linkage analysis, using screw system algebra. *Mechanism and Machine Theory*, **13**(3), pp. 333-343 (1978).

J.E. Baker. Limit positions of spatial linkages via connectivity sum reduction. *ASME Trans., Journal of Mechanical Design*, **101**(3), pp. 504-508 (1979).

J.E. Baker. Screw system algebra applied to special linkage configurations. *Mechanism and Machine Theory*, **15**(4), pp. 255-265 (1980).

J.E. Baker. Limiting positions of a Bricard linkage and their possible relevance to the cyclohexane molecule. *Mechanism and Machine Theory*, **21**(3), pp. 253-260 (1986).

J.E. Baker. A comparative survey of the Bennett-based, 6-revolute kinematic loops. *Mechanism and Machine Theory*, **28**(1), pp. 83-96 (1993a).

J.E. Baker. A geometrico-algebraic exploration of Altmann's linkage. *Mechanism and Machine Theory*, **28**(2), pp. 249-260 (1993b).

J. Duffy and M.J. Gilmartin. Limit positions of four-link spatial mechanisms - 1. Mechanisms having revolute and cylindric pairs. *Journal of Mechanisms*, **4**, pp. 261-272 (1969).

M.J. Gilmartin and J. Duffy. Limit positions of four-link spatial mechanisms - 2. Mechanisms having revolute, cylindric and prismatic pairs. *Journal of Mechanisms*, **4**, pp. 273-281 (1969).

K.H. Hunt. *Kinematic Geometry of Mechanisms*. Clarendon Press, Oxford (1978).

K. Wohlhart. Kinematotropic linkages. *Recent Advances in Robot Kinematics*, J. Lenarcic and V. Parenti-Castelli, eds., Kluwer, Netherlands, pp. 359-368 (1996).

CONSTRAINT SINGULARITIES AS $\mathcal{C}$-SPACE SINGULARITIES

Dimiter Zlatanov, Ilian A. Bonev and Clément M. Gosselin
Laboratoire de robotique, Université Laval, Québec, QC, Canada G1K 7P4
[zlatanov, bonev, gosselin]@gmc.ulaval.ca

Abstract This paper examines the phenomenon of constraint singularity of a parallel mechanism, as defined in a recent publication. We focus our attention on the fact that constraint singularities are always singular points of the configuration space of the kinematic chain. As such, they separate distinct configuration space regions and may allow transitions between dramatically different operation modes. All this is exemplified by a multi-operational parallel mechanism that can undergo a variety of transformations when passing through singular configurations.

1. Introduction

A *constraint singularity* is a configuration of a parallel mechanism with $n < 6$ dof, where both the mechanism as a whole and the moving platform have at least $n+1$ dof (Zlatanov et al., 2002). By definition, a constraint singularity is always an IIM-type singularity (Zlatanov et al., 1994) , i.e., a configuration with *increased instantaneous mobility* (IIM).

IIM-type (or *uncertainty*, Hunt, 1978) configurations are the singular points of the configuration space ($\mathcal{C}$-space) of a mechanism. The $\mathcal{C}$-space is the set of all possible configurations, i.e., the feasible arrays of joint-parameter values. For a closed-loop kinematic chain, the $\mathcal{C}$-space is a subset of the joint-space manifold, but not necessarily a submanifold; it may have points with no Euclidean neighbourhoods. These $\mathcal{C}$-space singularities are associated with what can be described as "branching" or "self-intersection" of the configuration space. A motion starting at such a point can end up in any one of the meeting $\mathcal{C}$-space "sheets".

This paper shows that the behaviours of the mechanism in bordering nonsingular regions can be very different. At a constraint singularity the adjoining $\mathcal{C}$-space regions can be expected to allow different systems of platform freedoms and, therefore, result in distinct modes of operation of the mechanism. We study how constraint and IIM singularities connect neighbouring $\mathcal{C}$-space regions. The example used is a parallel chain that undergoes fascinating transformations when moving through a constraint singularity from one 3-dof part of its $\mathcal{C}$-space to the next.

J. Lenarčič and F. Thomas (eds.), Advances in Robot Kinematics, 183–192.

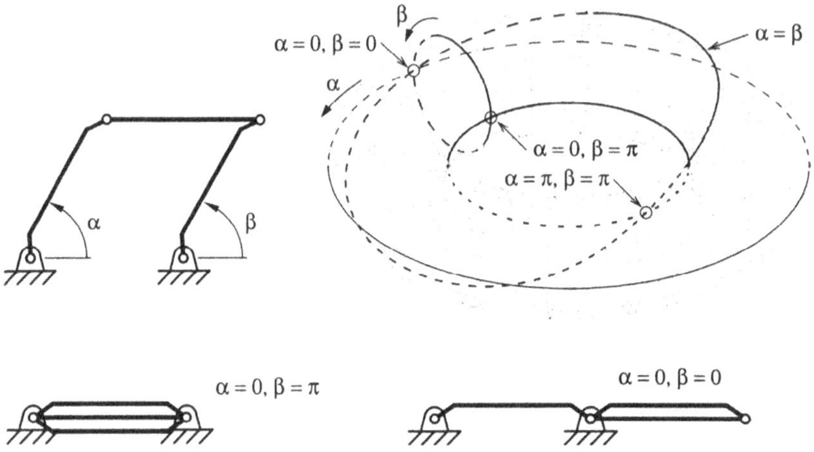

Figure 1. The equal-length four-bar mechanism, the $\mathcal{C}$-space and its singularities.

2. $\mathcal{C}$-space Singularities as Branching Points

The singularities of the $\mathcal{C}$-space are rarely mentioned explicitly in the literature. In (Zlatanov, 1998), $\mathcal{C}$-singularities were identified as one of the three main types of singular configurations of a closed-loop kinematic chain. A similar classification can be found in (Park and Kim, 1999).

Possibly the simplest interesting example of a $\mathcal{C}$-space singularity is a flattened four-bar linkage. As it was pointed out in (Zlatanov et al., 2002), this IIM configuration is a constraint singularity if the output link (the end-effector) is the *coupler* (since it has two transitory dof).

When all four bars have different lengths, the $\mathcal{C}$-space is homeomorphic to the figure eight, i.e., topologically it may be described as two circles joined by a single common point (the flattened configuration). The $\mathcal{C}$-space has two disjoint open connected nonsingular parts, and the mechanism can switch from one to the other (only) by passing through the IIM singularity. The two regions correspond to two different 1-dof sets of poses of the coupler. However, in both regions the allowed instantaneous coupler motion will be of the same type—an instantaneous rotation. (The screw system of the output freedoms remains the same.)

When all four bars are equal (Fig. 1), there are three constraint singularities, where the coupler has two instantaneous dof. The $\mathcal{C}$-space of a four-bar can be easily illustrated since it has a faithful projection on a torus, a configuration being defined in a unique way by (α, β), the base joint angles. Three smooth curves on the torus ($\alpha = 0$, $\beta = \pi$, and $\alpha = \beta$) form the $\mathcal{C}$-space; their intersection points are the singularities. These three points divide the $\mathcal{C}$-space into six non-singular regions.

In Fig. 1, two distinct regions of the $\mathcal{C}$-space may allow completely different coupler motions. While on one of the curves, $\alpha = \beta$, (i.e., in two of the $\mathcal{C}$-space regions) the coupler remains parallel to the base, on each of the other two curves (four regions) the coupler rotates about a point fixed in the base. At two singularities the mechanism has the options of entering two translational and two rotational regions, while at the third point the four-bar can choose among four rotational regions. We could say that the mechanism can be used both as a translational and as a rotational device and the transition between the two modes of operation may occur when the device goes through a constraint singularity.

Such "multi-functionality" is not limited to four-bars.

3. The DYMO Parallel Mechanism

The mechanism in Figs. 2–4 is a parallel chain with three *URU* legs. The centres (A, B, C) of the three platform U-joints form an equilateral triangle and so do the base U-joint centres. For each leg, we denote the revolute-joint screws by ϱ_i^P, $i = 1, \ldots, 5$, $P = A, B, C$. The three base joint axes, ϱ_1^P meet (to form a "Y") at the base centre, O, and so do the last R-axes, ϱ_5^P, at the platform centre, Q. The three intermediate R joints in each leg are parallel. The base frame is at O with a (vertical) z axis normal to the base plane π_b. We will refer to the manipulator as the *3-URU Double-Y Multi-Operational (DYMO)* parallel mechanism.

Since there are 15 joints, all revolute, the $\mathcal{C}$-space is a three-dimensional subset of a 15-torus. To avoid unnecessary complexity, we will not distinguish between configurations with the same pose of the platform and directions of the axes of ϱ_3^P (recall that $\varrho_2^P \parallel \varrho_3^P \parallel \varrho_4^P$). Given a generic platform pose, there are 4 different solutions for the posture of each leg, and hence 64 different configurations of the mechanism as a whole. The distinction between these will not be essential for our purposes. Therefore, we assume that the ambient manifold, where our $\mathcal{C}$-space is embedded, is the Cartesian product of $SE(3)$ and 3 copies of the projective line, $\mathbb{R}P^1$. (We do not distinguish between ϱ_5^P-joint angles that differ by 180°, hence the $\mathcal{C}$-space of the joint is $\mathbb{R}P^1$ rather than a circle.)

We will now proceed to examine the $\mathcal{C}$-space of the kinematic chain in a way similar to the analysis of the folding four-bar in Section 2. The goal is to identify the $\mathcal{C}$-space singularities and the non-singular $\mathcal{C}$-space regions. We want to know which $\mathcal{C}$-singularities are constraint singularities, what end-effector motions are possible in each non-singular region and what forms of transition exist between these regions.

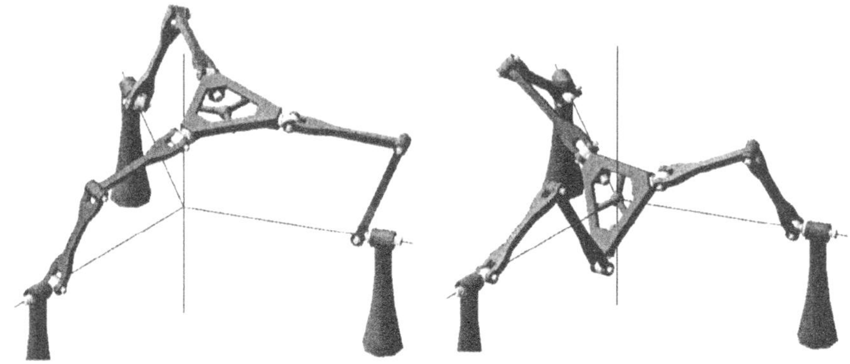
(a) (b)

Figure 2. The DYMO chain in (a) translation mode and (b) orientation mode.

4. Two Translational Mechanisms

It is no surprise that the mechanism can be translational in part of its $\mathcal{C}$-space (Fig. 2a). Translational manipulators of this type are well known (Appleberry, 1992, Tsai, 1996). Their singularities were studied by Di Gregorio and Parenti-Castelli, 1998.

When a leg, P, is nonsingular ($\boldsymbol{\varrho}_i^P$ are linearly independent) it imposes one constraint, $\boldsymbol{\zeta}^P$, on the platform. The wrench $\boldsymbol{\zeta}^P$ is either a force, $\boldsymbol{\varphi}^P$, parallel to $\boldsymbol{\varrho}_3^P$ and passing through the intersection of $\boldsymbol{\varrho}_1^P$ and $\boldsymbol{\varrho}_5^P$, or, if $\boldsymbol{\varrho}_1^P \parallel \boldsymbol{\varrho}_5^P$, a pure moment, $\boldsymbol{\mu}^P$, perpendicular to all $\boldsymbol{\varrho}_i^P$. The total constraint system is $\mathcal{W} = \text{Span}\,(\boldsymbol{\zeta}^P | P = A, B, C)$ and the platform freedom system is $\mathcal{T} = \mathcal{W}^\perp$. If a leg is singular, it imposes additional constraints. A constraint singularity occurs when $\dim \mathcal{W} < 3$. A $\mathcal{C}$-singularity also occurs when there are singular legs leading to d instantaneous dof in the legs *with a fixed platform* and $\dim \mathcal{W} < 3 + d$. We will not consider configurations where a leg is fully extended or folded, but will examine cases with $\boldsymbol{\varrho}_1^P = \boldsymbol{\varrho}_5^P$ (for exactly one or all three P) where a singular leg can freely spin when the platform is fixed.

Let the platform have zero orientation (parallel Ys), and no leg be singular, Fig. 2a. Then, $\boldsymbol{\zeta}^P = \boldsymbol{\mu}^P$, $\forall P$. It can be shown that the three moments $\boldsymbol{\mu}^P$ will remain linearly independent unless one of two things happens: (a) $Q \in Oz$; or (b) the platform plane, π_p, coincides with the base, $\pi_p \equiv \pi_b$, i.e., $Q \in \pi_b$. The intersection of (a) and (b) must be treated as a separate case (c) $Q \equiv O$. We can conclude that once the platform is assembled with zero orientation this will not change until a configuration defined by conditions (a) or (b) is reached.

Case (a) corresponds to a constraint singularity described in (Zlatanov et al., 2002). It is an IIM configuration where both the platform and the mechanism as a whole have five instantaneous dof; $\mathcal{T}$ includes all translations and the rotations with horizontal axes.

In case (b) ($\pi_p \equiv \pi_b$), $\boldsymbol{\mu}^P, P = A, B, C$, become coplanar and the platform acquires the capability to rotate about any vertical axis. The mechanism as well as the platform will generally have 4 dof. If $Q \in \varrho_3^P$ for one P, then the mechanism will still have 4 dof but the platform will have only 3 dof. Indeed, when Q is on the base Y, but not at O, ∞^1 configurations for a fixed platform are obtained while the leg spins. However, only one of those configurations, where $\varrho_3^P \parallel Oz$, $\forall P$, is a $\mathcal{C}$-space singularity. In this case the mechanism has 4 dof but the platform only three, two horizontal translations and a vertical rotation.

Case (c), when the two Ys coincide, corresponds to infinitely many translation-mode configurations. Indeed, all legs can spin while base and platform coincide forming a 3-torus of $\mathcal{C}$-space points. Not all are adjacent to translation-mode regions—these points form a 2-surface on T^3. This situation is further addressed in Section 7.

Note that there are $\mathcal{C}$-space singularities, of cases (b) and (c), for all positions of Q in π_b. Yet, this does not disconnect the set of nonsingular zero-orientation translation-mode configurations. There exist paths that avoid $\mathcal{C}$-space singularities and join configurations on both sides of the base plane. For such paths, Q passes through π_b in a point along the base Y (but not O). As we pointed out above, such configurations are not $\mathcal{C}$-space singularities as long as the single singular leg is not in the base plane. Moreover, note that each such configuration has a neighbourhood of non-singular $\mathcal{C}$-space points. Hence, the non-singular $\mathcal{C}$-points with zero platform orientation form a connected open set.

It is now time to observe that everything written since the beginning of the present section is valid if "zero orientation" becomes "180°-torsion orientation". Hence, there are two translation-mode regions (with different orientations). They have no common boundary points but, as we will see, there are continuous $\mathcal{C}$-space paths between them.

5. An Orientational Mechanism

In (Zlatanov et al., 2002) we discussed a 3-*UPU* parallel wrist proposed by Karouia and Hervé (2000) and analyzed by Di Gregorio (2001). The link-geometry condition is that the base *R* joints intersect in a common point and so do the platform *R* joints. DYMO satisfies this condition, therefore it has an orientation mode of operation (Fig. 2b).

Assume that $Q \equiv O$ and no leg is singular, Fig. 2b. For each P, the constraint $\boldsymbol{\zeta}^P = \boldsymbol{\varphi}^P$ is a force through O parallel to ϱ_3^P. When $\boldsymbol{\varphi}^P$, $P = A, B, C$, are linearly independent, the freedom system, $\mathcal{T}$, is the 3-system of all rotations about O. When the forces φ^P become coplanar the platform also acquires the freedom to translate in a direction orthogonal to the plane of the constraints. Such a configuration will then be a constraint singularity since the platform will have at least 4 dof (if all three forces are parallel, which can happen only if $\pi_p \equiv \pi_b$, the platform acquires *two* horizontal translations for a total of 5 dof).

Careful analysis reveals that the $\mathcal{C}$-space singularities adjacent to the orientation mode are the configurations with *zero-torsion* platform orientation, i.e., they are obtained from orientation zero by a single (finite) rotation with a horizontal axis (a tilt in the vertical plane of some azimuth). All such orientations correspond to $\mathcal{C}$-space singularities.

However, if the axis of the finite (non-zero) tilt is along ϱ_1^P, then there are infinitely many configurations (due to leg spinning) and not all of them lead to coplanar constraints and singularity. Only one of the orientations of the spinning leg (for every half-turn) is a constraint singularity, while for any other the configuration is non-singular.

The exact singularity identification for this mode was performed using the *Tilt-and-Torsion angles* (Bonev et al., 2002). This parameterization of orientation yielded simpler symbolic expressions and helped to establish that the IIM singularities divide the set of non-singular *orientations* of the platform into two disjoint connected components (with positive and negative torsion angle, respectively). However, the $\mathcal{C}$-singularity surface does not disconnect the set of non-singular *configurations* in the orientation mode. This is in close analogy to the translation mode and its singularities with $Q \in \pi_b$. There are non-singular orientation-mode paths that cross the singularity surface from positive to negative torsion via any configuration where the tilt axis is one of the base R-joints. Thus, the non-singular orientation-mode $\mathcal{C}$-space points form a connected set.

We can now resolve one transition question from the previous section. Assuming that the leg length is sufficient, it is possible to move continuously between two translation-mode configurations with opposite torsion angles via the orientation mode.We will see later that this is not the only route between the two translation modes.

6. Two Mechanisms for Planar Motion

In translation mode, we saw that if the platform is put in π_b it acquires as a fourth dof a vertical-axis rotation. In orientation mode, when φ^P, $P = A, B, C$, are vertical, the platform gains the horizontal translations.

In both cases, $\mathcal{T}$ includes as a subsystem the planar motions, $\mathcal{E}$. It is obvious that the capacity for planar motion is not transitory and that finite planar motions can originate from the described configurations.

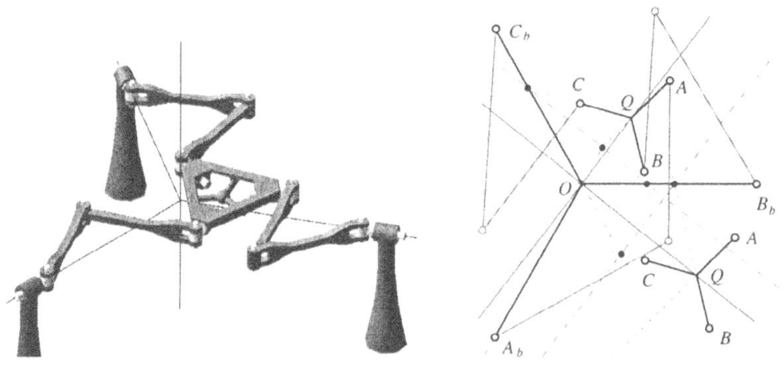

(a) (b)

Figure 3. The planar mode: (a) face up; (b) $\mathcal{C}$-singularities face down.

Indeed one glance at the mechanism should suffice to immediately recognize the features of a 3-*RRR* planar parallel manipulator (Fig. 3a).

If $\pi_p \equiv \pi_b$ and no leg is singular, $\boldsymbol{\varrho}_i^P$, $i = 2,3,4$, $P = A,B,C$ are vertical. Therefore, each $\boldsymbol{\zeta}^P$ is either a vertical force or a horizontal moment, and $\mathcal{T} \supset \mathcal{E} = \mathcal{E}^{\perp} \supset \mathcal{W}$. It must be observed that the above is true not only when the platform is in π_b "*face up*" but also "*face down*", although then the 3-*RRR* planar parallel mechanism will be less recognizable since the base and platform attachment points will be in reverse circular order. So there are two planar modes, which obviously do not have common adjacent configurations, yet it is still possible to go with a continuous motion from one to the other via the orientation mode. We will see later that there is another route as well.

If $O \not\equiv Q$, $\mathcal{W}$ degenerates when either: (a) $\boldsymbol{\zeta}^P = \boldsymbol{\varphi}^P$, $\forall P$, and the two Ys intersect in three collinear points; (b) $\boldsymbol{\zeta}^P = \boldsymbol{\mu}^P$, $\forall P$ (the Ys are parallel); (c) say $\boldsymbol{\zeta}^A = \boldsymbol{\mu}^A$ and the other two are forces whose plane is perpendicular to $\boldsymbol{\mu}^A$. Cases (a) and (c) require the two Ys to be mirror images and occur only when π_p is face-down, (b) is possible only face-up.

Therefore, the face-up $\mathcal{C}$-singularities occur when: (i) the platform orientation is $\sigma = 0$ or $\sigma = \pm\pi$, (constraint singularities adjoining the translation mode if no pair $(\boldsymbol{\varrho}_1^P, \boldsymbol{\varrho}_5^P)$ is aligned); (ii) $O \equiv Q$, ($\mathcal{C}$-singularities adjoining the orientational mode). The $\mathcal{C}$-singularities divide the face-up planar mode into two regions ($\sigma < 0$ and $\sigma > 0$).

It can be shown that the face-down singularities correspond to the following platform poses. For every orientation, the singular locations of Q are given by a pair of perpendicular lines through the origin. Fig. 3b shows such a cross at O and two singular poses ($A_bB_bC_b$ is the base). The orientation of this cross changes with the orientation of the platform, only twice slower. The singularity surface formed by the rotating cross divides the face-down planar mode into four component regions. In both (a) and (c) the platform acquires a rotation about a horizontal axis. This is both the line of the application points of the constraint forces and the axis of symmetry of the two Ys (the dash-dot lines for each pose, Fig. 4b). These singularities border neither translational nor orientation regions. They allow the mechanism to escape into a mixed mode of operation.

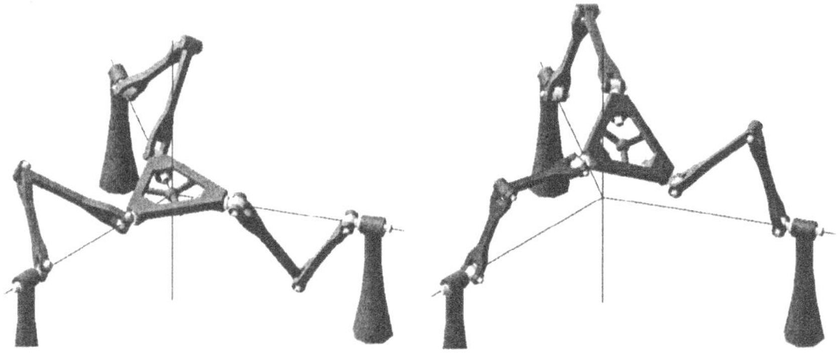

(a) (b)

Figure 4. The DYMO chain in (a) lockup mode and (b) mixed mode.

7. Two Lockup Poses

Let $Q \equiv O$, $\pi_p \equiv \pi_b$, face-up, with $\sigma = 0$ or $\sigma = \pi$ (Fig. 4a). All legs can spin independently, so there are two 3-tori of such configurations. For most, the mechanism has 3 dof but the platform cannot move. Each T^3 has ∞^3 dead-point configurations where the platform is "stuck." We can say that the tori are two separate "modes of operation," with $\mathcal{T} = 0$.

In a $\mathcal{C}$-singularity of this mode the platform acquires a freedom and can escape the lockup pose. Each leg imposes two constraints, a force, $\boldsymbol{\varphi}^P$, and a moment, $\boldsymbol{\mu}^P$. Two types of $\mathcal{C}$-singularities occur: (a) the $\boldsymbol{\mu}^P$ are coplanar and the platform can rotate into orientation mode; (b) the $\boldsymbol{\varphi}^P$ are in a plane and the platform can slide normally and into translation mode. There are two singularities where the platform acquires 3 dof, the whole mechanism has 6 dof, and both (a) and (b) are satisfied. When all $\boldsymbol{\varrho}_3^P$ are vertical, the platform has 3 dof and can escape with any planar motion. When $\boldsymbol{\varrho}_3^P$ are horizontal, the platform also has 3 dof: the rotations with axes in π_b and the vertical translation.

8. A New Mechanism with Mixed Freedoms

At several occasions we described singularities where DYMO has instantaneous motions which do not lead into the so-far listed operation modes. In Section 4, case (a) ($Q \in Oz, Q \not\equiv O$, zero orientation), the singularity allows two rotational freedoms but it is not adjacent to the orientational mode. Similarly, an orientation-mode constraint singularity allows translations with non-zero tilt of the platform (Section 5). In fact, these freedoms can be part of finite motions into $\mathcal{C}$-space regions with $Q \not\equiv O$ and $\pi_p \not\parallel \pi_b$, where the mechanism works as a new manipulator with three mixed freedoms, Fig. 4b. (Yet, a similar chain appears in (Hunt, 1973) as a constant velocity shaft coupling.)

It has been possible to describe completely the configurations in the mixed mode. The process is very interesting but the description cannot be made within the limits of this paper. Below we present an overview of the results and invite the reader to visit the ParalleMIC website, where a more detailed analysis as well as video demonstrations can be found (`http://www.parallemic.org/Reviews/Review008.html`).

A key in the analysis of the mixed mode is the observation that for any feasible configuration with $Q \not\equiv O$ and $\pi_p \not\parallel \pi_b$ the translation along OQ is a platform freedom. This is a finite freedom and it provides a route of entry from mixed into orientation mode. The freedom and constraint systems in a mixed-mode configuration are given by the same 4th special screw 3-system. It is composed of all zero-pitch screws in the plane bisecting π_p and π_b, as well as the infinite pitch screws perpendicular to it and parallel to OQ.

The mixed-mode configuration set is axisymmetric with respect to Oz. For every azimuth, there are configurations with arbitrary tilt between $0°$ and $180°$. For all configurations with given azimuth and tilt (ϕ, θ), Q is along a ray from O with azimuth and tilt $(\phi, \theta/2)$.

Thus, along the vertical axis we have tilt zero; along rays inclined at $45°$ are the configurations with tilt $90°$, i.e., with vertical platform facing away from the central axis; and along rays in π_b the tilt angle is $180°$, i.e. the platform is face down and entering the face-down planar mode. This construction can be performed both starting with zero orientation and with a platform twisted at $180°$ (about the platform normal). In other words for every feasible configuration of the mixed mode there is another with a platform rotated at $180°$ about its normal axis. Furthermore, there are poses on both sides of π_b.

The mixed-mode $\mathcal{C}$-space singularities occur for: (a) $Q \in Oz$, adjacent to the two translation modes; (b) $Q \equiv O$, on the orientation mode boundary; and (c) $Q \in \pi_b$, common with the face-down planar mode. Those in (c) disconnect the mixed-mode $\mathcal{C}$-space into four adjacent regions. The mixed mode provides connecting paths between the two translation modes, as well as the face-up and face-down planar modes.

9. Conclusions

The analyzed mechanism has a configuration space with multiple 3-dimensional regions separated by surfaces of $\mathcal{C}$-space singularities. These regions allow for five dramatically different types of platform motion, yet transition between the modes of operation is possible without disassembly. There is little doubt that this type of behaviour is exhibited by many other mechanisms with reduced freedoms.

10. Acknowledgements

The authors thank Mr Pierre-Luc Richard and Mr Thierry Laliberté for their work on the DYMO model. The financial support of the Canada Research Chair Programme and NSERC is also acknowledged.

References

Appleberry, W.T., 1992, "Anti-rotation positioning mechanism," *US Patent* 5, 156, 062.

Bonev, I.A., Zlatanov, D. and Gosselin, C.M., 2002, "Advantages of the Modified Euler Angles in the Design and Analysis of PKMs," *Parallel Kinematics Seminar*, Chemnitz, Germany, April 23-25.

Di Gregorio, R., and Parenti-Castelli, V., 1998, "A Translational 3-DOF Parallel Manipulator," *Advances in Robot Kinematics: Analysis and Control*, J. Lenarcic and M.L. Husty (eds.), Kluwer Academic Publishers, pp. 49-58.

Di Gregorio, R., 2001, "Statics and Singularity of the 3-UPU Wrist," *Proceedings of the IEEE/ASME International Conference on Advanced Intelligent Mechatronics*, Como, Italy, pp. 470-475, July 8-12.

Hunt, K. H., 1973, "Constant-velocity shaft couplings: a general theory," *Transactions of the ASME, Journal of Engineering for Industry*, Vol. 95B, pp. 455-464.

Hunt, K.H., 1978, *Kinematic Geometry of Mechanisms*, Oxford University Press.

Karouia, M., and Hervé, J.M., 2000,"A Three-DOF Tripod For Generating Spherical Rotation," *Advances in Robot Kinematics*, J. Lenarcic and M.M. Stanisic (eds.), Kluwer Academic Publishers, pp. 395-4020.

Park, F.C., and Kim, J.W., 1999, "Singularity analysis of closed-loop kinematic chains," *ASME Journal of Mechanical Design*, Vol. 121, pp. 32-38.

Tsai, L-W., 1996, "Kinematics of a Three-DOF Platform With Three Extensible Limbs," *Recent Advances in Robot Kinematics*, J. Lenarcic and V. Parenti-Castelli (eds.), Kluwer Academic Publishers, pp. 401–410.

Zlatanov, D., Benhabib, B. and Fenton, R.G., 1994, "Singularity Analysis of Mechanism and Robots Via a Velocity-Equation Model of the Instantaneous Kinematics," *Proceedings of the IEEE International Conference on Robotics and Automation*, San Diego, CA, pp. 986-991.

Zlatanov, D., 1998, "Generalized Singularity Analysis of Mechanisms," Ph.D. thesis, University of Toronto.

Zlatanov, D., Bonev, I.A., Gosselin, C.M., 2002, "Constraint Singularities of Parallel Mechanisms," accepted for publication, *Proceedings of the IEEE International Conference on Robotics and Automation*, Washington, DC, May 12-15.

HIERARCHICAL KINEMATIC ANALYSIS OF A REDUNDANT ROBOT

Daniel Martins
dmartins@das.ufsc.br

Raul Guenther
guenther@emc.ufsc.br
Departamento de Engenharia Mecnica
Universidade Federal de Santa Catarina
Caixa Postal 476 88040 900 - Florianpolis, SC, Brazil

Abstract This paper briefly reviews the hierarchical kinematic analysis, and apply it to a redundant manipulator. This application is done by extending the Jacobian matrix of the redundant manipulator with a careful selection of an extra row. This method is applied to the Roboturb manipulator, a redundant robot specifically designed to repair eroded hydraulic turbine blades.

Keywords: hierarchical kinematic analysis, redundant manipulators, graph theory, screw theory

1. Introduction

Robotic arms are defined as kinematically redundant when the number of joints is greater than the number of degrees of freedom needed to describe a specific task. The most general task of a manipulator is a general movement of the end-effector along the three-dimensional Cartesian space. This general movement requires 6 degrees of freedom ($d = 6$) *i.e.* 6 independent variables to be completely determined at every configuration. Therefore, every robot with more than 6 independent joints is a redundant robot, as each joint is responsible for an extra independent variable to the system.

The interest in redundant manipulators has growing in the last two decades. The main reason is that redundant manipulators can handle with movements in confined or cluttered spaces. Other reason is the extra ability that redundant manipulators have to avoid singularities, obstacles and joint limits. A third reason is that many specific applications are derived from, or may be significantly improved by, the use of redundant manipulators: Nenchev, 1989.

J. Lenarčič and F. Thomas (eds.), Advances in Robot Kinematics, 193–202.

The inverse kinematics is a crucial problem for redundant manipulators :Baillieul, 1985. The available methods lead either to high computational load or to algorithmic singularities. In this paper, we extend the Jacobian to apply the hierarchical kinematic analysis described by Martins and Guenther, 2001 and Martins and Guenther, 2002. The selection of the extra row require a special care to avoid unnecessary couplings, so diminishing the presence of algorithmic singularities.

This paper has the following organisation: section 2 shortly describes the graphical theoretical issues of how to transform a sparse matrix to its finest block-triangular form using simultaneous row and column permutation. Section 3 defines the kinematic structure matrix which enables obtaining the finest block-triangular form with *independent* row and column permutations. Section 4 discusses the sparsity of the Jacobian matrix and how it can be improved by coordinate system selection. Section 5 shortly describes the redundancy resolution, and section 6 applies the method to the Roboturb manipulator, a redundant robot specifically designed at the Universidade Federal de Santa Catarina to repair eroded hydraulic turbine blades.

2. Digraph Approach applied to the Analysis of Jacobian

The structure of a sparse matrix, also called structure matrix, is the relative position in the matrix of the null and non-null elements. The structure matrix can be modelled by directed graphs: Reinschke, 1988. The graphical representation of the structure matrix is also called the incidency matrix of the associated graph. Figure 1 shows a simple directed graph; its incidency matrix is shown in Fig. 2.

The incidency matrix $\bar{A} = [a_{ji}]$ is a graph theoretical concept: Reinschke, 1988. Their elements are defined by

$$a_{ji} \triangleq \begin{cases} 1 & \text{if there is an edge from vertex } i \text{ to vertex } j \\ 0 & \text{if, otherwise, there is no such edge.} \end{cases} \quad (1)$$

The incidency matrix $\bar{A} = [a_{ji}]$, eq. (1), is a Boolean matrix, which are here designated by Roman capitals with an overbar: $\bar{A}, \bar{R}, \bar{T}, \ldots$ A graph related to a incidency matrix $\bar{A}$ is designated by $G(\bar{A})$, or simply G for generic matrices. The properties of Boolean matrices and their relationship with directed graphs are fully discussed elsewhere, Reinschke, 1988 Jantzen, 1996.

A specific labelling of vertices in $G(\bar{A})$ is irrelevant to the properties of the graph itself. Relabelling the vertices has no impact on the directed graph behaviour, but produces a simultaneous row and column permu-

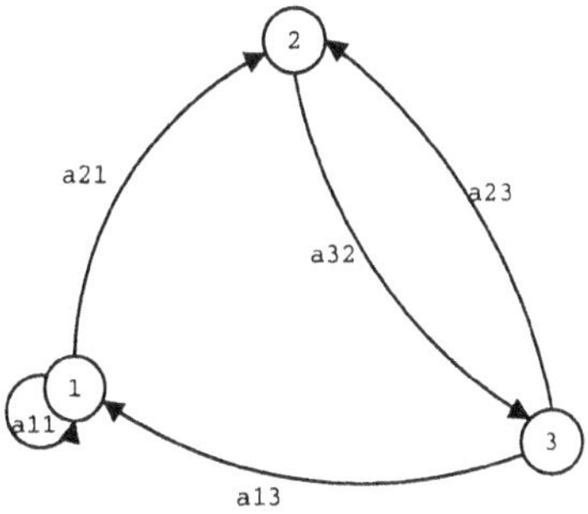

Figure 1. Directed graph (digraph)

$$\bar{A} = \begin{bmatrix} 1 & 0 & 1 \\ 1 & 0 & 1 \\ 0 & 1 & 0 \end{bmatrix} \quad A = \begin{bmatrix} a_{11} & 0 & a_{13} \\ a_{21} & 0 & a_{23} \\ 0 & a_{32} & 0 \end{bmatrix}$$

Figure 2. **Left:** incidency matrix $\bar{A}$ of digraph in Fig. 1. **Right:** matrix A with variables elements which has incidency matrix $\bar{A}$ and can be represented by the directed graph in Fig. 1

tation on the corresponding incidency matrix. Such permutations may be used to reorder the matrix, *i.e.* yielding a block triangular form. To obtain these permutation matrices we aply the concept of strong components of the associated graph.

A *strong component (or strong subgraph)* ia a maximal set of vertices of a graph that can reach and be reached one another: Reinschke, 1988. Let $K(i) \subset G$ be, by definition, the strongly connected subgraph that contains vertex i. A vertex $i \in G$ in a strongly connected set $K(i) \subset G$ is the initial vertex of a path that can reach any other vertex $j \in K(i)$. These strongly connected sets $K(i)$ are related to each diagonal block of the finest block-triangular form: Reinschke, 1988.

The matricial method to obtain this form is based on a series of matrices; namely the reachability matrix $\bar{R}$, the *symmetrical structure matrix* $\bar{S}$, and the *asymmetrical structure matrix* $\bar{T}$. With these matrices, which are discussed in Jantzen, 1996, we generate the asymmetry vector $\vec{w}$ which is responsible for the simultaneous row and column permutation of the matrix.

The asymmetry vector $\vec{w}$ is a permutation of the original vector $\vec{w} = P\,[1\ 2\ 3 \cdots n]^T$. Permutation matrix P converts the incidency matrix $\bar{A}$ to $\breve{A} = P\bar{A}P^T$ where $\breve{A}$ is block triangular: Reinschke, 1988.

An algorithm to obtain the block triangular form of the kinematic structure matrix Q is presented in Martins and Guenther, 2002. Basically this algorithm consists in obtaining sequentially the matrices $\bar{R}$, $\bar{S}$, $\bar{T}$, and the vector $\vec{w}$ starting from the structure matrix $\bar{Q}$. Vector $\vec{w}$ may be reordered to an ascending order by pre-multiplication by a permutation matrix P which will be the permutation matrix applied to Q.

It is important to remark that the row permutation (P) applied to Q is the same column permutation (P^T), *i.e.* the rearranged matrix is $\breve{Q} = P\bar{Q}P^T$.

3. Hierarchical Jacobian reordering

Simultaneous rows and columns permutations, as proposed by the algorithm cited above, are too restrictive to obtain the finest block-triangular form of a generally sparse Jacobian matrix. For this reason, digraph techniques, when directly applied to the Jacobian matrix, have poor results.

To overcome this deficiency, we define the *kinematic structure matrix* Q as

$$Q \triangleq \begin{pmatrix} 0 & J \\ J^{-1} & 0 \end{pmatrix} \Longrightarrow \begin{bmatrix} \dot{x} \\ \dot{q} \end{bmatrix} = Q \begin{bmatrix} \dot{x} \\ \dot{q} \end{bmatrix} \tag{2}$$

The matrix Q is part of a linear system with identical input and output vectors. The i-th column and i-th row are associated to the same variable in $\left[\dot{x}^T \; \dot{q}^T\right]^T$. This property turns the restrictive simultaneous rows and columns permutations on Q as general as independent rows and columns permutations of J.

The matrix Q can be reordered following the procedures described in Jantzen, 1996. These procedures basically rely on Boolean matrices products and sums, and may be performed in any numerical software.

The reordering of Q, eq. (2), is more properly a simultaneous reordering of both matrices J and J^{-1}. The matrix Q need not to be *a priori* known. In fact, only its structure matrix is needed

$$\bar{Q} \triangleq \begin{pmatrix} 0 & \bar{J} \\ \overline{J^{-1}} & 0 \end{pmatrix} \tag{3}$$

where $\bar{J}$ and $\overline{J^{-1}}$ are incidency matrices of the Jacobian and inverse Jacobian matrices, respectively.

While eq. (3) requires the prior knowledge of the incidency matrix of the inverse Jacobian, $\overline{J^{-1}}$, the knowledge of the exact values of J^{-1} are not necessary; therefore, the incidency matrix $\overline{J^{-1}}$ may be obtained without the symbolic inversion of J. Some methods exist to obtain $\overline{J^{-1}}$ without evaluation of J^{-1}; for instance, using the graphical interpretation of the determinant: Jantzen, 1996 or, numerically, by inverting a series of matrices whose incidency matrix is $\bar{J}$, see Martins, 2002.

Let $\check{Q}$ be the finest block-triangular form of the kinematic matrix Q, which is obtained by the algorithm cietd in section 2. Once $\check{Q}$ is obtained, we have at the same time the hierarchical canonical form of the Jacobian and also of the inverse Jacobian. For instance, to obtain the hierarchical canonical form of the Jacobian matrix is enough to drop the lines (rows and columns) of $\check{Q}$ which corresponds to the inverse Jacobian.

4. Sparsity of the Jacobian matrix

The rearrangement of matrices to their block-triangular form using graph techniques entail sparse matrices: Pissanetsky, 1984. An extreme situation is a full matrix, where any rearrangement is worthless. Nevertheless, this is not the only case; for instance, Fig. 2 shows two sparse matrices whose corresponding digraph, Fig. 1, is strongly connected. Any trial to find a block-triangular form from these matrices is, again, fruitless. The opposite extreme situation is an upper triangular matrix . In this case, the hierarchical canonical form is still obtained, so any trial to rearrange this matrix will yield either a worst block-triangular form or destroy this form.

The sparsity of the Jacobian matrix is strongly dependent upon the the choice of the coordinate system. The choice of the coordinate system using Denavit-Hartenberg method is less flexible than using screw theory methods. In fact, in screw theory methods the coordinate system can be freely chosen and then all joint velocities are represented in relation to this coordinate system. Furthermore, we can migrate from one coordinate system to another just using one matrix multiplication *i.e.* the screw transformation matrix: Hunt, 1987.

Some geometrical insight is required to select a good coordinate in order that the sparsity of the Jacobian be improved. Once this coordinate system is chosen, we obtain better results using graph theory techniques.

5. Redundancy resolution

After the path planning, the methods to deal with redundant and nonredundant manipulators differ significantly. The path planning is, in few words, a quest for a suitable path to do the desired task. Once established in Cartesian space, the path has to be converted to the joint space via inverse kinematics. This step for a redundant manipulator is sometimes called the *redundancy resolution* of the manipulator, *i.e.* the computation of the joint variables, as accurately as possible, to follow the path previously defined.

In general, the redundancy resolution yields an infinite number of joint solutions corresponding to a desired end-effector position/orientation. This multiplicity of solutions is the main difficulty of redundant manipulators. The redundancy resolution may be also described as the choice of one inverse kinematic solution among an infinitude of solutions.

Two approaches are suggested in the literature to the redundancy resolution: the use of generalised inverses, such as the Moore-Penrose pseudoinverse Klein and Huang, 1983, or the use of the extended Jacobian Baillieul, 1985, *i.e.* a Jacobian with extra rows.

The hierarchical kinematic analysis needs a linear system to be applied, so the first idea is to use the extended Jacobian method. However, general positional criteria, such as the joint availability criterion Klein et al., 1995, couples the associated digraph of the Jacobian more than necessary, and makes the use hierarchical kinematic analysis less beneficial. This coupling leads to the so-called algorithmic singularities, which normally appear unexpectedly during the movement of the robot.

While the hierarchical kinematic method can be applied to any extended Jacobian method, the results can be quite deceptive when the positional criteria tend to couple all, or almost all, of the kinematic variables. The consequences of this coupling are *per se* negative, *e.g.* increasing the presence of unexpected singularities. For this reason, we provided a variant of the extended Jacobian method which minimise this coupling and, as consequence, yields a more useful hierarchical kinematic analysis. For the Roboturb manipulator this extra row turns the redundancy to be resolved once the velocity of the first (prismatic) joint is given (see Guenther et al., 2000 and Martins, 2002).

6. Hierarchy of the singularities of the Roboturb robot

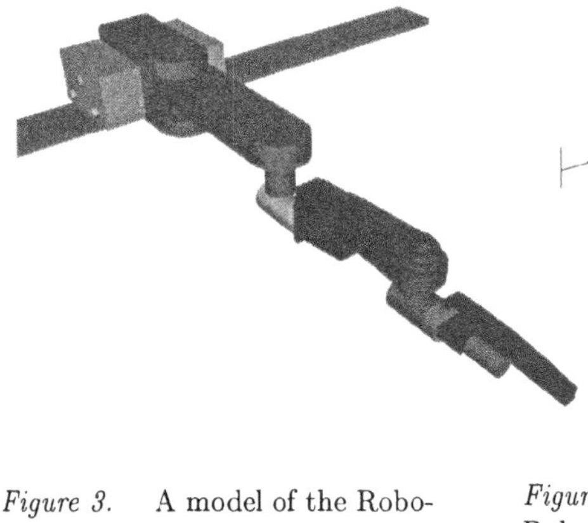

Figure 3. A model of the Roboturb manipulator at the initial position

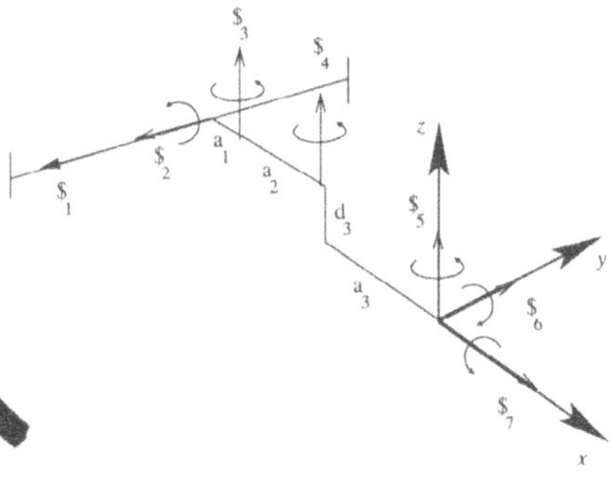

Figure 4. Coordinate system chosen to the Roboturb with the joint variables q_i at initial position.

The Roboturb robot is a serial manipulator with seven degrees of freedom originally conceived to repair turbine blades: Guenther et al., 2000. The first joint is prismatic, and the remaining 6 joints are rotative. The last three joint axes intersect one another in a single point forming a so-called spherical wrist. Figure 3 illustrates the manipulator that will be analysed in this work.

Figure 4 shows a sketch of the manipulator architecture, and the chosen coordinate system where screw theory methods are applied. This Figure remarks the initial configuration of the robot, *i.e.* all joint angles or displacements null: $q_i = 0, \quad \forall i = 1, 7$.

Screws $\$_i$ at joints are shown in Fig. 4. Position angles q_i are associated with each joint of the manipulator with the positive sense shown in Fig. 4. The first joint, represented by $\$_1$, is prismatic and has translational displacement along the direction of $\$_1$. The chosen coordinate system is represented by the triad x, y, z fixed on link 4. Figure 4 shows the coordinate system chosen and the manipulator at the null position, *i.e.* at the position where all joint variables are zero, $q_i = 0$ for $i = 1, 2, \cdots, 7$.

The Jacobian corresponding to Fig. 3 is

$$J = \begin{bmatrix} 0 & -s_{34} & 0 & 0 & 1 & 0 & c_6 \\ 0 & -c_{34} & 0 & 0 & 0 & -s_5 & c_5 s_6 \\ 0 & 0 & 1 & 1 & 0 & c_5 & s_5 s_6 \\ -s_{34} & d_4 c_{34} & a_2 s_4 & 0 & 0 & 0 & 0 \\ -c_{34} & d_4 c_{34} & x_{24} & a_3 & 0 & 0 & 0 \\ 0 & x_{14} & 0 & 0 & 0 & 0 & 0 \end{bmatrix} \tag{4}$$

where $s_i = \sin q_i$; $s_{ij} = \sin(q_i + q_j)$; $c_i = \cos q_i$; $c_{ij} = \cos(q_i + q_j)$, etc ... The symbols a_1, a_2, a_3 are distances shown in Figs. 3 and 4; $x_{14} = a_1 + a_2 c_3 + a_3 c_{34}$ and $x_{24} = a_3 + a_2 c_4$ represent distances along the x-axis from the spherical wrist centre to joints 1 and 2 axes, respectively.

The Jacobian of eq. (4) is extended by an extra row $\vec{n} = e_1$ due to the main positional movement of this manipulator is performed over the prismatic joint at the first joint.

$$J_e = \left[\begin{array}{ccccccc} 0 & -s_{34} & 0 & 0 & 1 & 0 & c_6 \\ 0 & -c_{34} & 0 & 0 & 0 & -s_5 & c_5 s_6 \\ 0 & 0 & 1 & 1 & 0 & c_5 & s_5 s_6 \\ -s_{34} & d_4 c_{34} & a_2 s_4 & 0 & 0 & 0 & 0 \\ -c_{34} & d_4 c_{34} & x_{24} & a_3 & 0 & 0 & 0 \\ 0 & x_{14} & 0 & 0 & 0 & 0 & 0 \\ \hline 1 & 0 & 0 & 0 & 0 & 0 & 0 \end{array}\right] \tag{5}$$

This decision is in accordance with the kinematic conception of the robot described in Guenther et al., 2000.

The structure of the inverse of the extended Jacobian, $\overline{J_e^{-1}}$, may be obtained numerically using the algorithm described in Martins, 2002. With the values of $\bar{J}_e$ and $\overline{J_e^{-1}}$ we mount matrix $\bar{Q}$ using eq. (3). Therefore, the hierarchically arranged linear system, obtained by applying Algorithm 1 to matrix $\bar{Q}$, becomes

$$
\begin{bmatrix} \omega_x \\ \dot{q}_5 \\ \hline \omega_y \\ \omega_z \\ \dot{q}_6 \\ \dot{q}_7 \\ \hline v_y \\ \dot{q}_4 \\ \hline v_x \\ \dot{q}_3 \\ \hline v_z \\ \dot{q}_2 \\ \hline r \\ \dot{q}_1 \end{bmatrix}
=
\left[\begin{array}{cc|cccc|cc|cc|cc|cc}
 & \times & & & & \times & & & & & & \times & & \\
\times & & \times & \times & & & \times & & \times & & \times & & \times & \\ \hline
 & & & & \times & \times & & & & & & \times & & \\
 & & & & \times & \times & & \times & & \times & & & & \\
 & & \times & \times & & & \times & & \times & & \times & & \times & \\
 & & \times & \times & & & \times & & \times & & \times & & \times & \\ \hline
 & & & & & & & \times & & \times & & \times & & \times \\
 & & & & & & \times & & \times & & \times & & \times & \\ \hline
 & & & & & & & & & \times & \times & & \times & \\
 & & & & & & & & & \times & & \times & & \times \\ \hline
 & & & & & & & & & & & \times & & \\
 & & & & & & & & & & \times & & & \\ \hline
 & & & & & & & & & & & & & \times \\
 & & & & & & & & & & & & \times &
\end{array}\right]
\begin{bmatrix} \omega_x \\ \dot{q}_5 \\ \hline \omega_y \\ \omega_z \\ \dot{q}_6 \\ \dot{q}_7 \\ \hline v_y \\ \dot{q}_4 \\ \hline v_x \\ \dot{q}_3 \\ \hline v_z \\ \dot{q}_2 \\ \hline r \\ \dot{q}_1 \end{bmatrix}
\qquad (6)
$$

where the end-effector screw $\dot{x} = \$_n = [\omega_x, \omega_y, \omega_z, v_x, v_y, v_z]^T$ is represented in coordinate system attached at link 4, see Fig. 4.

The hierarchically arranged linear system corresponding to eq. (6) is obtained using the techniques described in section 2. Extracting only the terms relative to the forward kinematics

Equation (7) below has a strictly upper block-triangular form. The only 2×2 diagonal block has determinant $-s_6$ which is associated to the so-called wrist singularity of the manipulator. From bottom to top the hierarchy of singularities is $(x_{14} = 0), (a_2 s_4 = 0), (-s_6 = 0)$. For instance $x_{14} = 0$ affects all screw coordinates of the end-effector while $s_6 = 0$ affects only $\omega_x, \omega_y, \omega_z$. The linear velocity components v_y, v_x, v_z of the end-effector are uniquely determined *even if* $s_6 = 0$. Physically, the spherical wrist singularity ($s_6 = 0$) affects the orientation of the end-effector, but not its position nor its linear velocity.

In the sequence, we show the direct kinematics and highlight the diagonal blocks of the inverse kinematics system

$$
\begin{bmatrix} \omega_x \\ \hline \omega_y \\ \hline \omega_z \\ \hline v_y \\ \hline v_x \\ \hline v_z \\ \hline r \end{bmatrix}
=
\underbrace{\left[\begin{array}{c|cc|c|c|c|c}
1 & & c_6 & & & -s_{34} & \\ \hline
 & -s_5 & c_5 s_6 & & & & \\
 & c_5 & s_5 s_6 & 1 & 1 & & \\ \hline
 & & & a_3 & x_{24} & -d_4 s_{34} & -c_{34} \\ \hline
 & & & & a_2 s_4 & d_4 c_{34} & -s_{34} \\ \hline
 & & & & & x_{14} & \\ \hline
 & & & & & & 1
\end{array}\right]}_{\check{J}_e}
\begin{bmatrix} \dot{q}_5 \\ \hline \dot{q}_6 \\ \dot{q}_7 \\ \hline \dot{q}_4 \\ \hline \dot{q}_3 \\ \hline \dot{q}_2 \\ \hline \dot{q}_1 \end{bmatrix}
\qquad (7)
$$

$$
\begin{bmatrix} \dot{q}_5 \\ \hline \dot{q}_6 \\ \dot{q}_7 \\ \hline \dot{q}_4 \\ \hline \dot{q}_3 \\ \hline \dot{q}_2 \\ \hline \dot{q}_1 \end{bmatrix} = \underbrace{\left[\begin{array}{c|cc|c|c|c|c} 1 & \times & \times & \times & \times & \times & \times \\ \hline & -s_5 & c_5 & \times & \times & \times & \times \\ & c_5/s_6 & s_5/s_6 & \times & \times & \times & \times \\ \hline & & & 1/a_3 & \times & \times & \times \\ \hline & & & & 1/(a_2 s_4) & \times & \times \\ \hline & & & & & 1/x_{14} & \\ \hline & & & & & & 1 \end{array}\right]}_{J_e^{-1}} \begin{bmatrix} \omega_x \\ \hline \omega_y \\ \omega_z \\ \hline v_y \\ \hline v_x \\ \hline v_z \\ \hline r \end{bmatrix} \quad (8)
$$

These results are quite similar to those obtained in Martins and Guenther, 2001 where the Puma robot in the coordinate system of Hunt, 1987 is analysed. In fact, the Roboturb has an architecture similar to the Puma robot, but with the shoulder eccentricity located at a different place. The Roboturb manipulator has d_3 located between joints 3 and 4, while the Puma manipulator has f (or d_3 according to the Denavit-Hartenberg convention) located between joints 1 and 4 axes.

Given the desired velocity of the manipulator over the trail, $\dot{q}_1(= r)$, the redundancy is solved with an analytical inverse by back-substitution. The most complex blocks to invert is the 2×2 diagonal block, so the computational gain over the classical extended Jacobian methods and mainly over the pseudoinverse methods is substantial.

7. Conclusions

This work reinforced that the hierarchical kinematic analysis can be successfully applied to redundant manipulators. The core of the problem is the use of an extended Jacobian and then using Algorithm 1 to reorder the extended (and square) Jacobian.

The inherent advantages of the hierarchical kinematic analysis are also observed when extending this method to redundant manipulators, *e.g.* the simplicity of the inverse kinematics, computational gain, rank of singularity, and flow of variables.

Other fact observed is that classical extended Jacobian methods tend to couple all, or almost all, variables without a clear need. With a judicious selection of the extra row, avoiding inutile couplings, a good hierarchical canonical form may be obtained, as shown for the Roboturb case, eqs (7)-(8). This selection can minimise or avoid completely the presence of algorithmic singularities.

8. Acknowledgements

This work was partially supported by CAPES Foundation, Brazil.

References

Baillieul, J. (1985). Kinematic programming alternatives for redundant manipulators. In *Proc. 1985 IEEE Int. Conf. Robotics & Automation*, pages 722–728.

Guenther, R., Simas, H., and de Pieri, E. R. (2000). Concepção cinemática de um manipulador para volumes de trabalho restritos. In *Proceedings of the Congresso Nacional de Engenharia Mecânica*, in CDROM. In English : *Kinematic conception of a manipulator for constrained workspaces.*

Hunt, K. H. (1987). Robot kinematics—a compact analytic inverse solution for velocities. *Trans. ASME, Journal of Mechanisms, Transmissions and Automation in Design*, 109:42–49.

Jantzen, J. (1996). *Digraph Analyses of Linear Control Systems.* Technical University of Denmark, Dept of Automation. publ. 96-H-838 (lecture notes).

Klein, C., Chu-Jenq, C., and Ahmed, S. (1995). A new formulation of the extended Jacobian method and its use in mapping algorithmic singularities for kinematically redundant manipulators.

Klein, C. A. and Huang, C. H. (1983). Review of pseudoinverse control for use with kinematically redundant manipulators. *IEEE Transactions on Systems Man and Cybernetics*, 13(3):245–250.

Martins, D. (2002). *Hierarchical Kinematic Analysis of Robot Manipulators.* PhD thesis, Universidade Federal de Santa Catarina- UFSC, Florianópolis, SC - Brazil.

Martins, D. and Guenther, R. (2001). Hierarchical singularity analysis of robots. In *Proceedings of the Nineth International Symposium on Dinamic Problems in Mechanics -IX DINAME*, pages 219–224, Florianopólis - Brazil. Brazilian Society of Mechanical Sciences.

Martins, D. and Guenther, R. (2002). Hierarchical kinematic analysis of robots. *Submitted to Mechanism and Machine Theory*

Nenchev, D. N. (1989). Redundancy resolution through local optimization: A review. *J. Robot. Syst.*, 6(6):769–798.

Pissanetsky, S. (1984). *Sparse Matrix Technology.* Academic Press, New York.

Reinschke, K. J. (1988). *Multivariable control : a graph-theoretic approach.* Springer-Verlag, Berlin.

MOTION PLANNING OF REDUNDANT MANIPULATORS FOR SPECIFIED TRAJECTORY TASKS

J. Alfonso Pámanes G.*
apamanes@itlalaguna.edu.mx

Philippe Wenger
IRCCyN - Ecole Centrale de Nantes
Rue de la Noë, BP 92101; 44321 Nantes – France
Philippe.Wenger@irccyn.ec-nantes.fr

José Luis Zapata D.*
* *Instituto Tecnológico de la Laguna*
Blvd. Revolución y Calz. Cuauhtémoc; Torreón, Coah. 27000; México
jlzapata@itlalaguna.edu.mx

Abstract A new approach is proposed in this paper for the motion planning of redundant manipulators in cluttered environments to accomplish specified trajectory tasks. The method is based on the use of feasibility maps, which identifies all the joint trajectories allowing the manipulator's end-effector to travel through the desired path. The procedure synthesizes a set of joint trajectories satisfying the feasibility conditions, even if singular configurations are necessary to achieve the task. The effectiveness of the method is illustrated in the motion planning of a planar 3-degree-of-freedom (dof) manipulator.

Keywords: Redundant manipulators, feasibility, motion planning, joint trajectories, singularities.

1. Introduction

In motion planning of a robotic manipulator the joint trajectories must be determined so that a desired motion of the end-effector can be efficiently achieved. We consider in this paper that such a motion is specified by a trajectory of the end-effector; i.e. a time function describing the history of positions and orientations of the end-effector during the task. Arc welding and laser cutting are industrial operations typically falling into this kind of robot's tasks.

Research supported by LAFMAA, IRCCyN of France, and COSNET of Mexico (Grant 635.01-P).

J. Lenarčič and F. Thomas (eds.), Advances in Robot Kinematics, 203–212.

For redundant manipulator generally an infinite number of joint trajectories allow to accomplish the desired task. To obtain a suitable solution in the planning process of this class of manipulators, a secondary task will be specified, which must be simultaneously achieved with the main task. The secondary task concerns the optimization of robot's kinematic performances (Yoshikawa, 1984) or obstacle avoidance (Maciejewski et al., 1985).

When the environment is too cluttered, the displacements of the robot are severely constrained and the motion planning problem can become complex. This problem has been studied by assuming that singular configurations of the robot are not necessary to achieve the task (Kuo et al., 1997, Pámanes et al., 1999). However, such kind of configurations may be essential to accede to a given task.

The possibility of using singular configurations in an augmented task space has been previously investigated by Wenger et al., 1993. In such reference a method is proposed to obtain feasibility maps which allows to observe all the possible solutions to the problem; however, the determination of suitable trajectories was not studied.

In the present paper we propose a method for the motion planning of redundant robotic manipulators to accomplish a desired trajectory of the end-effector in a cluttered environment. Singular configurations can be used to accede to the complete path with a continuous motion of the robot, and avoiding collisions.

2. Feasibility of tasks

A number of papers have been published which deal with the feasibility of tasks of manipulators (Latombe, 1991). In Wenger et al., 1991, 1993, Pamanes et al., 1995, conditions were established to characterize the feasibility of continuous tasks. To the authors' knowledge, only the approach presented in Wenger, 1993, allows to acquire the accessibility of a redundant manipulator by using singular configurations. We recall here some notions introduced in such reference.

2.1 The augmented task space

Let M be a n-dof redundant manipulator and $\boldsymbol{x}$ denote the vector of operational coordinates ($\boldsymbol{x} \in \mathbf{R}^m$) in the task space TS_m. We consider that r is the redundancy of M. We add r extra-parameters x_{m+1}, x_{m+2}, ..., x_{m+r} to the m-dimensional task space, such that the $m+r$ parameters are independent of each other.

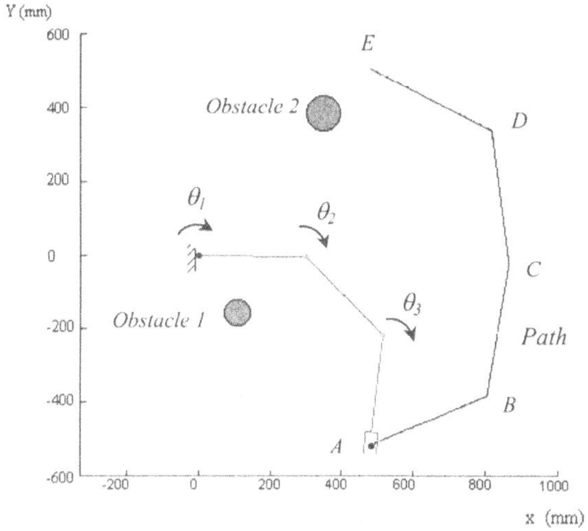

Figure 1. A planar 3R manipulator and desired path.

Therefore, we can define a n-dimensional augmented task space, TS_a, in which M is non redundant ($m + r = n$). In practice, the extra-parameters can be chosen as joint coordinates or as a supplementary coordinates defining the situation of the end-effector.

Thus, for the 3-dof planar manipulator of Fig. 1, whose end-effector situation in TS_m is only specified by its position ($m = 2$), a single extra-parameter denoted x_3 must be considered. We can define x_3 as θ_1 or θ_2 or θ_3. We can also choose x_3 as being the orientation of the last link in the plane; i.e. $x_3 = \theta_1 + \theta_2 + \theta_3$. The choice of the extra-parameter is not critical for the analysis. To illustrate here our choice we define $x_3 = \theta_1$.

2.2 Extended aspects

The extended Jacobian matrix of M in TS_a is the $n \times n$ matrix $\boldsymbol{J}_e(\boldsymbol{q})$ obtained by adding r rows to the original Jacobian matrix $\boldsymbol{J}(\boldsymbol{q})$ as follows:

$$\boldsymbol{J}_e(\boldsymbol{q}) = \begin{bmatrix} \boldsymbol{J}(\boldsymbol{q}) \\ \dfrac{\partial x_{m+k}(\boldsymbol{q})}{\partial q_j} \end{bmatrix} \tag{1}$$

where $k = 1,\ldots,r$; and $j = 1,\ldots,n$.

The r extra-parameters of $\boldsymbol{x}$ induce singularities related to the kinematics of M in the augmented task space (Baillieul, 1986). These singularities define extended aspects, denoted A_{ei}, in the joint space.

Such extended aspects are based on the determinant of the extended Jacobian matrix $\boldsymbol{J}_e(\boldsymbol{q})$. We recall here that an aspect of a manipulator is defined by Borrel and Liegois, 1986, as a connected subspace A_j enclosed in the admissible domain of configurations Q_a of the manipulator's joint space, such that $\forall \boldsymbol{q} \in A_j$ the determinant of any $m \times m$ matrix extracted from the manipulator's Jacobian matrix $\boldsymbol{J}(\boldsymbol{q})$ is not equal to zero, except if this minor is equal to zero everywhere in Q_a.

2.3 Feasibility maps

Let $\mathcal{T}$ be a given continuous trajectory of a manipulator M in TS_m, and S denote the traveled distance of the end-effector up to an arbitrary point on $\mathcal{T}$. A feasibility map, $FM_i(\mathcal{T})$, of one extended aspect A_{ei} of M, related to TS_a, is defined as the set of all the points $(S, x_{m+1}, x_{m+2}, \ldots, x_{m+r})$ in the $r+1$-dimensional space, such that the points which belong to $\mathcal{T}$ are reachable, without collision, by using only configurations of M in A_{ei}. The number of feasibility maps is equal to the number of extended aspects which contain configurations reaching at least one point of $\mathcal{T}$.

In Fig. 2 the feasibility maps are presented for the extended aspects associated with the path shown in Fig. 1 for the 3R manipulator. Such maps are modeled with quad-trees by using the approach of Wenger et al., 1993.

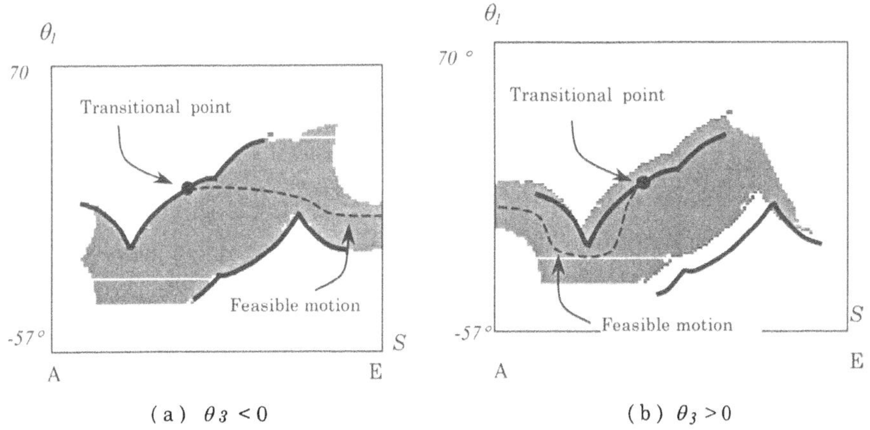

Figure 2. Feasibility maps of the manipulator and path in Fig. 1, associated with the first (a), and the second (b) extended aspects.

As pointed out in Wenger et al., 1993, a continuous trajectory $\mathcal{T}$ is feasible for a manipulator M if there exists a connected set of configurations of M capable to reach all the points on the trajectory in a continuous way, without collisions, and avoiding the joint limits. Therefore, the feasibility maps represent an interesting tool to identify all the joint trajectories allowing the end-effector to travel through the path.

The maps in Fig. 2 show that the desired path cannot be accomplished by using only one extended aspect of the manipulator. The doted line represents a possible trajectory which can be used. Thus, the task must be carried out by beginning with a configuration in the second extended aspect ($\theta_3 > 0$), and by finishing with a configuration in the first one ($\theta_3 < 0$).

To switch aspect, the doted line must attain a transitional point on a common border to both maps. Such borders are identified during the generation of the maps; they are indicated as bold lines on the maps of Fig. 2. A suitable transitional point must be chosen to facilitate the switching of the aspect.

The notion of feasibility maps referred in the preceding paragraphs can be easily extended to the domain of the time by relating the traveled distance S to time t with a function of the form

$$S = f_s(t) \tag{2}$$

The use of maps in the domain of time allows to synthesize the joint trajectories when the task is specified as a function of the time.

Eq. 2 defines a law of motion of the end-effector, which corresponds to a desired profile of velocity. Several standard laws can be applied to specify the motion of the end-effector. For instance, we use a cycloidal motion to produce zero velocity and acceleration at the bounds of the trajectory.

On the other hand, the operational coordinates are related to the traveled distance by a vectorial function $\boldsymbol{f}_x$ as follows:

$$\boldsymbol{x} = \boldsymbol{f}_x(S) \tag{3}$$

The derivatives of Eq. 2 define the magnitudes of the velocity and acceleration of the end-effector.

3. Motion planning of a redundant manipulator

The procedure proposed here for motion planning is composed of three parts. In the first one an exploration is carried out in the feasibility maps to determine whether the path can be achieved by using only one or several extended aspects of the robot.

In the second part, a connected path must be generated into the feasibility maps. Such a path will produce joint trajectories which accomplish the desired task, avoiding collisions and respecting the joint limits; however, it doesn't guarantee the smoothness of the trajectories and infinity joint accelerations could arise. To avoid such behavior, in the third part of the algorithm the trajectory must be rendered smooth.

The complete procedure will be presented based on the use of feasibility maps in the domain of time, and it will be illustrated by the motion planning of the manipulator in Fig. 1. The desired path, also shown there, must be traveled by using cycloidal motions in such a way that zero velocity be attained at the two boundaries of the four straight segments. Each segment must be traveled in 2 seconds. The joint variables θ_1, θ_2, θ_3 must range between -70° as minimum and 70° as maximum; and the link lengths are $l_1 = l_2 = l_3 = 300$ mm. It is noticed that two obstacles must be avoided.

The equation of the cycloidal law of motion at each straight segment is

$$S' = L'\left[\frac{t'}{T'} - \sin\left(\frac{2\pi t'}{T'}\right)\right] \tag{4}$$

where S' is the traveled distance at the instant t' on each straight segment; L' is the length of each segment; and T' is the required time to travel it. The values of such parameters are $L' = 340$ mm, $T' = 2$ sec. Therefore, the total time T to carry out the complete task is 8 seconds. By applying Eq. 4 we transform the maps of Fig. 2 to the domain of time. Such transformed maps were overlapped in Fig. 3 to obtain a map allowing to observe the complete feasible path.

3.1 Preliminary exploration

An elementary exploration of the feasible maps must be carried out to allow the user to know whether the trajectory is feasible or not. If it is feasible the user can also know whether the trajectory must be accomplished by using several extended aspects. In this case it is possible to arbitrarily choose a transitional point (θ_1*, t*) on a common border. In our example the task must be executed by using two extended aspects, as observed in Section 2. A segment of a common border of both extended aspects is shown in Fig. 3-a in which a transitional point is defined by $t^* = 3.8$ sec and $\theta_1^* = 22°$. Hence, a secondary goal of the robot is to obtain a configuration having $\theta_1^* = 22°$ at $t = 3.8$ sec into the aspect defined by $\theta_3 > 0$. This configuration will be also the initial one for the remaining segment of the path into the aspect of $\theta_3 < 0$.

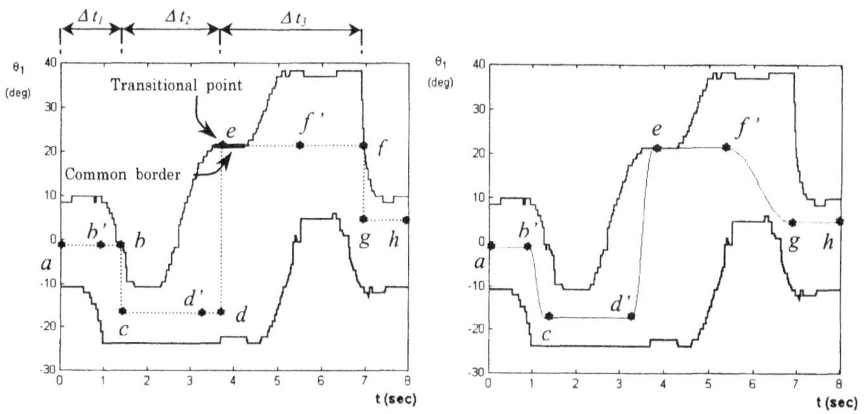

Figure 3. Preparatory (*a*) and smoothed (*b*) trajectories of θ_1 in the overlapped feasibility maps.

3.2 Preparatory connected trajectory

When the preliminary exploration has been concluded, a searching process is to be developed in order to determine a preparatory connected trajectory in the maps. This process is presented here for our example. The segments and points of the trajectory obtained are referred to in Fig. 3-a. The process is carried out in the following steps:

I. Establish a configuration allowing to start the task. The joint variable θ_1 is fixed at $t = 0$ into the initial feasibility map as far as possible from the boundaries (point *a*).

II. Increase the variable *t* by one time step (δt, arbitrarily defined by the user) and keep the last value of θ_1. Then repeat this process, step by step, as long as neither the boundary of the map nor the time values t^* or *T* are not reached. If some boundary of the feasibility map is attained (points *b* and *f*) then go to step III. If the time attains the value corresponding to the transitional point ($t = t^*$; point *d*) then go to step IV. If the end value T of the time is reached (point *h*) the task is completed; then go to step V.

III. Come back *t* to the preceding value and, preserving it, scan θ_1 in order to keep it as far as possible from the boundaries of the map (points *c* and *g*). Then return to step II and continue the process.

IV. Go to the transitional point defining the value of θ_1 as $\theta_1{}^*$ (point *e*). Then, preserving the value of θ_1, at the current value of *t*, go to the other feasible map allowing the robot to continue the task. This map was identified in the preliminary exploration. Then return to step II.

V. End of the process.

The trajectory obtained for θ_1 into the feasibility maps by applying the preceding procedure is shown in Figure 3-a.

3.3 Smoothing the trajectory

It is noticed that a trajectory obtained in the feasibility map as established in Section 3.2 will be connected but not necessarily smooth. Hence, infinity joint accelerations could arise in the manipulator. To avoid such unsuitable behavior, the trajectory must be rendered smooth. We can do that by using the procedure presented below, which is based on cycloidal laws.

I. Identify the points on the preparatory trajectory in which θ_1 must switch its values to avoid violation of a boundary or to go toward a transitional point (points b, d and f in Fig. 3-a).

II. Reduce to the half the time intervals Δt_k (Fig. 3-a; $k = 1,2,3$) of the preparatory trajectory just before the points where θ_1 is switched (points b, d, f in Fig. 3-a). The intervals reduced in this way, $\Delta t'_k$, displace the switching points (up to b', d', f'), such that intervals $\Delta t_k - \Delta t'_k$, greater than zero, are generated to allow to incorporate a cycloidal behavior for the shift of θ_1.

III. Modify the values of θ_1 on each time interval $\Delta t_k - \Delta t'_k$ by using a cycloidal time function for θ_1.

IV. Check whether the values of θ_1 obtained in step III violate or not any boundary of the feasibility map on any interval $\Delta t_k - \Delta t'_k$. If a violation is detected then reduce to the half the interval $\Delta t_k - \Delta t'_k$ and repeat the step III for such interval. If no violation is detected then go to the step V.

V. Determine the values of the other joint variables based on the trajectory obtained for θ_1 in the previous steps.

End of the process.

A computer program was developed in *Matlab©* to carry out both processes described in the preceding paragraphs, which use the data obtained from the feasibility maps. The maps were generated by applying the method described in Wenger, 1993.

The final trajectory of θ_1 into the overlapped feasibility maps is presented in Figure 3-b. Figure 4 shows the three joint trajectories finally obtained. Figure 5 displays a sample of configurations obtained by simulation of the task.

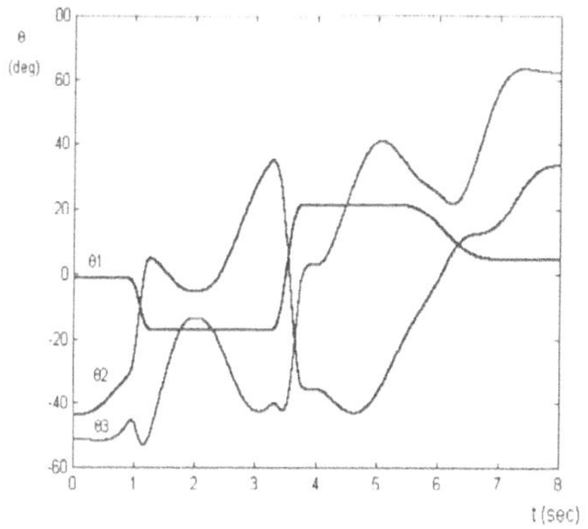

Figure 4. Joint trajectories.

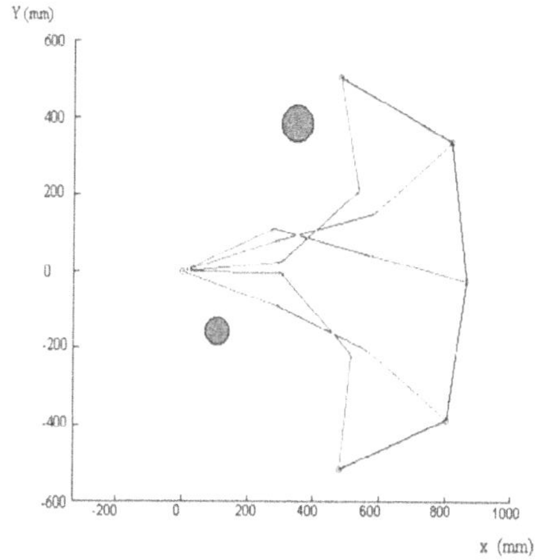

Figure 5. Simulation of the task.

5. Conclusion

A new approach for the motion planning of redundant robotic manipulators for specified trajectory tasks has been proposed in this paper. The proposed approach is based on the use of feasibility maps which allows the user to identify all the viable joint trajectories to achieve the desired path without collisions and avoiding the joint limits. Additionally, it is possible to know whether the use of singular configurations is essential to accomplish a desired task. The results obtained in the case studied here show the effectiveness of the proposed method.

A strategy was applied to generate the preparatory trajectory based on the idea that the displacements of the first joint would be minimum. Nevertheless, in feasibility maps having a more complex geometry, other techniques for path planning, like the A* algorithm, could be useful. Such algorithm could be also applied for a redundancy greater than one; this case will be studied in future work.

No discussion was considered here about the influence of the environment and joint limits on the geometry of the feasibility maps. A more detailed analysis of them can be found in Wenger, 1993.

It has been shown that cycloidal laws carry out an efficient gradual control on the shifts of joint variables into the feasibility maps. A simple approach was used here to locate the transitional point and to establish the amount of the time intervals reserved for the shifts. However, other procedures will be explored in future works to attempt the optimization of kinematic or dynamic performances of the manipulator.

Acknowledgments

This work was partially supported by the National Council of Technological Education -COSNET- (Grant 635.01-P) of Mexico. The subvention for cooperation between researchers of the ITLag of México and the IRCCyN of France was provided by the IRCCyN and the Laboratoire Franco-Mexicain d'Automatique Appliquée (LAFMAA).

References

Baillieul J. (1986), Avoiding obstacles and resolving kinematic redundancy, *Proceedings of the IEEE International Conference on Robotics and Automation.*

Borrel P. and Liegois A. (1986), A study of multiple manipulator inverse kinematic solution with applications to trajectory planning and workspace determination, *Proceedings of the IEEE International Conference on Robotics and Automation.*

Kuo J.A. and Sanger J. (1997), Task planning for serial redundant manipulators, *Robotica*, vol. 15, pp. 75-83.

Latombe J.C. (1991), Robot Motion Planning. Kluwer Academic Publishers, Boston, MA.

Maciejewski, A.A., and Klein C.A. (1985), Obstacle avoidance for kinematically redundant manipulators in dynamically varying environments, *International Journal of Robotics Research*, no. 3, vol. 4, pp. 109-117.

Pamanes G.J.A., Zeghloul S. (1995), On the task feasibility of non redundant manipulators, *Proceedings of the 9th World Congress on the Theory of Machines and Mechanisms*, Milan, Italy.

Pamanes G.J.A., Barron L.A., and Pinedo C. (1999), Constrained optimization in redundancy resolution of robotic manipulators, *Proceedings of the 10th World Congress on Theory of Machines and Mechanisms*, Oulu, Finland.

Wenger P., Chedmail P., and Reynier F. (1993), A global analysis of following trajectories by redundant manipulators in the presence of obstacles, *Proceedings of the IEEE International Conference on Robotics and Automation.*

Wenger P., Chedmail P. (1991), Ability of a robot to travel through its free workspace in an environment with obstacles, *The International Journal of Robotics Research*, no. 3, vol. 10, pp. 214-227.

Yoshikawa, T. (1984), Analysis and control of robot manipulators with redundancy, *Robotics Research: The First International Symposium*; M. Brady and R. Paul, Eds. Cambridge, MA. MIT Press.

REALTIME COORDINATED REDUNDANT MOTION OF A NONHOLONOMIC MOBILE MANIPULATOR

Günter Schreiber and Gerd Hirzinger
DLR Institute of Robotics and Mechatronics
82230 Wessling, Germany
Guenter.Schreiber@dlr.de

Abstract In this paper, the problem of coordinated motion control of a non-holonomic mobile platform with a serial manipulator mounted on top of the platform is addressed. The redundancy resolution algorithm bases on a constraint optimization approach, where the non-holonomic behaviour of the platform is modeled into inequality constraints leading to a convex optimization problem.

Keywords: Non-holonomy, Redundancy, Constraint Optimization, Realtime

1. Introduction

In service robotic applications the combination of a mobile platform with a serial manipulator is of big interest. Especially the coordinated motion of the two subsystems will be addressed within this paper, whereas the mobile platform has a non-holonomic behaviour. The most known applications about coordinated motions use a holonomic platform (Khatib et al., 1996) (Yamamoto & Yun, 1994) (Watanabe, Sato, Izumi, & Kunitake, 2000), which then allows the application of redundancy resolution algorithms based on the well known Moore-Penrose pseudo inverse (D.E.Whitney, 1969). Unfortunately in the case of non-holonomic behaviour, those redundancy resolution algorithms cannot be applied easily.

In our approach (section 3), the non-holonomic couplings of the joints are formulated into a convex optimization problem with linear inequality constraints, so that coordinated motion can be computed in realtime. This kind of algorithms base on (sequential) quadratic programming and constraint optimization problems, which have been studied by several authors (Schreiber, Otter, & Hirzinger, 1999) (Schlemmer, 1997) (Cheng, Chen, & Sun, 1994).

The paper is organized as follows: In the next section, the utilized platform is sketched. Then the redundancy resolution algorithm based

J. Lenarčič and F. Thomas (eds.), Advances in Robot Kinematics, 213–222.

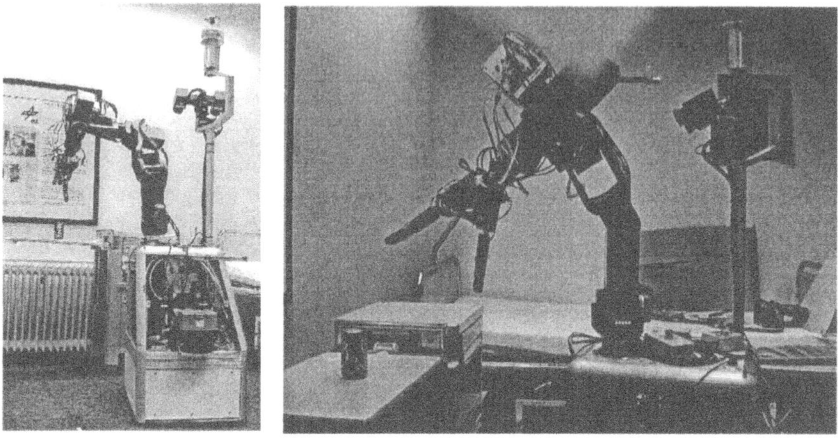

Figure 1. The DLR Lightweight Robot mounted on a mobile platform

on constraint optimization is outlined. Thereafter the linear inequality constraints are developed, and numerical results are presented.

2. Description and modelling of the mobile platform

The utilized mobile platform is "omni directional", since it is able to move in any direction, but not at any time (Saldic, Hanebeck, & Schmidt, 2000). The entire system is presented in figure 1. The platform itself is assembled with four wheelsets, which are located at the four corners of the platform. Those wheelsets are each controllable in two Degrees of Freedom (DoF). The fact, that the wheels have no offset between their steering axis and the point, where they touch the ground, is the reason for the non-holonomy of this setting. For our modelling purposes, we assume that the eight DoF are subject to low level control of the platform, and reduce the problem to three position coordinates $\boldsymbol{q_p} = (\xi, \zeta, \theta)^T$ which is shown in figure 2(a). Furtheron, as the entire manipulator system is considered, the joint vector $\boldsymbol{q} = \left(\boldsymbol{q_p}^T\ \boldsymbol{q_m}^T\right)^T$ is defined, which is constructed from the joint vector of the mobile platform $\boldsymbol{q_P}$ and the joint vector of the serial manipulator $\boldsymbol{q_m}$.

We assume, that the low level control of the platform chooses a configuration of the wheels so that motion will be possible. And furthermore that it computes a normalized direction vector $\boldsymbol{d}$ within the space of $\dot{\boldsymbol{q}}_P$ in which the wheels are facing. As the mobile platform is in motion, this will be nothing else than the normalized current speed in joint space of the platform.

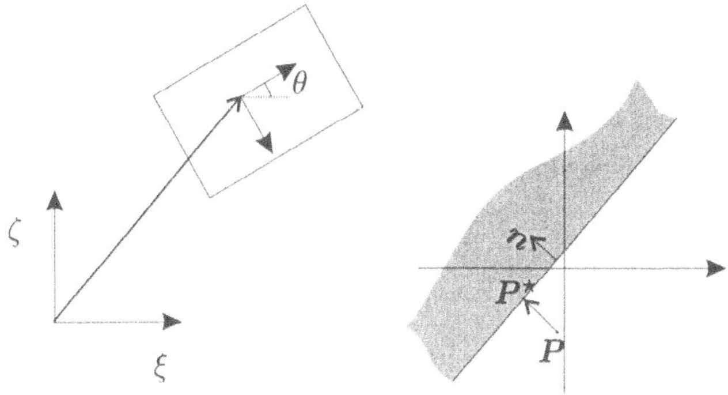

(a) Coordinate system of the mobile platform

(b) Inequality Constraints, the greyed area should be the feasible space. Backprojection, as a point lies outside

Figure 2. Coordinate system of the platform and Visualization of linear inequality constraints

3. Constraint optimization for redundancy resolution

In the sequel the problem of redundant inverse kinematics is reviewed and our current system based on constraint optimization is outlined. Various different authors dealt with the solution of the redundant inverse kinematic problem (see (Siciliano, 1990) for an overview). Cartesian position and orientation $\boldsymbol{x}$ of the end-effector can be described as a function of the vector of joint variables $\boldsymbol{q}$ of the manipulator.

$$\boldsymbol{x} = \boldsymbol{f}(\boldsymbol{q}) \tag{1}$$

While equation (1) can be obtained easily, the inverse problem is crucial. In the redundant case it is generally not possible to find an inverse mapping $\boldsymbol{f}^{-1}$. Instead of constructing an inverse function $\boldsymbol{g}(\boldsymbol{x})$ with $\boldsymbol{f}(\boldsymbol{g}(\boldsymbol{x})) = \boldsymbol{x}$ analytically, the problem is often reformulated in the velocities utilizing the partial derivative of $\boldsymbol{f}(\boldsymbol{q})$, which leads to the wellknown Jacobian matrix $\boldsymbol{J} = \frac{\partial \boldsymbol{f}(\boldsymbol{q})}{\partial \boldsymbol{q}}$.

$$\dot{\boldsymbol{x}} = \boldsymbol{J}(\boldsymbol{q})\dot{\boldsymbol{q}} \tag{2}$$

In our current system, a linear least square based constraint optimization is utilized (Schreiber et al., 1999; Cheng et al., 1994; Park, Chang, & Salisbury, 1997). The major advantage by utilizing the constraint optimization approach is that physical constraints, like joint speed limits or non-holonomic constraints, can be enforced in a natural and simple way via inequality constraints. Using the incremental formulation

$$\dot{\boldsymbol{q}} \approx \frac{\Delta \boldsymbol{q}}{T_a} \tag{3}$$

linear conditions of the form $\boldsymbol{A}\Delta\boldsymbol{q} = \boldsymbol{a}$ can be chosen for the nullspace motion. So the inverse kinematic problem can be formulated as an optimization problem $\min\limits_{\Delta \boldsymbol{q}} L_{IK}(\Delta\boldsymbol{q})$ for the cost function

$$L_{IK}(\Delta\boldsymbol{q}) = \|\boldsymbol{A}\Delta\boldsymbol{q} - \boldsymbol{a}\|_2 \tag{4}$$

with the equality constraint as main task

$$\Delta\boldsymbol{x} = \boldsymbol{J}\Delta\boldsymbol{q} = \boldsymbol{x}_d - \boldsymbol{x} \tag{5}$$

Herein the symbol $\|\cdot\|_2$ denotes the quadratic norm, $\boldsymbol{x}_d$ the desired cartesian, T_a is the sample time and the symbol Δ infront of a variables denotes an increment. Notice that the main task forms in fact the constraint of a linear optimization problem here. Additional constraints in form of inequalities can be added. Those are defined by the following wellknown expression of (hyper-) planes (see figure 2(b)):

$$\boldsymbol{n}^T\Delta\boldsymbol{q} \geq D \tag{6}$$

whereas $\boldsymbol{n}$ is the normal vector defining the plane and D is the shortest distance of the plane to the origin.

Also we define the rule, if point $\boldsymbol{P}$ is outside the allowed area, that means expression (6) does not hold, a new point $\boldsymbol{P}^\star$ will be defined such that $\boldsymbol{P}$ is pushed back on the shortest path into the allowed area:

$$\boldsymbol{P}^\star = \boldsymbol{P} + (\boldsymbol{n}^T\boldsymbol{P} - D)\boldsymbol{n} \tag{7}$$

This will be used to constrain the solutions for particular joints e.g. due to hardware restrictions as maximum joint speeds or joint limits.

$$\Delta\boldsymbol{q}_{min} \leq \Delta\boldsymbol{q} \leq \Delta\boldsymbol{q}_{max} \tag{8}$$

Any particular cost function $h(\boldsymbol{q})$ can be minimized locally by utilizing the $n \times n$–identity matrix $\boldsymbol{E}_{n \times n}$ and the gradient of the cost function, whereas n is the number of joints.

$$\boldsymbol{A} = \boldsymbol{E}_{n\times n}, \quad \boldsymbol{a} = -\frac{\partial h(\boldsymbol{q})}{\partial \boldsymbol{q}} \tag{9}$$

It is now possible to combine multiple behaviors for the null-space motion by choosing

$$\boldsymbol{A} = \begin{pmatrix} w_1\boldsymbol{A}_1 \\ \vdots \\ w_l\boldsymbol{A}_l \end{pmatrix}, \quad \boldsymbol{a} = \begin{pmatrix} w_1\boldsymbol{a}_1 \\ \vdots \\ w_l\boldsymbol{a}_l \end{pmatrix} \tag{10}$$

The particular equations $\boldsymbol{A}_i \Delta \boldsymbol{q} = \boldsymbol{a}_i$ are weighted by the factors w_i, which define how much an error in the ith equation affects the value of the cost function $L_{IK}(\Delta \boldsymbol{q})$. One very interesting feature of this formulation is that the main task can also be included into the cost function instead of constraining the optimization if the problem gets ill-posed, e.g. due to singular configurations (Schreiber et al., 1999). Other cost functions can be found in (Schreiber, Ott, & Hirzinger, 2001) and cost functions concerning interactive behaviour in (Schreiber & Hirzinger, 2000). Standard numerical algorithms exist to solve this special convex optimization problem in a finite number of steps in a reliable way, see e.g. (Lawson & Hanson, 1974; Björk, 1996) for details.

3.1 Metrics in coordinates

It should be noted, that in this system two different units (rotational and translational) in joint space are compared to each other, which is not obviously vaild. In fact the increments $\Delta \boldsymbol{q}$ are minimized with a quadratic norm. This implies that there is an implicit metric, with respect to joint space. This metric can be influenced by a non-singular weighting matrix $\boldsymbol{W}$, so that $\|\boldsymbol{W} \Delta \boldsymbol{q}\|_2$ is minimized. Therefore $\Delta \boldsymbol{q_W}$ is defined

$$\Delta \boldsymbol{q_W} = \boldsymbol{W} \Delta \boldsymbol{q} \tag{11}$$

$$\Delta \boldsymbol{q} = \boldsymbol{W}^{-1} \Delta \boldsymbol{q_W} \tag{12}$$

and used instead of $\Delta \boldsymbol{q}$ for the solution of the constraint optimization problem. This kind of metrics are considered to be applied in sequel.

4. Inequality Constraints for the non-holonomy

To embed the non-holonomic constraints into the previously introduced optimization algorithm, some linear inequality constraints have to be constructed according to (6). In principle the configuration of the mobile platform defines a one DoF manifold for an instantaneous motion within the three dimensional joint space $\boldsymbol{q_P}$. But under consideration of increments during the sample time T_a, the configuration can be changed, so that the feasible motion varies around the current direction $\boldsymbol{d}$ of the mobile platform (see figure 3(a), for the sake of simplicity, only two of the three dimensions will be outlined in the figures). This behaviour will be exploited to construct linear inequality constraints for generating coordinated motion in realtime. Due to linearization and convexity effects, the allowed area will be smaller than the feasible area, which the platform could reach during a time interval. But for (mechanical) stability reasons the maximum speeds and accellerations are reduced anyway.

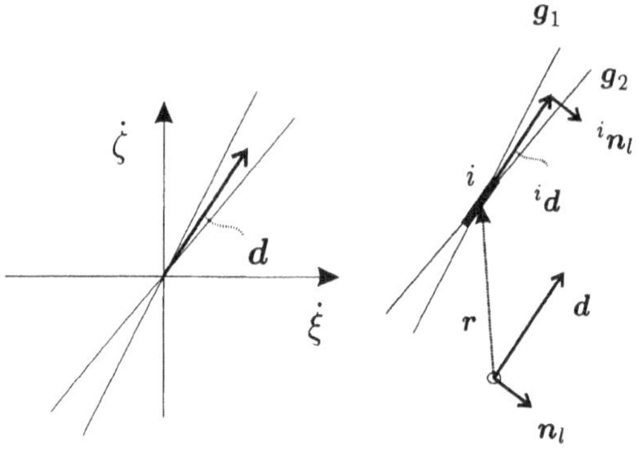

(a) The solution vector Δq_P is allowed around the direction d

(b) Velocity transformed direction and normal vectors for the i-th wheel

Figure 3. Inequality Constraints

The space of the feasible solution $\Delta q_{P,i}$ to the optimization problem can be thought of a pyramidal or conical shape around the current direction of the platform. The opening angle of that shape is restricted by the steering angular velocity of the wheels. So, this shape is approximated by a number of planes within Δq_P.

4.1 Inequality constraints when platform is in motion

As the mobile platform is moving, the new computed increment $\Delta q_{P,i}$ is dependend on the direction d. To construct a vector normal to d, we will use the vector product, since Δq_P is of dimension three. We will use the unit vectors $u_1 = (\,1 \;\; 0 \;\; 0)^T$, $u_2 = (\,0 \;\; 1 \;\; 0)^T$ and $u_3 = (\,0 \;\; 0 \;\; 1)^T$ as generators, whereas we select that unit vector u_j^T, which has the smallest scalar product to d, in order to guarantee that the generator is not parallel to d:

$$n_P = u_j \times d \tag{13}$$

To construct k pyramidically shaped inequality constraints, the vector n_P is rotated around the direction vector d and then rotated around vector u_j with the angle α, which is dependent on the maximum steering angular velocity twice for each of the sides. D of (6) is set to zero:

$$n_l = \left(\mathrm{rot}_{d}(\frac{l}{k}360°) n_P \right) \tag{14}$$

$$n_{l,\alpha} = \left(\mathrm{rot}_{\mathrm{rot}_{\boldsymbol{d}}(\frac{l}{k}360)\boldsymbol{u}_j}(\alpha_l)\boldsymbol{n}_l\right) \tag{15}$$

$$\boldsymbol{n}_{l,\alpha}^T \Delta \boldsymbol{q_P} \geq 0 \tag{16}$$

whereas for the counting index l holds $0 \leq l < k$. In our case, it is sufficient utilizing only $k = 4$ planes of the k-sided pyramid. $\mathrm{rot}_{\boldsymbol{u}_j}(\alpha)$ is a rotation around the axis of vector $\boldsymbol{u}_j$ with the angle α.

4.2 Computation of angle α_l

The angle α_l can be estimated from the current steering angle of each wheel. Therefore, the current direction $\boldsymbol{d}$ and the constructing normal vector $\boldsymbol{n}_l$ is velocity transformed into the i-th wheels coordinate system,

$$\begin{pmatrix} {}^i\dot{\xi} \\ {}^i\dot{\zeta} \end{pmatrix} = \begin{pmatrix} \dot{\xi} \\ \dot{\zeta} \end{pmatrix} + \dot{\theta} \begin{pmatrix} -r_2 \\ r_1 \end{pmatrix} \tag{17}$$

leading to ${}^i\boldsymbol{d}$ and ${}^i\boldsymbol{n}_l$ (see fig. 3(b)), whereas $\boldsymbol{r}$ is the vector connecting the platform coordinate system to the i-th wheel. Note that, due to the assumptions concerning the platform controller, in the resulting vectors only translational components are considered. For the i-th wheel, depending on its state, two lines $\boldsymbol{g}_{1,2}$ depending on its steering angle velocity can be computed, denoting the maximum variation of the steering angle within the time interval. Now a parameter $\lambda_i > 0$ is computed, which depicts the point, where the line $\boldsymbol{g}_{1,2}$ is intersected by ${}^i\boldsymbol{d} + \lambda_l^i \boldsymbol{n}_l$. From all wheels, the smallest $\lambda_i > 0$ is chosen to compute the angle $\alpha_l = \mathrm{atan}(\lambda_i)$.

4.3 Inequality constraints as platform has low speed

Special attention has to be paid, if the platform has with low speed or is at rest. In this case, the platform could physically revert it's direction. In this case at least two iterations will be needed (for details about iterative formulation see (Schreiber et al., 1999)) per sample time. A prismatic shaped constraint is used in the first iteration (see fig. 4). Therefore the opening angle α_l is set to zero, and the distance parameter D of equation (6) is set to a small value $\varepsilon_l > 0$, which correlates to the opening of the corresponding pyramid.

$$\boldsymbol{n}_l = \left(\mathrm{rot}_{\boldsymbol{d}}(\frac{l}{k}360^\circ)\boldsymbol{n_P}\right) \tag{18}$$

$$\boldsymbol{n}_l^T \Delta \boldsymbol{q_P} \geq -\varepsilon_l \tag{19}$$

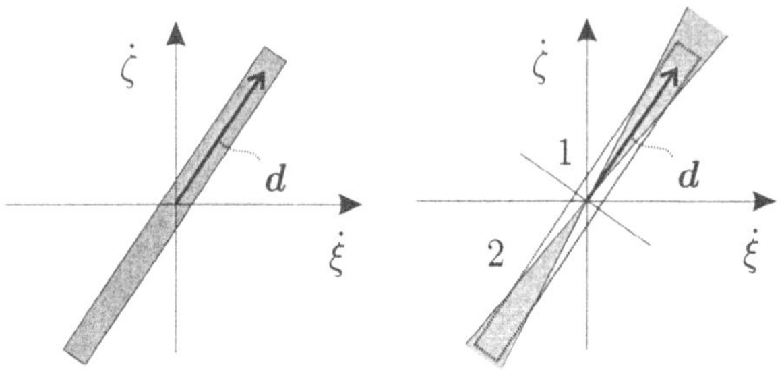

(a) First Iteration. The direction of the mobile platform may be reversed according to (20).

(b) The following iterations. Areas "1" and "2" are selected according to (20)

Figure 4. Inequality Constraints for low speed case

whereas $0 \leq l < k$. For the next iteration, the proper branch is selected according to the iterative solution $\Delta \boldsymbol{q_P}_i^0$. This is performed by the dot product:

$$\boldsymbol{d}^T \Delta \boldsymbol{q}_{\boldsymbol{P},i} \geq 0 \tag{20}$$

If condition (20) holds, the inequality constraints are constructed as in the previous section (see fig. 4(b) - denoted as area 1). If condition (20) is false, the direction vector $\boldsymbol{d}$ is reversed and the inequality constraints are constructed as in the previous section (see fig. 4(b) - denoted as area 2). The iterative solution vector $\Delta \boldsymbol{q}_i^0$, which is used as starting point for the next iteration is projected back into the feasible space utilizing equation (7).

5. Numerical results

The presented approach is implemented on our system. To demonstrate the non-holonomic effect best, the manipulator system has to follow a trajectory, for which the nonholonomic constraint is perpendicular to the requested motion. In figure 5, the manipulator is visualized during motion. In the upper left corner, the starting configuration is shown. The wheels are assumed to face along the greyed arrow, the trajectories of the endeffector and the position of the platform are displayed. If no non-holonomic constraint would be active, the mobile platform would follow a straight line, but with this constraint active, the mobile platform has to follow some curved trajectories due to the reconfiguration time of the wheels. This effect is obvious during the depicted motion. The motion falls into two phases. In the first phase, an endeffector motion is shown, in the second phase a nullmotion is performed to converge to a chosen configuration.

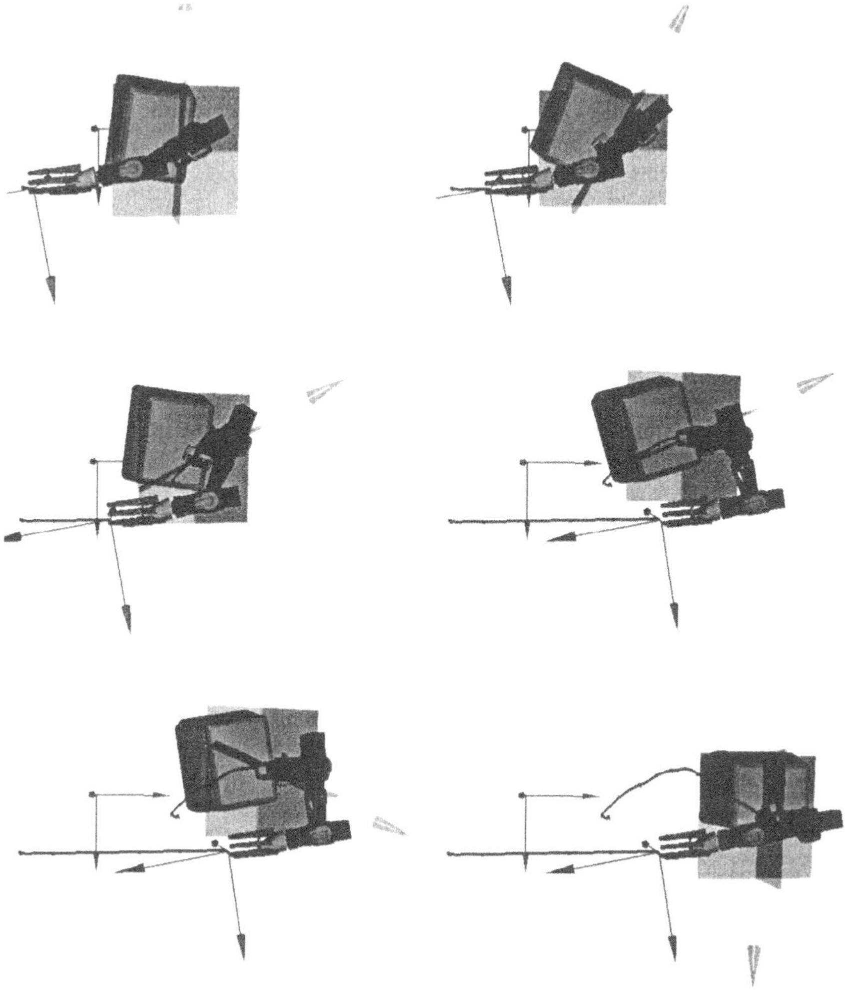

Figure 5. Snapshots of the sample trajectory. The manipulator's non-holonomic constraint faces along the greyed arrow. The linear inequality constraints are visualized as planes during the motion. One can see the curved motion of the first three axes as a consequence of the non-holonomic coupling of those joints. If no non-holonomic constraint were active, the trace of the platform would follow a straight line quite parallel to the trace of the endeffector. The first two rows of snapshots show an endeffector motion, the last row shows a subsequent nullmotion to a desired configuration.

6. Conclusion

Combined mobile manipulator systems, as the one presented, will be of great interest in service robotic applications. We have presented an approach to embed non-holonomic constraints into redundancy resolution systems for realtime purposes. The non-holonomic constraints are therefore squeezed into a convex optimization problem. The mobile manipulator utilizing this approach will be presented on Hannover Fare Exhibition 2002.

Acknowledgement

This work was funded by "BMBF-Leitprojekt MORPHA" Number ITL 0902 E.

References

Björk, A. (1996). *Numerical Methods for Least Squares Problems.* SIAM.

Cheng, F.-T., Chen, T., & Sun, Y.-Y. (1994). Resolving manipulator redundancy under inequality constraints. *IEEE Tr.act. on Robotics and Automation, 10*(1), 65–71.

D.E.Whitney. (1969). Resolved motion rate control of manipulators and human prostheses. *IEEE Tr.act. Man-Machine Systems, 10*, 47–53.

Khatib, O., Yokou, K., Chang, K., Ruspini, D., Holmberg, R., & Casal, A. (1996). Coordination and decentralized cooperation of multiple mobile manipulator. *J. Robotic Systems, 13*(11), 755–764.

Lawson, C., & Hanson, R. (1974). *Solving Least Squares Problems.* Prentice Hall.

Park, K. C., Chang, P. H., & Salisbury, J. K. (1997). A unified approach for local resolution of kinematic redundancy with inequality constraints and its application to nuclear power plant. In *ICRA, April 1997* (pp. 766–773).

Saldic, N., Hanebeck, U., & Schmidt, G. (2000). Systems for omnidirectional mobile robots. In *Proceedings 9 th International Workshop on Robotion in Alpe - Adria - Danube Region* (pp. 11–16).

Schlemmer, M. (1997). *Eine SQP-Rückwärtskinematik zur interaktiven, semi-autonomen Bahnführung kinematisch redundanter Manipulatoren.* Unpublished doctoral dissertation, Gerhard-Mercator-Universität, Gesamthochschule Duisburg.

Schreiber, G., & Hirzinger, G. (2000). An intuitive interface for nullspace teaching of redundant robots. In J. Lenarčič & M. Stanišić (Eds.), *Advances in Robot Kinematics* (pp. 209–216). Portoroz/ Slovenia: Kluwer Academic Publishers, Dordrecht.

Schreiber, G., Ott, C., & Hirzinger, G. (2001). Interactive redundant robotics: Control of the inverted pendulum with nullspace motion. In *IROS 2001* (pp. 158–164).

Schreiber, G., Otter, M., & Hirzinger, G. (1999). Solving the singularity problem of non-redundant manipulator by constraint optimization. In *IROS 1999, Kyongju, Korea* (pp. 1482–1488).

Siciliano, B. (1990). Kinematic Control of Redundant Robot Manipulators: A Tutorial. *Journal of Intelligent and Robotic Systems, 3*, 201–212.

Watanabe, K., Sato, K., Izumi, K., & Kunitake, Y. (2000). Analysis and control for an omnidirectional mobile manipulator. *J. of Int. and Rob. Systems, 27*, 3–20.

Yamamoto, Y., & Yun, X. (1994). Coordinting locomotion and manipulation of manipulator. *IEEE Tr.act. on Automation and Control, 39*(6), 1326–1332.

Methods in Kinematics

USING BODY FLEXIBILITY TO SIMPLIFY THE SOLUTION OF KINEMATIC EQUATIONS IN THE DYNAMIC ANALYSIS OF ROBOT MECHANISMS AND MULTIBODY SYSTEMS

P. Fanghella and C. Galletti
DIMEC - University of Genova
Via Opera Pia 15 A, 16145 Genova, ITALY
fanghell@mcae.dimec.unige.it

Abstract In this paper, it is shown how to derive systematically constraint equations for closed-loop mechanisms with flexible bodies as linear functions of deformation variables. The proposed approach is based on metric relations between kinematic constraints on a binary flexible body. The work describes how to evaluate the sensitivity of the closure equations to given deformation shapes that express the body flexibility. Several indexes and coefficients are introduced and rules are defined in order to verify the possibility of solving the linear constraint equations.

Keywords: flexible multibody, robot dynamics, modal shapes

1. Introduction

Robot mechanism analysis, design and control require the derivation and solution of kinematic and dynamic models of complex kinematic chains, often containing several functional and actuating closed loops and open subchains, with types of motion belonging to various displacement groups (families). The simultaneous solution of the algebraic and differential equations for the dynamic modelling of such constrained mechanical systems is a challenging task investigated by numerous researchers.

One class of methods to solve this problem is based on the idea of transforming the original DAEs into an ODE system, expressed in term of a properly adopted set of minimal, free coordinates. Various criteria have been suggested to select free coordinates: properties of the constraint tangent space are used by several authors, and different Jacobian partition or projection methods are applied (Wehage and Haug, 1982, Blajer et al., 1994, Jeng and Petzold, 1998). Other authors have proposed a way of selecting free coordinates in order to solve the constraint equations in closed, analytical form through a "*modular*" (recursive) decomposition of the kinematic problem (Hiller and

J. Lenarčič and F. Thomas (eds.), Advances in Robot Kinematics, 225–234.

Kecskeméthy, 1994, Fanghella and Galletti, 1993, Kovacs, G. Hommel 1999). The method presented in this paper deals with the solution of constraint equations in the context of dynamic simulation of a kinematic chain, with closed loops formed by rigid and flexible bodies. It is supposed that a minimal-coordinate approach and modular kinematics are adopted for formulating the dynamic equations. There are two main ways to obtain such modularization for a system with flexible bodies: one consists in applying known modularization algorithms to the system with all the bodies assumed to be rigid. In the case of a successful modularization, the vector of the free coordinates of the "rigid" system is augmented with the modal coordinates of all the flexible bodies, so the possibility of a modular solution is maintained also for the flexible system. In the case of complex chains, this strategy may fail, and a different, more powerful way of selecting free and dependent variables must be devised (Fanghella et al., 2001). In the following, we assume that the topology of the kinematic chain is such that a proper choice of the free coordinates yields a modular formulation of the kinematic model, i.e., constraint equations can be solved as a sequence of subsystems, one loop at a time. Therefore, we shall address the problem of solving constraint equations for a single loop containing both rigid and flexible bodies. The proposed approach can be applied only within a dynamic simulation, after an assembly of the mechanism in an undeformed configuration has been generated, and in which the small amplitude of elastic deformations is assured by the feedback of the elastic forces of flexible bodies. The resulting constraint equations will provide the single solution consistent with the constraints and the linear model of the flexible bodies.

2. Choice of free coordinates in a closed chain

Let us consider a single loop kinematic chain containing both rigid and flexible bodies (Fig. 1).

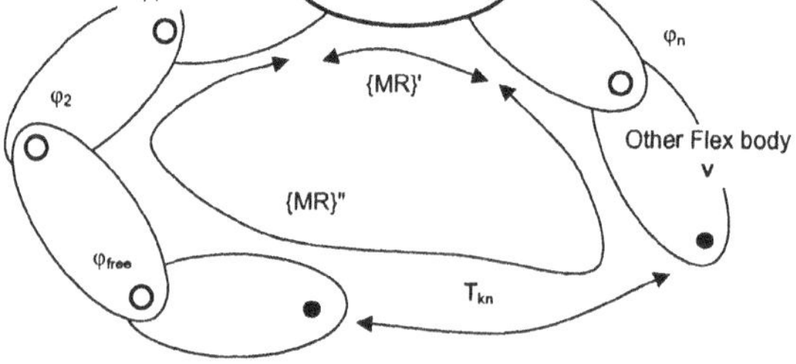

Figure 1. A single-loop kinematic chain

Closure equations for this chain can be obtained by metric relations between invariant properties of kinematic constraints, according to the approach discussed in Fanghella, C. Galletti, 1995 (see also next section and Hiller and Kecskeméthy, 1994). By choosing constraints KCa and KCb on the boundary of a flexible body ab, we obtain the equations:

$$\{MR(\mathbf{u})\}' = \{MR(\varphi, \varphi_{free}, \mathbf{v}, \mathbf{T}_{kn})\}'' \quad (1)$$

in which:

$\mathbf{u}$ = vector of modal coordinates of the flexible body ab between constraints KCa and KCb;

$\mathbf{v}$ = vector of modal coordinates of all other flexible bodies in the considered loop (it can be empty);

φ = vector of pair variables in the loop, excluded KCa and KCb, assumed to be dependent variables;

φ_{free} = vector of pair variables in the loop, excluded KCa and KCb, assumed to be free coordinates (it can be empty);

$\mathbf{T}_{kn}$ = known transformation between two bodies in the loop.

Under the hypotheses made in Fanghella and Galletti, 1995, the number n of equations (1) is equal to the dimension of the vector φ. Assuming the modal coordinates $\mathbf{u}$ and $\mathbf{v}$ to be free variables, the variables φ are solutions of a system of nonlinear algebraic equations.

A different way of solving (1) lies in changing the unknown variables: if we assume the pair variables φ to be free variables and the modal coordinates $\mathbf{u}$ of the body ab to be unknowns, under the hypothesis of small elastic displacements, it is possible to express the kinematic problem (1) as a set of linear algebraic equations. Its coefficient matrix is constant during the dynamic simulation and can be evaluated and inverted, once for all, before starting the simulation.

In order to formulate a well-conditioned problem, a suitable set of deformation modes for the flexible body ab must be selected, if any. Such a set must contain n modes, guarantee a good functional dependence (Di Gregorio and Parenti Castelli, 1999) of the metric relations MR used in (1) on the modal coordinates $\mathbf{u}$, and a good conditioning of the coefficient matrix of the position equations. A simple example is considered. Driven by a torque applied to pair A, the one-degree-of freedom chain in Fig. 2-a actuates joint E of an anthropomorphic robot arm. Body 4 is flexible, and its deformation depends on a modal coordinate u.

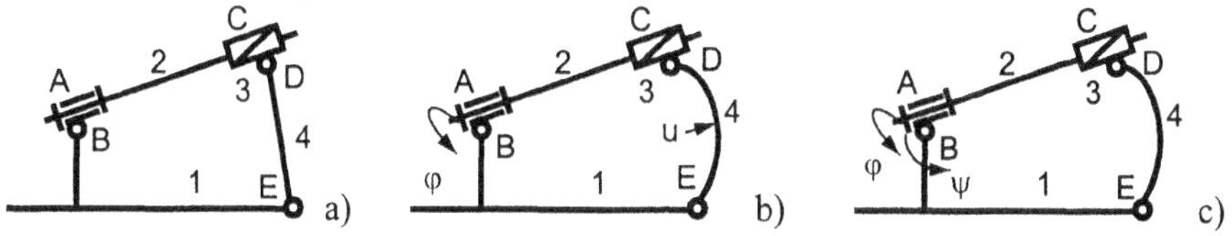

Figure 2. A flexible chain for actuating a robot joint

Assuming the rotation φ of revolute A and the modal coordinate u to be free coordinates (Fig. 2-b), the distance between points D and E can be computed for any value of u by using the knowledge of the deformation shape of body 4. Then the closure equations of the chain with the flexible body are the same as those of the rigid chain. If the rotation φ of revolute A and the rotation ψ of revolute B are assumed to be free coordinates (Fig. 2-c), the position of point D can be computed by solving the open chain of bodies 1, 2, and 3, then the distance between D and E is calculated. Finally, the dependent coordinate u is computed by using the distance DE and the deformation shape of the flexible body. This last computation involves the solution of one linear equation in the variable u. Obviously, this approach to simplifying the solution of the constraint equations can be adopted if the length of body 4 presents significant variations for non-zero values of u. In other words, the deformation shape of the body must be "suited" to providing a displacement of the whole chain. In the following sections, we discuss this possibility for spatial mechanisms, taking into account the shape of a generic binary body and the way of expressing constraint conditions.

3. Metric relations

The approach based on metric relations is very useful in deriving closure equations for a single-loop kinematic chain. This approach, which has been developed in a general way by Fanghella and Galletti, 1995, will be applied here to a generic binary body ab. The body is connected to two other bodies, a and b, by *kinematic constraints* - for instance, kinematic pairs - say KCa and KCb. A kinematic constraint is characterized by its *displacement group* GR and by a set of *invariant properties* IP of the group, i.e., by points and/or unit vectors that define geometric entities that do not vary when a relative displacement is given to the bodies connected by the constraint. *Metric relations* MR can be stated between IPs of displacement groups. Basically, 9 metric relations have to be considered to solve kinematic problems. Their symbolic expressions are given in Appendix 1.

Metric relations can be expressed for KCa and KCb on a body: MR_i(KCa, KCb), i=1,n, where n depends on the kind of groups GRa and GRb of the kinematic constraints on that body. For a system of rigid bodies forming a closed loop, each MR_i can be written twice: once for a subchain containing only the body ab, and once for a subchain containing the other bodies. The first set of MRs, say {MR}', is constant and depends only on the shape of the body ab, the second set of MRs, say {MR}", depends on the shapes of all the other bodies, on the variables φ and φ_{free} of the pairs connecting them (except those in KCa and KCb) and on other known time-varying transformations $\mathbf{T}_{kn}$. As MRs are scalar reference-independent quantities, by equating the two sets we obtain the non-linear closure equations for the closed chain

$$\{MR\}' = \{MR(\varphi, \varphi_{free}, \mathbf{T}_{kn})\}'' \qquad (2)$$

which contains as unknowns the variables φ of the pairs of the chain, except those in KCa and KCb and those assumed to be free coordinates φ_{free} (if any). A discussion of the applicability of this approach is given in Fanghella and Galletti, 1995.

Depending on GRa and GRb, their IPa and IPb are defined. For the purpose of this work, we assume that a point A and a unit vector $\mathbf{k}_A$, furnish all the information about IPa, and that a point B and a unit vector $\mathbf{k}_B$ furnish all the information about IPb. Their expressions are given in a generic reference frame fixed to the rigid body ab as a function of the constraint positions. Let $\mathbf{O}_{A0} = (O_{A0x}, O_{A0y}, O_{A0z})$ and $\mathbf{O}_{B0} = (O_{B0x}, O_{B0y}, O_{B0z})$ be the coordinates of A and B in this reference frame, and let

$$\mathbf{k}_{A0} = [\mathbf{R}_x(\theta_{A0x})][\mathbf{R}_y(\theta_{A0y})][0 \quad 0 \quad 1]^T = [\sin(\theta_{A0y}), \sin(\theta_{A0x})\cos(\theta_{A0y}), \cos(\theta_{A0x})\cos(\theta_{A0y})]^T$$

$$\mathbf{k}_{B0} = [\mathbf{R}_x(\theta_{B0x})][\mathbf{R}_y(\theta_{B0y})][0 \quad 0 \quad 1]^T = [\sin(\theta_{B0y}), \sin(\theta_{B0x})\cos(\theta_{B0y}), \cos(\theta_{B0x})\cos(\theta_{B0y})]^T$$

be the components of vectors $\mathbf{k}_A$ and $\mathbf{k}_B$, in the same reference.

By using these expressions, all the MRs in Appendix 1 can be made explicit functions of the actual values of the invariant properties. In Appendix 2 we show the resulting forms for relations Nos. 4 and 9.

The approach based on metric relations can also be used when the body ab is flexible. In this case, the {MR}' in (2) are not constant, but depend on the deformation of the body during its motion, and can be expressed as functions of the modal coordinates $\mathbf{u}$ of the body. For small deformations, the position $\mathbf{P}$ of a point in a deformed configuration, with respect to its position $\mathbf{P}_0$ in an initial undeformed configuration, can be determined with the modal superposition technique as

$$\mathbf{P} = \mathbf{P}_0 + \sum_{j=1}^{m} \mathbf{\Phi}^t_j u_j \qquad (3)$$

where $\mathbf{\Phi}^t_j$ is a three-component vector of translational degrees of freedom of deformation shapes, expressing the translational elastic displacement of that point for a unitary value of the j-th modal coordinate, and $\mathbf{u}=[u_1, u_2, \ldots u_m]$ is the vector of the modal coordinates. Moreover, an expression similar to (3) can be stated for small rotations. Such an expression contains the three-component vector $\mathbf{\Phi}^r_j$ of rotational degrees of freedom of deformation shapes. Relation (3) is written for points A and B, which express the invariant properties of constraints KCa and KCb

$$\mathbf{O}_A = \mathbf{O}_{A0} + \sum_{j=1}^{m} \mathbf{\Phi}^t_{Aj} u_j \qquad \mathbf{O}_B = \mathbf{O}_{B0} + \sum_{j=1}^{m} \mathbf{\Phi}^t_{Bj} u_j \qquad (4)$$

Inserting (4) and its rotational counterpart for $\mathbf{O}_A$, $\mathbf{O}_B$, $\mathbf{k}_A$ and $\mathbf{k}_B$ in the metric relations of Appendix 1, and taking into account that deformations are small, we obtain MR's expressions that are linear in the modal coordinates. For a generic MR_i we have

$$\mathrm{MR}_i = \sum_{j=1}^{m} c_{ij} u_j + D_i \tag{5}$$

where D_i has the same expression as that obtained for the body considered to be rigid. Every coefficient c_{ij} depends on the metric relation i, on the values $\mathbf{O}_{A0}$, $\mathbf{O}_{B0}$, $\mathbf{k}_{A0}$ and $\mathbf{k}_{B0}$ of the IPs in the undeformed state, and on the j-th deformation shape $\mathbf{\Phi}'^{t}_{jA}$, $\mathbf{\Phi}'^{t}_{jB}$ of points A and B and $\mathbf{\Phi}'^{r}_{jA}$, $\mathbf{\Phi}'^{r}_{jB}$ of vectors $\mathbf{k}_A$ and $\mathbf{k}_B$, i.e.,

$$c_{ij} = c_{ij}(\mathbf{O}_{A0}, \mathbf{O}_{B0}, \mathbf{k}_{A0}, \mathbf{k}_{B0}, \mathbf{\Phi}'^{t}_{jA}, \mathbf{\Phi}'^{t}_{jB}, \mathbf{\Phi}'^{r}_{jA}, \mathbf{\Phi}'^{r}_{jB}) \tag{6}$$

Therefore, the matrix **C** of the coefficients c_{ij} of all the modes for the chosen set of Metric Relations {MR} can be evaluated independently of the actual deformation of the body and of the position of the mechanism.

In Appendix 3, explicit expressions for a generic coefficient c for MRs Nos. 4 and 9 are provided.

4. Sensitivity of metric relation to deformations

The possibility of an effective use of a set of MRs on a flexible body to compute one or more modal coordinates u_j of that body depends on the properties of the coefficient matrix **C**. Obviously, if such a matrix is singular, the selected combination of deformation modes and MRs cannot be adopted. If the matrix is nearly singular or ill-conditioned, in theory the relations can be used, but errors in evaluating the coefficients of the unknowns and of the known terms of (1) will make the computation of the modal variables unacceptable. Therefore, the coefficient matrix must be, in a sense, "well conditioned." This means that the metric relations must be *sensitive* to the selected deformation modes, and that the resulting equations must be sufficiently independent.

Various possibilities of evaluating such sensitivity are discussed in the following part of the paper. The results of the discussion allow one to evaluate the suitability of each deformation shape in an MR independently of the others, and to verify the suitability of a set of deformation shapes.

4.1 Suitability of a deformation shape

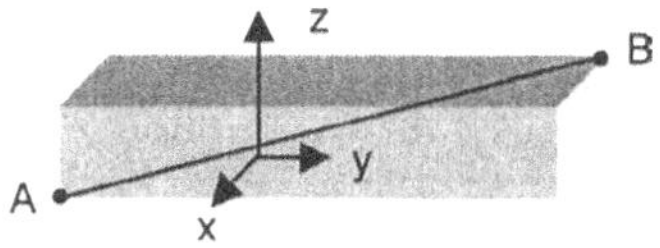

Figure 3. A body with to points A, B, expressing IPs of its constraints.

Two points, A and B, on the body are used to express the invariant properties of the two constraints that link this body to other bodies. Metric relation No. 9 gives the distance between these points. A generic coefficient c of matrix $\mathbf{C}$ for a generic mode in this relation is

$$c=2\left(O_{B0x}-O_{A0x}\right)\left(\Phi^{t}_{Bx}-\Phi^{t}_{Ax}\right)+2\left(O_{B0y}-O_{A0y}\right)\left(\Phi^{t}_{By}-\Phi^{t}_{Ay}\right)+$$
$$+2\left(O_{B0z}-O_{A0z}\right)\left(\Phi^{t}_{Bz}-\Phi^{t}_{Az}\right)=2(\mathbf{O}_{B0}-\mathbf{O}_{A0})\bullet(\mathbf{\Phi}^{t}_{B}-\mathbf{\Phi}^{t}_{A})$$

The coefficient has its maximum value when $\mathbf{\Phi}^{t}_{A}$ and $\mathbf{\Phi}^{t}_{B}$ are opposite and vectors $(\mathbf{O}_{B0}-\mathbf{O}_{A0})$ and $(\mathbf{\Phi}^{t}_{B}-\mathbf{\Phi}^{t}_{A})$ are parallel, i.e., when the deformation of the body makes points A and B be displaced in the direction that joins these points. The coefficient is zero when $\mathbf{\Phi}^{t}_{A}=\mathbf{\Phi}^{t}_{B}$, or $(\mathbf{O}_{B0}-\mathbf{O}_{A0})$ and $(\mathbf{\Phi}^{t}_{B}-\mathbf{\Phi}^{t}_{A})$ are perpendicular, i.e., when points A and B are displaced in such a way as to maintain a constant distance between them. Therefore, an appropriate index (besides the signs of $\mathbf{\Phi}^{t}_{A}$ and $\mathbf{\Phi}^{t}_{B}$) for evaluating the suitability of a shape is the cosine of the angle between $(\mathbf{O}_{B0}-\mathbf{O}_{A0})$ and $(\mathbf{\Phi}^{t}_{B}-\mathbf{\Phi}^{t}_{A})$, i.e.,

$$\frac{(\mathbf{O}_{B0}-\mathbf{O}_{A0})\bullet(\mathbf{\Phi}^{t}_{B}-\mathbf{\Phi}^{t}_{A})}{\left|(\mathbf{O}_{B0}-\mathbf{O}_{A0})\right|\left|\mathbf{\Phi}^{t}_{B}-\mathbf{\Phi}^{t}_{A}\right|} \tag{7}$$

In order to generalize (7) to any metric relation, we note that any coefficient c of the metric relations presents a linear dependence on $\mathbf{\Phi}^{t}_{A}$, $\mathbf{\Phi}^{t}_{B}$, $\mathbf{\Phi}^{r}_{A}$, and $\mathbf{\Phi}^{r}_{B}$ (see Appendix 3). We collect these vectors in two vectors with 6 terms, one with the translation shapes, $\mathbf{\Phi}^{t}$, and the other with the rotation shapes, $\mathbf{\Phi}^{r}$

$$\mathbf{\Phi}^{t}=\begin{bmatrix}\mathbf{\Phi}^{t}_{A}\\ \mathbf{\Phi}^{t}_{B}\end{bmatrix} \qquad \mathbf{\Phi}^{r}=\begin{bmatrix}\mathbf{\Phi}^{r}_{A}\\ \mathbf{\Phi}^{r}_{B}\end{bmatrix}$$

A second couple of vectors, $\mathbf{c}^{t}$ and $\mathbf{c}^{r}$, with 6 terms is formed by the expressions that multiply each term of $\mathbf{\Phi}^{t}$ and $\mathbf{\Phi}^{r}$ in the coefficient c of the metric relation. It results

$$c=\mathbf{c}^{t}\bullet\mathbf{\Phi}^{t}+\mathbf{c}^{r}\bullet\mathbf{\Phi}^{r} \tag{8}$$

Analogously to (7), we define 2 *sensitivity indexes* for the 2 terms in (8)

$$si^{t}=\frac{\mathbf{c}^{t}\bullet\mathbf{\Phi}^{t}}{\left|\mathbf{c}^{t}\right|\left|\mathbf{\Phi}^{t}\right|} \qquad si^{r}=\frac{\mathbf{c}^{r}\bullet\mathbf{\Phi}^{r}}{\left|\mathbf{c}^{r}\right|\left|\mathbf{\Phi}^{r}\right|}$$

with each si spanning from 0 (worst case) to 1 (best case). In other words, the coefficient c is obtained by projecting the vectors $\mathbf{\Phi}^{t}$ and $\mathbf{\Phi}^{r}$ of the deformation shapes on the vectors $\mathbf{c}^{t}$ and $\mathbf{c}^{r}$ that depend on the undeformed shape of the body and on the structure of the metric relation. The sensitivity indexes, which must be considered separately due to the different dimensions of the translations and rotations, define the functional dependence of the metric relation on the translational and rotational deformations.

For any metric relation and for any given deformation mode, the 4 numbers $\mathbf{c}^t \bullet \mathbf{\Phi}^t$, $\mathbf{c}^r \bullet \mathbf{\Phi}^r$, si^t, and si^r can be computed before starting a dynamic simulation. They furnish the information for evaluating the suitability of that mode for the considered metric relation. In order to establish if the mode is suited, it is necessary to consider all 4 numbers and to compare them. Three aspects must be taken into account: a) the metric relation must have a strong functional dependence on either the translation or the rotation (or both) deformations; b) if only one kind of deformation affects the MR, it must constitute the most significant component of the coefficient c; c) the contributions of the two kinds of deformations do not cancel each other. Condition a): a large si^t (or si^r) - e.g., with a value greater that 0.5 - indicates that a translational (or rotational) deformation substantially affects the metric relation; condition b): if this is true, the corresponding component of the coefficient c, $\mathbf{c}^t \bullet \mathbf{\Phi}^t$ (or $\mathbf{c}^r \bullet \mathbf{\Phi}^r$), must be significant with respect to the other $\mathbf{c}^r \bullet \mathbf{\Phi}^r$ (or $\mathbf{c}^t \bullet \mathbf{\Phi}^t$); otherwise, the corresponding deformation gives a negligible contribution to c in (8); condition c): if the signs of $\mathbf{c}^t \bullet \mathbf{\Phi}^t$ and $\mathbf{c}^r \bullet \mathbf{\Phi}^r$ are different, one of the two components must be significantly greater than the other.

From these observations, it is possible to assess the suitability of a mode by using the following rules:

1. At least one of the sensitivity indexes must be large.
2. If both si^t and si^r are large the mode can be considered.
3. If only si^t is large and $\mathbf{c}^t \bullet \mathbf{\Phi}^t$ is large as compared with $\mathbf{c}^r \bullet \mathbf{\Phi}^r$, the, mode can be considered.
4. If only si^r is large and $\mathbf{c}^r \bullet \mathbf{\Phi}^r$ is large as compared with $\mathbf{c}^t \bullet \mathbf{\Phi}^t$, the mode can be considered.
5. If the mode can be considered according to items 1-4 and
   a) if $\mathbf{c}^t \bullet \mathbf{\Phi}^t$ and $\mathbf{c}^r \bullet \mathbf{\Phi}^r$ have the same sign, the mode is suitable;
   b) if $\mathbf{c}^t \bullet \mathbf{\Phi}^t$ and $\mathbf{c}^r \bullet \mathbf{\Phi}^r$ have different signs, the mode is suitable only if $\mathbf{c}^t \bullet \mathbf{\Phi}^t$ and $\mathbf{c}^r \bullet \mathbf{\Phi}^r$ have substantially different values (the coefficient c is not small with respect to the largest of its components).

4.2 Using a set of deformation shapes

In general, for a flexible body it is possible to find a desired number of deformation modes. FEM-based software codes allow one to obtain them in a straightforward way. From this set of modes, it is possible, following the approach outlined in the previous section, to verify if and how many suitable modes exist. In order to obtain the closure equations (1), n suited modes must be selected. If only a smaller number of modes is available, the method cannot be applied. If a larger number is at hand, it is possible to select the best ones. This can be accomplished by comparing the values of the coefficients c_{ij} of the different modes in the MRs of (1). In any case, when n modes are accepted for a flexible body, a system of n equations (1) can be obtained in which the left-hand sides {MR}' of the equations are in the form (5) with $n=m$

$$\mathbf{Cu} = \{MR(\varphi, \varphi_{free}, \mathbf{v}, \mathbf{T}_{kn})\}'' - \mathbf{D} \tag{9}$$

Then, for a given set of free coordinates φ, φ_{free}, the system must be solved for the n unknown modal variables $\mathbf{u}$.

As said before, an interesting feature of the proposed formulation of the closure equations is that, under the small-deformation hypothesis, the system (9) is linear. This is useful to solve the equations and to perform a further test of the suitability of the deformation shapes. To this end, well-known techniques for evaluating the conditioning of linear systems can be applied. The condition number of the coefficient matrix **C** of (9), which quantifies the sensitivity of the algebraic problem, must be small. Moreover, as small deformations are to be expected in the motion of the mechanism, the determinant of **C** must be great with respect to the known terms of (9) in order to ensure small values of the unknown modal coordinates. If more than n suited deformation modes are available, all of them can be tested in order to find their best combination.

At any rate, it is important to note that the coefficients in **C**, the indexes si , the condition number and the determinant of the coefficients can be evaluated numerically starting from the information that does not depend either on the structure of the multibody system, or on the actual displacement of the flexible body during its motion. Then, all those terms can be computed and the suitability of the modes evaluated only once for any flexible body of the system.

References

Blajer, W., Schielen, W., and Schirm, W, (1994) A Projective Criterion to the Coordinate Partitioning Method for Multibody Dynamics, *Archive of Applied Mech.*, Vol. 64, pp. 86-98.

Di Gregorio, R., and Parenti Castelli, V., (1999) Influence of Leg Flexibility on the Kinetostatic Behaviour of a 3-dof Fully Parallel Manipulator, *X World Congress on the Theory of Machines and Mechanisms*, pp. 1091-1098.

Fanghella, P., and Galletti, C., (1993) A Modular Method for Computational Kinematics, in *Computational Kinematics*, Kluwer, pp. 275-284.

Fanghella, P., and Galletti, C., (1995) Metric Relations and Displacement Groups in Mechanism and Robot Kinematics, *Trans. ASME, J. Mech. Design*, Vol. 117, pp. 470-478.

Fanghella, P., Galletti, C., and Mezzadra, V., (2001) Choice of Free Coordinate to Simplify the Dynamic Equations of Multibody Systems with Flexible Bodies (in Italian), Paper n. SP_ME_26, *XV AIMETA Conference.*

Hiller, M., and Kecskeméthy, A., (1994) Dynamics of Multibody Systems with Minimal Coordinates, in *Computer-Aided Analysis of Rigid and Flexible Mechanical Systems*, NATO ASI Series, Vol. 268, pp. 61-100.
Jeng, Y., and Petzold L. R., (1998) An Efficient Newton-Type Iteration for the Numerical Solution of Highly Oscillatory Constrained Multibody Dynamic Systems, *SIAM J. Scient. Comput.*, Vol. 19, No. 5, pp. 1513-1534.
Kovacs, P., and Hommel, G., (1999) An algorithm for the kinematic analysis and optimal input joint assignment of recursive mechanisms, *Mech. Mach. Theory*, Vol. 34, pp. 615-635.
Wehage, R. A., and Haug, E. J. (1982), Generalized Coordinate Partitioning for Dimension Reduction in Analysis of Constrained Dynamic Systems, *Trans. ASME, J. Mech. Design*, Vol. 134, pp. 247-255.

Appendix 1 Metric relations between displacement groups

Given two displacement groups and their invariant properties, defined by points $\mathbf{O}_A$ and $\mathbf{O}_B$ and unit vectors $\mathbf{k}_A$, $\mathbf{k}_B$, the following metric relations (MRs) that arise among the IPs of the two groups are considered

1) $\mathbf{k}_A \bullet \mathbf{k}_B$; 2) $(\mathbf{O}_A - \mathbf{O}_B)\bullet(\mathbf{k}_A \times \mathbf{k}_B)$; 3) $(\mathbf{O}_A - \mathbf{O}_B)\bullet\mathbf{k}_A$; 4) $(\mathbf{O}_B - \mathbf{O}_A)\bullet\mathbf{k}_B$;
5) $((\mathbf{O}_B - \mathbf{O}_A)\times\mathbf{k}_A)^2$; 6) $((\mathbf{O}_B - \mathbf{O}_A)\times\mathbf{k}_B)^2$; 7) $((\mathbf{O}_A - \mathbf{O}_B)\times\mathbf{k}_A)\bullet(\mathbf{k}_B \times \mathbf{k}_A)$;
8) $((\mathbf{O}_B - \mathbf{O}_A)\times\mathbf{k}_B)\bullet(\mathbf{k}_A \times \mathbf{k}_B)$; 9) $(\mathbf{O}_B - \mathbf{O}_A)\bullet(\mathbf{O}_B - \mathbf{O}_A)$

Appendix 2 Metric relations 4 and 9 (rigid body)

$$MR_4 \equiv (O_{B0y} - O_{A0y})\sin\vartheta_{Bx}\cos\vartheta_{By} - (O_{B0x} - O_{A0x})\sin\vartheta_{By} - (O_{B0z} - O_{A0z})\cos\vartheta_{Bx}\cos\vartheta_{By}$$

$$MR_9 \equiv O_{B0x}^2 + O_{B0y}^2 + O_{B0z}^2 - 2O_{A0x}O_{B0x} - 2O_{A0y}O_{B0y} - 2O_{A0z}O_{B0z} + O_{A0x}^2 + O_{A0y}^2 + O_{A0z}^2$$

Appendix 3 Coefficients of metric relations 4 and 9 (flex. body, index j is omitted)

$$\begin{aligned} c_4 \equiv &-((B_{0x} - A_{0x})\sin\vartheta_{B0x}\cos\vartheta_{B0y} + (B_{0y} - A_{0y})\sin\vartheta_{B0y})\Phi^r_{Bz} + \\ &+((B_{0z} - A_{0z})\sin\vartheta_{B0y} - (B_{0x} - A_{0x})\cos\vartheta_{B0x}\cos\vartheta_{B0y})\Phi^r_{By} + \\ &+((B_{0y} - A_{0y})\cos\vartheta_{B0x}\cos\vartheta_{B0y} + (B_{0z} - A_{0z})\sin\vartheta_{B0x}\cos\vartheta_{B0y})\Phi^r_{Bx} + \\ &-(\Phi^t_{Bx} - \Phi^t_{Ax})\sin\vartheta_{B0y} + (\Phi^t_{By} - \Phi^t_{Ay})\sin\vartheta_{B0x}\cos\vartheta_{B0y} - (\Phi^t_{Bz} - \Phi^t_{Az})\cos\vartheta_{B0x}\cos\vartheta_{B0y} \end{aligned}$$

$$c_9 \equiv 2(B_{0x} - A_{0x})(\Phi^t_{Bx} - \Phi^t_{Ax}) + 2(B_{0y} - A_{0y})(\Phi^t_{By} - \Phi^t_{Ay}) + 2(B_{0z} - A_{0z})(\Phi^t_{Bz} - \Phi^t_{Az})$$

Acknowledgment: This work has been developed under a grant of Italian MIUR (progetto METAFORE)

SUBDIVISION ALGORITHMS FOR MOTION DESIGN BASED ON HOMOLOGOUS POINTS

M. Hofer and H. Pottmann
Institute of Geometry
Vienna University of Technology, Vienna, Austria
hofer@geometrie.tuwien.ac.at, pottmann@geometrie.tuwien.ac.at

B. Ravani
Department of Mechanical & Aeronautical Engineering
University of California, Davis, USA
bravani@ucdavis.edu

Abstract We present two algorithms for the interpolation of given positions of a moving body by a smooth and fair motion, such that chosen feature points of the moving system run on smooth and fair paths. We outline algorithms which rely on known interpolatory variational subdivision for curves and on registration techniques from Computer Vision. For the numerical solution of the arising optimization problems we propose a geometric method which is based on instantaneous kinematics.

Keywords: motion design, interpolation, variational subdivision, kinematics, registration

1. Introduction

An important topic in motion design are *smooth* motions with interpolation or approximation constraints. Motions of that type, which make use of NURBS techniques, have been investigated in several publications (see e.g. Jüttler and Wagner, 1996; Röschel, 1998). Recently, Hyun et al., 2001 constructed affine rational spline motions with minimal distortion. For motion design based on the quaternion representation of the spherical component and nonlinear extensions of spline constructions in affine spaces to the sphere see Fang et al., 1998.

The present paper provides alternatives for motion design using *subdivision algorithms.* This has several advantages over previous approaches: It is easier to deal with *variational motion design*, it leads in a natural way to a *multiresolution representation* for the designed motion, and it can be extended to motion design in the *presence of obstacles. Homol-*

J. Lenarčič and F. Thomas (eds.), Advances in Robot Kinematics, 235–244.

ogous points are the different locations of a single feature point as the moving body takes several positions Our work emphasizes optimality with respect to some fairness criterion based on trajectories of chosen feature points. As a rigid body motion is a curve in the Lie group SE(3), variational motion design could be based on the fairness of that curve (see Park and Ravani, 1997). However, for the applications we have in mind, we believe that it is most important that the body to be moved behaves well.

We present two *subdivision algorithms based on homologous points* for the design of fair and smooth motions. Consider the given positions of the moving body as a coarse approximation of the motion we want to design. Then we refine this motion by iteratively inserting more and more intermediate positions. The resulting dense set of discrete positions of the moving body might be sufficient for the application in mind, or it can easily be interpolated with other motion design techniques without caring about the fairness of the motion anymore. The main difference of the two algorithms is how intermediate positions of the moving body are inserted to refine the motion.

The paper is organized as follows. In Sec. 2, 3, 4 we present the main ingredients for the two motion design algorithms, *interpolatory variational subdivision*, *instantaneous kinematics* and *registration*. In Sec. 5 we present the algorithm *subdivision for motion design I*, which is actually a transfer principle from curve to motion design. A related, but different approach, is the algorithm *subdivision for motion design II*, presented in Sec. 6. We conclude the paper in Sec. 7 with an outlook on possible extensions of the presented algorithms.

2. Interpolatory variational subdivision

Variational subdivision has been introduced by Kobbelt, 1996 and involves the minimization of some energy functional to control the fairness of the curves (or surfaces) that are constructed.

We start with an *open* polygon and in each iteration step we insert new points $\mathbf{q}_i, i = 1, \ldots, N-1$ between the points $\mathbf{p}_i, i = 1, \ldots, N$ of the polygon to be refined such that the second forward differences are minimized (see Fig. 1, left). The position of the new points is thus found by minimizing the objective function

$$F(\mathbf{q}_1, \ldots, \mathbf{q}_{N-1}) = \sum_{i=1}^{N-1} \|\mathbf{m}_i - \mathbf{q}_i\|^2 + \sum_{i=2}^{N-1} \|\mathbf{n}_i - \mathbf{p}_i\|^2 \qquad (1)$$

where $\mathbf{m}_i = (\mathbf{p}_i + \mathbf{p}_{i+1})/2$ and $\mathbf{n}_i = (\mathbf{q}_{i-1} + \mathbf{q}_i)/2$. F is *quadratic* in the unknowns $\mathbf{q}_i \in \mathbb{R}^d$ and thus the minimization of F leads to the following

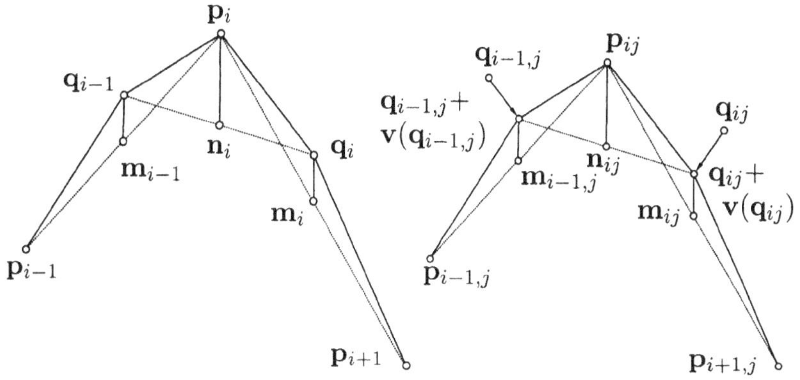

Figure 1. Squared distances that are minimized in Kobbelt's interpolatory variational subdivision (left) and in the simultaneous variational subdivision on the K sequences of homologous points of motion design with subdivision II (right).

tridiagonal linear system of equations, which can be solved efficiently using e.g. sparse matrix techniques,

$$\begin{pmatrix} 5 & 1 & & & \\ 1 & 6 & 1 & & \\ & \ddots & \ddots & \ddots & \\ & & 1 & 6 & 1 \\ & & & 1 & 5 \end{pmatrix} \begin{pmatrix} \mathbf{q}_1 \\ \mathbf{q}_2 \\ \vdots \\ \vdots \\ \mathbf{q}_{N-2} \\ \mathbf{q}_{N-1} \end{pmatrix} = \begin{pmatrix} 2\mathbf{p}_1 + 4\mathbf{p}_2 \\ 4(\mathbf{p}_2 + \mathbf{p}_3) \\ \vdots \\ \vdots \\ 4(\mathbf{p}_{N-2} + \mathbf{p}_{N-1}) \\ 4\mathbf{p}_{N-1} + 2\mathbf{p}_N \end{pmatrix} \tag{2}$$

This refinement scheme is global, i.e., every new point depends on all points of the polygon to be refined. Interpolation is guaranteed since the old points belong to the newly calculated finer version. Kobbelt, 1996 has shown that this scheme generates at least C^2 curves. The scheme corresponds to an underlying *uniform* parametrization and can be extended to a *non-uniform* parametrization, cf. Kobbelt and Schröder, 1998.

3. Instantaneous kinematics

Consider a smooth one-parameter rigid body motion in Euclidean 3-space. Introducing Cartesian coordinate systems in the moving system Σ and in the fixed system Σ_0, the time dependent position $\mathbf{x}_0(t)$ of a point $\mathbf{x} \in \Sigma$ in the fixed system is given by

$$\mathbf{x}_0(t) = \mathbf{a}(t) + M(t) \cdot \mathbf{x}. \tag{3}$$

Here, the time dependent orthogonal matrix $M(t)$ represents the spherical component of the motion, and $\mathbf{a}(t)$ describes the trajectory of the origin of the moving system. All arising functions shall be C^1. By differentiation we get the velocity vectors. It is well-known that the velocity vector field is linear at any time instant and that it has the form

$$\mathbf{v}(\mathbf{x}) = \bar{\mathbf{c}} + \mathbf{c} \times \mathbf{x}, \tag{4}$$

where $\bar{\mathbf{c}}$ represents the velocity vector of the origin, and $\mathbf{c}$ is the so-called Darboux vector (vector of angular velocity). Only very special one-parameter motions have a time-independent velocity vector field. These motions are a *translation* with constant velocity (if $\mathbf{c} = \mathbf{0}$), a uniform *rotation* about an axis (if $\mathbf{c} \cdot \bar{\mathbf{c}} = 0$) and a uniform *helical motion* (if $\mathbf{c} \cdot \bar{\mathbf{c}} \neq 0$). The most general case is that of a uniform helical motion, which is the superposition of a rotation with constant angular velocity about an axis A and a translation with constant velocity parallel to A. If the moving body rotates about an angle α, the translation distance is $p \cdot \alpha$. We compute axis A, pitch p and angular velocity ω from $\mathbf{c}, \bar{\mathbf{c}}$ by

$$\mathbf{a} = \frac{\mathbf{c}}{\|\mathbf{c}\|}, \qquad \bar{\mathbf{a}} = \frac{\bar{\mathbf{c}} - p\mathbf{c}}{\|\mathbf{c}\|}, \qquad p = \frac{\mathbf{c} \cdot \bar{\mathbf{c}}}{\mathbf{c}^2}, \qquad \omega = \|\mathbf{c}\|, \tag{5}$$

where the axis A is parallel to $\mathbf{a}$ and $\bar{\mathbf{a}} = \mathbf{p} \times \mathbf{a}$ is the *moment vector* of A, which is independent of the choice of the point $\mathbf{p}$ on A (see e.g. Pottmann and Wallner, 2001).

4. Registration

Consider two clouds X, Y of *corresponding* points $\mathbf{x}_i$, $i = 1, \ldots, N$ and $\mathbf{y}_i$, $i = 1, \ldots, N$, respectively. The problem of applying to one cloud, say X, a Euclidean motion m which brings each $\mathbf{x}_i$ as close as possible to $\mathbf{y}_\mathrm{i}$ is well studied. Formulating it in a least squares sense, i.e., minimizing

$$F = \sum_{i=1}^{N} \|m(\mathbf{x}_i) - \mathbf{y}_i\|^2, \tag{6}$$

the solution amounts to an eigenvalue problem (see e.g. Horn, 1987).

Although there is a direct way to solve the minimization of F in Eq. 6, we describe the following *iterative* algorithm which is based on instantaneous kinematics. It has the advantage that it can be applied directly to the simultaneous alignment of more than two clouds of corresponding points, which we use in the algorithm motion design with subdivision II (see Sec. 6). To each point $\mathbf{x}_i$ attach the velocity vector $\mathbf{v}(\mathbf{x}_i) = \bar{\mathbf{c}} + \mathbf{c} \times \mathbf{x}_i$ of the still unknown velocity vector field of an instantaneous helical motion which can be described by a pair $\mathbf{C} = (\mathbf{c}, \bar{\mathbf{c}}) \in \mathbb{R}^6$. The objective function we minimize in each iteration step,

$$F(\mathbf{C}) = \sum_i \|\mathbf{x}_i + \mathbf{v}(\mathbf{x}_i) - \mathbf{y}_i\|^2, \tag{7}$$

is quadratic in $\mathbf{c}, \bar{\mathbf{c}}$ and thus the solution can be computed using a linear system of equations.

Note that the transformation which maps $\mathbf{x}_i$ to $\mathbf{x}_i + \mathbf{v}(\mathbf{x}_i)$ is an affine map and not a Euclidean transformation. Therefore we use the uniform helical motion that is uniquely determined by the velocity vector field represented by the pair $(\mathbf{c}, \bar{\mathbf{c}})$. Applying a rotation about the axis of the helical motion through an angle of $\alpha = \arctan(\|\mathbf{c}\|)$ and a translation parallel to this axis by the distance $p \cdot \alpha$ brings the points close to the tips $\mathbf{x}_i + \mathbf{v}(\mathbf{x}_i)$ of the velocity vectors used in Eq. 7.

The algorithm iteratively minimizes F in Eq. 7 and updates the position of X with the corresponding helical motion.

5. Motion design with subdivision I

Given are N_0 positions $\Sigma_1, \ldots, \Sigma_{N_0}$ of a moving body Σ which are to be interpolated by a fair and smooth motion. We take a sample of $K \geq 4$ feature points $\mathbf{p}_1, \ldots, \mathbf{p}_K$ from the moving body Σ. In the m-th iteration step we have $N_m = 2^{m-1}(N_0 - 1) + 1$ positions $\Sigma_1, \ldots, \Sigma_{N_m}$ and we insert $N_m - 1$ new positions, one between each two adjacent positions from the previous iteration step. This is done by repeating procedure 1 and 2 outlined below.

Procedure 1. Insert (affinely) distorted copies of the moving body: Apply interpolatory variational subdivision separately to the K sequences of homologous points $\mathbf{p}_{1j}, \ldots, \mathbf{p}_{N_m j}, j = 1, \ldots, K$ (see Fig. 2, left). This gives $N_m - 1$ initial intermediate *affine* copies Σ'_i of the moving body (see Fig. 3, left), which follows from the *linearity* of variational subdivision (Eq. 1). Therefore, it would be sufficient to run the variational subdivision only with four non-coplanar feature points of the moving body. From that one could compute affine maps from an initial position to the inserted affine copies, and use these affine maps to compute the remaining homologous locations of the other $K - 4$ feature points.

Procedure 2. Replace distorted copies of the moving body with Euclidean ones using registration: To each affinely distorted intermediate position Σ'_i we individually register (see Sec. 4) the moving body Σ to find an intermediate position resulting from a rigid body motion. These Euclidean copies of Σ together with those of the previous step are the input to the next iteration step (see Fig. 3, right).

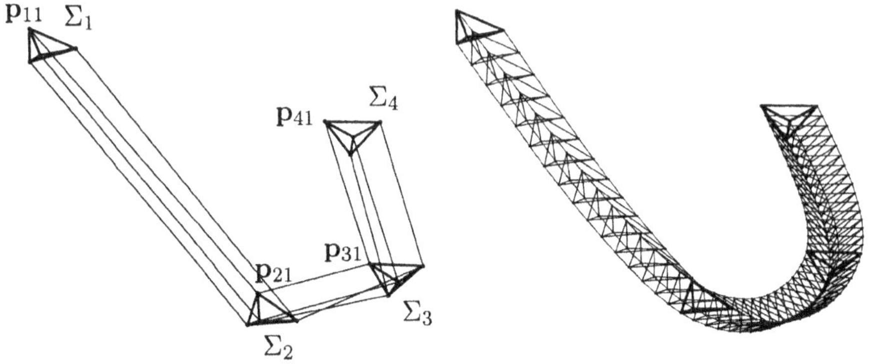

Figure 2. Motion design with subdivision I: The input to the algorithm are four given positions of a moving body and the connecting polygons of the four sequences of homologous points (left). The output after four iterations consists of 49 discrete positions of the moving body (right).

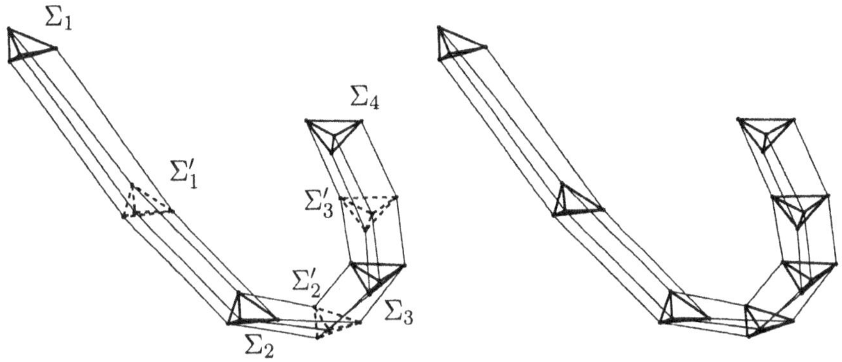

Figure 3. Motion design with subdivision I: To the input we apply procedure 1 of the algorithm (left) and then procedure 2 (right).

Procedure 1 is a global operation that works separately on each sequence of homologous points. It controls the fairness and smoothness of the path of each feature point. Using only procedure 1 we would get an *affine* motion. Therefore, in procedure 2 we use registration to find the best fit, in the least squares sense, of the rigid moving body to each affinely distorted intermediate position. Procedure 2 is a local operation that works on every intermediate position separately.

We know that the variational subdivision scheme produces at least C^2 curves (for a proof we refer to Kobbelt, 1996). Numerical tests give

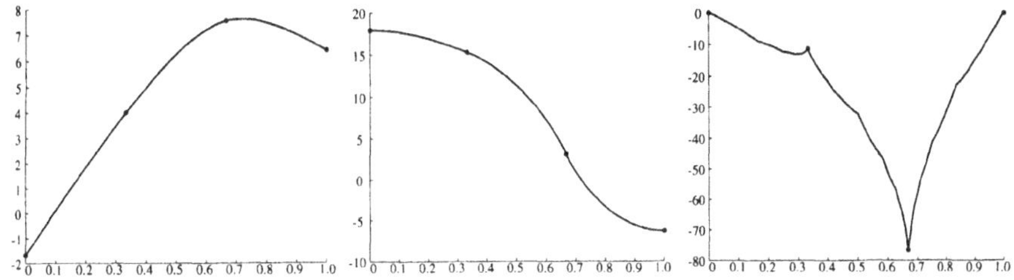

Figure 4. Motion design with subdivision I: Plot of the functions $x(t)$, $x'(t)$, $x''(t)$ where $x(t)$ is the function that assumes over *uniform* parameter values from the interval $[0, 1]$ the x-values of the path of one feature point of the motion.

evidence that using Kobbelt's variational subdivision together with the registration step results in C^2 paths of the feature points and in a C^2 motion (Fig. 4).

The computational cost of the algorithm in the m-th iteration step is the following. First we have to solve K tridiagonal systems of size $N_m - 1$ for the refinement of the paths of the feature points. Then, if we use the iterative registration described by Eq. 7, we need to solve $M(N_m - 1)$ linear systems for 6 unknowns. With a good starting position (centroid of distorted body and moving body coincide), the number of iterations M needed in the registration step is usually 3.

To reduce the computational cost one may stop the algorithm described above after 4-5 iterations, since then a variational scheme is not necessary anymore. One may switch to a local subdivision scheme (see e.g. Warren and Weimer, 2001) and insert as many further positions as needed.

Remarks. Any linear subdivision scheme applied to sequences of homologous points will result in affine intermediate positions. Insertion of intermediate points may also be done with splines, i.e. cubic C^2 splines which are evaluated at intermediate positions. This is one way to address the time distribution of the motion. Another way would be the use of the subdivision scheme corresponding to a *non-uniform* parametrization, see Kobbelt and Schröder, 1998. We may also use nonlinear schemes, but then we do no longer get affine copies and thus have to apply subdivision to *all* sequences of homologous points. In the subsequent registration step, it does not matter whether we have an affine copy or not. But in any case, one has to use all K feature points for the registration.

6. Motion design with subdivision II

Similar to the first method we use K chosen feature points $\mathbf{p}_1, ..., \mathbf{p}_K$ from the moving body Σ. In the m-th iteration step we consider the homologous points $\mathbf{p}_{ij}, i = 1, \ldots, N_m, j = 1, \ldots, K$. Our calculations also use the *centroid* (center of mass) $\mathbf{s}$ of the moving body Σ. In this second algorithm we first insert initial Euclidean copies Σ'_i between each pair of adjacent positions Σ_i and Σ_{i+1}. Then we reposition $\Sigma'_1, \ldots, \Sigma'_{N_m-1}$ *simultaneously.* Thereby, the fairness of the result is envoked by using a simultaneous variational subdivision on the K sequences of homologous points. The iterative algorithm repeats the following two procedures.

Procedure 1. Insert intermediate Euclidean copies of the moving body: First, we apply to the homologous positions $\mathbf{s}_1, \ldots, \mathbf{s}_{N_m}$ of the *centroid* $\mathbf{s}$ one iteration step of the interpolatory variational subdivision to obtain intermediate positions $\mathbf{s}'_1, \ldots, \mathbf{s}'_{N_m-1}$. Second, we compute the angle ϕ_i of the rotational part A_i of the unique helical motion $\mathbf{x} \mapsto A_i\mathbf{x} + a_i$ that transforms Σ_i to the adjacent system Σ_{i+1}, for $i = 1, \ldots, N_m - 1$. The initial intermediate position Σ'_i between Σ_i and Σ_{i+1} is chosen such that its centroid is $\mathbf{s}'_i$ and a rotation through $\frac{\phi_i}{2}$ is applied to Σ_i (see Fig. 5, left).

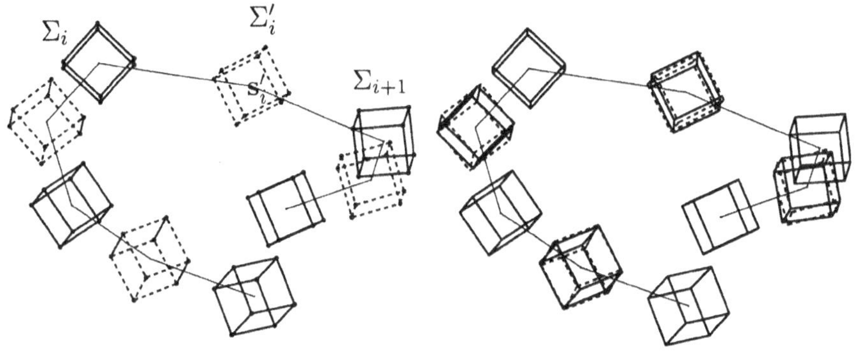

Figure 5. Motion design with subdivision II. The result after procedure 1 (left) and then procedure 2 (right) have been applied to the input data once.

Procedure 2. Simultaneously reposition the inserted Euclidean copies with respect to a fairness criterion: For the simultaneous optimization (with respect to fairness) of the intermediate positions we use linearized motions (velocity fields). To the feature points in each single position Σ'_i we attach vectors belonging to the linear velocity vector field of an instantaneous helical motion (see Eq. 4) described by the pair $(\mathbf{c}_i, \bar{\mathbf{c}}_i)$. The velocity vectors are used for first order estimates of the new positions. Simultaneous variational subdivision on the K sequences of homologous points amounts to the minimization of the following objective function

$$F(\mathbf{C}) = \sum_{j=1}^{K} \left(\sum_{i=1}^{N_m-1} \|\mathbf{m}_{ij} - \mathbf{q}_{ij} - \mathbf{v}(\mathbf{q}_{ij})\|^2 + \sum_{i=2}^{N_m-1} \|\mathbf{n}_{ij} - \mathbf{p}_{ij}\|^2 \right) \tag{8}$$

where $\mathbf{C}$ is the $(2N_m - 2)$-tupel $(\mathbf{c}_1, \bar{\mathbf{c}}_1, \ldots, \mathbf{c}_{N_m-1}, \bar{\mathbf{c}}_{N_m-1})$,

$$\mathbf{m}_{ij} = (\mathbf{p}_{ij} + \mathbf{p}_{i+1,j})/2, \quad \mathbf{v}(\mathbf{q}_{ij}) = \bar{\mathbf{c}}_i + \mathbf{c}_i \times \mathbf{q}_{ij} \tag{9}$$

$$\mathbf{n}_{ij} = [\mathbf{q}_{i-1,j} + \mathbf{v}(\mathbf{q}_{i-1,j}) + \mathbf{q}_{ij} + \mathbf{v}(\mathbf{q}_{ij})]/2, \tag{10}$$

and $\mathbf{q}_{ij}, i = 1, \ldots, N_m - 1, j = 1, \ldots, K$ are the initial intermediate positions of the chosen feature points (see e.g. Fig. 1, right). F is quadratic in the $2N_m - 2$ unknowns $\mathbf{c}_1, \bar{\mathbf{c}}_1, \ldots, \mathbf{c}_{N_m-1}, \bar{\mathbf{c}}_{N_m-1}$, the minimization leads to the solution of a banded linear system of equations which can be solved efficiently.

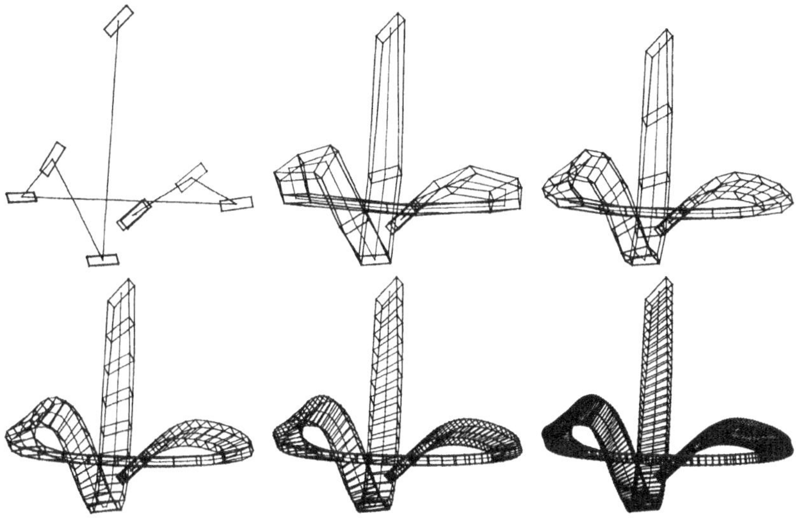

Figure 6. Example of motion design with subdivision II. Shown are the input to the algorithm and the first to fifth iteration step.

Now the initial intermediate positions Σ_i' are corrected using the helical motion that is uniquely determined by the velocity vector field represented by the pair $(\mathbf{c}_i, \bar{\mathbf{c}}_i)$. This means that we apply to Σ_i' a rotation through an angle of $\alpha_i = \arctan(\|\mathbf{c}_i\|)$ and a translation parallel to the axis $(\mathbf{a}_i, \bar{\mathbf{a}}_i)$ by the distance $p_i \cdot \alpha_i$. It brings the points $\mathbf{q}_{ij}$ close to the tips $\mathbf{q}_{ij} + \mathbf{v}(\mathbf{q}_{ij})$ of the velocity vectors used in Eq. 8.

Remarks. The smoothness analysis of the resulting motion is analogous to the one performed in Sec. 5. Numerical tests again suggest that the resulting motion is C^2. The computational cost of the algorithm in the m-th iteration step is the following. We have $N_m = 2^{m-1}(N_0 - 1) + 1$ positions and insert $N_m - 1$. Thus, first we have to solve one linear system of equations of size $N_m - 1$ to refine the sequence of homologous positions of the centroid $\mathbf{s}$ of the moving body. Then, for the simultaneous repositioning of the inserted Euclidean copies of the moving body we have to solve a banded linear system of $6(N_m - 1)$ equations.

7. Conclusions

We have presented two subdivision algorithms for motion design that generate C^2 motions with respect to some optimality criterion. Our approach deals with variational motion design. Both algorithms can be modified for the design of motions approximating given input positions. Algorithm II is extendable to motion design in the presence of obstacles.

8. Acknowledgements

This work has been carried out within the K plus Competence Center *Advanced Computer Vision* and was funded from the K plus program. We would like to thank Bert Jüttler and Stefan Leopoldseder for their insightful and valuable comments on an earlier version of this paper.

References

Fang, Y.C., Hsieh, C.C., Kim, M.J., Chang, J.J., Woo, T.C. (1998), Real time motion fairing with unit quaternions. *Computer-Aided Design*, vol. 30, no. 3, pp. 191–198.

Horn, B.K.P. (1987), Closed form solution of absolute orientation using unit quaternions. *Journal of the Optical Society A*, vol. 4, no. 4, pp. 629–642.

Hyun, D.E., Jüttler, B., Kim, M.S. (2001), Minimizing the distortion of affine spline motions. *Proceedings of Pacific Graphics 01*, Tokyo, Japan, pp. 50–59.

Jüttler, B., Wagner, M. (1996), Computer Aided Design with spatial rational B-spline motions, *ASME Journal of Mechanical Design*, vol. 118, no. 2, pp. 193–201.

Kobbelt, L. (1996), A variational approach to subdivision. *Computer Aided Geometric Design*, vol. 13, no. 8, pp. 743–761.

Kobbelt, L., Schröder, P. (1998), A multiresolution framework for variational subdivision. *ACM Transactions on Graphics*, vol. 17, no. 4, pp. 209–237.

Park, F.C., Ravani, B. (1997), Smooth invariant interpolation of rotations. *ACM Transactions on Graphics*, vol. 16, no. 3, pp. 277–295.

Pottmann, H., Wallner, J. (2001), *Computational Line Geometry*, Springer-Verlag.

Röschel, O. (1998), Rational motion design - a survey. *Computer-Aided Design*, vol. 30, no. 3, pp. 169–178.

Warren, J., Weimer, H. (2001), *Subdivision Methods for Geometric Design: A Constructive Approach*, Morgan Kaufmann Series in Computer Graphics.

GROUP THEORY CAN EXPLAIN THE MOBILITY OF PARADOXICAL LINKAGES

José M. Rico
Instituto Tecnológico de Celaya, Dept. de Ingeniería Mecánica
Celaya, Gto. 38000, México
mrico@itc.mx

Bahram Ravani
University of California, Davis
Dept. of Mechanical and Aeronautical Engineering
Davis, California 95616, U.S.A.
bravani@ucdavis.edu

Abstract The present contribution shows that a closer look into the applications of group theory to the kinematics of spatial linkages provides a concise, simple and correct explanation of the mobility of several linkages classified as overconstrained and paradoxical. The analysis readily yields the Bennett linkage as well as families of R-C-R-C and H-C-H-C linkages, all of them overconstrained and paradoxical. Moreover, the analysis not only yields the geometric characteristics of the linkages but also the relationships between the displacements of the different kinematic pairs. Furthermore, these relationships easily provide, when the motion of the linkages starts from a special configuration, the input-output equations. The study of paradoxical linkages are important to robot kinematics because they are inherently more rigid that trivial linkages and they might provide better architectures for serial or parallel manipulators.

Keywords: Group theory, mobility, paradoxical linkages, overconstrain

1. Introduction

One of the few remaining unsolved problems of classical kinematics is that of determination and classification of overconstrained and paradoxical linkages. Obviously, this topic has close ties with the development of mobility criteria, in particular the Kutzbach-Grübler criterion. Indeed, the overconstrained linkages were defined by comparing its mobility with that predicted by the Kutzbach-Grübler criterion. For example, Waldron, 1979, after presenting the Kutzbach-Grübler criterion, states "...it comes as a shock to most students of mechanism science to discover that

J. Lenarčič and F. Thomas (eds.), Advances in Robot Kinematics, 245–254.

there are linkages that do not obey equation (1) – the Kutzbach-Grübler criterion –. These exceptions, called overconstrained or overclosed...".

In 1978, Hervé, 1978 resorting to group theory proposed a classification of linkages based on the subgroups of the Euclidean group associated with the relative motions between its links and the intersection between some of these subgroups. Hervé, 1978 classified the linkages into trivial, exceptional and paradoxical. Trivial linkages satisfy the Kutzbach-Grübler criterion or one of its special forms. Exceptional linkages owe their mobility to the non-trivial intersection of two subgroups, where none of the subgroups is a subgroup of the other. Unfortunately, Hervé, 1978 did not propose an alternative mobility criterion for exceptional linkages. Paradoxical linkages must satisfy special constraints involving the links dimensions in order to be mobile. None of the exceptional and paradoxical linkages satisfy the Kutzbach-Grübler criterion. Angeles, 1982 made the following comment "... Finally, it –a linkage or kinematic chain– is said to be paradoxical if its mobility cannot be analyzed with the sole aid of the theory of groups. Experience shows that paradoxical chains possess certain axes of symmetry which produce their mobility...".

Recently, Rico and Ravani, 2001 building upon the notions introduced by Hervé, 1978 and adding some newer concepts due to Fanghella, 1988 and Fanghella and Galletti, 1995 were able to find a mobility criterion applicable to exceptional linkages such that the Kutzbach-Grübler criterion is a special case. Thus, paradoxical linkages appear to be the last obstacle to obtain a comprehensive group theory based mobility analysis of linkages. In this contribution, it will be shown that, contrary to Angeles' comment, group theory is all what is needed to explain the mobility of at least a good portion of the paradoxical linkages. Due to space limitations, most results are stated without proof.

2. Conjugacy and Symmetric Spherical Linkages

Rico and Ravani, 2000 showed that conjugacy, a notion of group theory, allows the synthesis of symmetric spherical linkages. More precisely, if $g_1 = C\frac{\theta_1}{2} + S\frac{\theta_1}{2}\hat{u}_1$ and $g_2 = C\frac{\theta_2}{2} + S\frac{\theta_2}{2}\hat{u}_2$, are the quaternion representations of a pair of rotational motions around a fixed point, there is a constant unit quaternion, $g \in \Re^3 \subset H_u$, that represents a reflection around a unit vector $\hat{u}$ parallel to $tan\frac{\theta_1}{2}\hat{u}_1 + tan\frac{\theta_2}{2}\hat{u}_2$, such that[1]

$$g_1 g_2 = g^-(g_2 g_1)g = g(g_2 g_1)g^- = (g g_2 g^-)(g g_1 g^-) = g_4 g_3, \qquad (1)$$

where

$$g_4 = g g_2 g^- \qquad \text{and} \qquad g_3 = g g_1 g^-.$$

Here, g_4, and g_3 are the image motions of g_2, and g_1 respectively under the conjugation by g. It should be noted that the unit vectors $\hat{u}$, $\hat{u}_1$, and $\hat{u}_2$ are coplanar. The original axes and the conjugate axes are shown in Figure 1.

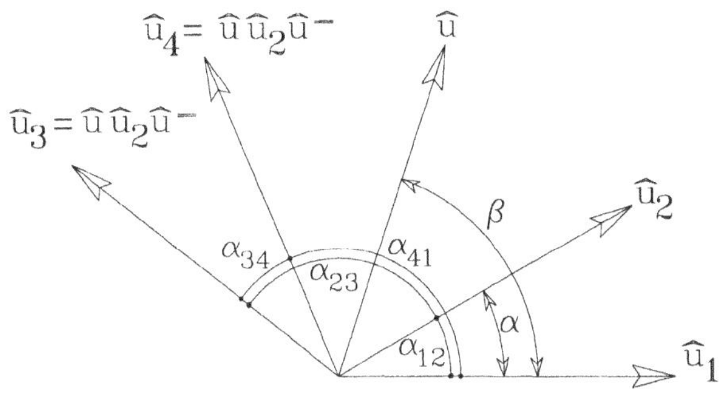

Figure 1. Rotation and Reflection Axes in the Resulting Spherical Linkage in a "Flattened Configuration".

Further, it can be shown that the angular displacements are related by

$$tan\frac{\theta_2}{2} = -tan\frac{\theta_1}{2}\frac{S\beta}{S(\beta - \alpha)}. \qquad (2)$$

Moreover, it is also shown that

$$\alpha_{12} = \alpha_{34} = \alpha \qquad \text{and} \qquad \alpha_{23} = \alpha_{41} = 2\beta - \alpha.$$

Therefore

$$\alpha = \alpha_{12} \qquad \text{and} \qquad \beta = \frac{\alpha_{12} + \alpha_{41}}{2}.$$

Thus, equation (2) can be written as

$$tan\frac{\theta_2}{2} = -tan\frac{\theta_1}{2}\frac{S\left(\frac{\alpha_{41}+\alpha_{12}}{2}\right)}{S\left(\frac{\alpha_{41}-\alpha_{12}}{2}\right)}. \qquad (3)$$

Equation (3) relates the angular displacements θ_1 around $\hat{u}_1$, and θ_2 around $\hat{u}_2$. Furthermore, due to equation (1), the remaining angular displacements are $\theta_3 = \theta_1$ around $\hat{u}_3 = \hat{u}\hat{u}_1\hat{u}^-$ and $\theta_4 = \theta_2$ around $\hat{u}_4 = \hat{u}\hat{u}_2\hat{u}^-$. Despite the similarity with already published results, equation (3) is not, properly speaking, the input-output equation of the spherical symmetric linkage. However, if the initial position of the four bar linkage is that indicated in Figure 1, it follows that equation (3) also provides the required input-output equation.

3. Conjugacy and Symmetric Spatial Linkages

In this section, the results obtained for symmetric spherical four-bar linkages will be extended to the spatial case. The spatial case will provide two sets of equations, the first set involves the orientation of the kinematic chain pairs, and it will be the same set of equations already solved in the corresponding spherical case. The second set involves not only the orientation but also the translational location of the linkage's pairs.

It should be noted that the symmetric spherical linkages have the property that the four revolute axes together with the reflection axis are, at some position, coplanar. Therefore, all the links of the spatial linkage at the same position will be perpendicular to the common plane of the revolute and reflection axes; i.e. any spatial linkage obtained from a symmetric spherical linkage will have, at this position, all their links in line. The following analysis will take advantage of this property.

In a very formal way, the following problem will be investigated. Let $\hat{g}_1(\theta_1, s_1)$, $\hat{g}_2(\theta_2, s_2)$ be the biquaternion representation of a pair of two-parameter motions, each one associated with a cylindric pair. This section investigates the conditions under which

$$\hat{g}(\hat{g}_1(\theta_1, s_1)\hat{g}_2(\theta_2, s_2))\hat{g}^- = \hat{g}_2(\theta_2, s_2)\hat{g}_1(\theta_1, s_1), \tag{4}$$

where $\hat{g}$ is a constant Euclidean displacement, is satisfied. As indicated previously, the dual equation (4) is composed of two parts. The real part of the equation is related to the spherical problem already solved in Section 2. The dual part of the equation involves both, spherical and translational, motions and its solution is the main challenge of this section.

Proposition. Assume that the screw axes associated with the spatial motions $\hat{g}_1, \hat{g}_2$, and the reflection with a possible attached translation $\hat{g}$ are given by $(\hat{u}_1, \vec{r}_1), (\hat{u}_2, \vec{r}_2)$, and $(\hat{u}, \vec{r})$ respectively, where $\hat{u}_i$ is a unit vector along the screw axis and $\vec{r}_i$ is the position vector from an origin to the respective screw axis. Then

$$\hat{g}(\hat{g}_1(\theta_1, s_1)\hat{g}_2(\theta_2, s_2))\hat{g}^- = \hat{g}_2(\theta_2, s_2)\hat{g}_1(\theta_1, s_1),$$

if

1 θ_1 and θ_2 satisfy the relationship

$$tan\frac{\theta_2}{2} = -tan\frac{\theta_1}{2}\frac{S\beta}{S(\beta-\alpha)}.$$

2 $\hat{g}$ is a line reflection around $(\hat{u}, \vec{r})$.

3 There is a line, see Figure 2, called the common normal, that intersects perpendicularly the three axes $(\hat{u}_1, \vec{r}_1)$, $(\hat{u}_2, \vec{r}_2)$, and $(\hat{u}, \vec{r})$.[2]

From these common results, it is possible to distinguish the following subcases.

Case 1

1-a $s_2 = s_1 = 0$.

1-b The distance along the common normal between $(\hat{u}_1, \vec{r}_1)$ and $(\hat{u}, \vec{r})$, denoted z, and the distance along the common normal between $(\hat{u}_1, \vec{r}_1)$ and $(\hat{u}_2, \vec{r}_2)$, denoted a, are related by

$$z = \frac{S\beta}{S\alpha} C(\beta - \alpha)\, a.$$

This case leads to the Bennett linkage. An overconstrained and paradoxical, class II, kinematic chain, Rico and Ravani, 2001.

Case 2

2-a $s_2 = 0$, $s_1 = k_1 a\, S\theta_1$, where $k_1 \in \Re$.

2-b The distance along the common normal between $(\hat{u}_1, \vec{r}_1)$ and $(\hat{u}, \vec{r})$, denoted z, and the distance along the common normal between $(\hat{u}_1, \vec{r}_1)$ and $(\hat{u}_2, \vec{r}_2)$, denoted a, are related by

$$z = \frac{S\beta}{S\alpha}[C(\beta - \alpha) + k_1\, S(\beta - \alpha)]a.$$

This case leads to a C-R-C-R linkage. An overconstrained and paradoxical, class II, kinematic chain, see Rico and Ravani, 2001.

Case 3

3-a $s_2 - s_1 = k_2\, a\, S\theta_1$, where $k_2 \in \Re$.

3-b The angles α and β satisfy the condition

$$\alpha = 2\beta.$$

3-c The distance along the common normal between $(\hat{u}_1, \vec{r}_1)$ and $(\hat{u}, \vec{r})$, denoted z, and the distance along the common normal between $(\hat{u}_1, \vec{r}_1)$ and $(\hat{u}_2, \vec{r}_2)$, denoted a, are related by

$$z = \frac{1 + k_2\, tan\beta}{2} a.$$

This case leads to a parallel R-C-R-C linkage and other variants,including a H-C-H-C linkage. All these linkages are overconstrained and paradoxical, class I, kinematic chains, see Rico and Ravani, 2001.

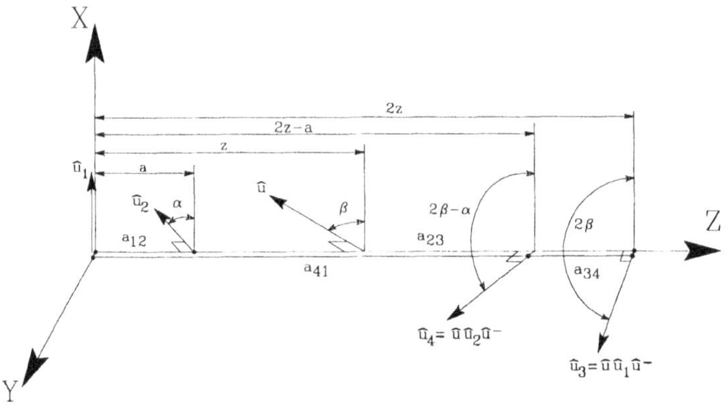

Figure 2. Screw and Reflection Axes in the Second Symmetry Condition for Spatial Linkages.

Proof: From the analysis of the corresponding spherical equation, see Section 2, it follows the relationship between θ_1 and θ_2, and the result that $\hat{g}$ must be a reflection; although, the location of the reflection axis and the possibility of a translation along the reflection axis are still undetermined. All these results will be discovered in the course of this proof.

Without loss of generality, assume that the screw axes of the motions $\hat{g}_1$ and $\hat{g}_2$ are given by

$$(\hat{u}_1 = (1,0,0), \vec{r}_1 = (0,0,0)) \quad \text{and} \quad (\hat{u}_2 = (C\alpha, S\alpha, 0), \vec{r}_1 = (0,0,a)),$$

respectively. Finally assume that the screw axis of the constant Euclidean displacement $\hat{g}$ is given by $(\hat{u} = (C\beta, S\beta, 0), \vec{r}_1 = (0, y, z))$. Then

$$\hat{g}_1 = g_1 + \frac{1}{2}\epsilon \vec{t}_1 g_1$$

where

$$g_1 = C\frac{\theta_1}{2} + S\frac{\theta_1}{2}\hat{i} \quad \text{and} \quad \vec{t}_1 = s_1 \hat{i},$$

$$\hat{g}_2 = g_2 + \frac{1}{2}\epsilon \vec{t}_2 g_2,$$

where

$$g_2 = C\frac{\theta_2}{2} + S\frac{\theta_2}{2}(C\alpha\hat{\imath} + S\alpha\hat{\jmath}),$$

and

$$\vec{t}_2 = (-aS\alpha S\frac{\theta_2}{2} + s_2 C\alpha)\hat{\imath} + (-aC\alpha S\frac{\theta_2}{2} + s_2 S\alpha)\hat{\jmath} + S\frac{\theta_2}{2}\hat{k}.$$

Finally

$$\hat{g} = g + \frac{1}{2}\epsilon\vec{t}g$$

where

$$g = C\frac{\phi}{2} + S\frac{\phi}{2}(C\beta\hat{\imath} + S\beta\hat{\jmath}) \quad \text{and} \quad \vec{t} = (s_u C\beta)\hat{\imath} + (s_u S\beta + 2y)\hat{\jmath} + 2z\hat{k}.$$

Here, s_u indicates a possible translation along the reflection axis, $(\hat{u}, \vec{r})$, associated with $\hat{g}$. Finally, the $\vec{t}_1, \vec{t}_2$, and $\vec{t}$ are the translation vectors of the origin O. Then, substituting $\hat{g}_1, \hat{g}_2$, and $\hat{g}$, and solving equation (4), the real part yields the condition (see equation (2), in Section 2)

$$tan\frac{\theta_2}{2} = -tan\frac{\theta_1}{2}\frac{S\beta}{S(\beta - \alpha)},$$

together with the result that $\hat{g}$ must be a line reflection, although, at this point, a translation, s_u, along the reflection axis, $(\hat{u}, \vec{r})$, is still possible. The dual part of equation (4) yields

$$y = 0.$$

Thus, the existence of a common normal intersecting the three axes is proved. Furthermore, the spatial linkage has a position where all their links are colinear. Finally, the location of the reflection axis along the common normal is given by

$$\begin{aligned} z &= \frac{S\beta}{S\alpha}\left[C(\beta - \alpha)\,a + \frac{1 + {x_1}^2}{2x_1}S(\beta - \alpha)s_1\right] - \frac{1}{2}x_1\, s_u\, S\beta \\ &+ \left[\frac{1 + {x_1}^2}{2x_1}\frac{S^2\beta}{S\alpha} + \frac{S\alpha - 2S\beta\, C(\beta - \alpha)}{2x_1}\right] s_2. \end{aligned} \tag{5}$$

Since z and s_u must be constants, the only solutions are those indicated in the statement of the proposition.

Case 1. In addition to $s_u = 0$, also assume that $s_1 = s_2 = 0$. Then, the location of the reflection axis is given by

$$z = \frac{S\beta}{S\alpha}C(\beta - \alpha)\,a.$$

Since, we have already shown, see equation (2) in Section 2, that the relationship between θ_1 and θ_2 is the one corresponding to the Bennett linkage, and the links angles satisfy

$$\alpha_{12} = \alpha_{34} = \alpha, \quad \text{and} \quad \alpha_{23} = \alpha_{41} = 2\beta - \alpha,$$

it suffices to show that the link lengths also satisfy

$$a_{12} = a_{34}, \quad \text{and} \quad a_{23} = a_{41}$$

and the Bennett condition

$$\frac{a_{12}}{S\alpha_{12}} = \frac{a_{41}}{S\alpha_{41}}.$$

From Figure 2, it is evident that

$$a_{12} = a_{34} = a, \quad \text{and} \quad a_{23} = a_{41} = 2z - a.$$

Then consider

$$\begin{aligned} a_{41} &= 2z - a = \left[2\frac{S\beta}{S\alpha}C(\beta-\alpha) - 1\right]a = \frac{a}{S\alpha}\left[2S\beta C(\beta-\alpha) - S\alpha\right] \\ &= \frac{a_{12}}{S\alpha_{12}}\left[2S\beta C(\beta-\alpha) - S\alpha\right] \end{aligned}$$

It is straightforward to show that

$$S\alpha_{41} = S(2\beta - \alpha) = 2S\beta\, C(\beta - \alpha) - S\alpha. \tag{6}$$

Thus, the Bennett condition has been proved.

Case 2. In addition to $s_u = 0$, also assume that $s_2 = 0$ and

$$s_1 = k_1 a \frac{2x_1}{1 + x_1^2} = k_1 a\; S\theta_1,$$

where, $k_1 \in \Re$. Then, the location of the reflection axis is given by

$$z = \frac{S\beta}{S\alpha}\left[C(\beta - \alpha) + k_1 S(\beta - \alpha)\right] a.$$

Finally, the geometric characteristics of the corresponding linkage will be presented. The linkage is a C-R-C-R where the cylindrical pairs are associated with the axes 1 and 3. Moreover, the links angles are

$$\alpha_{12} = \alpha_{34} = \alpha, \quad \text{and} \quad \alpha_{23} = \alpha_{41} = 2\beta - \alpha,$$

and the link lengths are

$$a_{12} = a_{34} = a, \quad \text{and} \quad a_{23} = a_{41} = 2z - a.$$

Unfortunately, the dependence of the link lengths upon the arbitrary value of $k_1 \in \Re$ prevents further analysis.

Case 3. In addition to $s_u = 0$, also assume that, in equation (5),

$$S\alpha - 2S\beta\, C(\beta - \alpha) = 0.$$

However, equation (6), shows that

$$S(2\beta - \alpha) = -[S\alpha - 2\, S\beta\, C(\beta - \alpha)].$$

Thus, the assumption implies that

$$2\beta - \alpha = \begin{cases} 0 \\ \pi \end{cases} \quad \text{or} \quad \alpha = \begin{cases} 2\beta \\ 2\beta + \pi \end{cases}$$

In the rest of this analysis, it will be assumed that $\alpha = 2\beta$, the other possibility yields the same results. Then, from Figure 1, it follows that

$$\hat{u}_3 = \hat{u}\hat{u}_1\hat{u}^- = \hat{u}_2,$$

and

$$\hat{u}_4 = \hat{u}\hat{u}_2\hat{u}^- = \hat{u}_1.$$

Hence, the result is a linkage with a pair of parallel dyads represented by $\hat{u}_4 - \hat{u}_1$, and $\hat{u}_2 - \hat{u}_3$ respectively. Whatever linkage that can be obtained from these partial results will be an overconstrained paradoxical kinematic chain of class I, since the linkage can be obtained from the intersection of a pair of Schönflies subgroups, whose rotation axes coincide with the rotation axes of the parallel dyads obtained here.

Furthermore, substituting the assumptions into equation (5), it follows that

$$z = \frac{a}{2} + \frac{1 + x_1^2}{2x_1}\frac{1}{2}\, tan\beta\, (s_2 - s_1).$$

Thus, if

$$s_2 - s_1 = k_2\, a\, \frac{2x_1}{1 + x_1^2} = k_2\, a\, S\theta_1,$$

where $k_2 \in \Re$. It follows that

$$z = \frac{1 + k_2\, tan\beta}{2} a.$$

Many different variants of this parallel dyad linkage can be obtained since the condition over the displacements s_1 and s_2 must be satisfied, unlike case 2, by the difference $s_2 - s_1$. These variants include a family of H-C-H-C linkages.

4. Conclusion

This contribution has shown a simple group theoretical explanation for the paradoxical mobility of some linkages. The obvious goal is to prove that the mobility of the remaining paradoxical linkages can be also explained in a similar way. The authors believe that these results are important for the systematic design of manipulator architectures.

5. Dedication and Acknowledgements

This work is dedicated to the memory of Prof. Joseph Duffy, doctoral advisor of the first author. This work was supported by the California Department of Transportation (Caltrans) under the Basic Research Component of Advanced Highway Maintenance and Construction Tecnology (AHMCT) Research Center at UC-Davis, the Consejo del Sistema Nacional de Educación Tecnológica and the Consejo de Ciencia y Tecnología del Estado de Guanajuato.

Notes

1. Since $g \in H_u$, conjugation and inverse are equivalent; i.e. $g^- = g^{-1}$. Moreover, since $g \in \Re^3$, $g^- = -g$. Thus $g^{-1} = g^- = -g$.
2. Because a symmetric spherical linkage has a "flattened" configuration, it is ensured that the corresponding symmetric spatial linkage has a position such as the indicated in Figure 2.

References

Angeles, J., (1982), *Spatial Kinematic Chains*, New York: Springer Verlag.

Fanghella, P., (1988), Kinematics of Spatial Linkages by Group Algebra : A Structure-Based Approach, *Mechanism and Machine Theory*, vol. 23, pp. 171-183.

Fanghella, P. and Galletti, C., (1995), Metric Relations and Displacement Groups in Mechanism and Robot Kinematics", ASME *Journal of Mechanical Design*, vol. 117, pp. 470-478.

Hervé, J.M., (1978), Analyse Structurelle des Mécanismes par Groupe des Déplacements, *Mechanism and Machine Theory*, vol. 13, pp. 437-450.

Rico, J.M., and Ravani, B. (2000), Designing Linkages with Symmetric Motions: The Spherical Case, in *Advances in Robot Kinematics*, J. Lenarčič, and M.M. Stanišić eds. Dordrecht: Kluwer Academic Publishers.

Rico, J.M., and Ravani, B., (2001), On Mobility Analysis of Linkages Using Group Theory, Accepted for Publication, *ASME Journal of Mechanical Design*.

Waldron, K. J., (1979), Overconstrained Linkages, *Environment and Planning B*, Vol. 6, pp. 393-402.

SOLVING MULTI-LOOP LINKAGES BY ITERATING 2D CLIPPINGS

J.M. Porta, L. Ros, F. Thomas and C. Torras
Institut de Robòtica i Informàtica Industrial (CSIC-UPC)
Llorens i Artigas 4-6, 08028-Barcelona, Spain
{jporta,llros,fthomas,ctorras}@iri.upc.es

Abstract

A multi-loop linkage can be viewed as a graph of kinematic constraints, and its solution entails finding joint configurations satisfying all constraints simultaneously. In this paper, a general algorithm to obtain all solution configurations is presented. Each loop provides a matrix equation in homogeneous coordinates, which is handled as a constraint on the involved joint variables. The algorithm uses one constraint at a time to reduce the valid ranges for variables, by iteratively applying two simple geometric operations: segment-trapezoid and circle-rectangle clippings. The obtained experimental results qualify this approach as a valuable alternative to other interval-based strategies.

Keywords: Kinematic constraints, constraint graph, geometric constraint systems, multilinear equations, clipping.

1. Introduction

The resolution of graphs of kinematic constraints has aroused interest not only within Kinematics (analysis of multi-loop linkages, simulation of deployable mechanisms), but also within related disciplines such as Robotics (path planning of closed-loop kinematic chains, contact analysis, assembly specification) and Computer Graphics (constraint-based sketching and design, geometric modeling for CAD). In the latter contexts, the problem is formulated as a system of geometric constraints between objects (Kramer, 1992), whose resolution entails finding object positions and orientations that satisfy all constraints simultaneously.

Solving a constraint graph amounts to solving its underlying system of loop-closure equations. Thus, the general methods developed for finding all the roots of systems of algebraic equations can be (and have been) applied to this problem. Three types of methods can be distinguished: algebraic geometry, homotopy and interval-based ones.

Our group has been exploring the latter type of algorithms, first using interval Newton methods (Castellet and Thomas, 1998) and, afterwards,

J. Lenarčič and F. Thomas (eds.), Advances in Robot Kinematics, 255–264.

the subdivision and convex-hull properties of the Bernstein polynomials (Bombín et al., 2000). An interesting feature of this second technique is that it does not require the computation of any Jacobian. Along this line, we propose here a remarkably simpler branch-and-bound algorithm that only relies on two basic 2D clipping operations.

2. From a constraint graph to a system of equations

When reference frames are attached to the objects, the constraints can be expressed as sets of allowed transformations between these reference frames. Although these transformations are usually described by means of homogeneous coordinates and Denavit-Hartemberg parameters, other representations are possible but this does not modify the discussion that follows.

We will only consider transformations with independent translational and rotational degrees of freedom. This includes most problems of practical interest but excludes, for example, those including screw motions because the translational and rotational motions are linked by the pitch of the screw.

Each loop or cycle c in the constraint graph can be expressed as a sequence of homogeneous transformations, where each transformation is either a constant matrix or a translation/rotation matrix including a single variable. In a solution configuration, the result of composing the transformations in the loop must equal the identity. Thus, each cycle leads to a matrix equation:

$$c(x_1, \ldots, x_n, \theta_1, \ldots, \theta_m) = I, \quad (1)$$

where $x_1, \ldots, x_n$ are variables involved in translation transformations, $\theta_1, \ldots, \theta_m$ are rotational variables, and I is the identity matrix.

For multi-loop graphs, not all resulting matrix equations are independent, but only those belonging to a cycle basis of the graph. In the current implementation, we use a basis of fundamental cycles (derived from a spanning tree), but this is clearly a point susceptible of improvement, as discussed in Section 7.

Each matrix equation leads to 12 scalar equations, one for each row i (but the last, which has constant values) and column j in the transformation matrix:

$$f_{i,j}(x_1, \ldots, x_n, \theta_1, \ldots, \theta_m) = \delta_{i,j}, \quad (2)$$

where $\delta_{i,j}$ is 1 if i is equal to j, and 0 otherwise.

Obviously, the system of matrix equations defined by the cycle basis of the graph can be solved by working on the resulting set of scalar equations. However, the presence of rotational variables in the scalar functions (2) is quite troublesome because it leads to trigonometric expressions. The tangent-half-angle substitution has been the usual choice to transform them into a system of algebraic equations, but it might introduce extraneous roots because of its singularity at $\theta_k = \pm\pi$.

As an alternative, it is possible to introduce the changes $y_k = \cos\theta_k$ and $z_k = \sin\theta_k$ that present no discontinuity. Using this variable substitution, the scalar equations become:

$$f_{i,j}(x_1, \ldots, x_n, y_1, \ldots, y_m, z_1, \ldots, z_m) = \delta_{i,j}. \tag{3}$$

The drawback of this variable substitution is that we have m additional variables and, thus, the search space is considerably enlarged. The advantage is that now all $f_{i,j}$ functions are multilinear and, consequently, the equations are simpler to solve.

In the next sections, we describe how these equations can be solved using simple 2D clipping techniques and how the circle-based restriction ($y_k^2 + z_k^2 = 1$) existing between some of the variables is used to ensure that only valid roots are found.

3. Bounding multilinear functions

Because of its application in the field of nonlinear programming, a great deal of effort has been devoted to finding piecewise-linear bounds for multilinear functions. It is well-known that the convex envelope of one such function on the unit hypercube is polyhedral and that finding this envelope is an NP-hard problem.

In this paper, we will use a much looser and easy-to-compute bound, namely the projections of the convex envelope on the planes $(x_k, f(\mathbf{x}))$ for all k. To this end, we will make use of the following result, which is a direct consequence of Theorem 1.1 in (Rikun, 1997):

The point $(\mathbf{x}, f(\mathbf{x})) \in \Re^{n+1}$, where f is a scalar multilinear function and $\mathbf{x} = (x_1, \ldots, x_n) \in [x_1^l, x_1^u] \times \cdots \times [x_n^l, x_n^u]$ is contained in the convex hull of the 2^n points $\{(\mathbf{x}, f(\mathbf{x})) | x_k \in \{x_k^l, x_k^u\}\}$.

This convex hull property leads to a simple clipping strategy for bounding the solutions of a system of multilinear equations, as described in Section 4.1.

4. Two clipping operations

The algorithm we present takes an initial box $\mathcal{B} \subset \Re^n$ in configuration space and uses two clipping operations —*segment-trapezoid clipping* and *circle-rectangle clipping*— to iteratively cut off portions of $\mathcal{B}$ containing no root. For simplicity, we here explain how they work on a system of just one equation in two unknowns. The extension to multiple equations and any number of variables will then become straightforward, as summarized in Section 5.

4.1 Segment-trapezoid clipping

Assume we want to find all solutions of a multilinear equation $f(\mathbf{x}) = \delta$, $\mathbf{x} = (x_1, x_2)$, for $\mathbf{x}$ in the box $\mathcal{B} = [x_1^l, x_1^u] \times [x_2^l, x_2^u] \in \Re^2$. Since the graph of $(\mathbf{x}, f(\mathbf{x}))$ must lie within the convex hull of the $2^2 = 4$ points $\{(\mathbf{x}, f(\mathbf{x}))|\ \mathbf{x} \in \{x_1^l, x_1^u\} \times \{x_2^l, x_2^u\}\}$, we can compute the convex hull of these points in $\Re^3$ and intersect it with the plane $f(\mathbf{x}) = \delta$ to obtain a polygon whose rectangular hull gives a better bound for the solutions. Although this method seems inefficient for a high number of variables, Bombín et al. (2000) show that the explicit computation of the convex hull may be avoided, and that the overall procedure can be reduced to solving $2n$ linear programming problems. Nevertheless, the method can be further simplified by accepting looser bounds through the following variation, illustrated in Fig. 1. Instead of computing the intersection of the convex hull with the $f(\mathbf{x}) = \delta$ plane, we simply start by projecting the hull onto each coordinate plane as depicted in Fig. 1a, and intersecting each of the resulting trapezoids with the $f(x) = \delta$ line, as shown in Fig. 1b. Usually, these segment-trapezoid clippings reduce the ranges of some variables, and the Cartesian product of them all gives a box smaller than $\mathcal{B}$ still bounding the root locations (the black rectangle in Fig. 1a). The experiments show that, although the latter strategy produces less pruning than the former, a root finding process using it is faster due to the lower cost of its simpler operations.

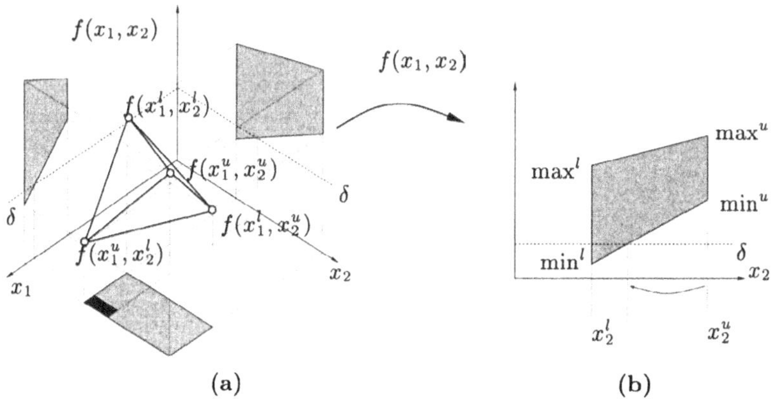

Figure 1. Segment-trapezoid clipping.

4.2 Circle-rectangle clipping

When two variables of a same rotation, say y_k and z_k, are constrained by an equation $y_k^2 + z_k^2 = 1$, their ranges can be further reduced with a circle-rectangle clipping operation, as illustrated in Fig. 2. Suppose that after all possible segment-trapezoid clippings we end up with a certain box $\mathcal{B}$ of possible values for y_k and z_k (see Fig. 2a). Since only the values lying on $y_k^2+z_k^2 = 1$ are consistent with the equations, we can clip this box with the circle, yielding the smaller shaded rectangles in Fig. 2b. Note that, in general, clipping against the unit circle may produce more than one clipped box (Srinivasan, 1992). Although this frequently prunes large portions of the search space, it would also produce a great deal of branching in the search algorithm below, and it is often preferable to take the rectangular hull of all clipped boxes as the output of this clipping operation, rather than the individual boxes themselves. This also permits a simpler box-handling strategy in the search algorithm described next.

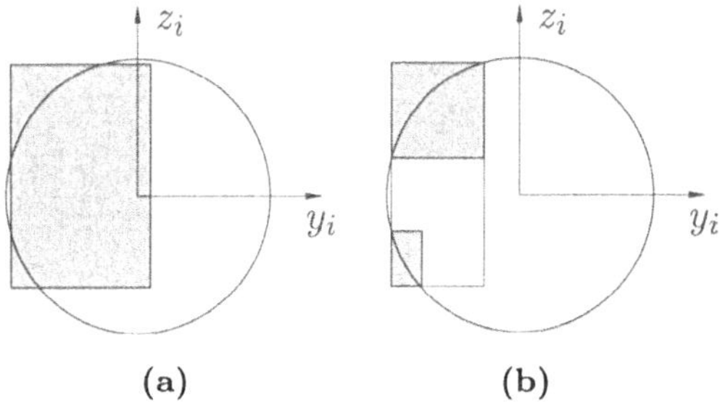

Figure 2. Circle-rectangle clipping.

5. The algorithm

The proposed algorithm, which we have named *Cuik* (Fig. 3), aims at determining a set S of small boxes that include the solutions for a given kinematic problem. The valid ranges for the variables (including those coming from the *sine* and *cosine* substitutions) are used to set up an initial list of boxes. This list constitutes the search space where to look for solution boxes. In this algorithm, a box is a solution when it has a size below a user-provided threshold (σ) and it fulfills all the possible segment-trapezoid and rectangle-cycle clipping tests. This simple criterion may classify as solution boxes those that are very close to a solution point, but that do not actually include the point itself. Avoiding these pseudo-solutions calls for more strict acceptance criteria, as mentioned in Section 7.

```
The Cuik Algorithm
  Input: A kinematic graph description
  Output: A set of solution boxes (S)
  Process:
    Compute a cycle basis of the graph
    S ← ∅
    L ← Initial list of boxes
    while not empty(L)
        B ← first box(L)
        do
            s ←size(B)
            Reduce_Box(B)
        until empty(B) or size(B)< σ or size(B)/s > ρ
        if not empty(B) then
          if size(B)≤ σ then
            S ← S ∪ {B}
          else
            Split B into two sub-boxes: B1, B2
            Add B1 and B2 to L
          endif
        endif
    endwhile
```

Figure 3. The Cuik algorithm.

To determine the set of solution boxes, each one of the initial boxes is reduced as much as possible by the successive application of the trapezoid and circle clippings described in the previous section (see Fig. 4). Three possible outcomes are possible from the box-reduction process:

- **The box becomes empty:** In any of the clippings a variable interval can become empty because there is no intersection between the segment and the trapezoid or between the circle and the rectangle. This indicates that the analyzed box does not contain solution points and we can simply stop the exploration in the search space delimited by this box.
- **The box becomes small:** Then we add it to the solution set.
- **The box can not be significantly reduced:** If the reduction ratio of the box is above a user-provided threshold (ρ), the box reduction is stopped and the box is split into two sub-boxes. In the current version of the algorithm, we simply split the box along its larger dimension. However, as commented in section 7 this

```
Reduce_Box(B)
  Input: A box defined as a set of intervals:
        B = {[x_1^l, x_1^u], ..., [x_n^l, x_n^u]}
  Output: The same box but eventually resized
  Process:
    for each cycle c
      V ← {v_0, ..., v_k} (Set of indexes of variables involved in c)
      C = {c(x) | x = (x_{v_0}, ..., x_{v_k}) ∈ {x_{v_0}^l, x_{v_0}^u} × ... × {x_{v_k}^l, x_{v_k}^u}}
      for each v ∈ V
        for each i ∈ [1,3] and j ∈ [1,4]
          min^l ← min{c_{i,j} | c ∈ C, x_v = x_v^l}
          max^l ← max{c_{i,j} | c ∈ C, x_v = x_v^l}
          min^u ← min{c_{i,j} | c ∈ C, x_v = x_v^u}
          max^u ← max{c_{i,j} | c ∈ C, x_v = x_v^u}
          if i = j then δ = 1
          else δ = 0
          Trapezoid_Clipping(x_v^l, x_v^u, min^l, max^l, min^u, max^u, δ)
          if x_v is a rotation variable then
            w ← Index of the circle-related variable of x_v
            Circle_Clipping(x_v^l, x_v^u, x_w^l, x_w^u)
          endif
        endfor
      endfor
    endfor
```

Figure 4. The **Reduce_Box** function. $c(x_0, \ldots, x_k)$ refers the evaluation of the cycle matrix equation at point $(x_0, \ldots, x_k)$, and $c_{i,j}$ denotes the element at row i and column j of matrix c.

point requires further consideration. After the split, the circle clipping process is used to transmit the interval reduction to the corresponding circle-related variable. If the two new boxes are added to the beginning of the list, the Cuik algorithm implements a depth-first search, and if they are added to the end of the list the search is breadth-first.

Two factors have to be taken into account to determine the cost of any interval-based algorithm: the number of processed boxes and the cost to process each box. In the case of the Cuik algorithm, the emphasis is put in efficiently processing each box, even at the cost of having to process more boxes than if a more costly box reduction process were used.

Note that the algorithm in Fig. 4 projects 4 times the 2^k cycle evaluations for each cycle c_l ($l \in [1..n]$), for each variable x_v ($v \in [1..k]$), and for each element of the matrix equation. This amounts to $4 \cdot n \cdot k \cdot 12 \cdot 2^k$

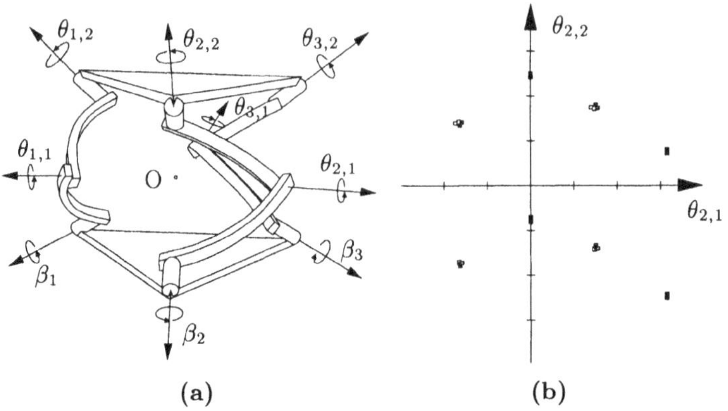

Figure 5. **(a)** The Gosselin platform. **(b)** The solution boxes determined by the Cuik algorithm with $\beta_1 = -0.2266$, $\beta_2 = -0.7103$ and, $\beta_3 = 0.073$.

point projections and, hence, even for few variables per cycle, this is the most expensive step of the box-reduction process.

One advantage of the Cuik algorithm is that box-reductions are completely independent of one another and, therefore, the algorithm can be readily parallelized with the consequent increment in efficiency.

6. Experimental results

Fig. 5a shows the platform presented in (Gosselin and Gagné, 1995). In this mechanism, the β parameters are the inputs for the platform actuators and the θ's are the joint values to be determined by the Cuik algorithm. This platform has eight valid configurations, four attainable for any value of the actuators parameters, and four particular for each set of inputs.

Observe that this problem involves six rotational variables and, consequently, the Cuik algorithm works in a twelve-dimensional space.

The kinematic graph for this platform has two fundamental cycles, each involving four variables (two of them shared with the other cycle).

The initial range for the θ variables is $[0, 2\pi]$. With $\sigma = 0.1$, these initial ranges conform a search space where approximately $4 \cdot 10^{15}$ solution boxes can be defined. Despite this large search space, the problem is solved by an optimized version of the Cuik algorithm in less than 1.25 seconds on a Pentium III at 700MHz, using the parameters $\sigma = 0.1$ and $\rho = 0.8$.

In this example, the algorithm starts with a single box and it examines up to 679 boxes, only 58 of which are considered solutions. Fig. 5b shows the solution boxes projected on a plane defined by a pair of prob-

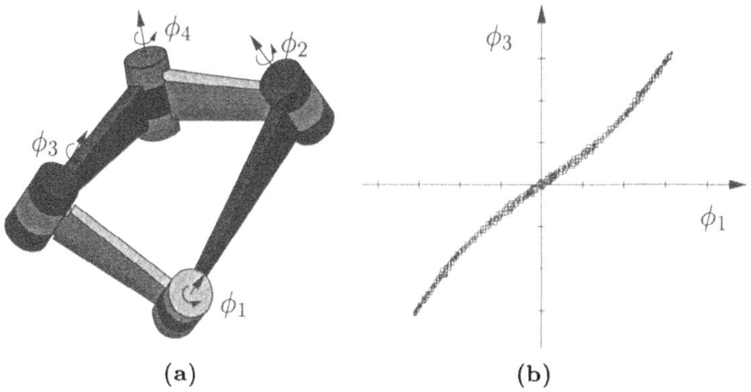

Figure 6. (**a**) The Bennett linkage. (**b**) The discretization of the one-dimensional space of solutions provided by the Cuik algorithm.

lem variables. We see that, as expected, the solution boxes accumulate around the eight solution points of the proposed problem.

If we need more precision in the solution boxes, we can run the program with a lower value for the parameter σ. For instance, for $\sigma = 0.01$ (10 times smaller than before), the search space is enlarged up to $4 \cdot 10^{27}$, but the execution time is increased in less than 0.2 seconds.

Fig. 6a represents the Bennett's linkage, a 4R spatial closed chain with 1-d.o.f. mobility. Fig. 6b shows the discretization of the one-dimensional solution space provided by the Cuik algorithm. This discretization includes about 300 boxes and is obtained in less than 1 second (with $\sigma = 0.1$ and $\rho = 0.8$).

We have also tested our implementation with more complex problems. As a reference, all the valid solutions of a 6R manipulator are found by the algorithm in about 5 minutes.

7. Conclusions and envisaged enhancements

In this paper, we have shown the feasibility of efficiently solving complex kinematic problems using very simple 2-D clipping techniques.

The presented Cuik algorithm can be seen as the skeleton of a software package to solve arbitrary graphs of kinematic constraints. Currently, the algorithm deals with cycles, variables and matrix components in an exhaustive, uninformed way. Future work will try to develop good heuristics for all choice points in the algorithm. For example, it is not clear what the most suitable cycle basis is, or even if the addition of redundant cycles could speed up the process. Likewise, first solving the cycles with discrete solutions may prune drastically the search space. It seems equally advisable to begin by processing variables shared between many cycles. Another point that also deserves consideration is that of deciding how to split boxes whose size can not be further reduced. Solution accuracy should also be improved by using more elaborate criteria to determine whether a small box contains a root.

The results we have obtained show that the presented algorithm is specially adequate for kinematic problems with many loops but with few variables per loop (up to 4 or 5). To efficiently confront problems with more variables per loop, the algorithm should be extended with other techniques for solving algebraic equations.

Acknowledgements

This research has been partially supported by the Catalan Research Commission, through the "Robotics and Control" group, and the Spanish CICYT under contracts TAP99-1086-C03-01 and TIC2000-0696.

References

Bombín, C., Ros, L., and Thomas, F. (2000), A concise Bézier clipping technique for solving inverse kinematics problems, in *Advances in Robot Kinematics,* J. Lenarcic and M.M. Stanisic (eds.), pp. 53-60, Kluwer Academic Publishers.

Castellet, A., and Thomas, F. (1998), An algorithm for the solution of inverse kinematic problems based on an interval method, in *Advances in Robot Kinematics,* M. Husty and J. Lenarcic (eds.), pp. 393-403, Kluwer Academic Publishers.

Gosselin C. M. and Gagné M. (1995), A close-form solution for the direct kinematics of a special class of spherical three-degree-of-freedom parallel manipulators, in *Computational Kinematics*, J.-P. Merlet and B. Ravani (eds.), pp. 231-240, Kluwer Academic Publishers.

Kramer, G.A. (1992), *Solving Geometric Constraint Systems*, Cambridge, The MIT Press.

Rikun, A.D. (1997), A convex envelope formula for multilinear functions, *J. of Global Optimization*, vol. 10, pp. 425-437.

Srinivasan, R.V. (1992), A fast circle clipping algorithm, in *Graphics Gems III*, pp. 182-187, Academic Press.

REVISITING PLÜCKER COORDINATES IN VISION-BASED CONTROL

Nicolas Andreff
LaRAMA-UBP,
Institut Français de Mécanique Avancée
BP 265, 63175 Aubière Cedex, France
Nicolas.Andreff@ifma.fr

Bernard Espiau
INRIA Rhône-Alpes
38334 Saint-Ismier Cedex, France
Bernard.Espiau@inrialpes.fr

Abstract Plücker Coordinates are largely used in kinematics. What is perhaps much less known is the fact that such a representation of lines exhibits very interesting properties in some classes of control problems. In this paper we start from classical Plücker Coordinates and show that, by adding a supplementary normality constraint, we obtain a representation which is very efficient for controlling a system from images of lines obtained through a perspective projection. Furthermore, the physical interpretation of the approach is very easy. Having derivated the related equations of motion, we then design a control scheme the stability of which can be proved. Some simulation and experimental results illustrate the approach.

Keywords: Plücker Coordinates, visual servoing, control

1. Introduction

In computer vision, when considering industrial parts, the extraction and the tracking of lines, generally seen as intersecting planes, is natural. Knowing that, in vision, control properties are strongly linked to the parametrization of the manifolds in the 3D space from which come the projected features, the question which occurs is now: what representation for these lines?

Coming back to geometrical aspects, it should be recalled that the 3D Euclidean lines form a 4-dimensional manifold the properties of which are described by the Grassman's geometry. One of the most noticeable property is the absence of a natural (Riemannian) distance. Since such a distance is somewhere needed for defining a feedback control, the choice

J. Lenarčič and F. Thomas (eds.), Advances in Robot Kinematics, 265–274.

of the parametrisation is crucial. Among the classical ones, the Plücker coordinates look particularly interesting since they come from the embedding of the line manifold in a projective space in which things are linear. More, recalling that the camera transformation is also projective, we understand intuitively that some nice properties could be expected. In this framework, the paper is organized as follows. In a first step, we discuss a restriction on classical Plücker coordinates, which looks very pertinent in visual servoing. The second part of the paper presents a control law on this submanifold, which is then applied to an orthogonal trihedron. Some results end the paper.

2. Binormalized Plücker coordinates

2.1 Recalls on Plücker coordinates

The (projective) Plücker coordinates (Plücker, 1865; Bottema and Roth, 1979; Faugeras, 1993; Semple and Kneebone, 1952) of a line in space, or 3D line, $(\mathcal{L})$ are the 6 coordinates of a point in the 5-dimensional projective space P^5. This point is sometimes called the *Plücker point* (Goodman and O'Rourke, 1997). Its coordinates are usually noted as a couple of two 3-vectors defined up to a common scale factor:

$$\mathcal{L} = \begin{pmatrix} \mathbf{u} \\ \mathbf{v} \end{pmatrix} \tag{1}$$

where $\mathbf{u} \neq \mathbf{0}$ denotes the orientation of $(\mathcal{L})$ and $\mathbf{v}$ contains the coefficients of the *interpretation plane* equation, the latter being the plane passing by the origin and $(\mathcal{L})$. Since physical lines in space form a 4-dimensional manifold, called the *Klein quadric* (Pottmann *et al.*, 1998) or the *Plücker hypersurface* (Chazelle *et al.*, 1996), $\mathbf{u}$ and $\mathbf{v}$ satisfy the following constraint:

$$\mathbf{u}^T\mathbf{v} = 0 \tag{2}$$

which expresses, via the so-called *Grassmannian*, that a line intersects itself.

The Euclidean Plücker coordinates are obtained by normalizing $\mathbf{u}$:

$$\mathcal{L} = (\underline{\mathbf{u}}, \mathbf{h}) \;\; ; \;\; \underline{\mathbf{u}}^T\mathbf{h} = 0 \; ; \; \underline{\mathbf{u}}^T\underline{\mathbf{u}} = 1 \tag{3}$$

where $\underline{\mathbf{u}} = \mathbf{u}/\|\mathbf{u}\|$ and $\mathbf{h} = \mathbf{v}/\|\mathbf{u}\|$. Hence, they can be seen as *normalized Plücker coordinates* (Pottmann *et al.*, 1998). Notice that, due to the Euclidean structure, the constraint $\underline{\mathbf{u}}^T\mathbf{h} = 0$ expresses that $\mathbf{h}$ is orthogonal to the interpretation plane of $(\mathcal{L})$. It can moreover be shown that, for any point P on $(\mathcal{L})$:

$$\mathbf{h} = P \times \underline{\mathbf{u}} \tag{4}$$

Remark 1 $\mathbf{h} = 0$ *iff* $(\mathcal{L})$ *passes through the origin.*
Since the case where a line passes through the center of projection is degenerate in computer vision, we will assume in the remainder of this paper that it never occurs.

Remark 2 *Notice that* $(\underline{\mathbf{u}}, \mathbf{h})$ *is a bivector as McCarthy (MacCarthy, 1990) defined it to represent kinematic screws. Consequently (Murray et al., 1994, p. 427), there does not exist a bi-invariant Riemannian metric on the manifold of 3D lines.*

The Plücker coordinates (both projective and Euclidean) give trivially the image projection (ℓ) of $(\mathcal{L})$, since (ℓ) is simply the intersection of the interpretation plane and the image plane. One can verify that (4) defines an implicit orientation of the 2D space, and that (ℓ) being thus oriented can be uniquely defined by:

$$\ell = \underline{\mathbf{h}} \tag{5}$$

where $\underline{\mathbf{h}} = \mathbf{h}/\|\mathbf{h}\|$. With this new notation and introducing $h = \|\mathbf{h}\|$, the Euclidean Plücker coordinates of a line $(\mathcal{L})$ can be rewritten as:

$$\mathcal{L} = (\underline{\mathbf{u}}, h\underline{\mathbf{h}}) \quad ; \quad \underline{\mathbf{u}}^T\underline{\mathbf{h}} = 0 \; ; \; \underline{\mathbf{u}}^T\underline{\mathbf{u}} = 1$$

2.2 Binormalized Plücker coordinates

Definition. From all the preceeding, we are now ready to define the binormalized Plücker coordinates, as follows:

Definition 1 *The binormalized Plücker coordinates of the pencil of all the 3D lines oriented by the unit vector* $\underline{\mathbf{u}}$ *and lying in the plane of unit normal vector* $\underline{\mathbf{h}}$ *are the couple* $(\underline{\mathbf{u}}, \underline{\mathbf{h}})$ *with* $\underline{\mathbf{u}}^T\underline{\mathbf{h}} = 0 \;\; ; \;\; \underline{\mathbf{u}}^T\underline{\mathbf{u}} = \underline{\mathbf{h}}^T\underline{\mathbf{h}} = 1$

By extension, we call binormalized Plücker coordinates of a 3D line the binormalized Plücker coordinates of the pencil it defines. Consequently:

Proposition 1 (3D line representation) *A 3D line* $(\mathcal{L})$ *is uniquely defined by its binormalized Plücker coordinates* $(\underline{\mathbf{u}}, \underline{\mathbf{h}})$ *and its depth* h*:*

$$\mathcal{L} = (\underline{\mathbf{h}}^T, \underline{\mathbf{u}}^T, h)^T \tag{6}$$

Notice that this is why we consider $(\mathcal{L})$ lying in a pencil of parallel lines that h is not only a distance but also a depth.

Geometrical interpretation. Our representation yields a sequential interpretation suited to the case of vision (Figure 1): $(\mathcal{L})$ is the line in the interpretation plane of (ℓ) (defined by its normal vector $\underline{\mathbf{h}}$) with orientation $\underline{\mathbf{u}}$ and lying at depth h. This interpretation corresponds to the usual sequence followed in image analysis to determine a 3D line

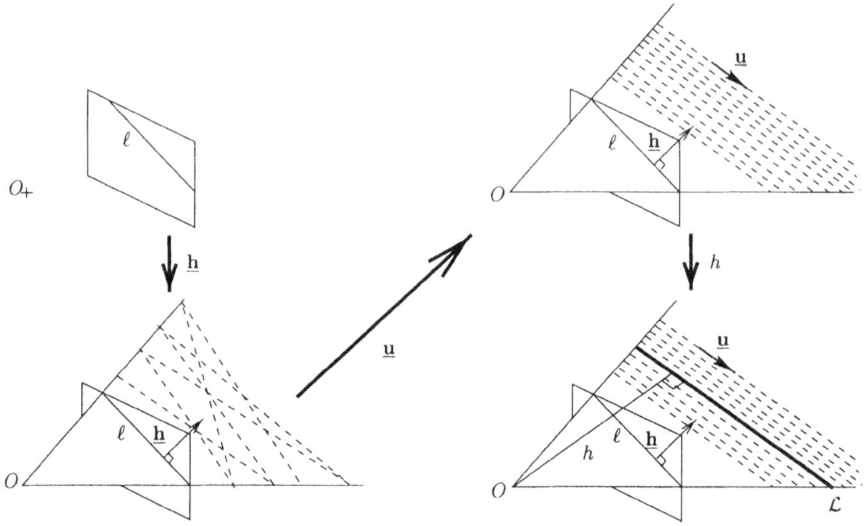

Figure 1. The binormalized Plücker coordinates have a sequential interpretation: from image measurements to the estimation of 3D components.

from one or several images: extraction of (ℓ) from the image, then computation of the orientation of $(\mathcal{L})$, and finally, computation of the 3D position of $(\mathcal{L})$.

Motion equations. Now, the instantaneous motion of a camera relatively to a fixed world frame is defined by its velocity screw $\tau = (V, \Omega)$, where Ω is the instantaneous angular velocity and V the instantaneous linear velocity of a given point.

Then, the apparent motion of the normalized Plücker coordinates $(\underline{\mathbf{u}}, \mathbf{h})$ of a fixed 3D line is (Rives and Espiau, 1987):

$$\dot{\underline{\mathbf{u}}} = \Omega \times \underline{\mathbf{u}} \tag{7}$$

$$\dot{\mathbf{h}} = \Omega \times \mathbf{h} + V \times \underline{\mathbf{u}} \tag{8}$$

Using now binormalized Plücker coordinates, let $\mathcal{L} = (\underline{\mathbf{h}}^T, \underline{\mathbf{u}}^T, h)^T$ be a 3D line. Then, it can be shown (Andreff, 1999) that its motion writes as:

$$\dot{\underline{\mathbf{u}}} = \Omega \times \underline{\mathbf{u}} \tag{9}$$

$$\dot{\underline{\mathbf{h}}} = \Omega \times \underline{\mathbf{h}} - \frac{V^T \underline{\mathbf{h}}}{h} (\underline{\mathbf{u}} \times \underline{\mathbf{h}}) \tag{10}$$

$$\dot{h} = V^T (\underline{\mathbf{u}} \times \underline{\mathbf{h}}) \tag{11}$$

2.3 Discussion

The parameterization we propose, based on binormalized Plücker coordinates, has some advantages, from the control point of view, over the parameterizations found in the litterature (Andreff, 1999):

- Our representation is continuous and covers the whole space, except the case of lines going through the optical center.
- It is coherent in the 3D space and the image, as it has been explained in the geometrical interpretation. Furthermore, the extraction of $\underline{\mathbf{h}}$ from the image of a line is trivial and there exists accurate methods for the computation of $\underline{\mathbf{u}}$.
- It is unique provided that some constraints, easy to satisfy, hold.
- As shown by equations (9), (10), it exhibits a partial decoupling between rotation and translation, which is very useful for designing a stable control scheme.

3. Control Issues

3.1 Introduction

Let us consider a fixed reference frame F_r and a moving object with associated frame F_0 (here the camera frame). We define a goal frame F^* in F_r and the control problem we consider is to drive F_0 from its initial value $F_0(0)$ to F^* in a stable way from image information. The representation we use allows to extract directly 2 parameters from the image, i.e. $\underline{\mathbf{h}}$, the normal to the interpretation plane. The remaining 2 parameters may be considered as 3D ones: the depth h and one component of $\underline{\mathbf{u}}$. Finally, instead of the frame F^* itself, our goal can therefore be defined as the set $\mathcal{L}_i^* \;:\; \{\underline{\mathbf{u}}_i^*, \underline{\mathbf{h}}_i^*, h_i^* \;;\; i = 1 \ldots n\}$, provided that the set contains at least 2 independent lines.

We will firstly demonstrate the principle of the control scheme we propose by using a single line. Before, let us set the assumptions which will be used in all the following.

- The set of goal lines $\mathcal{L}_i^*$ corresponds to a fixed achievable configuration of the system.
- All 3D trajectories of the camera which are needed are feasible (no obstacles, no joint limits neither geometrical singularities are reached by the device which moves the camera).
- During the motion, no line goes through the camera optical center.

- The camera is fully calibrated.
- All the needed variables **but the depth** can be measured.
- The velocity screw (Ω, V) of the camera is the control variable.

3.2 Case of a Single Line

Let us consider a single target line specified by the set $\{\underline{\mathbf{u}}^*, \underline{\mathbf{h}}^*, h^*\}$ and forget the depth h. The state equations of this subsystem are given by eqs (9) and (10) only. We may remark that the orientation part is independent from the translation part, which allows to consider that this system is in cascaded form. Since the manifold of lines does not exhibit natural metric we have to define convenient errors. By exploiting the above mentioned partial decoupling, we set

$$\mathbf{e_u} = \underline{\mathbf{u}} \times \underline{\mathbf{u}}^* \tag{12}$$

where $\mathbf{e_u}$ is the classical error on the sphere. We can therefore set the following cascaded control:

$$\begin{cases} \Omega = \mu_u \mathbf{e_u} \\ V = -\mu_h \mathbf{e_h}|_{\Omega=0;\underline{\mathbf{u}}=\underline{\mathbf{u}}^*} \end{cases} \tag{13}$$

where $\mu_u > 0$, $\mu_h > 0$. The last expression shows that the choice made for defining the error $\mathbf{e_h}$ allows $\underline{\mathbf{h}}$ to move on a sphere using translational degrees of freedom only.

Note that, since we do not consider h, the null space of the system is 3-dimensional. It is given by

$$\Omega_{ns} = \lambda \underline{\mathbf{u}} \tag{14}$$

$$V_{ns} = \mu \underline{\mathbf{u}} + \lambda h \underline{\mathbf{h}} + \nu(\underline{\mathbf{u}} \times \underline{\mathbf{h}}) \tag{15}$$

where μ, λ and ν are real numbers. Any motion within this subspace will not affect the stabilization. Considering the latter remark, we notice that the translation control in (13) may have components in the two dimensions of the null space (μ,ν) that are independent from the rotation control. These motions are useless disturbances unless they are used for complementary tasks. It is therefore pertinent to project the linear velocity on the only relevant component, i.e. $\underline{\mathbf{h}}$ when $\Omega = 0$. Hence, we obtain the final cascaded control for one line:

$$\begin{cases} \Omega = \mu_u \mathbf{e_u} \\ V = -\mu_h \epsilon \underline{\mathbf{h}}|_{\Omega=0;\underline{\mathbf{u}}=\underline{\mathbf{u}}^*} \end{cases} \tag{16}$$

where $\epsilon = (\underline{\mathbf{u}} \times \underline{\mathbf{h}})^T \underline{\mathbf{h}}^*$, $\mu_u > 0$, $\mu_h > 0$.

We have now the following result, the proof of which is easy using $\frac{1}{2}||\underline{\mathbf{u}}-\underline{\mathbf{u}}^*||^2$ and $\frac{1}{2}||\underline{\mathbf{h}}-\underline{\mathbf{h}}^*||^2$ as Lyapunov functions respectively (Andreff, 1999; Andreff *et al.*, 2002):

Proposition 2 *The control (16) is such that:*
1- if $\underline{\mathbf{u}}(t=0) \neq -\underline{\mathbf{u}}^*$, *then* $\underline{\mathbf{u}}(t)$ *asymptotically converges to* $\underline{\mathbf{u}}^*$*;*
2- if $\underline{\mathbf{h}}(t=0|_{\Omega=0;\underline{\mathbf{u}}=\underline{\mathbf{u}}^*}) \neq -\underline{\mathbf{h}}^*$, *then* $\underline{\mathbf{h}}(t)$ *asymptotically converges to* $\underline{\mathbf{h}}^*$.

3.3 Case of an Orthogonal Trihedron

The trihedral configuration of lines is of particular interest since it is very often present in structured environments. However, from the image point of view, it has been shown (Dhome *et al.*, 1989) that any pencil of lines (i.e. with a common intersection) is degenerated for pose computation. Indeed, for example, it is intuitive that observing an ideal infinite trihedron with a single camera doesn't allow to recover its depth. Let us now study the control without taking the depth into account. The final equilibrium will therefore have a single degree of freedom left.

Let us thus consider a set of three 3D lines with the following constraints:

- $\underline{\mathbf{u}}_i^T \underline{\mathbf{u}}_j = 0 \ \forall \ i \neq j$(orthogonality)
- $\mathbf{h}_i^T \underline{\mathbf{u}}_j + \mathbf{h}_j^T \underline{\mathbf{u}}_i = 0 \ \forall \ i, j$ (intersection)

Taking these constraints into account leads to the following results (Andreff, 1999; Andreff *et al.*, 2002):

Theorem 1 (Control in rotation) *If* $\underline{\mathbf{u}}_i(t=0) \neq -\underline{\mathbf{u}}_i^*$, $\forall i = 1,\ 2,\ 3$, *then, the control*

$$\Omega = \mu_u \sum_{i=1}^{3} \underline{\mathbf{u}}_i \times \underline{\mathbf{u}}_i^*, \quad \mu_u > 0 \tag{17}$$

asymptotically stabilizes the equilibrium of the system,(9):

$$\underline{\mathbf{u}}_i = \underline{\mathbf{u}}_i^*, \quad i = 1,\ 2,\ 3 \tag{18}$$

The proof comes in two steps. First, show that $\frac{1}{2}\sum_{i=1}^{3}||\underline{\mathbf{u}}_i - \underline{\mathbf{u}}_i^*||^2$ is a Lyapunov function with stationnary points of the form: $\sum_{i=1}^{3}\underline{\mathbf{u}}_i \times \underline{\mathbf{u}}_i^*$, then, using the rigidity constraint, that the latter points verify $\underline{\mathbf{u}}_i = -\underline{\mathbf{u}}_i^*$, $\forall i = 1,\ 2,\ 3$ and are maxima of the Lyapunov function.

Theorem 2 (Control in translation) *If* $\underline{\mathbf{h}}_i(t=0) \neq -\underline{\mathbf{h}}_i^*, \forall i = 1, 2, 3$ *then the control*

$$V = -\mu_h \sum_{i=1}^{3} \epsilon_i \underline{\mathbf{h}}_i \ |_{\Omega=0;\underline{\mathbf{u}}_i=\underline{\mathbf{u}}_i^* \forall i}, \quad \mu_h > 0 \tag{19}$$

asymptotically stabilizes, whatever the depth, the equilibrium of the system (10):

$$\underline{\mathbf{h}}_i = \underline{\mathbf{h}}_i^*, \quad i = 1,\ 2,\ 3, \tag{20}$$

The proof shows that $\frac{1}{2}\sum_{i=1}^{3} ||\underline{\mathbf{h}}_i - \underline{\mathbf{h}}_i^*||^2$ is a Lyapunov function. The only subtlety in the calculus (Andreff *et al.*, 2002) is to notice that $\frac{1}{2}(\mathbf{I}_9 - U \otimes U)$ is a projection operator, where $U = (\underline{\mathbf{u}}_1^T, \underline{\mathbf{u}}_2^T, \underline{\mathbf{u}}_3^T)^T$

4. Experimental Results

This work was performed in the framework of a european project VIGOR, 1998–2001. A major concern of the project was to improve the automatic welding of ship parts. In particular, it was required to accurately move a welding tool with respect to a ship part, the position of which is not exactly known. Since 75% of the parts to be welded are composed of three intersecting orthogonal planes, we put the emphasis of the application of our studies to this junction of three orthogonal lines.

With respect to the theoretical control schemes previously presented, the following problems had to be solved in order to achieve the implementation:

1. how to extract and track the lines in the image in a fast and robust way? To achieve that along the motion, we used an adapted version of the 2D-3D model-based approach proposed in (Marchand *et al.*, 1999).
2. how to compute the orientations $\underline{\mathbf{u}}_i$? We obtained an analytical solution by adapting to the trihedron lines the method proposed in (Horaud *et al.*, 1989) for the perspective 4 points problem.
3. how to cope with the missing depth? The solution we chose was to use a commercial laser pointer that we rigidly linked to the camera (figure 2). In this configuration the image of the laser spot moves on a given straight line whatever the observed scene and the motion of the set {camera-laser}. The system was **not** calibrated. We designed the associated control law in such a way that the related motion always took place in the null space of the binormalized Plücker coordinates motion.

The control law proposed in this paper was largely tested both in simulation and experimentally (Andreff, 1999). We present here one simulation result (Figure 3) showing the control behavior when the three controls (rotation, translation, depth) are cascaded. Notice the expected exponential decay in rotation control and error. Since the translation control is not immediately activated, rather large errors in the image occur. Finally, depth control is activated when both orientation and image errors are small enough.

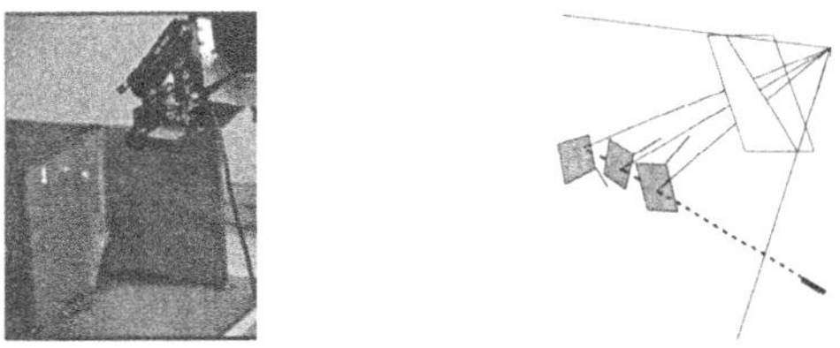

Figure 2. A laser pointer is rigidly linked to the camera

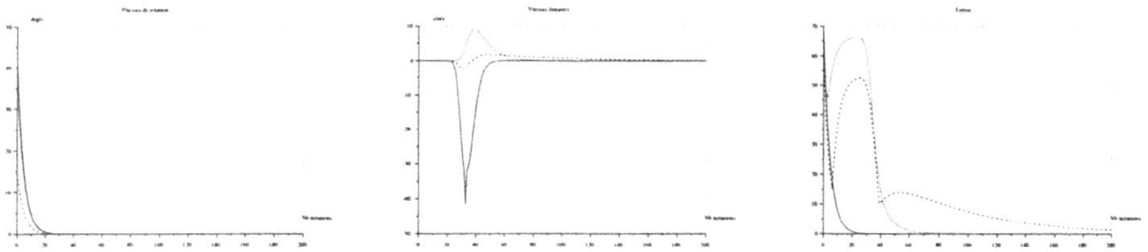

Figure 3. Simulation result with cascaded controls in rotation, translation and depth. **Left:** Rotation control. **Center:** Translation control. **Right:** Errors in rotation (-), in the image (–) and in depth (..).

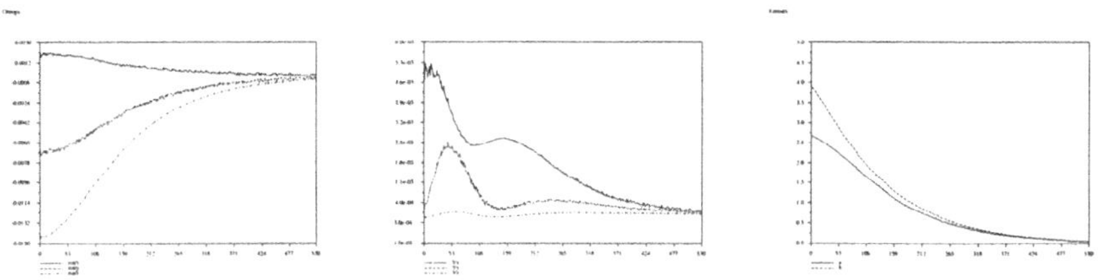

Figure 4. Experimental result with simultaneous control in rotation and translation. **Left:** Rotation control. **Center:** Translation control. **Right:** Errors in rotation (-) and in the image (–).

We present also one experimental result (Figure 4) showing the performance of simultaneous control in rotation and translation in a challenging initial situation. Indeed, the motion to achieve to reach the desired line configuration is composed of a translation of some 50 cm and a rotation of about 80 deg. However, both errors in rotation and in the image converge to 0, despite the translation control is perturbated by the rotation control.

5. Summary

In this work, we defined a new representation, the so-called binormalized Plücker coordinates, which can be easily interpreted geometrically and has some advantages from the control point of view. We could therefore define a control law which realizes an image line alignment with some very interesting properties. First, the proposed law can be seen as an analytic inversion of the motion equations. Second, it shows a partial decoupling between translation and rotation. Third, global asymptotic stability conditions could be found. This control law was validated both in simulation and experimentally in the case of an orthogonal trihedron.

References

Andreff, N., (1999), Asservissement visuel à partir de droites et auto-étalonnage pince-caméra, Thèse de Doctorat, Institut National Polytechnique de Grenoble, 1999.

Andreff, N., Espiau, B., Horaud, R. (2002), Visual Servoing from Lines, *submitted to Int. J. of Robotics Research*

Bottema, O., and Roth, B., (1979), Theoretical Kinematics, Dover Publications.

Chazelle, B., Edelsbrunner, H., Guibas, L. J., Sharir, M. and Stolfi, J., (1996), Lines in Space : Combinatorics and Algorithms, *Algorithmica*, vol. 15, pp. 428–447, 1996.

Dhome, M., Richetin, M., Lapresté, J.T. and Rives, G., (1989), Determination of the Attitude of 3D Objects from a Single Perspective View, IEEE *Transactions on Pattern Analysis and Machine Intelligence*, no. 12, vol. 11, pp. 1265–1278, december 1989.

Faugeras, O., (1993), Three-Dimensional Computer Vision - A Geometric Viewpoint, MIT Press, Cambridge, MA, USA.

Goodman, J. and O'Rourke, J., (1997), Handbook of Discrete and Computational Geometry, CRC Press.

Horaud, R., Conio, B., Leboulleux, O. and Lacolle, B., (1987), An Analytic Solution for the Perspective 4-Point Problem, *Computer Vision, Graphics and Image Processing*, vol. 47, pp. 33–44, 1989.

MacCarthy, J.M., (1990), Introduction to Theoretical Kinematics, MIT Press.

Marchand, E., Bouthémy, P., Chaumette, F. and Moreau, V., (1999), Robust real-time visual tracking using a 2D-3D model-based approach IEEE *International Conference on Computer Vision*, vol. 1, pp. 262–268, Kerkira, Greece, September 1999.

Murray, R. M., Li, Z. and Sastry, S. S., (1994), A Mathematical Introduction to Robotic Manipulation, CRC Press.

Navab, N., (1993), Visual motion of lines and cooperation between motion and stereo, Thèse de Doctorat, Université Paris-Sud, 1993.

Plücker, J., (1865), On a New Geometry of Space, *Philosophical Transactions of the Royal Society of London*, vol. 155, pp. 725–791.

Pottman, H., Peternell, M. and Ravani, B., (1998), Approximation in Line Space – Applications in Robot Kinematics and Surface Reconstruction, *Advances in Robot Kinematics: Analysis and Control* J. Lenarčič and M. L. Husty, eds, Kluwer Academic Publishers, pp. 403–412, 1998.

Rives, P. and Espiau, B., (1987), Estimation récursive de primitives 3D au moyen d'une caméra mobile, *Traitement du Signal*, vol. 4, pp. 259–272, 1987.

Semple, J.G. and Kneebone, G.T., (1952), Algebraic Projective Geometry, Oxford Science Publication.

VIGOR, (1998–2001), Visually Guided Robots Using Uncalibrated Cameras, ESPRIT-IV reactive LTR project number 26247, `http://www.inrialpes.fr/VIGOR`

ON DISPLACING A SCREW ABOUT A SCREW

I.A. Parkin
School of Information Technologies
The University of Sydney, N.S.W. 2006, Australia
ian@it.usyd.edu.au

J.E. Baker
School of Mechanical and Manufacturing Engineering
The University of New South Wales, NSW 2052, Australia
jebaker@it.usyd.edu.au

Abstract We derive, using compact vector methods which do not rely on Kotel'nikov's principle, the dualised "Rodrigues formula" which is a well-established expression for the screw that has suffered a screw displacement from a known initial location. An alternative perception – that the screw has suffered rigid-body rotation through a dual angle – is examined by expressing the operation as a transformation of the initial screw by a 3×3 dual-valued orthogonal matrix. We draw attention to the often unrecognised extent of Rodrigues's contributions in this area.

Keywords: kinematics, screw theory, screw displacement, Rodrigues

1. Background

In books and articles which deal with the rotation of a point about an axis through an origin of co-ordinates, it is common to encounter "Rodrigues's formula" as the means of determining the displaced vector. In extending the treatment, such a work might lay down a more general result for the screw displacement of a vector about its unique axis of motion. If this authority is concerned with more advanced matters, it might even cite Kotel'nikov's principle of transference in stating the "dualised Rodrigues formula" for displacing an arbitrarily disposed screw about an arbitrarily disposed axis. Within these various equations there appears the multiplier $\tan \frac{1}{2}\theta$, or some function of it, θ being the angle of rotation. By way of introduction to the present article, we have two comments to make on the development just described.

First, in fact, Rodrigues (Rodrigues, 1840; Baker and Parkin, 2002) presents as his leading formula one which takes into account screw mo-

J. Lenarčič and F. Thomas (eds.), Advances in Robot Kinematics, 275–280.

tion with axis generally configured. Hence, the typical designation given earlier does him a disservice by neglect. Second, the presence of $\tan\frac{1}{2}\theta$ is an artefact of the method of derivation which Rodrigues employs. Appealing though it may be, it is not an essential feature of the transformation, as we now set out to demonstrate.

2. Notation and Basic Geometry

Throughout this paper a *screw* will typically be written as a 3-tuple of dual numbers

$$\hat{\mathbf{G}}=|\hat{\mathbf{G}}|\,(1+\varepsilon\,p)\,\hat{\mathbf{g}},\quad \hat{\mathbf{g}}=\mathbf{l}+\varepsilon\,\mathbf{M},\quad \hat{\mathbf{g}}^2=\mathbf{l}^2+\varepsilon\,2\,\mathbf{l}\cdot\mathbf{M}=1+\varepsilon\,0,\quad \mathbf{l}\times\mathbf{M}=\mathbf{R}. \tag{1}$$

in which ε is a quasi-scalar satisfying $\varepsilon^2=0$ and such that for all real a, b, c, d, $(a+\varepsilon\,b=c+\varepsilon\,d)\Leftrightarrow(a=c)$ *and* $(b=d)$. Bold letters represent 3-vectors, lower case bold letters indicate *unit* vectors, and the overwritten 'hat' symbol indicates dual quantities. $|\hat{\mathbf{G}}|$ is the *real magnitude* and p is the *pitch* of the screw $\hat{\mathbf{G}}$ which is located spatially by its *normalised line* $\hat{\mathbf{g}}$, of unit magnitude (and zero pitch), with direction vector $\mathbf{l}=(l,\ m,\ n)$ and *moment* vector $\mathbf{M}=(P,\ Q,\ R)$ which together determine its *radius* vector $\mathbf{R}$. The values l, m, n, P, Q, R are *Plucker* coordinates of that line.

Two screws $\hat{\mathbf{G}}_1=\mathbf{G}_1+\varepsilon\,\mathbf{G}_{\mathbf{p}_1}$, $\hat{\mathbf{G}}_2=\mathbf{G}_2+\varepsilon\,\mathbf{G}_{\mathbf{p}_2}$ have a *scalar product*

$$\hat{\mathbf{G}}_1\cdot\hat{\mathbf{G}}_2=\mathbf{G}_1\cdot\mathbf{G}_2+\varepsilon\,\hat{\mathbf{G}}_1\otimes\hat{\mathbf{G}}_2\quad \textit{where}\quad \hat{\mathbf{G}}_1\otimes\hat{\mathbf{G}}_2=\mathbf{G}_1\cdot\mathbf{G}_{\mathbf{p}_2}+\mathbf{G}_2\cdot\mathbf{G}_{\mathbf{p}_1},$$

in which $\hat{\mathbf{G}}_1\otimes\hat{\mathbf{G}}_2$ is the *mutual moment* of the screws $\hat{\mathbf{G}}_1$, $\hat{\mathbf{G}}_2$. Two screws $\hat{\mathbf{G}}_1$, $\hat{\mathbf{G}}_2$ are said to be *perpendicular* if $\mathbf{G}_1\cdot\mathbf{G}_2=0$, to be *reciprocal* if $\hat{\mathbf{G}}_1\otimes\hat{\mathbf{G}}_2=0$. We shall call them *orthogonal* if both of these are true, *i.e.* if $\hat{\mathbf{G}}_1\cdot\hat{\mathbf{G}}_2=0$, which implies that each intersects the other at right angles. Their *cross product* screw $\hat{\mathbf{G}}_1\times\hat{\mathbf{G}}_2$ is sited in the *common perpendicular line* of $\hat{\mathbf{G}}_1$ and $\hat{\mathbf{G}}_2$.

3. Displacing a Screw about a Screw

We here determine, from fundamentals, the location $\hat{\mathbf{G}}_f$ into which a general screw $\mathcal{G}$, initially at $\hat{\mathbf{G}}_i$, is carried when displaced through dual angle $\hat{\theta}$, where $\hat{\theta}=\theta+\varepsilon\,\sigma$, about a unit *screw-axis* $\hat{\mathbf{s}}$, with $\hat{\mathbf{s}}^2=1$. Such a displacement may be represented by the *sine*-form finite displacement screw, $\hat{\mathbf{S}}=\sin\frac{1}{2}\hat{\theta}\,\hat{\mathbf{s}}$, or by the corresponding *tangent*-form screw, $\hat{\mathbf{T}}=\tan\frac{1}{2}\hat{\theta}\,\hat{\mathbf{s}}$.

For the purpose of analysis we temporarily introduce new xyz-axes which are embedded in the screw-pair consisting of $\hat{\mathbf{s}}$ and $\mathcal{G}$ in such a way that the x-axis lies on their common perpendicular and the z-axis

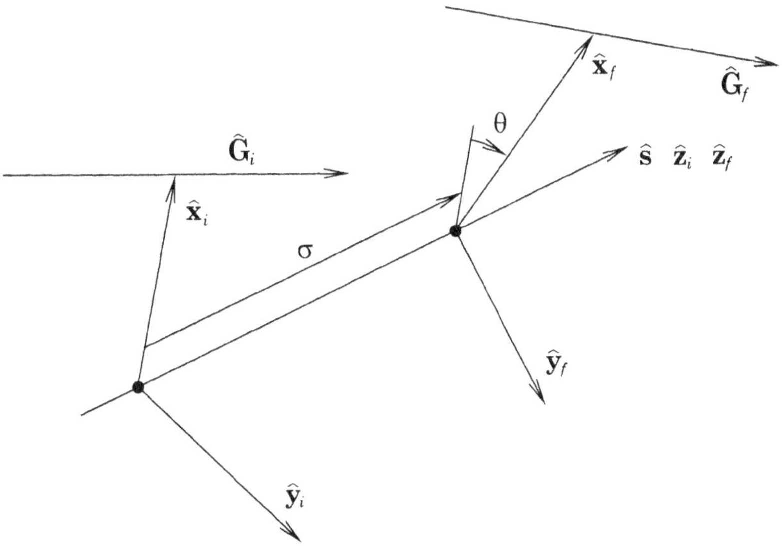

Figure 1. Shows the displacement of a general screw from location $\hat{\mathbf{G}}_i$ to location $\hat{\mathbf{G}}_f$ by rotation and translation through dual angle $\hat{\theta} = \theta + \varepsilon\,\sigma$ about screw-axis $\hat{\mathbf{s}}$. To aid analysis, a triad of xyz-axes, which displaces with the screw, is constructed so that the z-axis is sited in the screw-axis $\hat{\mathbf{s}}$ and the x-axis lies on the common perpendicular of that axis and the displacing screw. The lines of those axes are shown in their initial locations $\hat{\mathbf{x}}_i$, $\hat{\mathbf{y}}_i$, $\hat{\mathbf{z}}_i$ and their final locations $\hat{\mathbf{x}}_f$, $\hat{\mathbf{y}}_f$, $\hat{\mathbf{z}}_f$.

is sited in $\hat{\mathbf{s}}$ (Fig. 1). Thus, in their *initial* locations $\hat{\mathbf{x}}_i$, $\hat{\mathbf{y}}_i$, $\hat{\mathbf{z}}_i$, these normalised axial lines are expressed in the world-frame by

$$\hat{\mathbf{x}}_i = \frac{\hat{\mathbf{s}} \times \hat{\mathbf{G}}_i}{\sqrt{(\hat{\mathbf{s}} \times \hat{\mathbf{G}}_i)^2}}, \quad \hat{\mathbf{y}}_i = \hat{\mathbf{s}} \times \hat{\mathbf{x}}_i, \quad \hat{\mathbf{z}}_i = \hat{\mathbf{s}} \quad \textit{where} \quad \hat{\mathbf{s}}^2 = 1. \tag{2}$$

The coordinates of $\mathcal{G}$ – *i.e.* of $\hat{\mathbf{G}}_i$ – in this xyz-frame are

$$\begin{bmatrix} \hat{\mathbf{x}}_i^T \\ \hat{\mathbf{y}}_i^T \\ \hat{\mathbf{z}}_i^T \end{bmatrix} \hat{\mathbf{G}}_i = \begin{bmatrix} 0 \\ \hat{\mathbf{s}} \times \hat{\mathbf{x}}_i \cdot \hat{\mathbf{G}}_i \\ \hat{\mathbf{s}} \cdot \hat{\mathbf{G}}_i \end{bmatrix} = \begin{bmatrix} 0 \\ -\sqrt{(\hat{\mathbf{s}} \times \hat{\mathbf{G}}_i)^2} \\ \hat{\mathbf{s}} \cdot \hat{\mathbf{G}}_i \end{bmatrix}. \tag{3}$$

These coordinates do not change when $\mathcal{G}$, together with this xyz-frame, is displaced to a new location.

When displacement has proceeded through the dual angle $\hat{\theta}$ from the initial to the final location, the xyz-axes constitute a *final-location* frame whose axes $\hat{\mathbf{x}}_f$, $\hat{\mathbf{y}}_f$, $\hat{\mathbf{z}}_f$, expressed in the world-frame, are

$$\begin{bmatrix} \hat{\mathbf{x}}_f^T \\ \hat{\mathbf{y}}_f^T \\ \hat{\mathbf{z}}_f^T \end{bmatrix} = \begin{bmatrix} \cos\hat{\theta} & \sin\hat{\theta} & 0 \\ -\sin\hat{\theta} & \cos\hat{\theta} & 0 \\ 0 & 0 & 1 \end{bmatrix} \begin{bmatrix} \hat{\mathbf{x}}_i^T \\ \hat{\mathbf{y}}_i^T \\ \hat{\mathbf{z}}_i^T \end{bmatrix} = \begin{bmatrix} \cos\hat{\theta}\,\hat{\mathbf{x}}_i^T + \sin\hat{\theta}\,\hat{\mathbf{y}}_i^T \\ -\sin\hat{\theta}\,\hat{\mathbf{x}}_i^T + \cos\hat{\theta}\,\hat{\mathbf{y}}_i^T \\ \hat{\mathbf{z}}_i^T \end{bmatrix}. \tag{4}$$

Relative to this final-location frame, the displaced screw $\mathcal{G}$ has the same coordinates as those written in eqn. (3). So, on substituting from eqns. (2, 3, 4) and expanding in terms of the dual half-angle $\frac{1}{2}\hat{\theta}$, the final-location screw $\hat{\mathbf{G}}_f$ may be expressed in the world-frame progressively as

$$\begin{aligned} \hat{\mathbf{G}}_f &= \sqrt{(\hat{\mathbf{s}} \times \hat{\mathbf{G}}_i)^2}\,(\sin\hat{\theta}\,\hat{\mathbf{x}}_i - \cos\hat{\theta}\,\hat{\mathbf{y}}_i) + \hat{\mathbf{s}} \cdot \hat{\mathbf{G}}_i\,\hat{\mathbf{z}}_i \\ &= 2\sin\frac{1}{2}\hat{\theta}\cos\frac{1}{2}\hat{\theta}\,\hat{\mathbf{s}} \times \hat{\mathbf{G}}_i - (1 - 2\sin^2\frac{1}{2}\hat{\theta})\,\hat{\mathbf{s}} \times (\hat{\mathbf{s}} \times \hat{\mathbf{G}}_i) + \hat{\mathbf{s}} \cdot \hat{\mathbf{G}}_i\,\hat{\mathbf{s}} \\ &= \hat{\mathbf{G}}_i + 2\sin\frac{1}{2}\hat{\theta}\cos\frac{1}{2}\hat{\theta}\,\hat{\mathbf{s}} \times \hat{\mathbf{G}}_i + 2\sin^2\frac{1}{2}\hat{\theta}\,\hat{\mathbf{s}} \times (\hat{\mathbf{s}} \times \hat{\mathbf{G}}_i). \end{aligned} \tag{5}$$

In particular, when expressed in terms of the *sine*-form finite displacement screw, $\hat{\mathbf{S}} = \sin\frac{1}{2}\hat{\theta}\,\hat{\mathbf{s}}$,

$$\hat{\mathbf{G}}_f = \hat{\mathbf{G}}_i + 2\hat{\mathbf{S}} \times (\cos\frac{1}{2}\hat{\theta}\,\hat{\mathbf{G}}_i + \hat{\mathbf{S}} \times \hat{\mathbf{G}}_i), \tag{6}$$

and, when expressed in terms of the *tangent*-form finite displacement screw, $\hat{\mathbf{T}} = \tan\frac{1}{2}\hat{\theta}\,\hat{\mathbf{s}}$,

$$\hat{\mathbf{G}}_f = \hat{\mathbf{G}}_i + 2\cos^2\frac{1}{2}\hat{\theta}\,\hat{\mathbf{T}} \times (\hat{\mathbf{G}}_i + \hat{\mathbf{T}} \times \hat{\mathbf{G}}_i). \tag{7}$$

These are the desired representations of the displaced screw $\hat{\mathbf{G}}_f$. To obtain the actual formula presented by Rodrigues, we suppress the dual parts of $\hat{\mathbf{G}}_i$ and $\hat{\mathbf{G}}_f$ in the foregoing, thereby finding, after some manipulation,

$$\mathbf{R}_f - \mathbf{R}_i = \sigma\,\mathbf{s} + 2\tan\frac{1}{2}\theta\,\mathbf{s} \times [\frac{1}{2}(\mathbf{R}_i + \mathbf{R}_f) - \mathbf{R}_s]$$

where $\mathbf{R}_i$ and $\mathbf{R}_f$ are respectively the radius vectors of $\hat{\mathbf{G}}_i$ and $\hat{\mathbf{G}}_f$, and $\mathbf{R}_s$ is the radius vector of the screw axis $\hat{\mathbf{s}}$ whose unit direction vector is $\mathbf{s}$. This equation has been expressed in vectorial format, a conciseness not available at the time of Rodrigues's article.

In the special case $\frac{1}{2}\hat{\theta} = \pi/2$, in which $\hat{\mathbf{S}} = \hat{\mathbf{s}}$, $\cos\frac{1}{2}\hat{\theta} = -\varepsilon\frac{1}{2}\sigma$, the *tangent*-form screw $\hat{\mathbf{T}}$ attains infinite magnitude and cannot be readily used. The result for the *sine*-form finite displacement screw $\hat{\mathbf{S}}$ is then:

$$\hat{\mathbf{G}}_f = \hat{\mathbf{G}}_i + 2\hat{\mathbf{s}}\times(-\varepsilon\frac{1}{2}\sigma\,\hat{\mathbf{G}}_i + \hat{\mathbf{s}}\times\hat{\mathbf{G}}_i) = -\hat{\mathbf{G}}_i + 2(\hat{\mathbf{s}}\cdot\hat{\mathbf{G}}_i\,\hat{\mathbf{s}} - \varepsilon\frac{1}{2}\sigma\,\hat{\mathbf{s}}\times\hat{\mathbf{G}}_i)$$

which, for a pure half-turn rotation without translation ($\sigma = 0$), shows the expected line-mirror symmetry of $\hat{\mathbf{G}}_f$ and $\hat{\mathbf{G}}_i$ about the line $\hat{\mathbf{S}} = \hat{\mathbf{s}}$.

4. Orthogonal Matrix Formulation

The evaluations represented by eqns. (5, 6, 7) may appropriately be applied to individual screws $\mathcal{G}$; but, when several screws are to be made subject to the same displacement, it is more efficient to construct a matrix for (rigid-body) transformation of each initial screw $\hat{\mathbf{G}}_i$ into its corresponding final value $\hat{\mathbf{G}}_f$. Creation of such a matrix embeds, just once, a number of calculations which must otherwise be repeated for each screw which is transformed.

The appropriate matrix, $\hat{D}$ below, may be determined from our formulation as follows. Consider the finite displacement screw $\hat{\mathbf{T}}$ which appears in eqn. (7) to be expressed by $\hat{\mathbf{T}} = (\hat{T}_1, \hat{T}_2, \hat{T}_3)$. Then the cross-product operations in that equation may be represented by the 3×3 dual matrices

$$\hat{\mathbf{T}}\times = \begin{bmatrix} 0 & -\hat{T}_3 & \hat{T}_2 \\ \hat{T}_3 & 0 & -\hat{T}_1 \\ -\hat{T}_2 & \hat{T}_1 & 0 \end{bmatrix}, \quad \hat{\mathbf{T}}\times\hat{\mathbf{T}}\times = \begin{bmatrix} \hat{T}_1^2 - \hat{\mathbf{T}}^2 & \hat{T}_2\hat{T}_1 & \hat{T}_3\hat{T}_1 \\ \hat{T}_1\hat{T}_2 & \hat{T}_2^2 - \hat{\mathbf{T}}^2 & \hat{T}_3\hat{T}_2 \\ \hat{T}_1\hat{T}_3 & \hat{T}_2\hat{T}_3 & \hat{T}_3^2 - \hat{\mathbf{T}}^2 \end{bmatrix} \tag{8}$$

which are respectively skew-symmetric and symmetric, the latter being the square of the former. In these terms, and on using the identities

$$\hat{\mathbf{T}}^2 = \hat{\mathbf{T}}\cdot\hat{\mathbf{T}} = \hat{T}_1^2 + \hat{T}_2^2 + \hat{T}_3^2 \quad \textit{and} \quad 1 + \hat{\mathbf{T}}^2 = 1 + \tan^2\frac{1}{2}\hat{\theta} = 1/\cos^2\frac{1}{2}\hat{\theta},$$

the transformation of $\hat{\mathbf{G}}_i$ to $\hat{\mathbf{G}}_f$ of eqn. (7) may be immediately transcribed into the matrix notation

$$\hat{\mathbf{G}}_f = \hat{D}\hat{\mathbf{G}}_i \quad \textit{where} \quad \hat{D} = \hat{I} + \frac{2}{1+\hat{\mathbf{T}}^2}\Big[\hat{\mathbf{T}}\times + \hat{\mathbf{T}}\times\hat{\mathbf{T}}\times\Big], \tag{9}$$

in which the matrix $\hat{D}$ embodies the displacement and $\hat{I}$ is the identity matrix. We readily confirm that the matrix $\hat{D}$ is *orthogonal* – representing a rigid-body displacement – by demonstrating that

$$\hat{D}\hat{D}^T = \hat{I} + \frac{4}{(1+\hat{\mathbf{T}}^2)^2}\hat{\mathbf{T}}\times\hat{\mathbf{T}}\times\Big[\hat{\mathbf{T}}^2\hat{I} + \hat{\mathbf{T}}\times\hat{\mathbf{T}}\times\Big] = \hat{I}. \tag{10}$$

For the second of eqns. (8) shows that $\hat{\mathbf{T}} \times \hat{\mathbf{T}} \times = \hat{A} - \hat{\mathbf{T}}^2 \hat{I}$ where $\hat{A}$ is the matrix defined by $\hat{A}_{ij} = \hat{T}_i \hat{T}_j$ with the property:

$$(\hat{A}^2)_{ij} = \sum_{k=1}^{3} \hat{T}_i \hat{T}_k \hat{T}_k \hat{T}_j = \hat{\mathbf{T}}^2 \hat{A}_{ij} \quad i.e. \quad \hat{A}^2 = \hat{\mathbf{T}}^2 \hat{A} \, ;$$

so the term added to the identity in eqn. (10) vanishes in the form

$$\hat{\mathbf{T}} \times \hat{\mathbf{T}} \times \left[\hat{\mathbf{T}}^2 \hat{I} \; + \; \hat{\mathbf{T}} \times \hat{\mathbf{T}} \times \right] = \left[\hat{A} - \hat{\mathbf{T}}^2 \hat{I} \right] \hat{A} = \hat{A}^2 - \hat{\mathbf{T}}^2 \hat{A} = 0.$$

5. Closing Remarks

Although the *sine*-screw and the *tangent*-screw may be regarded merely as alternatives, we believe that the former offers advantages in computation and visualisation. Indeed, Rodrigues draws upon the analogy between planar and spherical triangles, whereby the side (or half-side) of the former corresponds with the *sine* of the half-angle of the latter. Consequently, it appears to be appropriate to regard pitch based upon the *sine*-screw (Parkin, 1992) as the natural form for this entity. The *sine/cosine*-half-angle formulation also appears in Rodrigues's crucial expressions for compounding rotations. It has been pointed out by Baker (Baker, 1998) that dualising these relationships leads directly to corresponding formulas for compounding finite displacements.

6. Acknowledgment

The authors acknowledge the assistance of the School of Information Technologies, The University of Sydney, in providing computing facilities used in preparation of this paper.

References

O. Rodrigues. Des lois géometriques qui régissent les déplacements d'un système solide dans l'espace, ... *Journal de Mathématiques Pures et Appliquées de Liouville* **5**, pp. 380-440 (1840).

J.E. Baker and I.A. Parkin. *Fundamentals of Screw Motion: Seminal Papers by Michel Chasles and Olinde Rodrigues,* (transls. J.E. Baker and I.A. Parkin) (2002).

I.A. Parkin. A third conformation with the screw systems: finite twist displacements of a directed line and point. *Mechanism and Machine Theory* **27** (2), pp. 177-188 (1992).

J.E. Baker. On compounding finite-displacement screws. *Proc. Eighth International Conference on Engineering Design Graphics and Descriptive Geometry,* Austin, Texas, USA, July 31 - August 3, **1**, pp. 109-113. (1998)

ALGEBRAIC SOLUTION OF INVERSE KINEMATICS REVISITED

W. Korb and W. Schlegel
German Cancer Research Center, Dept. for Medical Physics
Im Neuenheimer Feld 280, D-69120 Heidelberg, Germany.
w.korb@dkfz.de.

J.P. Schlöder and H.G. Bock
Interdisciplinary Center for Scientific Computing (IWR) of the University of Heidelberg
Im Neuenheimer Feld 368, D-69120 Heidelberg, Germany.

Abstract Different methods are known for giving all the solutions to the inverse kinematics problem for general serial-link manipulators. One of these is Raghavan and Roth's algebraic algorithm. An eigenvalue problem based on this, which is numerically robust, was formulated by various authors. Practical implementation requires algebraic precomputation, because the underlying equations are computationally involved. This can be done by "computer algebra systems" (CAS). We show here a notational concept that avoids the use of CAS. Therefore the algorithm can be implemented on every personal computer by every programming language; even for those who have no access to CAS. But even more importantly, the solution process is much faster.

Keywords: Inverse Kinematics, Polynomial Systems, Serial Manipulators, Generalized Eigenvalue Problem, Surgical Robot.

1. Introduction

For general serial link manipulators, the inverse kinematics problem is nonlinear and has up to 16 real solutions, Primrose, 1986. One of these solutions can be determined by using the Jacobi-matrix, see Sciavicco and Siciliano, 2000. The solution is dependent on the inital value of the iteration. In many applications (e.g. medical robotics) but also for theoretical considerations, it is desired to determine all the solutions. Therefore different methods have been developed. An early work in this direction was done by Paul and Zhang, 1986. But Paul's approach is only suitable for special configurations. Other authors presented solutions for robots of general geometry. A numerical procedure was described by Tsai and Morgan, 1985 and algebraic methods were given by Lee and Liang, 1988; Raghavan and Roth, 1990; Raghavan and Roth,

J. Lenarčič and F. Thomas (eds.), Advances in Robot Kinematics, 281–290.

1991. The drawback of the first is that it is slow; the others are numerically ill-conditioned, Manocha and Zhu, 1994. To overcome these problems, Raghavan and Roth's algorithm was advanced by transforming the system to a generalized eigenvalue problem (Manocha and Zhu, 1994; Kohli and Osvatic, 1992b; Ghazvini, 1993). To get the eigenvalue formulation, symbolical precomputation is needed. This can be done by "computer algebra systems" (CAS) as *Maple* or *Mathematica*, see Manocha and Zhu, 1994; Mavroidis et al., 1994. We give here a new notational concept that avoids the use of CAS. This has three main advantages. First the complete solution algorithm of inverse kinematics can be implemented in any programming language (only some library routines such as eigenvalue determination, etc. are required). Even those engineers who have no access to CAS can implement it. Second, the precomputation phase is accelerated, because only *numerical* matrix manipulation is needed. Third, we get no complex equations compared to CAS, which often gives tangled solutions. This leads to more insight into the solution process.

The *tangent-substitution* of variable θ by $x = \tan(\frac{\theta}{2})$ does not work for configurations with a solution $\theta = \pm\pi$, because $\tan(\pm\frac{\pi}{2})$ is not defined. This can be solved by homogeneous tangens transform, which was (theoretically) introduced in Kovacs, 1993. We show here, how to implement this practically.

We show an application of the algorithm in surgical robotics: the "stereotactic-neurosurgery-robot" is a 5-DOF-manipulator that is used for highly precise interventions, Korb et al., 2001. In surgery it is crucial to implement controllers with predictive behaviour. Therefore iterative methods must be discarded, and algebraic methods, such as discussed here, must be used. Further, in medical applications, the robot and the surgeon work interactively. This means, for example, that the control system determines all the solutions for the inverse kinematics problem. Then the surgeon (possibly supported by the computer system) decides which one of the solutions to use. The inverse kinematics algorithm should work for different parameter sets because robot calibration is performed. Robot calibration is necessary because the surgical manipulator must perform with high precision.

2. Notational Preliminaries

This section gives a short introduction to the terms and notation used. The manipulator is modelled by a series of rigid bodies (*links*) connected by means of kinematic pairs (*joints*). For the sake of simplicity, in our case only revolute joints are considered. The joint variables are called

$\theta_1, \ldots, \theta_6$. A coordinate system (*frame*) is attached to each link. A translation vector $\mathbf{t}_i$ and a rotational matrix $\mathbf{D}_i$ (with $\mathbf{D}_i^{-1} = \mathbf{D}_i^T$ and $\det(\mathbf{D}_i) = 1$) give the position and orientation of the $(i+1)$-th frame in relation to the i-th frame. The **translational part** is given by a_i (*length of link* i), and d_i (*offset* along joint i) as $\mathbf{t}_i := (a_i,\, 0,\, d_i)^T$. The **rotational part** is given by α_i (*twist angle* between the axes of joints i and $i+1$) and the *joint variable* θ_i (rotation angle about joint axes i) as $\mathbf{D}_i := \mathbf{C}_i \cdot \mathbf{E}_i$ with

$$\mathbf{C}_i := \begin{pmatrix} c_i & -s_i & 0 \\ s_i & c_i & 0 \\ 0 & 0 & 1 \end{pmatrix} \qquad \mathbf{E}_i := \begin{pmatrix} 1 & 0 & 0 \\ 0 & \lambda_i & -\mu_i \\ 0 & \mu_i & \lambda_i \end{pmatrix}$$

while $c_i := \cos(\theta_i)$, $s_i := \sin(\theta_i)$, $\lambda_i := \cos(\alpha_i)$, $\mu_i := \sin(\alpha_i)$. a_i, d_i, α_i and θ_i are called *Denavit-Hartenberg (DH) parameters.* This leads to the *homogeneous transformation matrices*

$$\mathbf{A}_i := \left(\begin{array}{c|c} \mathbf{C}_i \cdot \mathbf{E}_i & \mathbf{C}_i \cdot \mathbf{t}_i \\ \hline \mathbf{0} & 1 \end{array}\right); \text{ thus } \mathbf{A}_i^{-1} = \left(\begin{array}{c|c} \mathbf{E}_i^T \cdot \mathbf{C}_i^T & -\mathbf{E}_i^T \cdot \mathbf{t}_i \\ \hline \mathbf{0} & 1 \end{array}\right). \quad (1)$$

$\in \mathbb{R}^{4\times 4}$. A more detailed description is given in textbooks (e.g. Sciavicco and Siciliano, 2000). The end-effector frame in terms of the base frame is given by

$$\mathbf{T} := \left(\begin{array}{c|c} \mathbf{D}_7 & \mathbf{t}_7 \\ \hline \mathbf{0} & 1 \end{array}\right); \text{ thus } \mathbf{T}^{-1} = \left(\begin{array}{c|c} \mathbf{D}_7^T & -\mathbf{D}_7^T \mathbf{t}_7 \\ \hline \mathbf{0} & 1 \end{array}\right). \quad (2)$$

Solving the *inverse kinematics* problem means solving the closure equation

$$\mathbf{T} = \mathbf{A}_1 \cdot \mathbf{A}_2 \cdot \mathbf{A}_3 \cdot \mathbf{A}_4 \cdot \mathbf{A}_5 \cdot \mathbf{A}_6 \quad (3)$$

for the variables $\theta_1, \ldots, \theta_6$. As stated before, the problem is nonlinear and has up to 16 solutions.

Without loss of generality we set $\alpha_6 = a_6 = d_6 = 0$. Otherwise, we adapt the end-effector frame to

$$\mathbf{T} = \left(\begin{array}{c|c} \mathbf{D}_7 & \mathbf{t}_7 \\ \hline \mathbf{0} & 1 \end{array}\right) \left(\begin{array}{c|c} \mathbf{E}_6 & \mathbf{t}_6 \\ \hline \mathbf{0} & 1 \end{array}\right)^{-1} = \left(\begin{array}{c|c} \mathbf{D}_7 \cdot \mathbf{E}_6^T & -\mathbf{D}_7 \cdot \mathbf{E}_6^T \cdot \mathbf{t}_6 + \mathbf{t}_7 \\ \hline \mathbf{0} & 1 \end{array}\right) \quad (4)$$

3. Solving Inverse Kinematics without Computer Algebra Systems (CAS)

We use the basic methods of Ghazvini, 1993. The main goal of this work is to provide a notation that gives an algorithm without algebraic precomputation and therefore no need for CAS.

Because we have $\alpha_6 = a_6 = d_6 = 0$, we can determine in (3)

$$\hat{\mathbf{A}}_6 := \mathbf{A}_6 \cdot \mathbf{T}^{-1} = \left(\begin{array}{c|c} \mathbf{C}_6 & 0 \\ \hline \mathbf{0} & 1 \end{array}\right) \cdot \left(\begin{array}{c|c} \mathbf{D}_7^T & -\mathbf{D}_7^T \mathbf{t}_7 \\ \hline \mathbf{0} & 1 \end{array}\right)$$

When identifying $\mathbf{E}_6 = \mathbf{D}_7^T$ and $\mathbf{t}_6 = -\mathbf{D}_7^T \mathbf{t}_7$, the frame $\hat{\mathbf{A}}_6$ has the structure of (1) like $\mathbf{A}_1, \ldots, \mathbf{A}_5$. Therefore the closure equation can be rewritten. We do this in six different forms:

$$\begin{array}{lcll} \mathbf{A}_1\mathbf{A}_2\mathbf{A}_3 & = & \hat{\mathbf{A}}_6^{-1}\mathbf{A}_5^{-1}\mathbf{A}_4^{-1} & (5) \\ \mathbf{A}_2\mathbf{A}_3\mathbf{A}_4 & = & \mathbf{A}_1^{-1}\hat{\mathbf{A}}_6^{-1}\mathbf{A}_5^{-1} & (6) \\ \mathbf{A}_3\mathbf{A}_4\mathbf{A}_5 & = & \mathbf{A}_2^{-1}\mathbf{A}_1^{-1}\hat{\mathbf{A}}_6^{-1} & (7) \end{array} \qquad \begin{array}{lcll} \mathbf{A}_4\mathbf{A}_5\hat{\mathbf{A}}_6 & = & \mathbf{A}_3^{-1}\mathbf{A}_2^{-1}\mathbf{A}_1^{-1} & (8) \\ \mathbf{A}_5\hat{\mathbf{A}}_6\mathbf{A}_1 & = & \mathbf{A}_4^{-1}\mathbf{A}_3^{-1}\mathbf{A}_2^{-1} & (9) \\ \hat{\mathbf{A}}_6\mathbf{A}_1\mathbf{A}_2 & = & \mathbf{A}_5^{-1}\mathbf{A}_4^{-1}\mathbf{A}_3^{-1} & (10) \end{array}$$

In the general case any of these equations can be used. In some cases not all equations are suitable (compare Mavroidis et al., 1994). We can write eq. (5) – (10) in a general form:

$$A_{(x)}A_{(a)}A_{(b)} = A_{(c)}^{-1}A_{(d)}^{-1}A_{(e)}^{-1} \tag{11}$$

Correspondingly we name the joint variables $\theta_{(x)}, \theta_{(a)}, \theta_{(b)}, \theta_{(c)}, \theta_{(d)}$ and $\theta_{(e)}$. In detail the left hand side of (11) is

$$\left(\begin{array}{c|c} \mathbf{C}_{(x)}\mathbf{E}_{(x)} & \mathbf{C}_{(x)}\mathbf{t}_{(x)} \\ \hline \mathbf{0} & 1 \end{array}\right) \left(\begin{array}{c|c} \mathbf{C}_{(a)}\mathbf{E}_{(a)} & \mathbf{C}_{(a)}\mathbf{t}_{(a)} \\ \hline \mathbf{0} & 1 \end{array}\right) \left(\begin{array}{c|c} \mathbf{C}_{(b)}\mathbf{E}_{(b)} & \mathbf{C}_{(b)}\mathbf{t}_{(b)} \\ \hline \mathbf{0} & 1 \end{array}\right) \tag{12}$$

Note that only the matrices $\mathbf{C}_{(\cdot)}$ are dependent on the variables $\theta_{(x)}$, $\theta_{(a)}$ and $\theta_{(b)}$. The other matrices and vectors are constant. The next step is casting up (12) by matrix multiplication. The third and forth columns of the resulting matrix (without the homogeneous entry) can be written as

$$\begin{aligned} \mathbf{z}_l &= \mathbf{C}_{(x)}\,\mathbf{E}_{(x)}\mathbf{C}_{(a)}\,\mathbf{E}_{(a)}\mathbf{C}_{(b)}\,\mathbf{E}_{(b)}\,\mathbf{e}_3 && (13) \\ \mathbf{p}_l &= \mathbf{C}_{(x)}\,\{\mathbf{t}_{(x)} + \mathbf{E}_{(x)}\mathbf{C}_{(a)}\,\mathbf{t}_{(a)} + \mathbf{E}_{(x)}\mathbf{C}_{(a)}\,\mathbf{E}_{(a)}\mathbf{C}_{(b)}\,\mathbf{t}_{(b)}\} && (14) \end{aligned}$$

Analogously $\mathbf{p}_r$ and $\mathbf{z}_r$ can be determined by casting up the right hand side of (11). Note that then the inverses $\mathbf{A}_{(\cdot)}^{-1}$ of the matrices are used. This leads to

$$\mathbf{z}_r = \mathbf{E}_{(c)}^T\,\mathbf{C}_{(c)}^T\mathbf{E}_{(d)}^T\,\mathbf{C}_{(d)}^T\mathbf{E}_{(e)}^T\,\underbrace{\mathbf{C}_{(e)}^T\,\mathbf{e}_3}_{=\mathbf{e}_3} \tag{15}$$

$$\mathbf{p}_r = -\mathbf{E}_{(c)}^T\,\{\mathbf{t}_{(c)} + \mathbf{C}_{(c)}^T\mathbf{E}_{(d)}^T\,\mathbf{t}_{(d)} + \mathbf{C}_{(c)}^T\mathbf{E}_{(d)}^T\,\mathbf{C}_{(d)}^T\mathbf{E}_{(d)}^T\,\mathbf{t}_{(e)}\} \tag{16}$$

Eq. (13) – (16) are independent of $\theta_{(e)}$; this means that the system of equations is reduced to 5 variables.

We can now set up the 14 **Raghavan-and-Roth (RR) equations** (see Raghavan and Roth, 1990): $\mathbf{p}_l = \mathbf{p}_r$, $\mathbf{z}_l = \mathbf{z}_r$, $\mathbf{p}_l^T\mathbf{p}_l = \mathbf{p}_r^T\mathbf{p}_r$, $\mathbf{p}_l^T\mathbf{z}_l = \mathbf{p}_r^T\mathbf{z}_r$, $\mathbf{p}_l \times \mathbf{z}_l = \mathbf{p}_r \times \mathbf{z}_r$ and $\mathbf{p}_l \times (\mathbf{p}_l \times \mathbf{z}_l) + (\mathbf{p}_l^T\mathbf{z}_l)\mathbf{p}_l = \mathbf{p}_r \times (\mathbf{p}_r \times \mathbf{z}_r) + (\mathbf{p}_r^T\mathbf{z}_r)\mathbf{p}_r$. These are sufficient to solve the inverse kinematics problem.

The left sides of these 14 equations can be determined as follows. Note that the appearing matrices $\mathbf{N}$ are orthogonal, hence $\mathbf{N}^T\mathbf{N}$ is identity.

$$\mathbf{z}_l = \mathbf{C}_{(x)} \underbrace{\mathbf{E}_{(x)}\mathbf{C}_{(a)}}_{\hat{\mathbf{A}}} \underbrace{\mathbf{E}_{(a)}\mathbf{C}_{(b)}}_{\hat{\mathbf{B}}} \underbrace{\mathbf{E}_{(b)}\mathbf{e}_3}_{\hat{e}_3} \tag{17}$$

$$\mathbf{p}_l = \mathbf{C}_{(x)}\{\mathbf{t}_{(x)} + \hat{\mathbf{A}}\mathbf{t}_{(a)} + \hat{\mathbf{A}}\hat{\mathbf{B}}\mathbf{t}_{(b)})\} \tag{18}$$

$$\begin{aligned}\mathbf{p}_l^T\mathbf{p}_l = {} & \mathbf{t}_{(x)}^T\mathbf{t}_{(x)} + 2\mathbf{t}_{(x)}^T\hat{\mathbf{A}}\mathbf{t}_{(a)} + \\ & + 2\mathbf{t}_{(x)}^T\hat{\mathbf{A}}\hat{\mathbf{B}}\mathbf{t}_{(b)} + \mathbf{t}_{(a)}^T\mathbf{t}_{(a)} + 2\mathbf{t}_{(a)}^T\hat{\mathbf{B}}\mathbf{t}_{(b)} + \mathbf{t}_{(b)}^T\mathbf{t}_{(b)}\end{aligned} \tag{19}$$

$$\mathbf{p}_l^T\mathbf{z}_l = \underbrace{\mathbf{t}_{(x)}^T\hat{\mathbf{A}}\hat{\mathbf{B}}\hat{\mathbf{e}}_3}_{n_1} + \underbrace{\mathbf{t}_{(a)}^T\hat{\mathbf{B}}\hat{\mathbf{e}}_3}_{n_2} + \underbrace{\mathbf{t}_{(b)}^T\hat{\mathbf{e}}_3}_{n_3} \tag{20}$$

$$\underbrace{(\mathbf{p}_l \times \mathbf{z}_l)}_{\mathbf{w}_l} = \mathbf{C}_{(x)}\{\underbrace{(\mathbf{t}_{(x)} \times \hat{\mathbf{A}}\hat{\mathbf{B}}\hat{e}_3)}_{\mathbf{u}_1} + \hat{\mathbf{A}}\underbrace{(\mathbf{t}_{(a)} \times \hat{\mathbf{B}}\hat{e}_3)}_{\mathbf{u}_2} + \hat{\mathbf{A}}\hat{\mathbf{B}}\underbrace{(\mathbf{t}_{(b)} \times \hat{e}_3)}_{\mathbf{u}_3}\} \tag{21}$$

$$\begin{aligned}\mathbf{v}_l = {} & \mathbf{C}_{(x)}\{(\mathbf{t}_{(x)} \times \mathbf{u}_1) + n_1\mathbf{t}_{(x)} + 2[(\mathbf{t}_{(x)} \times \hat{\mathbf{A}}\mathbf{u}_2) + \\ & + n_2\mathbf{t}_{(x)}] + 2[(\mathbf{t}_{(x)} \times \hat{\mathbf{A}}\hat{\mathbf{B}}\mathbf{u}_3) + n_3\mathbf{t}_{(x)}] + \\ & + \hat{\mathbf{A}}(\mathbf{t}_{(a)} \times \mathbf{u}_2) + n_2\hat{\mathbf{A}}\mathbf{t}_{(a)} + 2[\hat{\mathbf{A}}(\mathbf{t}_{(a)} \times \hat{\mathbf{B}}\mathbf{u}_2) + \\ & + n_3\hat{\mathbf{A}}\mathbf{t}_{(a)}] + \hat{\mathbf{A}}\hat{\mathbf{B}}(\mathbf{t}_{(b)} \times \mathbf{u}_3) + n_3\hat{\mathbf{A}}\hat{\mathbf{B}}\mathbf{t}_{(b)}\}\end{aligned} \tag{22}$$

with $\mathbf{v}_l = \mathbf{p}_l \times \mathbf{w}_l + (\mathbf{p}_l^T\mathbf{z}_l)\mathbf{p}_l$. The right hand sides are determined analogously.

The next step is **tangens substitution** of the joint variable $\theta_{(x)}$ by $x = \tan(\theta_{(x)}/2)$ what leads to $\cos(\theta_{(x)}) = (1 - x^2)/(1 + x^2)$ and $\sin(\theta_{(x)}) = (2x)/(1 + x^2)$. When using the equivalent matrix notation

$$\mathcal{X}^-\mathbf{C}_{(x)} = \mathcal{X}^+ \quad \mathcal{X}^+ := \begin{pmatrix} x & 1 & 0 \\ 1 & -x & 0 \\ 0 & 0 & 1 \end{pmatrix} \; \mathcal{X}^- := \begin{pmatrix} -x & 1 & 0 \\ 1 & x & 0 \\ 0 & 0 & 1 \end{pmatrix} \tag{23}$$

tangens substitution means just to multiply the RR equations by $\mathcal{X}^-$ from the left. This leads to $\mathcal{X}^+\hat{\mathbf{A}}\hat{\mathbf{B}}\hat{e}_3 = \mathcal{X}^-\mathbf{p}_r$ and so forth.

In the following we show how to transform the 14 RR-equations into the form

$$\boxed{\left(x\hat{\mathcal{L}} + \mathcal{L}\right)\mathbf{w}_l = \left(x\hat{\mathcal{R}} + \mathcal{R}\right)\mathbf{w}_r} \tag{24}$$

with

$$\begin{aligned} \mathbf{w}_l &:= (s_a s_b, s_a c_b, c_a s_b, c_a c_b, s_a, c_a, s_b, c_b, 1)^T \\ \mathbf{w}_r &:= (s_c s_d, s_c c_d, c_c s_d, c_c c_d, s_c, c_c, s_d, c_d, 1)^T \end{aligned}$$

where $s_a := \sin(\theta_{(a)})$, $c_a := \cos(\theta_{(a)})$, ... and $\hat{\mathcal{L}}$, $\mathcal{L}$, $\hat{\mathcal{R}}$, $\mathcal{R}$ are (14×9)-matrices. For this transformation we do not use CAS, but a simple notational concept. Therefore we set

$$\mathbf{C}_{(\cdot)} = \sin(\theta_{(\cdot)})\,\mathbf{D}_s + \cos(\theta_{(\cdot)})\,\mathbf{D}_c + \mathbf{D}_u \tag{25}$$

$$\mathcal{X}^+ = \mathbf{D}_p + x\mathbf{D}_x \tag{26}$$

$$\mathcal{X}^- = \mathbf{D}_p - x\mathbf{D}_x \tag{27}$$

with

$$\mathbf{D}_s := \begin{pmatrix} 0 & -1 & 0 \\ 1 & 0 & 0 \\ 0 & 0 & 0 \end{pmatrix} \quad \mathbf{D}_c := \begin{pmatrix} 1 & 0 & 0 \\ 0 & 1 & 0 \\ 0 & 0 & 0 \end{pmatrix} \quad \mathbf{D}_u := \begin{pmatrix} 0 & 0 & 0 \\ 0 & 0 & 0 \\ 0 & 0 & 1 \end{pmatrix}$$

and

$$\mathbf{D}_p := \begin{pmatrix} 0 & 1 & 0 \\ 1 & 0 & 0 \\ 0 & 0 & 1 \end{pmatrix} \quad \mathbf{D}_x := \begin{pmatrix} 1 & 0 & 0 \\ 0 & -1 & 0 \\ 0 & 0 & 0 \end{pmatrix}$$

To show the idea, we determine $\mathbf{X}^-\mathbf{z}_l = \mathbf{X}^+\hat{\mathbf{A}}\hat{\mathbf{B}}\hat{\mathbf{e}}_3 = x\mathbf{D}_x\mathbf{q} + \mathbf{D}_p\mathbf{q}$ with $\mathbf{q} = \hat{\mathbf{A}}\hat{\mathbf{B}}\hat{\mathbf{e}}_3 = \mathbf{E}_{(x)}(s_a\mathbf{D}_s + c_a\mathbf{D}_c + \mathbf{D}_u)\mathbf{E}_a(s_b\mathbf{D}_s + c_b\mathbf{D}_c + \mathbf{D}_u)\mathbf{E}_b\hat{\mathbf{e}}_3$ $= \{\mathbf{E}_{(x)}\mathbf{D}_s\mathbf{E}_{(a)}\mathbf{D}_s\mathbf{E}_{(b)}\hat{\mathbf{e}}_3\}\ s_a\ s_b + \{\mathbf{E}_{(x)}\mathbf{D}_s\mathbf{E}_{(a)}\mathbf{D}_c\mathbf{E}_{(b)}\hat{\mathbf{e}}_3\}\ s_a\ c_b + \ldots +$ $\{\mathbf{E}_{(x)}\mathbf{D}_u\mathbf{E}_{(a)}\mathbf{D}_u\mathbf{E}_{(b)}\hat{\mathbf{e}}_3\}$. This shows how the elements of $\hat{\mathcal{L}}$ and $\mathcal{L}$ can be evaluated:

$$\begin{aligned} (l_{1,1}|l_{2,1}|l_{3,1})^T &= \mathbf{D}_p\mathbf{E}_{(x)}\mathbf{D}_s\mathbf{E}_{(a)}\mathbf{D}_s\mathbf{E}_{(b)}\hat{\mathbf{e}}_3 \\ (l_{1,2}|l_{2,2}|l_{3,2})^T &= \mathbf{D}_p\mathbf{E}_{(x)}\mathbf{D}_s\mathbf{E}_{(a)}\mathbf{D}_c\mathbf{E}_{(b)}\hat{\mathbf{e}}_3 \\ (l_{1,3}|l_{2,3}|l_{3,3})^T &= \mathbf{D}_p\mathbf{E}_{(x)}\mathbf{D}_c\mathbf{E}_{(a)}\mathbf{D}_s\mathbf{E}_{(b)}\hat{\mathbf{e}}_3 \\ &\vdots \\ (\hat{l}_{1,1}|\hat{l}_{2,1}|\hat{l}_{3,1})^T &= \mathbf{D}_x\mathbf{E}_{(x)}\mathbf{D}_s\mathbf{E}_{(a)}\mathbf{D}_s\mathbf{E}_{(b)}\hat{\mathbf{e}}_3 \\ (\hat{l}_{1,2}|\hat{l}_{2,2}|\hat{l}_{3,2})^T &= \mathbf{D}_x\mathbf{E}_{(x)}\mathbf{D}_s\mathbf{E}_{(a)}\mathbf{D}_c\mathbf{E}_{(b)}\hat{\mathbf{e}}_3 \\ (\hat{l}_{1,3}|\hat{l}_{2,3}|\hat{l}_{3,3})^T &= \mathbf{D}_x\mathbf{E}_{(x)}\mathbf{D}_c\mathbf{E}_{(a)}\mathbf{D}_s\mathbf{E}_{(b)}\hat{\mathbf{e}}_3 \\ &\vdots \end{aligned}$$

In a similar way $\hat{\mathcal{R}}$ and $\mathcal{R}$ are determined. It is important that the matrices $\hat{\mathbf{A}}$ and $\hat{\mathbf{B}}$ occur only once in the summands of (17) – (22)

(namely the terms are completely factorized). "Factorizing completely" is in fact the same as "complete" simplification of every equation by $\sin^2(\theta_{(\cdot)}) + \cos^2(\theta_{(\cdot)}) = 1$ ("trigonometric pythagorean theorem").

Based on (24), the eigenvalue problem can be formulated analogously to Kohli and Osvatic, 1992b or Ghazvini, 1993. We use the latter one, because implementation of both showed that the first has numerical problems in some special cases (similar problems as reported by Mavroidis et al., 1994). Ghazvini introduces another 6 equations, which are based on the RR equations, but are linearly independent from these ($\mathbf{a}^3$ means the third entry of vector $\mathbf{a}$): $x\mathbf{z}_l^3 + \mathbf{p}_l^3 = x\mathbf{z}_r^3 + \mathbf{p}_r^3$, $x\mathbf{p}_l^3 + \mathbf{z}_l^3 = x\mathbf{p}_r^3 + \mathbf{z}_r^3$, $x(\mathbf{p}_l^T\mathbf{p}_l) + (\mathbf{p}_l^T\mathbf{z}_l) = x(\mathbf{p}_r^T\mathbf{p}_r) + (\mathbf{p}_r^T\mathbf{z}_r)$, $x(\mathbf{p}_l^T\mathbf{z}_l) + (\mathbf{p}_l^T\mathbf{p}_l) = x(\mathbf{p}_r^T\mathbf{z}_r) + (\mathbf{p}_r^T\mathbf{p}_r)$, $x\mathbf{w}_l^3 + \mathbf{z}_l^3 = x\mathbf{w}_r^3 + \mathbf{z}_r^3$ and $x\mathbf{v}_l^3 + \mathbf{p}_l^3 = x\mathbf{v}_r^3 + \mathbf{p}_r^3$.

These 20 equations can be formulated analogous to (24) as

$$\begin{matrix}\text{16 equations } \{ \\ \text{4 equations } \{\end{matrix} \quad \underbrace{\left(\begin{array}{c|c} \hat{\mathcal{L}}_1 & \mathcal{L}_1 \\ \hline \hat{\mathcal{L}}_2 & \mathcal{L}_2 \end{array}\right)}_{\text{18 columns}} \begin{pmatrix} x\,\mathbf{w}_l \\ \mathbf{w}_l \end{pmatrix} = \underbrace{\left(\begin{array}{c|c} \hat{\mathcal{R}}_1 & \mathcal{R}_1 \\ \hline \hat{\mathcal{R}}_2 & \mathcal{R}_2 \end{array}\right)}_{\text{16 columns}} \begin{pmatrix} x\,\mathbf{w}_r \\ \mathbf{w}_r \end{pmatrix} \tag{28}$$

Note that all the constant terms of the equations are summerized on the left side, therefore $\mathbf{w}_r = (s_c s_d, s_c c_d, c_c s_d, c_c c_d, s_c, c_c, s_d, c_d)^T$ here. The equations are arranged so that $\left(\begin{array}{c|c} \hat{\mathcal{R}}_1 & \mathcal{R}_1 \end{array}\right)$ is a full-rank matrix. Therefore we can determine $(x\,\mathbf{w}_r^T, \mathbf{w}_r^T)^T = \mathbf{P} \cdot (x\,\mathbf{w}_l^T, \mathbf{w}_l^T)^T$ with $\mathbf{P} = \left(\begin{array}{c|c} \hat{\mathcal{R}}_1 & \mathcal{R}_1 \end{array}\right)^{-1} \cdot \left(\begin{array}{c|c} \hat{\mathcal{L}}_1 & \mathcal{L}_1 \end{array}\right)$. This can be substituted back into (28), which leads to

$$\left[(\hat{\mathcal{L}}_2 - \hat{\mathcal{R}}_2\mathbf{P})\,x + (\mathcal{L}_2 - \mathcal{R}_2\mathbf{P})\right]\mathbf{w}_l = \mathbf{0} \tag{29}$$

This is a representation of 4 equations in 9 variables. The next steps are the tangens substitution $t_{(\cdot)} = \tan(\theta_{(\cdot)}/2)$ in $\mathbf{w}_l$ and then the *dialytic elimination* by multiplication of each equation by $t_{(a)}$, $t_{(b)}$ and $t_{(a)}t_{(b)}$. The resulting eigenvalue value problem has at most 16 eigenvalues and therefore 16 solutions of x. The joint values can now easily be determined by backsubstitution into appropriate equations. For details see Ghazvini, 1993.

Computational Performance. The implementation of the presented algorithm in *Matlab* showed that the avoidance of CAS leads to a much better performance. For comparison we implemented the numerical approach and the CAS approach. For the latter we used the *Symbolic Toolbox* within the environment of *Matlab* (which is based on the *Maple* kernel) and the strategies of Manocha, 1992. Because the terms that are manipulated by the symbolic part of the algorithm are very complicated, the performance is very poor for the CAS approach. Esspecially the simplification command takes a lot of time for some terms. The performance of the CAS approach could be increased by symbolic precomputation and storing of the matrices needed. But then at least six different sets of matrices would be needed. And even more such precomputed matrices would be needed if the algorithm is generalized to manipulators with revolute and prismatic joints.

4. Homogeneous tangens transform

As stated before, a solution of $\theta_{(x)} = \pi/2$ causes problems because of the tangens substitution used. Therefore we use the *homogeneous tangens substitution* (HTS) instead of the plain tangens substitution given by formula (23), see also Kovacs, 1993. The HTS is given by $\tilde{\mathcal{X}}^- \mathbf{C}_{(x)} = \tilde{\mathcal{X}}^+$ with

$$\tilde{\mathcal{X}}^+ := \tilde{\mathcal{X}}^+(x,z) = \begin{pmatrix} -x & z & 0 \\ z & x & 0 \\ 0 & 0 & 1 \end{pmatrix} \tag{30}$$

and

$$\tilde{\mathcal{X}}^- := \tilde{\mathcal{X}}^-(x,z) = \begin{pmatrix} x & z & 0 \\ z & -x & 0 \\ 0 & 0 & 1 \end{pmatrix} \tag{31}$$

where $(x,z) \in \{\mathbb{R} \times \{0,1\}\} \setminus (0,0)$, which is the *one-dimensional projective space*. For $z = 1$ we get the standard tangens substitution, which does not work for $\theta = \pm\pi$, as stated before. For $z = 0$ it follows that $x = 1$ and $\theta = \pm\pi$. Practically we implement the above given algorithm with two different cases. First we use $\tilde{\mathcal{X}}^+(x,1)$ and $\tilde{\mathcal{X}}^-(x,1)$ and get a eigenvalue problem in x. All solutions are valid except $x = \pm\inf$. The second case is done by using $\tilde{\mathcal{X}}^+(1,z)$ and $\tilde{\mathcal{X}}^-(1,z)$ which leads to an eigenvalue problem in z. Only the solutions $z = 0$ are valid (see definition of HTS). Those are the solutions for $\theta_{(x)} = \pm\pi$. The other joint values can be determined by backsubstitution as in the standard case. Of course, an analogous procedure is necessary for the tangens substitution in $\mathbf{w}_l$. This leads together to 8 cases, which must be distinguished.

5. Application to the 5R neurosurgery manipulator

As stated above, a 5R manipulator is used for neurosurgery. This is reasonable, because the position of the end effector and the position of the needle tip are important, but the rotation angle around the needle axis is arbitrary. The goal in this section is to determine a closure equation like (3). Given are the DH parameters, the end-effector position $\mathbf{p}_1$ and the needle-tip position $\mathbf{p}_2$. The translation vector and rotation matrix of the end effector frame $\mathbf{T}$ can be determined as $\mathbf{t}_7 = \mathbf{p}_1$ and $\mathbf{D}_7 = \mathbf{D}_7(\vartheta)$. The angle ϑ gives the rotation angle about the rotation axis $\mathbf{v} = (\mathbf{p}_2 - \mathbf{p}_1)$. It must be suitably chosen, so that $\mathbf{T} = \mathbf{A}_1 \cdot \mathbf{A}_2 \cdot \mathbf{A}_3 \cdot \mathbf{A}_4 \cdot \mathbf{A}_5$ is solvable. In the following $\mathbf{R}_{\mathfrak{x}}(\gamma)$ means the rotation matrix with the angle γ around the axis given by $\mathfrak{x}$. To determine $\mathbf{D}_7(\vartheta)$ for a given $\mathbf{v}$ we compute $N = \sqrt{\mathbf{v}(1)^2 + \mathbf{v}(2)^2}$ and

$$\sin(\alpha) = \frac{\mathbf{v}(2)}{N} \qquad \cos(\alpha) = \frac{\mathbf{v}(1)}{N} \qquad \sin(\beta) = N \qquad \cos(\beta) = \mathbf{v}(3)$$

first, and then

$$\begin{aligned}\mathbf{D}_7(\vartheta) = \mathbf{R}_\mathbf{v}(\vartheta)\mathbf{R}_\mathbf{z}(\alpha)\mathbf{R}_\mathbf{y}(\beta) = \\ \mathbf{R}_\mathbf{z}(\alpha)\mathbf{R}_\mathbf{y}(\beta)\mathbf{R}_\mathbf{z}(\vartheta)\mathbf{R}_\mathbf{y}(-\beta)\mathbf{R}_\mathbf{z}(-\alpha)\mathbf{R}_\mathbf{z}(\alpha)\mathbf{R}_\mathbf{y}(\beta) \\ \mathbf{R}_\mathbf{z}(\alpha)\mathbf{R}_\mathbf{y}(\beta)\mathbf{R}_\mathbf{z}(\vartheta)\end{aligned}$$

This leads to

$$\mathbf{A}_1 \cdot \mathbf{A}_2 \cdot \mathbf{A}_3 \cdot \mathbf{A}_4 \cdot \mathbf{A}_5 \begin{pmatrix} \mathbf{R}_\mathbf{z}(\vartheta)^T & \mathbf{0} \\ \mathbf{0} & 1 \end{pmatrix} = \begin{pmatrix} \mathbf{R}_\mathbf{z}(\alpha)\mathbf{R}_\mathbf{y}(\beta) & \mathbf{p}_1 \\ \mathbf{0} & 1 \end{pmatrix}$$

which means that we have reduced the inverse kinematics problem of a 5R manipulator to solving eq. (3) and therefore to the inverse kinematics problem of a 6R manipulator.

6. Conclusion

The new notational concept avoids involved symbolic precomputation and allows the algorithm to be implemented without use of CAS. The algorithm does work for "general" and "special" 6R-manipulators. The method can easily be expanded to general serial-link manipulators with revolute and prismatic joints, Kohli and Osvatic, 1992a. The main advantage is the saving in computation time. This makes the algorithm feasible for implementation in practical applications, such as surgical robotics.

References

Ghazvini, M. (1993). Reducing the inverse kinematics of manipulators to the solution of a generalized eigenproblem. In Angeles, J., Hommel, G., and Kovacs, P., editors, *Computational Kinematics*, pages 15–26. Kluwer Academic Publishers.

Kohli, D. and Osvatic, M. (1992a). Inverse kinematics of general 4r2p, 3r3p, 4rlc, 2r2c and 3c serial manipulators. In *Robotics, Spatial Mechanisms and Mechanical*

Systems, DE-Vol. 45, volume 47 of *Flexible Mechanisms, Dynamics and Analysis*, pages 619 - 627. ASME.

Kohli, D. and Osvatic, M. (1992b). Inverse kinematics of general 6r and 5r,p serial manipulators. In *Flexible Mechanisms, Dynamics and Analysis, DE-Vol. 47*, volume 47 of *Flexible Mechanisms, Dynamics and Analysis*, pages 619 - 627. ASME.

Korb, W., Barthold, S., Bendl, R., Echner, G., Grosser, K., Pastyr, O., Treuer, H., Sturm, V., and Schlegel, W. (2001). Entwicklung eines Hochpräzisionsmanipulators für die stereotaktische Neurochirurgie. In Wörn, H., Mühling, J., Vahl, C., and Meinzer, H.-P., editors, *Proc. of "Rechner- und Sensorgestützte Chirurgie"*, pages 336 - 343, Heidelberg/Germany. DFG.

Kovacs, P. (1993). *Rechnergestützte Symbolische Kinematik*. Fortschritte der Robotik. Vieweg, Braunschweig, Wiesbaden.

Lee, H.-Y. and Liang, C.-G. (1988). Displacement analysis of the general spatial 7-link 7r mechanism. *Mechanism and Machine Theory*, 23(3):219–226.

Manocha, D. (1992). *Algebraic and Numeric Techniques for Modelling and Robotics*. PhD thesis, Dept. of Electrical Engineering and Computer Science, Univ. of California at Berkeley. chap. 6.

Manocha, D. and Zhu, Y. (1994). A fast algorithm and system for the inverse kinematics of general serial manipulators. In *Proc. Of IEEE Conf. On Robotics and Automation*, pages 3348–3354, San Diego, CA.

Mavroidis, C., Ouezdou, F., and Bidaud, P. (1994). Inverse kinematics of six-degree of freedom "general" and "special" manipulators using symbolic computation. *Robotica*, 12:421–430.

Paul, R. and Zhang, H. (1986). Computationally efficient kinematics for manipulators with spherical wrists based on the homogeneous transformation representation. *The International Journal of Robotics Research*, 5(2):32–44.

Primrose, E. (1986). On the input-output equation of the general 7r-mechanism. *Mechanism and Machine Theory*, 21(6):509–510.

Raghavan, M. and Roth, B. (1990). Kinematic analysis of the 6r manipualtor of general geometry. In Miura, H. and Arimoto, S., editors, *Proc. of the 5th Internat. Symp. on Robotics Research*, pages 263–269, Cambridge. MIT Press.

Raghavan, M. and Roth, B. (1991). A general solution for the inverse kinematics of all series chains. In *Proc. of the 8th CISM-IFToMM Symp. on Robots and Manipulators, Cracow, Poland, 1990*, pages 24 - 31, Warsaw. Univ. of Technology.

Sciavicco, L. and Siciliano, B. (2000). *Modelling and Control of Robot Manipulators*. Advanced Textbooks in Control and Signal Processing. Springer, London, 2 edition.

Tsai, L.-W. and Morgan, A. (1985). Solving the kinematics of the most general six- and five-degree-of-freedom manipulators by continuation methods. *Journal of Mechanisms, Transmissions and Automation in Design*, 107:189—200.

A VARIANT OF A 6–RKS HUNT–TYPE PARALLEL MANIPULATOR TO EASILY USE INSENSITIVITY POSITION CONFIGURATIONS

I. Zabalza, J. Ros, J. J. Gil, J. M. Pintor and J. M. Jiménez
Public University of Navarra. Mechanical Engineering Dep.
Campus Arrosadia, 31006 Pamplona, Spain
izavi@unavarra.es

Abstract In this paper, the "insensitivity position configurations" (IPCs) of a 6–RKS Hunt–type parallel manipulator are analysed. The advantages of the aforementioned manipulator IPCs in positioning tool or pieces in being manufacturing processes are discussed. The difficulty to design such manipulator showing IPCs in some predefined positions is studied. To simplify this design process a variant of 6–RKS parallel manipulator is presented.

Keywords: Kinematics, Singularity, Insensitivity, Multibody, Robotics, Parallel Manipulator, Machine Theory.

1. Introduction

The planar slider–crank and crank–rocker mechanisms are used in many automatisms: the first one when a rectilinear output movement is needed and the second when a revolving movement is sought. Both are alternative movements obtained from a continuous input revolving movement at the crank. In the slider–crank mechanism the output will be the linear alternative slider motion and in the crank–rocker, the alternative rotation of the rocker.

Both mechanisms are extensively used mainly because each one has two singular configurations (also known as limit or dead points) in which the speed of the slider, or the rocker speed vanish for any value of the speed of the input crank. In the slider–crank mechanism, two singular configurations are reached when the crank and the connecting rod are aligned (Fig. 1); analogously, in the crank–rocker mechanism, the singular configurations are obtained when the crank and the coupler are aligned (Fig. 2).

In this paper, those configurations are called "insensitivity position configurations" (IPCs) as they are configurations at which the position of the output link is set with very high precision. In those configurations small perturbations in the position of the input element have no significant effect

J. Lenarčič and F. Thomas (eds.), Advances in Robot Kinematics, 291–300.

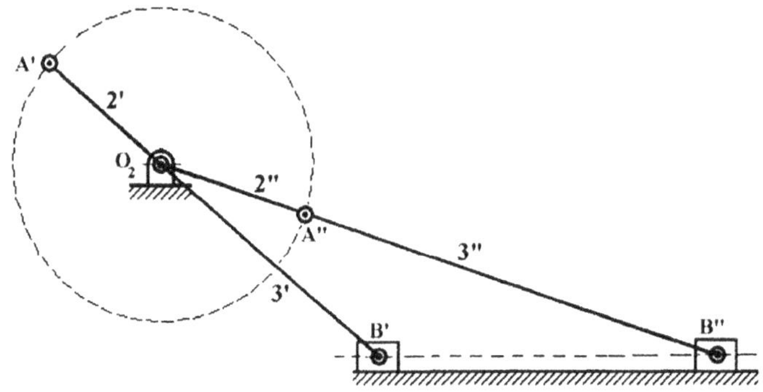

Figure 1. IPCs in the slider–crank mechanism

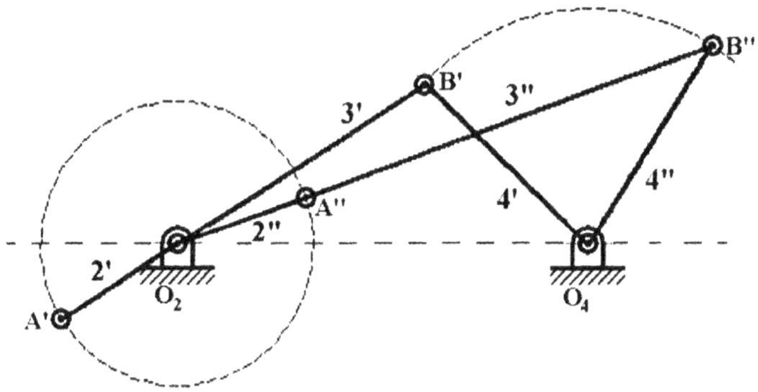

Figure 2. IPCs in the crank–rocker mechanism

on the position of the output link. It can be said that the output links position is not 'sensible' to the position of the input element.

The null speed of the output link in the IPCs leads to a great advantage when this kind of mechanisms are used to make a stroke that ends in a halt. In this way, the mechanism is stopped in the IPCs, when the output link has null speed, thus dramatically reducing the values of inertia forces. Note that at this configuration, even if the mechanism is brought quickly to a halt, the acceleration of the output link will be the same as if passing through this same configuration without stopping. However, the abrupt halt of the mechanism in any other no–IPC configuration, where the speed of the output link is not null, will produce a very high acceleration.

Also, at the mentioned IPCs, if the masses of the coupler and crank links are neglected, the input torque in the crank will be null independent of the forces or torques applied to the output link. This is because the reaction that appears in the crank and coupler link union articulation has the direction of the crank.

It becomes clear from the previous paragraphs that the described IPCs are advantageous or positive characteristics of the slider–crank and crank–rocker mechanisms. This is also demonstrated by the fact that these mechanisms IPCs have been widely used in automatism and machines.

2. IPCs in a 6–RKS parallel manipulator

The 6–RKS parallel manipulator (Fig. 3) was introduced by Hunt, 1983. This parallel robot is composed of two triangular platforms, one of them fixed to the ground. On the fixed platform there are six rotative actuators (R) with their axis located on the edges of the triangle. These actuators are the input elements on which the motors are located. The crank of each actuator is linked to the moving platform through a rod. In each rod, one tip is linked to an actuator crank through a universal joint (K) and the other tip is connected to the moving platform by means of a spherical joint (S), as depicted in Fig. 3. A couple of rods converge on each spherical joint of the moving platform.

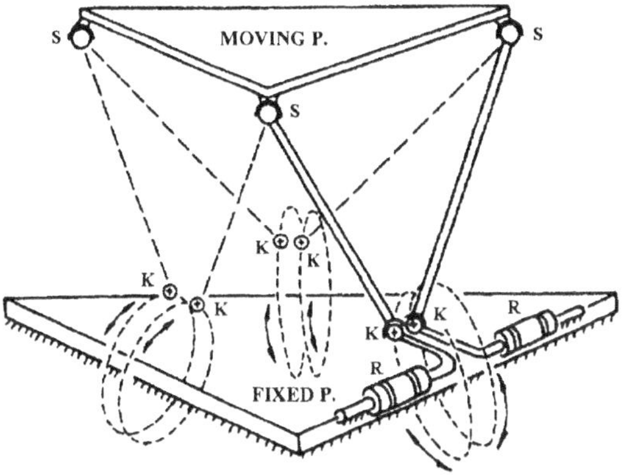

Figure 3. 6–RKS parallel manipulator

This manipulator, if it has suitable dimensions, can reach 64 different IPCs. At each one of these singular configurations the axis of the actuator, the crank and the rod of each one of the kinematic chains between the platforms are in a particular plane (Benea, 1996). In these configurations, the moving platform remains fixed independently of the movement introduced through the actuators.

Each kinematic chain actuator–crank–rod can reach 2 IPCs, one when the crank and the rod are nearly superimposed and another one when they are nearly aligned, since the platforms are coupled with 6 chains, the total number of IPCs is $2^6 = 64$.

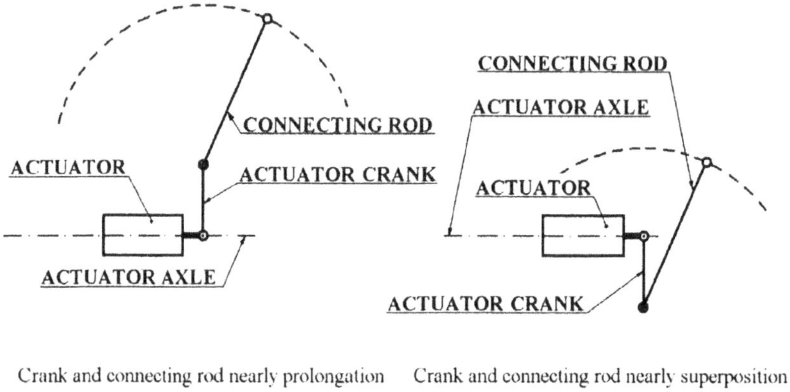

Figure 4. Two IPCs for each kinematic chain

As a generalization of the slider–crank and crank–rocker planar mechanisms IPCs, Zabalza et al., 1999, exposed the advantages of these configurations in parallel robots: great precision in positioning the moving platform, null torque in the actuators to balance the forces and torques applied on the moving platform and null speeds of this platform in the IPCs.

When the mechanism approaches an IPC, the speed of the moving platform decreases progressively until it vanishes. In this position small perturbations in the input elements have nearly no effect on the position of the moving platform. For this reason, these mechanisms can be used as halting mechanisms, taking advantage of the IPCs. In these configurations, although the mechanism is stopped abruptly using asynchronous motors with brake, the accelerations of the moving platform, and therefore the inertia forces, will have the same values as if were passing through this configuration without stopping.

However, if the mechanism is stopped in any other configuration, the accelerations will be inversely proportional to the stop time. In this case accelerations will be very high and it will be needed to use controlled speed actuators that could halt the robot progressively.

Zabalza et al., 2000, proposed parallel manipulator IPC use for pieces or tool positioning in manufacturing processes, taking advantage of their kinematic properties.

3. Design of a 6–RKS parallel manipulator using its IPCs

To take advantage of the IPCs kinematic characteristics, the parallel manipulator dimensions must be chosen so that the desired platform halting positions in a given manufacturing process happen at some manipulator IPC subset.

Theoretically it would be possible to design a 6–RKS parallel manipulator with a maximum of 64 IPCs, each one coinciding with a precision position in the manufacturing process. Actually it is very difficult to make this happen at more than three manufacturing precision positions.

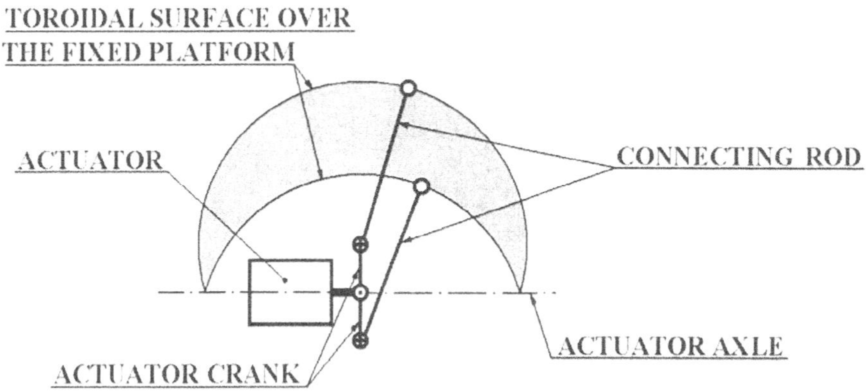

Figure 5. Section of the toroide

When only one of the six kinematic chains attaching the triagular platforms meets the IPC condition, (Fig. 5), the rod tip attached to the moving platform must be placed on a toroidal surface, Zabalza et al., 1999. As two rod tips meet at each of the three triangular platform vertex, the manipulator will be only in IPC if each of vertex of the platform is at the intersection of the two toroidal surfaces corresponding to their kinematic chains.

It can be easily understood that if more than three precision positions are sought the resulting problem is rather geometrical complex problem that would have nearly no feasible solutions.

4. Variant of the 6–RKS parallel manipulator

As described in the previous paragraphs, the use of the kinematic IPC characteristics in a 6–RKS parallel manipulator may lead to several advantages. However, it is difficult to design a robot in which more than three IPCs can be used in practice.

With the purpose of taking advantage of IPCs overcoming the difficulties of the manipulator design, a variant of the 6–RKS parallel manipulator is introduced. In this variant each of the actuators and their corresponding cranks are replaced by the output link of a crank–rocker mechanism, as depicted in Fig. 6.

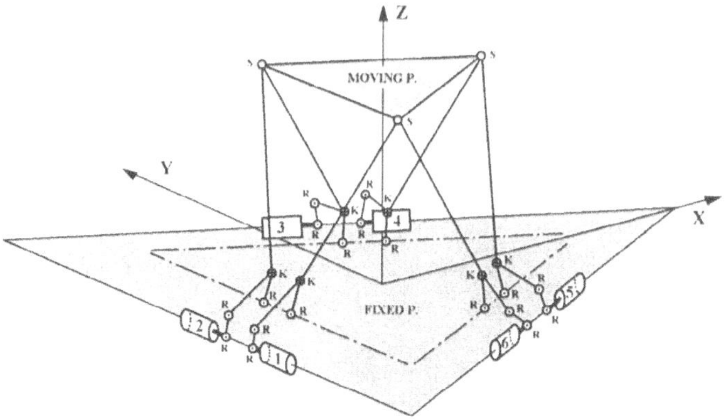

Figure 6. Variant of the 6–RKS parallel manipulator

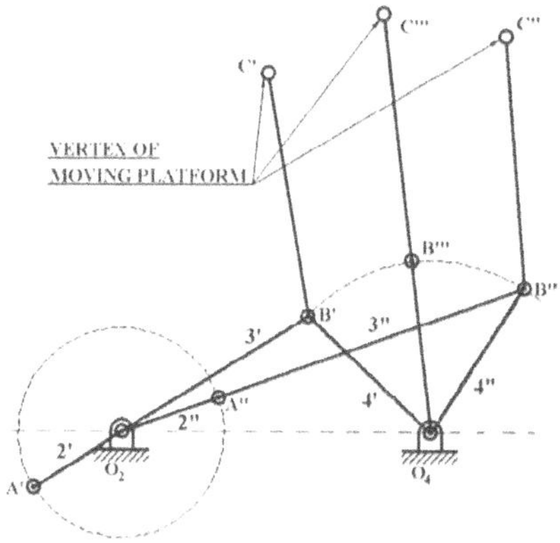

Figure 7. IPCs in each kinematic chain

In Fig. 7 is depicted how each of the attaching kinematic chains will have three (instead of the original two) possible IPCs: one when the axis of the revolving pair O_4, the rocker and the rod are in the same plane (one of the IPCs of the primitive manipulator) and two additional IPCs corresponding to the ones of the crank–rocker mechanism alone. In this way, the new manipulator will have $3^6 = 729$ IPCs instead of the 64 of the primitive manipulator

In this design each kinematic chain loses one of the primitive manipulator IPCs in favor of two new ones coming from the new crank–rocker architecture. It can be seen in Fig. 7 that the primitive manipulator IPC in which the crank tip would be under the fixed platform is lost.

With this variant of the mechanism the same design techniques explained by Zabalza et al., 1999, are easily applicable to the IPC subset corresponding to the B"' configurations in Fig. 7. Nevertheless the complexity of such design can make it discardable in favour to the design towards the new B' and B" type configurations depicted in Fig 7.

In this case, taking the crank, coupler and fixed link lengths and the angle that the fixed link sustains with the platform as free design variables for the crank-rocker mechanisms, the volume contained by the intersection of the above mentioned toroidal surfaces could be a valid position of the corresponding mobile platform vertex. Once the position of the vertex is fixed the coupler length, the mentioned angle or both can be conveniently chosen so that the position becomes a IPC. This implies a enormous design flexibility in comparison with the limitations of the primitive manipulator in which the vertex positions need to be in the intersecting edges of the corresponding toroidal surfaces.

5. Example

In this section a graphical example of a modified 6–RKS manipulator at 6 different IPC is presented. Fig. 8 schematically represents the dimensions of the manipulator used in this example.

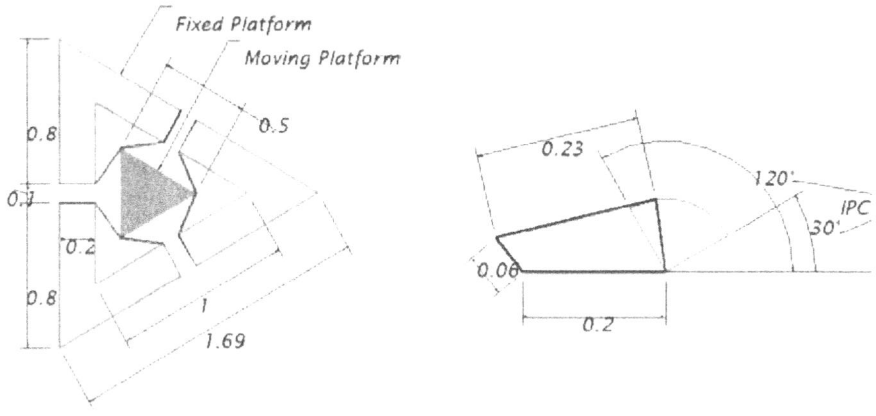

Figure 8. Example manipulator dimensions

Figure 9 schematically represent three B'–B" type insensitivity positions. It is easily observable that the complexity of IPC determination reduces to that of six simple crank–rocker mechanisms, that can be easily determined numerically or using 2D graphical techniques.

Mechanism at three B"' type IPCs is depicted in Fig. 10. Positions are much more difficult to compute and to design for as demonstrated by Zabalza et al., 1999, and can be considered less interesting in the context of the modified manipulator.

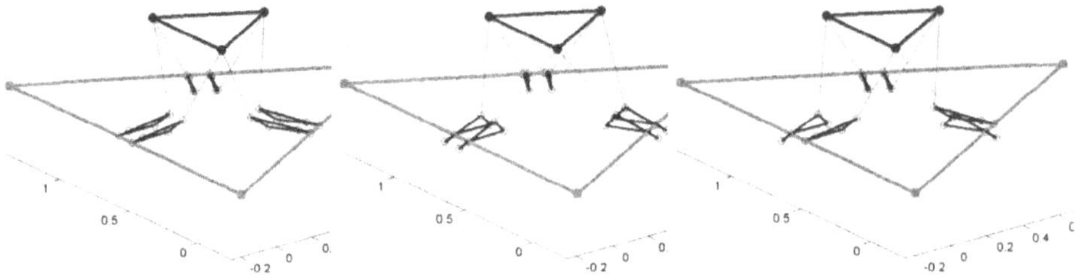

Figure 9. B' and B" type IPCs

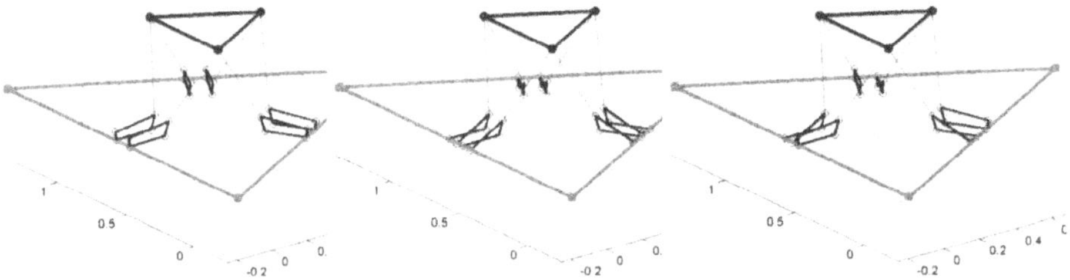

Figure 10. B"' type IPCs

It should be noted that the figures presented in this example do not consider any variability in the crank, coupler and fixed link lengths nor in the angles that the fixed link of any of the 6 crank–rocker mechanism can sustain with the fixed platform.

6. Case study

A simplistic case study is presented in this section. A 6–RKS parallel manipulator of the type shown in Fig. 6 is designed so that the manipulator achieves the following two IPC positions:

1 The position obtained after turning the moving platform 5 degrees around an axis parallel to the Y axis that passes through its gravity centre, starting from a horizontal position with gravity centre coordinates (0, 0, 0.6) and with the moving platform edges parallel to the actuator axes.

2 The same position as before, but now turning the same amount in the opposite direction.

The design variables considered are the crank and the coupler lengths. The parallel manipulator fixed dimensions are such:

- The axes of the actuators form an equilateral triangle having one meter edge length.

- Each actuator pair is symmetrically placed respect to the center of the associated triangle edge. The distance between the pair cranks is 0.1 meters.

- All actuators cranks are 0.1 meters in length.

- All connecting rods are 0.6 meters in length

- The movable platform is an equilateral triangle whose edges are 0.5 meters in length.

- The crank–rocker mechanisms fixed links all equal to 0.2 meters in length.

The problem is solved using a Matlab program. Basically a system of equations exhibiting the system coordinates constraints and the two IPC constraints is solved for the design variables chosen.

The computed design variables in the above presented example are shown in Table 1. In general several solutions might arise and then additional criteria must be used to choose the most appropriate one.

Table 1. Crank and coupler lengths

Actuator	Crank length in meters	Coupler length in meters
1	0.0128	0.1543
2	0.0128	0.1543
3	0.0128	0.1538
4	0.0260	0.1564
5	0.0260	0.1564
6	0.0128	0.1538

7. Future work

More work is still needed regarding to the possible optimisations that can still affect the proposed mechanism design. For example, it may result obvious from the example figures, that the working space of the mechanism is greatly reduced if B"' type positions (Fig. 9) are along the path between B' an B" type positions. Also due to the intrinsic characteristics of the crank–rocker mechanism the IPC approach and returning velocities can be very different, making some of different possible design configurations preferable.

The ideas exposed are evidently extendible by changing the proposed crank–rocker mechanisms with different planar or spatial mechanisms that present some sort of IPCs. Other possible input mechanism and the systematisation of design criteria are currently under study.

8. Conclusions

The "Insensitivity position configurations" (IPCs) of a 6–RKS type parallel manipulator have been analyzed by mean of a generalization of the slider–crank and crank–rocker mechanisms IPCs. The advantages that manipulator IPCs present in part or tool positioning to obtain great position precision, null moving platform speed and null torque in the actuators have been exposed.

Having in mind the complexity involved in the design of such a manipulator, needing to fulfil the IPC conditions at specified working positions, a variant of the 6–RKS parallel manipulator that simplifies the design is introduced.

With this variant, and taking the crank, coupler and fixed link lengths and the angle that the fixed link sustains with the platform as free design variables of the crank–rocker mechanisms, the space of the positions that each of the platform vertex can attain in IPC is greatly extended from the edges of the primitive torus intersection to the volume contained between. The design variables are then trivially chosen. The mechanism also presents a bigger design flexibility because the increase in the number of attainable IPCs.

References

Benea R., (1996), *Contribution à L'Ètude des Robots Pleinement Parallèles de Type 6R–RR–S*, Doctoral Thesis, Université de Savoie, Annecy, France.

Hunt K. H., (1983), Structural Kinematics of In–Parallel–Actuated Robot-Arms, *J. of Mech. Trans. and Auto. in Design*, no. 4, vol. 105 , pp. 705–712.

Zabalza I., Pintor J. M., Ros J. y Jiménez J. M., (1999), Evaluation of the 64 Insensitivity Positions for a 6–RKS Hunt–Type Parallel Manipulator, *Tenth World Congres on the Theory of Machine and Mechanisms*, Oulu, Finland, pp. 1152–1157.

Zabalza I., Ros J., Pintor J. M. y Jiménez J. M., (2000), Aplicaciones de los Manipuladores Paralelos con Actuadores Giratorios, (X–RKS), XIV Congreso Nacional de Ingeniería Mecánica, Leganés, España, pp. 2367–2372.

Kinematic Design

ON THE KINEMATICS OF PARALLEL MECHANISMS WITH BI-STABLE POLYMER ACTUATORS

Andreas Wingert,
Department of Mechanical Engineering
Massachusetts Institute of Technology, Cambridge, MA 02139 USA
arw@mit.edu

Matthew D. Lichter,
Department of Mechanical Engineering
Massachusetts Institute of Technology Cambridge, MA 02139 USA
lichter@mit.edu

Steven Dubowsky
Department of Mechanical Engineering
Massachusetts Institute of Technology Cambridge, MA 02139 USA
dubowsky@mit.edu

Abstract Binary robotic devices have been proposed to perform complex tasks. These systems do not require feedback and are very easy to control. Their kinematic performance approaches that of continuous devices as their degrees of freedom becomes large. To date, high DOF binary systems have not been demonstrated, largely due to actuator limitations. Recently, significant advances have been made with dielectric polymer actuators. Here it is shown that such actuators have the potential to make binary robotic devices practical. Two designs of a Binary Robotic Articulated Intelligent Device (BRAID) using dielectric polymer actuators are presented. The kinematic performance based on laboratory results and analysis is predicted.

Keywords: Parallel Mechanism, Binary, Bi-Stable, Dielectric Polymer Actuators, Double-Octahedral Variable Geometry Truss, BRAID

1 Introduction

Proposed future tasks for robotic systems such as for missions to Mars will require robot subsystems that are lightweight, inexpensive, and easy to control (Weisbin et al., 1999, Sujan et al., 2001). Conventional robotic technology does not meet these requirements. A key limitation is actuator performance. Recently, important progress has been made in

J. Lenarčič and F. Thomas (eds.), Advances in Robot Kinematics, 303–310.

two areas that have the potential to meet this challenge. The first is hyper redundant binary robotic devices, and the second is dielectric polymer actuators (Pelrine et al., 2001). Here these two technologies are brought together in a concept called BRAID – Binary Robotic Articulated Intelligent Device. The BRAID is a lightweight, hyper-redundant binary manipulator with a large number of embedded actuators (Sujan et al. 2001). The kinematic performance is predicted based on measured results of actuator prototypes. Performance is also shown for projected performance of these actuators. Analysis suggests that dielectric polymer actuators have the potential of overcoming current actuator limitations and serving as a key component for high degree of freedom binary element manipulators. Such systems have the potential to be important in future robotic systems

2 Background and Literature

In the 1960's and 70's the concept of binary and sensorless robotics was introduced (Anderson et al., 1967, Roth et al., 1973). More recently increased computation power made the analysis, control, and planning for binary robots feasible (Chirikjian, 1994, Lees et al., 1996). Some experimental work has been done on binary redundant manipulators (Chirikjian, 1994). An example is a large variable geometry truss (VGT) manipulator that was constructed using pneumatic actuators. This implementation, while acceptable for systems with few DOF, cannot readily be extended to develop practical systems with very large DOF. To date, little work has been done to develop simple, lightweight, robust binary design concepts.

Development of dielectric polymer actuators started in the late 1990s (Pelrine et al., 1997). These actuators are one of a large class of electro active polymers (EAP) that has been studied (Madden et al., 2000). Recent progress has resulted from the identification of new and more effective dielectric materials, making it one of the most promising new actuator technologies, with relative strains of up to 380% having been reported (Pelrine et al., 2000, Pelrine et al., 2001). Applications such as robots, acoustic speakers, and solid-state optical devices have been proposed, including a snake-like manipulator using a small number antagonistic pairs of dielectric polymers, controlled in a conventional analog fashion (Pelrine et al., 2001).

To date, very high DOF binary robotic systems implemented with polymer actuators have yet to be explored. This is the subject of this paper. The focus is on the expected system kinematic performance given the demonstrated and anticipated actuator performances.

3 The BRAID

Figure 1 shows the kinematics for two designs of 3 degree-of-freedom parallel platforms. The first one is a revolute-revolute- spherical (RRS) configuration. The second one is a double-octahedral variable geometry truss (DOVGT) with a revolute-spherical-revolute (RSR) configuration (Rhodes et al., 1985). Each degree of freedom is binary-actuated, allowing a single stage to reach $2^3=8$ discrete states. The combination of identical stages forms the BRAID.

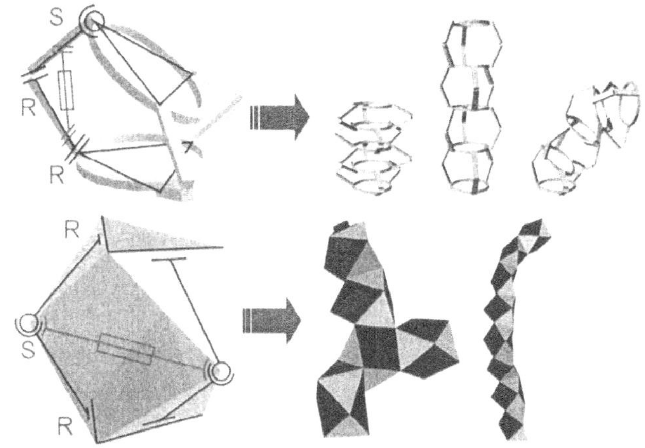

Figure 1 BRAID kinematics; top) RRS; bottom) RSR

As the number of stages of the BRAID increases, its performance approaches that of a continuous system (Lichter et al., 2002). The major advantage over conventional systems is that due to the on/off nature of each actuator and the two-state latching no feedback is required, which greatly simplifies control. The device also has good disturbance rejection and graceful failure modes. Figure 2 shows potential applications for space exploration of a 7-stage BRAID, such as a manipulator and a 6-legged walking rover.

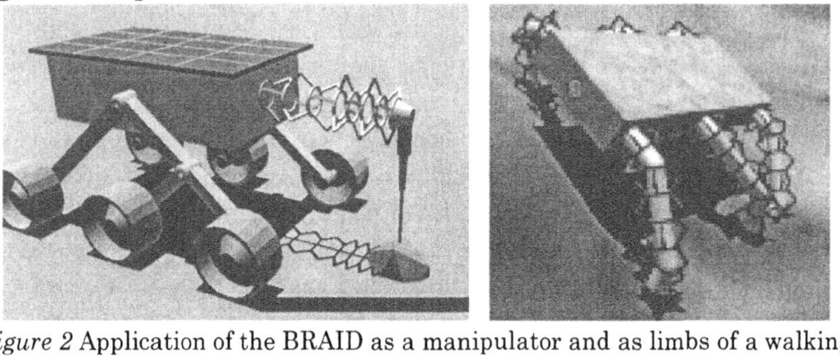

Figure 2 Application of the BRAID as a manipulator and as limbs of a walking robot

Implementing such systems as the BRAID with conventional actuators is not practical (Lichter et al., 2000, Chirikjian, 1994).

A first generation BRAID was built that used electromagnetic actuators (Hafez et al., 2002). It was found that the low force to weight ratio of the electromagnets greatly limited the number of stages, and hence the number of DOF. It is shown that dielectric polymer actuators have potential to be effective for this application. Figure 3 shows a sketch of how these actuators can be used into an RRS-BRAID.

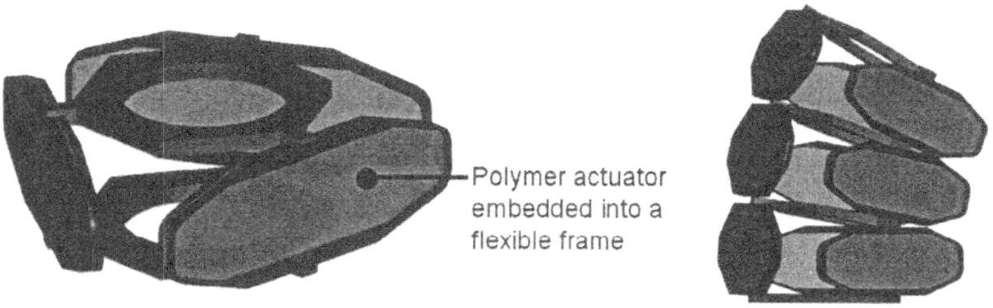

Figure 3 Polymer-actuated BRAID

4 Actuator Design

4.1 Dielectric polymer actuators

The operating principle of the dielectric polymer actuators is simple. An elastomeric film is coated on both sides with compliant electrodes. As a voltage is applied across the electrodes, the electrostatic charges will force the film to compress in thickness and expand in area (Pelrine et al., 1997). This expansion in area can be used to actuate mechanical systems.

4.2 Actuator implementation

For most elastomeric film materials, the response is much larger if the film is pre-stretched in both planar directions (Kofod et al., 2001). For application in the BRAID, the film is placed between two flexible frames, which have a degree of freedom in the intended direction of motion. The frame ensures the desired boundary conditions on the film. It further supplies a restoring force to the stretched film, forming an actuator that can work under both tension and compression. The restoring force is also tuned by a passive bi-stable element, which effectively cancels the stiffness of the film and frame over a certain range. This increases the range of motion and ensures a nearly constant force throughout the actuator stroke. Figure 4 shows an actuator assembly before and after actuation. The relative strain of the active film is about 50%. The actuator assemblies serve both as actuator and structural components when integrated into the RRS-BRAID (Figure 3).

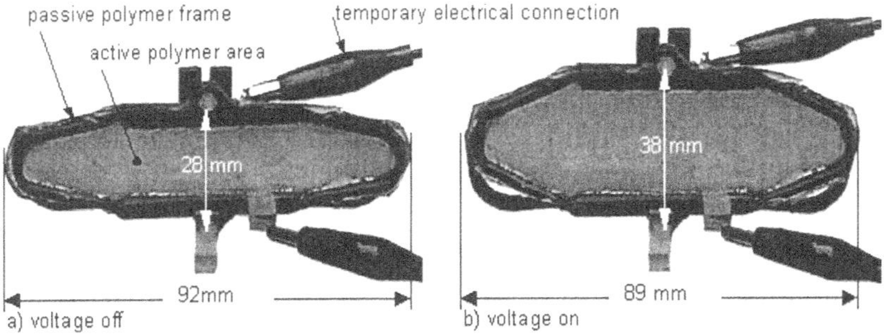

Figure 4 Photograph of embedded polymer actuator before and after actuation

4.3 Actuator Performance

The actuation forces are proportional to the cross sectional area of the film and can therefore be increased by combining multiple layers of the actuator material. Since a single layer of stretched dielectric material is only about 50μm thick, adding additional layers increases the actuation forces without significantly changing the geometry of the actuator. For an actuator with the same geometry shown in Figure 4, the actuation force is about 0.75N per layer.

The mass of a complete actuator unit is approximately 8 grams. Most of this mass is due to the frame and bi-stable element. For a double-layer actuator, the mass of the dielectric polymer actuator itself is less than 0.1 grams, only a fraction of what the actuator is capable of lifting. The performance of the actuator material is also independent of scale. In contrast to voice-coil actuators, which have poor performance on small scales, these actuators lend themselves for miniaturization (Pelrine et al., 2001). This suggests that the proposed BRAID can find applications on the micro- scale.

The actuator requires high voltage (5kV), but only draws currents of about 50μA, hence they can be powered by conventional miniature batteries. New thinner films will result in lower voltages.

5 BRAID Kinematics and Workspace

To illustrate the current and predicted capabilities of a dielectric polymer powered BRAID, a 4-stage (12 DOF) device is considered. The workspace is a function of the geometry of the individual stages as well as the strokes of the actuators.

Figure 5a shows the workspace cloud of an RRS-BRAID that can be achieved readily with actuator performance demonstrated in laboratory experiments (Figure 4) (Wingert et al., 2002). The actuator overall length changes from 28mm to 38mm.

The size of the workspace is on the order of a single stage of the BRAID. The high density of the workspace would make such a design suitable for a micro-positioning device where high resolution is required rather than a large range of motion.

Dielectric polymer actuators have shown strains that are much larger under laboratory conditions than in the self-contained modular frame design described above. This suggests that the dominant limitation to system performance is not the actuator material itself. Optimization studies currently being performed on the frame and passive elements suggest that actuator strokes that are double those cited above will be demonstrated shortly. Figure 5b shows the significantly larger workspace, using an actuator that expands from 28mm to 48mm. The workspace diameter would then be on the order of the stowed RRS-BRAID height. Such a system lends itself to a variety of applications, such as a camera placement device for planetary exploration, see Figure 2.

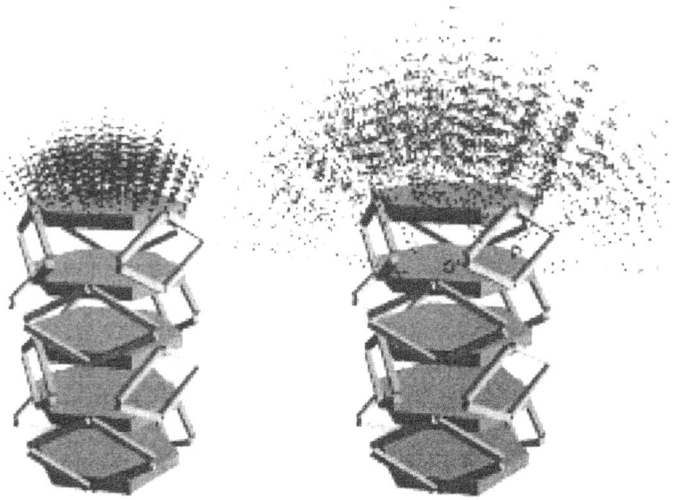

Figure 5 4-stage RRS-BRAID a) workspace with current actuator performance b) workspace with predicted actuator performance

A single stage of the RSR-BRAID has 14 faces, 8 of which have a fixed triangular geometry. The remaining 6 faces form diamond shapes that are occupied by dielectric polymer actuators (Figure 1). The RSR-stage is actuated by changing the diagonal lengths of the diamond-shaped areas with the dielectric polymer. Actuators enclosed in a diamond-shaped frame have been successfully demonstrated.

One advantage of the RSR-BRAID over the RRS-BRAID is that the actuators are directly integrated into the structure, reducing the number of components. This comes at the cost of adding some complexity to the design process, as actuators and structure can no longer be tuned independently. The joints of the RSR-BRAID could be designed to provide the desired restoring forces. As with the previous design, the actuation forces can easily be increased by adding layers of actuator film.

Taking the currently achieved performance of the frame actuators and applying them to the RSR-BRAID results in a workspace shown in Figure 6a. If the effective motion of each actuator were to be doubled, a significantly larger workspace would be achieved (Figure 6b).

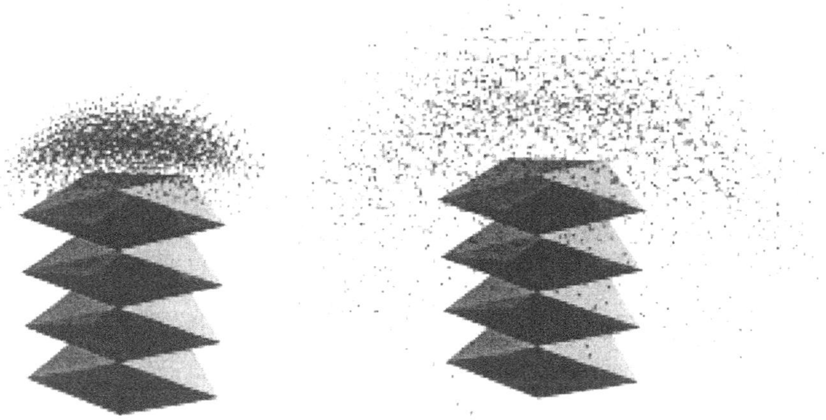

Figure 6 4-stage RSR - BRAID workspace cloud a) with current actuator performance b) with predicted actuator performance

6 Conclusion

It has been shown that dielectric polymer actuators have the potential of overcoming the key limitations of traditional actuators in making high-degree of freedom binary parallel robots practical. The simplicity of the actuators allows for a BRAID design that is virtually all-plastic, inexpensive, lightweight, and easy to control. Research in this program is continuing to develop devices that fully exploit the dielectric polymer actuators in very high DOF practical binary systems.

7 Acknowledgements

The support of the NASA Institute for Advanced Concepts (NIAC) for this research is acknowledged.

8 References

Anderson, V., Horn, R. (1967) Tensor Arm Manipulator Design. *ASME paper*, 67-DE-57

Chirikjian, G. (1994) A Binary Paradigm for Robotic Manipulators, *Proc. of IEEE International Conference on Robotics and Automation*, Vol. 4, pp. 3063-3069

Hafez, M., Lichter, M., Dubowsky, S. (2002) Optimized Binary Modular Reconfigurable Robotic Devices, *IEEE International Conf. on Robotics and Automation*. Washington, D.C.

Kofod, G., Kornbluh, R., Pelrine, R., Sommer-Larsen, P. (2001) Actuation response of polyacrylate dielectric elastomers, *Proc. of SPIE Vol. 4329*, pp. 141-147

Lees D., Chirikjian, G. (1996) A Combinatorial Approach to Trajectory Planning for Binary Manipulators, *Proc. of the IEEE International Conf. on Robotics and Automation*, Minneapolis, Mn

Lichter, M., Sujan, V., Dubowsky, S. (2000) Experimental Demonstrations for a New Paradigm in Space Robotics, *Proc. of the Seventh International Symposium on Experimental Robotics, ISER '00. Hawaii,* pp. 225-234.

Lichter, M., Sujan, V., Dubowsky, S., (2002) Computational Issues in the Planning and Kinematics of Binary Robots, *ICRA*, Washington D.C.

Madden, J., Cush, R., Kanigan, T., et al. (2000) Fast-contracting Polypyrrole Actuators, *Synthetic Metals,* 113: pp.185-193

Pelrine, R., Kornbluh, R., Joseph, J., Chiba S.,. (1997) Electrostriction of Polymer Films for Microactuators, *Proc. IEEE Tenth Annual International Workshop on Micro Electro mechanical Systems*, Nagoya, Japan, pp. 238-243

Pelrine, R., Kornblu, R., Pei, Q., Joseph, J. (2000) High Speed Electrically Actuated Elastomers with Strain Greater Than 100%, *Science*, Vol. 287, pp. 836-839

Pelrine, R., Sommer-Larsen, P., Konrbluh, R., Heydt, R., Kofod, G. Pei, P. Gravesen, Q. (2001) Applications of Dielectric Elastomer Actuators, *Proc. of SPIE* Vol. 4329

Rhodes, M., Mikulas M., Jr,. (1985) Deployable Controllable Geometry Truss Beam, *NASA Technical Memorandum.*

Roth, B., Rastegar, J., Scheinman, V. (1973) On the Design of Computer Controlled Manipulators, *First CISM-IFTMM Symposium on Theory and Practice of Robots and Manipulators*, pp. 93-113

Sujan, V., Lichter, M., Dubowsky, S., (2001) Lightweight Hyper-redundant Binary Elements for Planetary Exploration Robots, *Proc. IEEE/ASME Int. Conf. on Advanced Intelligent Mechatronics,* July 2001, Como, Italy.

Weisbin, C., Rodriguez, G., Schenker, P., et al (1999) Autonomous Rover Technology for Mars Sample Return, *International Symposium on Artificial Intelligence, Robotics and Automation in Space* (i-SAIRAS'99), 1-10.

Wingert, A., Lichter, M., Hafez, M., Dubowsky, S. (2002) Plastic Hyper Redundant Robot Manipulators Actuated by Optimized Binary Electrostrictive Polymers, *SPIE Vol. 4695*, San Diego.

THE KINETOSTATIC DESIGN OF A SCHÖNFLIES-MOTION GENERATOR

K. Al-Widyan and J. Angeles
Department of Mechanical Engineering & Centre for Intelligent Machines
McGill University, 817 Sherbrooke St. West, Montreal, Canada H3A 2K6
widyan@cim.mcgill.ca, angeles@cim.mcgill.ca

Abstract The kinetostatic design of a four-degree-of-freedom manipulator of the SCARA type, i.e., a Schönflies-motion generator, is reported here. The optimum design is obtained by means of kinetostatic isotropy, which thus leads to robustness against measurement and round-off errors. It is shown that a kinetostatically isotropic design is possible.

Keywords: Kinetostatic design, isotropy, SCARA, Schönflies motions, robustness

1. Introduction

The design of robotic mechanical systems (RMS) can be formulated as an optimization problem. To this end, an objective function is called for to quantify the system performance. Over the last twenty years or so, many performance indices have been proposed in the literature, which are based on the properties of the Jacobian matrix $\mathbf{J}$ mapping the joint-rate vector into the Cartesian-velocity vector. The ellipsoid associated with this mapping is a key element in assessing the kinetostatic performance. Two schools are found in this context: that using the determinant of $\mathbf{J}$ and that using the condition number of $\mathbf{J}$ to evaluate the kinetostatic performance. The determinant is related to the volume, while the condition number is a measure of the eccentricity of the ellipsoid. The former aims at minimizing the volume, while the latter aims at minimizing the eccentricity, Salisbury and Craig, 1982; Yoshikawa, 1985; and Ranjbaran et al., 1996. A serious drawback of all indices pertaining to the determinant is that those indices cannot tell a large end-effector (EE) from a small one Li, 1990.

While the condition number does not entail the foregoing drawback, its application to the manipulator Jacobian for positioning and orienting tasks is not trouble-free. Indeed, the entries of the Jacobian matrix corresponding to rotations are dimensionless, whereas those corresponding to translations have units of length. Consequently, an ordering of the singular values of this matrix from smallest to largest, needed to evaluate its condition number, is not possible. As a means to cope with

J. Lenarčič and F. Thomas (eds.), Advances in Robot Kinematics, 311–318.

this dimensional inhomogeneity, the concept of characteristic length was proposed by Angeles and López-Cajún, 1992.

An optimally kinetostatic design can be achieved by means of kinetostatic isotropy, a concept that guarantees positioning robustness to errors due to: joint-encoder noisy readouts; actuator noisy response; and manufacturing and assembly inaccuracies.

In this paper we focus on the kinetostatic design of a Schönflies-Motion Generator (SMG) based on isotropy. Schönflies motions constitute a set of four-degree-of-freedom (four-dof) displacements endowed with the algebraic structure of a group, Hervé, 1997. The first mechanical realization of this displacement subgroup of the six-dof displacement group was achieved by means of what is known as a SCARA (Selective-Compliance Assembly Robot Arm) system, Furuya et al., 1983. The SMG proposed here comprises two cascaded driving units, each producing a tilt-pan motion.

2. Kinetostatic Model

The kinetostatic model of a general *serial* manipulator can be described as

$$\mathbf{t} = \mathbf{J}\dot{\mathbf{q}}, \quad \boldsymbol{\tau} = \mathbf{J}^T\mathbf{w} \tag{1}$$

where $\mathbf{t}$ is the twist of the end-effector (EE), $\dot{\mathbf{q}}$ is the vector of the actuated joint rates, $\mathbf{J}$ is the Jacobian of the manipulator, $\mathbf{w}$ denotes the wrench acting on the EE, and $\boldsymbol{\tau}$ is the vector of joint forces and torques. A twist is an array containing both the angular and translational velocities of the EE, while a wrench is an array containing both the resultant force and the resultant moment acting at the EE.

If all the actuated joints are revolutes and the twist $\mathbf{t}$ involves both translational and angular velocities, then $\mathbf{J}$ is bound to have some entries with units of length and some dimensionless. We rely on the system characteristic length, which enables us to render $\mathbf{J}$ dimensionless, and hence, to compute its condition number.

3. A Multi-Modular-Manipulator System

The system at hand is a multi-modular manipulator (M^3) system capable of long-reach accurate tasks. It has a redundant architecture with 11 degrees of freedom. Its intended applications comprise the servicing and maintenance of aircraft and space structures. An example of such a manipulator is Lufthansa's Skywash robot, which is supplied with a cylindrical brush rotating about its axis, for fuselage cleaning. This task is quite tolerant in terms of positioning accuracy, other tasks required in the servicing and maintenance of aircraft being more demanding, e.g., stripping and painting.

The M^3 system is designed as a cascade of three modules: (*i*) the *Schönflies-Motion Generator* a.k.a. the *Macro-manipulator*; (*ii*) the *CuatroArm* and; (*iii*) the *Agile Wrist*, all displayed in Fig. 1. The *Schönflies-*

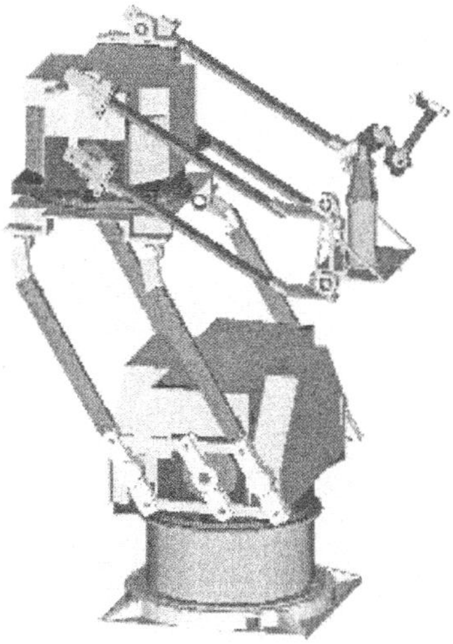

Figure 1. The M^3 system

Motion Generator (SMG) is the ground-base module of Fig. 1, its end-effector being the horizontal platform carrying the smaller manipulator. The SMG generates a four-degree-of-freedom (four-dof) SCARA-type long-reach motion, of about 3.5m. It is responsible for the gross positioning and orientation of the two other modules. This motion generator is built upon two similar architectures, the *proximal* and the *distal* parallelepipeds. Each has two actuated revolute axes and two opposite kinematic pairs of the class that Hervé, 1997, has termed Π pairs. The kinetostatic design of this module is the subject of the paper.

The *Cuatro Arm* and the *Agile Wrist* constitute the two other modules. They are responsible for the fine positioning and orientation of the end-effector. The *Cuatro Arm* has a four-dof serial architecture, the *Agile Wrist* having a three-dof parallel architecture, as depicted in Figs. 2. The conceptual design of the SMG was reported in Angeles et al., 2000. This paper is concerned with applying the concept of kinetostatic isotropy to its optimum design .

The purpose of the M^3 system is to perform fine manipulation tasks, for example: painting a curve on the fuselage of an airplane. These tasks will be performed by the seven-dof subsystem, i.e., the assembly of the *Cuatro Arm* and the *Agile Wrist*, termed the *distal modules*, while maintaining the four actuated joints of the SMG locked.

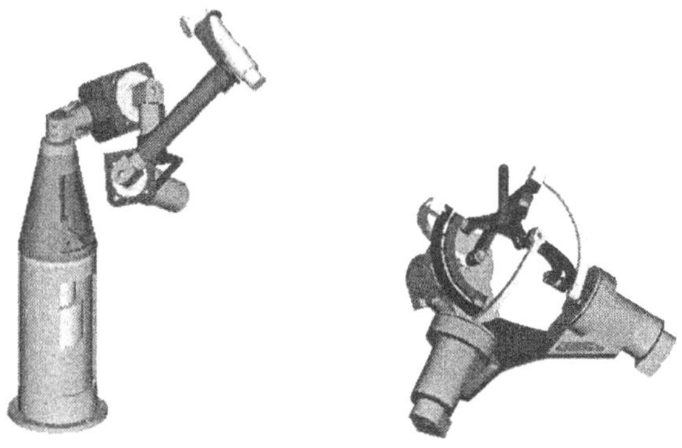

(a) (b)

Figure 2. The seven-dof subsystem: (a) Cuatro Arm ; (b) Agile Wrist

The twist $\mathbf{t} = [\dot{\phi}, \quad \dot{\mathbf{p}}^T]^T$ of the end-effector of the gross manipulator has four components, namely, the three components of the translational velocity vector $\dot{\mathbf{p}}$ of the operation point P shown in Fig. 3, and the scalar angular velocity $\dot{\phi}$ of the same. The 4×4 Jacobian matrix $\mathbf{J}$ of the gross manipulator of Fig. 3, relating the four joint rates $\{\dot{\theta}_i\}_1^4$ to the twist of the end-plate, takes the form, Angeles et al., 2000

$$\mathbf{J} = \begin{bmatrix} 1 & 0 & 1 & 0 \\ \mathbf{e}_1 \times \mathbf{p}_1 & \mathbf{e}_2 \times \mathbf{r}_2 & \mathbf{e}_3 \times \mathbf{p}_3 & \mathbf{e}_4 \times \mathbf{r}_4 \end{bmatrix} \tag{2}$$

Here, $\mathbf{e}_1$ and $\mathbf{e}_3$ are unit vectors pointing upwards in the vertical direction, while $\mathbf{e}_2$ and $\mathbf{e}_4$ are unit vectors normal to the vertical planes of the two parallelograms of Fig. 3, where the projections λ_1 and λ_2 are defined as: $\lambda_1 = l_1\sqrt{1 - s_1^2 c_1^2}$ and $\lambda_2 = l_2\sqrt{1 - s_{13}^2 c_4^2}$, with $c_i = \cos(\theta_i)$, $c_{ij} = \cos(\theta_i + \theta_j)$, and similar definitions for s_i and s_{ij}. Moreover, l_1 and l_2 are the lengths of the tilting beams of the the parallelogram. These vectors are given by

$$\mathbf{e}_1 = \mathbf{e}_3 = \begin{bmatrix} 0 \\ 0 \\ 1 \end{bmatrix}, \quad \mathbf{e}_2 = \begin{bmatrix} s_1 \\ -c_1 \\ 0 \end{bmatrix}, \quad \mathbf{e}_4 = \begin{bmatrix} s_{13} \\ -c_{13} \\ 0 \end{bmatrix} \tag{3}$$

The two vectors $\mathbf{p}_1$ and $\mathbf{p}_3$, directed from O_0 and O_3, respectively, to

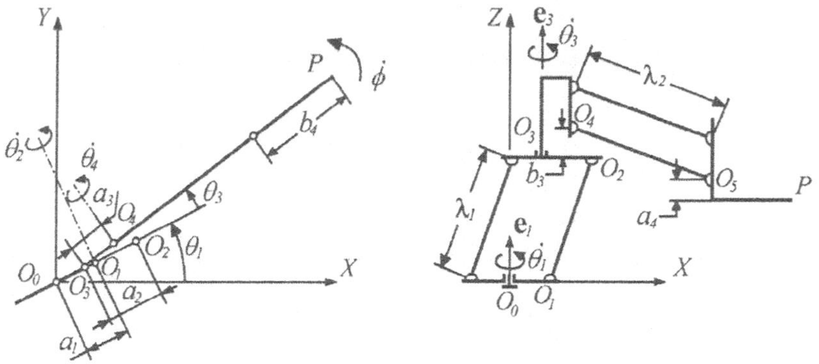

Figure 3. Kinematic Chain of Gross Manipulator, $l_1 = \overline{O_1O_2}$, $l_2 = \overline{O_4O_5}$

P, are defined as

$$\mathbf{p}_1 = \begin{bmatrix} Ac_1 + Bc_{13} \\ As_1 + Bs_{13} \\ l_1s_2 + b_3 - l_2c_4 - a_4 \end{bmatrix}, \quad \mathbf{p}_3 = \begin{bmatrix} (a_3 + l_2s_4 + b_4)c_{13} \\ (a_3 + l_2s_4 + b_4)s_{13} \\ b_3 - l_2c_4 - a_4 \end{bmatrix} \tag{4}$$

in which A and B are defined as: $A \equiv a_1 + l_1c_2 - a_2$ and $B \equiv a_3 + l_2s_4 + b_4$. Moreover, $\mathbf{r}_2$ and $\mathbf{r}_4$ are defined as the vectors directed from O_1 to O_2 and from O_4 to O_5, correspondingly. These vectors are given by

$$\mathbf{r}_2 = \begin{bmatrix} l_1c_1c_2 \\ l_1s_1c_2 \\ l_1s_2 \end{bmatrix}, \quad \mathbf{r}_4 = \begin{bmatrix} l_2c_{13}c_4 \\ l_2s_{13}c_4 \\ l_2s_4 \end{bmatrix} \tag{5}$$

It is clear that the above form of the Jacobian matrix is dimensionally inhomogeneous. In order to render it dimensionless, the last three rows containing components with dimensions of length are divided by what is called the characteristic length L, thus defining the normalized Jacobian $\overline{\mathbf{J}}$. The optimum kinetostatic design at hand is based on the condition under which the product $\overline{\mathbf{J}}\overline{\mathbf{J}}^T$ is proportional to the 4×4 identity matrix. The kinetostatic isotropy condition for the Jacobian matrix $\overline{\mathbf{J}}$ is, then,

$$\overline{\mathbf{J}}\overline{\mathbf{J}}^T = \sigma^2 \begin{bmatrix} 1 & \mathbf{0}^T \\ \mathbf{0} & \mathbf{1}_{3\times3} \end{bmatrix} \tag{6}$$

where σ^2 is the quadruple singular value of $\overline{\mathbf{J}}$, $\mathbf{0}$ is the three-dimensional zero vector, and $\mathbf{1}_{1\times3}$ is the 3×3 identity matrix. By developing the above expression we obtain

$$\overline{\mathbf{J}}\overline{\mathbf{J}}^T = \begin{bmatrix} 2 & (\mathbf{p}_1 + \mathbf{p}_3)^T\mathbf{E}^T/L \\ \mathbf{E}(\mathbf{p}_1 + \mathbf{p}_3)/L & \mathbf{C}/L^2 \end{bmatrix} \tag{7}$$

in which the 3×3 matrix $\mathbf{C}$ is defined as

$$\mathbf{C} = \mathbf{E}(\mathbf{p}_1\mathbf{p}_1^T + \mathbf{p}_3\mathbf{p}_3^T)\mathbf{E}^T + \mathbf{A}\mathbf{r}_2\mathbf{r}_2^T\mathbf{A}^T + \mathbf{B}\mathbf{r}_4\mathbf{r}_4^T\mathbf{B}^T \tag{8}$$

with matrices $\mathbf{E}$ and $\mathbf{A}$ are given by

$$\mathbf{E} = \begin{bmatrix} 0 & -1 & 0 \\ 1 & 0 & 0 \\ 0 & 0 & 0 \end{bmatrix}, \quad \mathbf{A} = \begin{bmatrix} 0 & 0 & -c_1 \\ 0 & 0 & -s_1 \\ c_1 & s_1 & 0 \end{bmatrix} \tag{9}$$

Matrix $\mathbf{B}$ is defined with the same form as $\mathbf{A}$, but with c_{13} and s_{13} instead of c_1 and s_1. From the above discussion it is apparent that, in order for $\overline{\mathbf{J}}\overline{\mathbf{J}}^T$ to be proportional to the identity matrix, we must have

$$\sigma^2 = 2 \tag{10}$$

$$\mathbf{E}(\mathbf{p}_1\mathbf{p}_1^T + \mathbf{p}_3\mathbf{p}_3^T\mathbf{E}^T + \mathbf{A}\mathbf{r}_2\mathbf{r}_2^T\mathbf{A}^T + \mathbf{B}\mathbf{r}_4\mathbf{r}_4^T\mathbf{B}^T = \sigma^2 L^2 \mathbf{1}_{3\times 3} \tag{11}$$

$$\mathbf{E}(\mathbf{p}_1 + \mathbf{p}_3) = \mathbf{0} \tag{12}$$

which means that the quadruple singular value σ of $\overline{\mathbf{J}}$ is $\sqrt{2}$. Moreover, by taking the trace of both sides of Eq. (11), we obtain

$$\mathrm{tr}[\mathbf{E}(\mathbf{p}_1\mathbf{p}_1^T + \mathbf{p}_3\mathbf{p}_3^T)\mathbf{E}^T + \mathbf{A}\mathbf{r}_2\mathbf{r}_2^T\mathbf{A}^T + \mathbf{B}\mathbf{r}_4\mathbf{r}_4^T\mathbf{B}^T] = 3\sigma^2 L^2 \tag{13}$$

Upon simplifying Eq. (13), we further obtain

$$L^2 = \frac{1}{6}(\|\mathbf{e}_1 \times \mathbf{p}_1\|^2 + \|\mathbf{e}_3 \times \mathbf{p}_3\|^2 + \|\mathbf{e}_2 \times \mathbf{r}_2\|^2 + \|\mathbf{e}_4 \times \mathbf{r}_4\|^2) \tag{14}$$

Further, from Eq. (12), $\mathbf{p}_1 + \mathbf{p}_3$ must lie in the nullspace $\mathcal{N}(\mathbf{E})$ of $\mathbf{E}$, which is spanned by $\mathbf{e}_1$, and hence,

$$\mathbf{p}_1 + \mathbf{p}_3 = \alpha \mathbf{e}_1 \tag{15}$$

From the geometry of Fig. 3,

$$\overrightarrow{O_0O_3} + \mathbf{p}_3 = \mathbf{p}_1 \quad \Longrightarrow \quad 2\mathbf{p}_1 + \overrightarrow{O_0O_3} = \alpha \mathbf{e}_1 \tag{16}$$

Substituting Eqs. (10) and (12) into Eq. (11), we obtain

$$2\mathbf{E}\mathbf{p}_1\mathbf{p}_1^T\mathbf{E}^T + \mathbf{A}\mathbf{r}_2\mathbf{r}_2^T\mathbf{A}^T + \mathbf{B}\mathbf{r}_4\mathbf{r}_4^T\mathbf{B}^T = 2L^2\mathbf{1}_{3\times 3} \tag{17}$$

Upon expansion, Eq. (17) leads to

$$2(\mathbf{e}_1 \times \mathbf{p}_1)(\mathbf{e}_1 \times \mathbf{p}_1)^T + (\mathbf{e}_2 \times \mathbf{r}_2)(\mathbf{e}_2 \times \mathbf{r}_2)^T + (\mathbf{e}_4 \times \mathbf{r}_4)(\mathbf{e}_4 \times \mathbf{r}_4)^T = 2L^2\mathbf{1}_{3\times 3} \tag{18}$$

One simple way to satisfy Eq. (18) is by rendering the three vectors $\sqrt{2}(\mathbf{e}_1 \times \mathbf{p}_1)$, $\mathbf{e}_2 \times \mathbf{r}_2$, and $\mathbf{e}_4 \times \mathbf{r}_4$ orthogonal and of equal Euclidean norm. Shown in Fig. 4 is one configuration satisfying these conditions and producing the locus $\mathcal{L}$ of isotropy. In this configuration, θ_1 is arbitrary; $\theta_3 = 0$; $\theta_4 = \theta_2$; and $\mathbf{e}_1 \times \mathbf{p}_1 = \mathbf{e}_3 \times \mathbf{p}_3$. Figure 5 depicts the isotropy surface, swept by $\mathcal{L}$ as the manipulator makes a full turn about the z_1 axis.

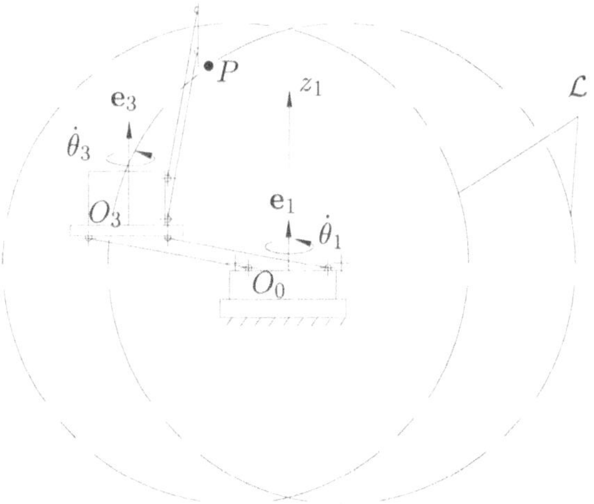

Figure 4. Isotropic configuration and isotropy locus

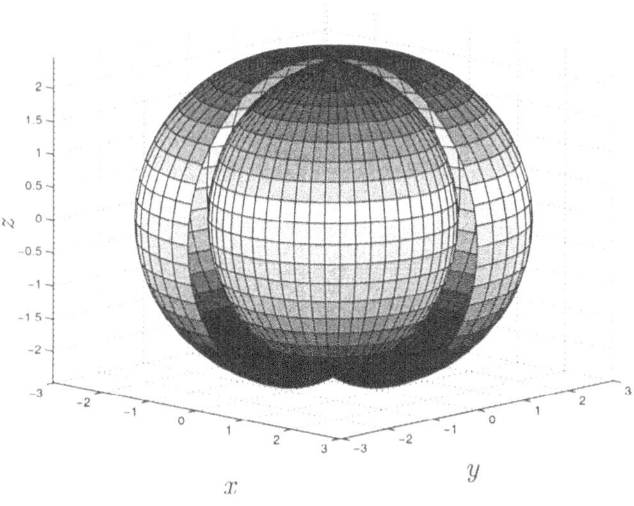

Figure 5. Isotropy surface

4. Conclusions

The concept of kinetostatic isotropy, as applied to a Schönflies-Motion Generator, was used here as a design criterion to produce a robust and accurate design.

The isotropic posture is an optimum configuration at which the manipulator is maximally robust in terms of its kinetostatic performance.

To improve a manipulator performance during task execution, the task should be planned within a region of the workspace where the Jacobian attains a low condition number.

5. Acknowledgements

The work reported here is supported by NSERC (Canada's Natural Sciences and Engineering Research Council) under Strategic Project 215729-98.

References

Angeles, J. and López-Cajún, C., "Kinematic isotropy and conditioning index of serial robotic manipulators," *The International J. Robotics Res.*, Vol. 11, No. 6, pp. 560–571, 1992.

Angeles, J., Morozov, A., and Navarro, O., "A Novel Manipulator Architecture for the Production of SCARA Motions," *Proc. IEEE Int. Conf. Robotics & Automation,* San Francisco, pp. 2370–2375, 2000.

Furuya, N., Soma, K., Chin, E., and Makino H., "Research and development of Selective Compliance Assembly Robot Arm. II. Hardware and Software of SCARA Controller," *J. Japan Society of Precision Engineering/Seimitsu Kogaku Kaishi,* Vol. 49, No. 7, pp. 835–841, 1983.

Hervé, J.M. 1997, "The Lie group of rigid body displacements, a fundamental tool for mechanism design," *NATO ASI on Computational Methods in Mechanisms*, Varna (Bulgaria).

Golub, G. and Van Loan, C., *Matrix Computations*, The Johns Hopkins University Press, Baltimore, 1989.

Lee, J., Duffy, J. and Hunt, K., 1998, "A Practical Quality Index Based on the Octahedral Manipulator", *The International J. Robotics Res.*, Vol. 17, No. 10, pp. 1081–1090.

Li, Z., "Geometrical considerations of robot kinematics," *The Int. J. Robotics and Automation,* Vol. 5, No. 3, pp. 139–145, 1990.

Ranjbaran, F., Angeles, J., and Kecskeméthy, A., "On the kinematic conditioning of robotic manipulators," *Proc. IEEE Int. Conf. Robotics & Automation*, pp. 3167-3172, Minneapolis, 1996.

Salisbury, J. and Craig J., "Articulated hands: force control and kinematics issues," *The Int. J. Robotics Res.*, Vol. 1, No. 1, pp. 4–17, 1982.

Yoshikawa, T., "Manipulability of robotic mechanisms," *The Int. J. Robotics Res.* Vol. 4, No. 2, pp. 3–9, 1985.

DESIGN OF 2-DOF PARALLEL MECHANISMS FOR MACHINING APPLICATIONS

Félix Majou
Institut de Recherches en Communications et Cybernétique de Nantes[1],
1 rue de la Noë, 44321 Nantes, FRANCE
Felix.Majou@irccyn.ec-nantes.fr

Philippe Wenger
Institut de Recherches en Communications et Cybernétique de Nantes,
1 rue de la Noë, 44321 Nantes, France
Philippe.Wenger@irccyn.ec-nantes.fr

Damien Chablat
Institut de Recherches en Communications et Cybernétique de Nantes,
1 rue de la Noë, 44321 Nantes, FRANCE
Damien.Chablat@irccyn.ec-nantes.fr

Abstract A design method based on velocity amplification factors analysis is presented in this paper, through the design of two simple two-degree-of-freedom Parallel Kinematic Mechanisms (PKM) dedicated to machining applications. The largest square useful Cartesian workspace is achieved for given kinetostatic performances. Its orientation and position are chosen to avoid singularities and to produce the best ratio between Cartesian workspace size and mechanism size. The size of each designed mechanism is used as a comparative criterion.

Keywords: Parallel Kinematic Machine Tool, Velocity Amplification Factors, Optimal Workspace Design.

1. Introduction

Most industrial Machine Tools (MT) have a serial kinematic architecture: each axis supports the following one, including its actuators and joints. High Speed Machining (HSM) highlights some drawbacks of such architectures: heavy moving parts require high stiffness from the

[1] IRCCyN : UMR n° 6597 CNRS, Ecole Centrale de Nantes, Université de Nantes, Ecole des Mines de Nantes

J. Lenarčič and F. Thomas (eds.), Advances in Robot Kinematics, 319–328.

machine structure to limit bending problems that lower the machine accuracy and limit the dynamic performances of the feed axes.

Parallel Kinematic Machines (PKM) attract more and more researchers and companies, because they are claimed to offer several advantages over their serial counterparts, like high structural rigidity and high dynamic capacities. Indeed, the parallel kinematic arrangement of the links provides higher stiffness and lower moving masses that reduce inertia effects. Thus, PKM have better dynamic performances, which is interesting for HSM.

However, most existing PKM have a complex geometrical workspace shape and highly non linear input/output relations. For most PKM, the Jacobian matrix which relates the joint velocities to the output velocities is not constant. Consequently, the performances may vary significantly for different points in the workspace and for different directions at one given point, which is a serious drawback for machining applications, Kim et al., 1997. To satisfy the needs of machining applications, a parallel kinematic architecture should preserve good workspace properties such as a regular shape and homogeneous kinetostatic performances throughout.

The design method presented in this paper is conducted for two-degree-of-freedom (2-DOF) mechanisms. Each mechanism is defined by a set of three design variables. The notion of useful workspace is then explained. Given prescribed kinetostatic performances, the link dimensions and actuated joint ranges of each mechanism are calculated for the largest square useful workspace. The orientation and position of the useful workspace are chosen to avoid singularities and to produce the best ratio between useful workspace and Cartesian workspace. Then, the size of the resulting mechanisms are compared.

The organization of this paper is as follows: the next section presents the kinematics of the studied mechanisms, sections 3 and 4 are devoted to the design of the two mechanisms through the velocity amplification factors analysis, and the last section concludes this paper.

2. Kinematic study

2.1 Description of the mechanisms

The two mechanisms under study have two constant length struts gliding along fixed linear actuated joints with different relative orientation. The two struts are of equal lengths L. Figure 1 presents the Biglide and Figure 2 shows the 2-DOF Orthoglide, Wenger et al. 2001.

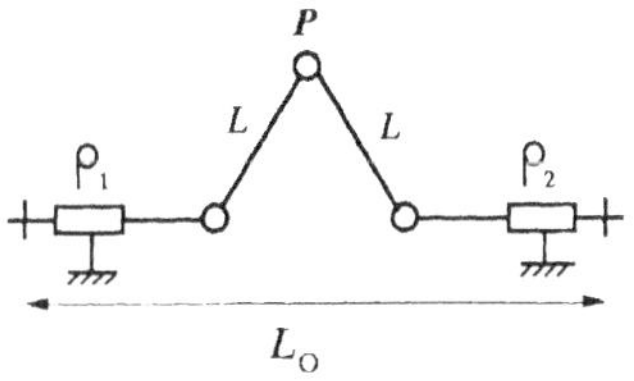

Figure 1. The Biglide mechanism

The joint variables are ρ_1 and ρ_2 associated with the two actuated prismatic joints and the output variables are the Cartesian coordinates of the tool center point $P = [x, y]^T$. Parameters characterizing each mechanism are the lengths L_0, L, and the actuated joint ranges $\Delta\rho$.

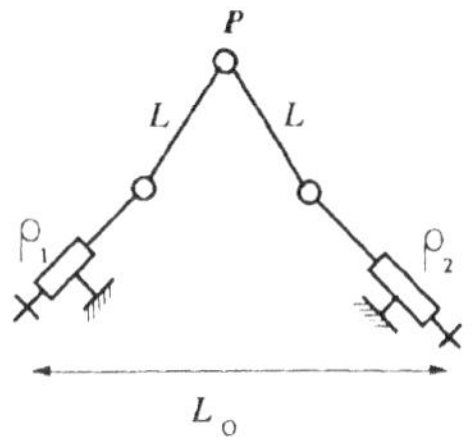

Figure 2. The 2-DOF Orthoglide

These two mechanisms are convenient for machining applications because they fit the technological constraints that a convenient PKM architecture for machining should respect as explained in Majou et al., 2002:

- only one DOF in each lower pair, for a simple design and a low cost;
- actuators fixed on the frame, to reduce inertia effects;
- actuated prismatic joints to allow linear motors;
- similar legs, for a low cost.

2.2 Velocity analysis

The Jacobian matrix relates the velocity vector $\dot{\mathbf{t}}$ of the tool point P to the velocity vector $\dot{\boldsymbol{\rho}}$ of the prismatic joints.

$$\dot{\boldsymbol{\rho}} = \begin{bmatrix} \dot{\rho}_1 \\ \dot{\rho}_2 \end{bmatrix} \quad \text{and} \quad \dot{\mathbf{t}} = \begin{bmatrix} \dot{x} \\ \dot{y} \end{bmatrix}$$

The vectors $\dot{\mathbf{t}}$ and $\dot{\boldsymbol{\rho}}$ are related by

$$\mathbf{A}\dot{\mathbf{t}} = \mathbf{B}\dot{\boldsymbol{\rho}}$$

where **A** and **B** are the parallel and serial Jacobian matrices, Gosselin and Angeles, 1990. When **A** and **B** are not singular, the following relations are obtained:

$$\dot{\mathbf{t}} = \mathbf{J}\dot{\boldsymbol{\rho}} \text{ with } \mathbf{J} = \mathbf{A}^{-1}\mathbf{B} \text{ and } \dot{\boldsymbol{\rho}} = \mathbf{J}^{-1}\dot{\mathbf{t}} \text{ with } \dot{\mathbf{t}} = \mathbf{J}\dot{\boldsymbol{\rho}}$$

The inverse Jacobian matrix $\mathbf{J}^{-1}$ is used for more simplicity.

2.3 Singularity analysis

Six types of singularities can arise in a mechanism, Zlatanov et al. 1996, but focusing on the three common ones, Gosselin and Angeles, 1990, is enough for the purpose of the work presented here.

The first type occurs when the determinant of matrix **A** vanishes, i.e. when det(**A**) = 0, Gosselin and Angeles, 1990. This type of singularity is called a parallel singularity (Fig. 3 and Fig. 4).

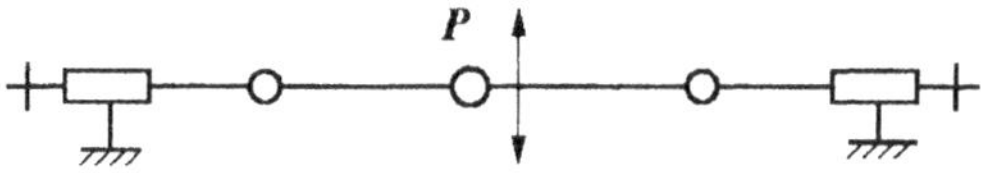

Figure 3. Biglide parallel singularity

In this configuration, it is possible to move the tool center point whereas the actuated joints are locked, thus the control of the tool point P is lost. These singularities have to be eliminated from the Cartesian workspace to prevent damaging the mechanism.

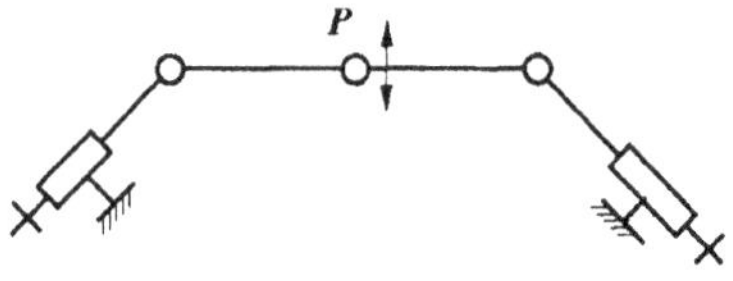

Figure 4. 2-DOF Orthoglide parallel singularity

The second type of singularity occurs when the determinant of matrix **B** vanishes, i.e. when det(**B**) = 0, Gosselin and Angeles, 1990. This type of singularity is called a serial singularity (Fig. 5 and Fig. 6). In this configuration, there exists a direction along which no velocity can be produced. For a PKM, serial singularities define the boundary of the Cartesian workspace, Merlet, 1997. Because the two struts are of equal lengths, the serial singularity is also a structural singularity and P can freely rotate around the two coincident revolute centers, Gosselin and Angeles, 1990.

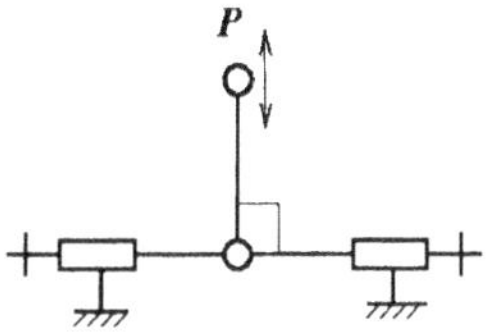

Figure 5. Biglide serial singularity

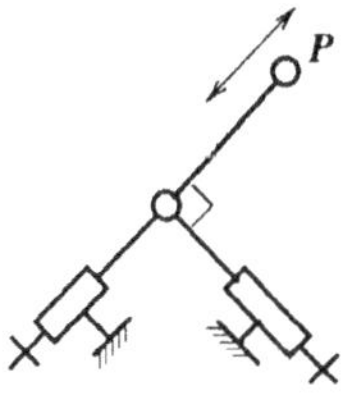

Figure 6. 2-DOF Orthoglide serial singularity

3. Shape, position and orientation of the useful workspace

3.1 Velocity amplification factor boundaries

In order to keep reasonable and homogeneous kinetostatic properties inside the Cartesian workspace, the manipulability ellipsoids of velocity defined by the inverse Jacobian matrix $\mathbf{J}^{-1}$ are studied, Yoshikawa, 1985. The $\mathbf{JJ}^{-1}$ eigenvalues square roots, γ_1 and γ_2, are the lengths of the semiaxes of the ellipse that define the two Velocity Amplification Factors (from now on called VAF) between the actuated joints velocities and the velocity vector $\dot{\mathbf{t}}$, $\lambda_1 = 1/\gamma_1$ and $\lambda_2 = 1/\gamma_2$. To limit the variations of these factors inside the Cartesian workspace, the following constraints are set

$$1/3 \leq \lambda_i \leq 3$$

This means that for a given joint velocity, the output velocity is at most three times larger or, at least, three times smaller. These constraints also permit to limit the loss of rigidity (velocity amplification lowers rigidity) and of accuracy. The boundaries on VAF were chosen as an example and should be revised depending on the machining tasks (accuracy needs for example).

3.2 Useful workspace shape

The Cartesian workspace (from now on called C-workspace) is the manipulator's workspace defined in Cartesian space. The useful workspace (from now on called u-workspace) is defined as a part of the C-workspace. Its shape and size are a design parameters and have to be defined. Furthermore, inside the u-workspace, VAF remain under the prescribed values.

The u-workspace shape of the two mechanisms should be similar to the one of classical serial MT, which is parallelepipedic if the machine has three translational degrees of freedom for instance. The u-workspace of a serial three axis MT is equivalent to the C-workspace because the input/output relations are linear. Therefore, a square u-workspace is prescribed here. And it must be a t-connected region, i.e. it must be free of serial and parallel singularities, Chablat and Wenger, 1998.

3.3 Isotropy continuum of the 2-DOF Orthoglide

The 2-DOF Orthoglide mechanism, extended to three DOF in Wenger and Chablat, 2000, was designed to have an isotropic configuration for which the VAF are unitary. But this mechanism also provides an isotropy continuum which is a straight line (Fig. 9).

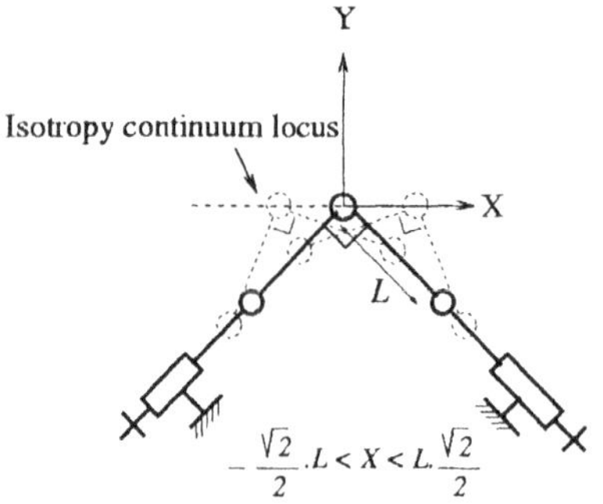

Figure 9. 2-DOF Orthoglide isotropy continuum locus

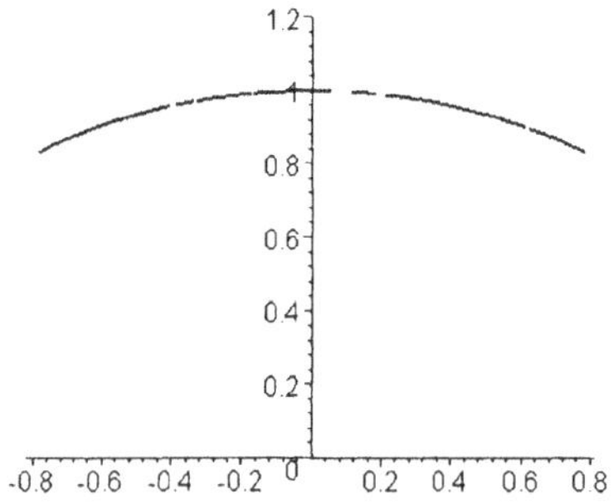

Figure 10. VAF along the isotropy continuum locus

The two VAF are equal along the continuum, but not constant, Angeles, 1997. It means that $\lambda_1 = \lambda_2$, and therefore cond(**J**) = 1. The variation of the VAF along the isotropy continuum is limited (Fig. 10), which is interesting as it shows that isotropy brings homogeneousness to kinetostatic performances, which is prefered for this application.

The Biglide isotropy continuum is not studied here because it has few consequences on the VAF homogeneousness inside the u-workspace. See section 4.1 for more details on its location.

3.4 Useful workspace orientation

The 2-DOF Orthoglide u-workspace is first arbitrarily centered on the point S where the VAF are equal to 1 (Fig. 11). Changing the u-workspace center position will be discussed in section 4. Two possible u-workspace orientations are studied (Fig. 11) and it appears that orientation A has a bad ratio between the u-workspace and the C-workspace, which yields a poor machine compactness because of the larger joint ranges.

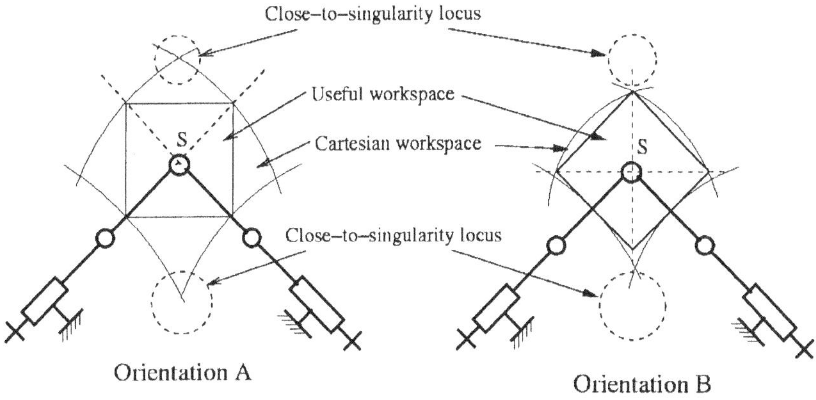

Figure 11. Two orientations for the 2-DOF Orthoglide u-workspace

Furthermore in the case of orientation A, singular configurations may appear inside the C-workspace, which is not acceptable. Indeed, the u-workspace is used for the machining task, but the C-workspace can be used for changing the tool position between two machining operations. Singularities are then strictly prohibited. Thus orientation B is selected for the 2-DOF Orthoglide.

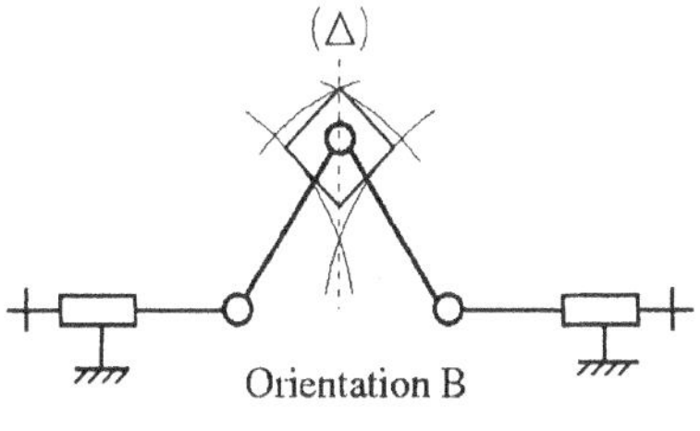

Figure 12. Orientation of the Biglide u-workspace

Same comments about compactness and singularities avoidance can be made for the Biglide, thus orientation B is chosen (Fig. 12).

4. Optimal useful workspace design

This section explains how the u-workspace is designed: first the best workspace center locus is found by computing the VAF along the u-workspace sides. Then the u-workspace is sized so that the VAF are inside the boundaries defined in section 3.1.

4.1 Useful workspace center locus

To find the best u-workspace center locus, we shift the u-workspace perpendicularly to (Δ) and along (Δ) (Fig. 13) and the VAF are computed for each configuration.

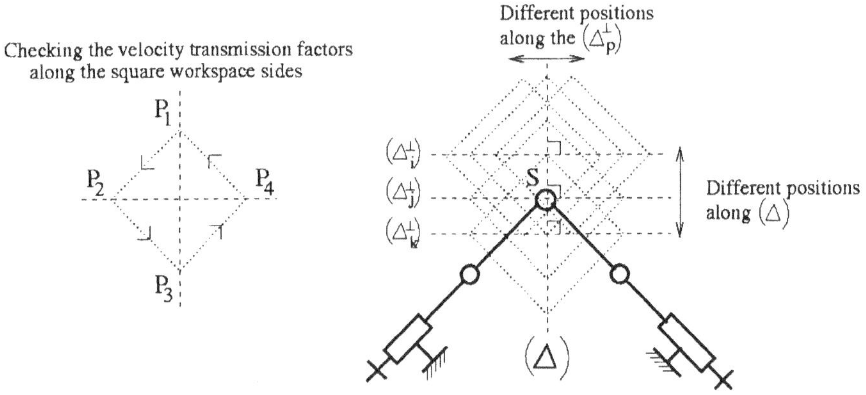

Figure 13. Looking for the best u-workspace center locus of 2-DOF Orthoglide

In each case, VAF extrema are located along the sides P_iP_j: they start from 1 at point S, then they vary until they reach prescribed boundaries on VAF (Fig. 15, section 4.2). Computing the VAF (which analytical expressions λ_i (Xp,Yp) have been obtained with Maple) along the 4 sides of the square takes only 5 sec. with a Pentium II class PC.

As Biglide configurations are identical along every horizontal line orthogonal to (Δ) (Fig. 12), VAF are constant along these lines. Consequently, the workspace position will only be discussed along (Δ). This corresponds to the u-workspace sizing process described in section 4.2.

4.2 Useful workspace size

To size the u-workspace, one initial point is chosen on the locus found in section 4.1 then the u-workspace is grown until the VAF meet their limits (Fig. 14). It appears that the VAF limits are met at points P1 and P3 (Fig. 15). For the 2-DOF Orthoglide, the initial u-workspace center is point S. It appears that the first limit met is the upper one ($\lambda_i < 3$), and that it is met by λ_2, simultaneously at points P1 and P3, therefore point S remains the final u-workspace center. We see on Fig. 15 that λ_1 does not vary much compared to λ_2.

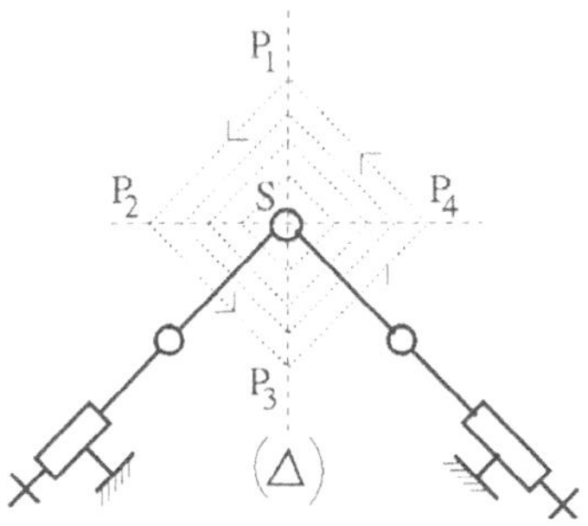

Figure 14. Sizing the 2-DOF Orthoglide u-workspace

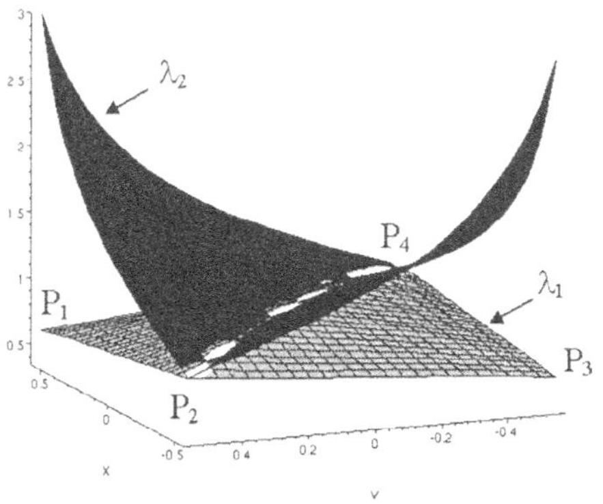

Figure 15. VAF values of the 2-DOF Orthoglide inside its u-workspace

For the Biglide, the two boundary lines are found (Fig. 16) and the distance between them define the diagonal of the u-workspace.

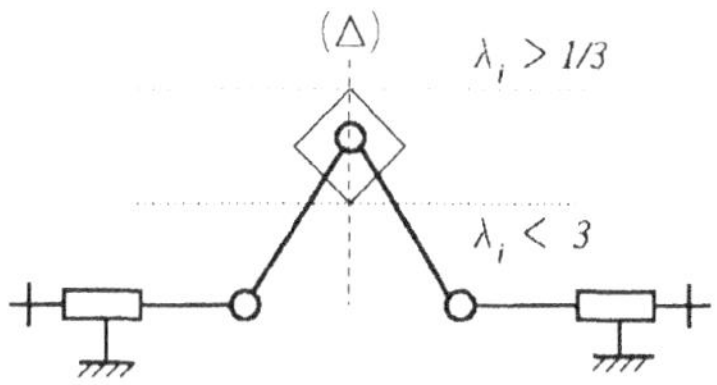

Figure 16. Sizing the Biglide u-workspace

4.3 Comparison of the mechanisms envelope size

For a square u-workspace of 1m^2, the design parameters (L_0, L, $\Delta\rho$) computed by Maple are given in Tb. 1. Obviously, the 2-DOF Orthoglide is more compact than the Biglide, Wenger et al. 2001.

Table 1. Design parameters for the Biglide and for the 2-DOF Orthoglide

	L_0 (m)	L (m)	$\Delta\rho$ (m)	Mechanism enveloppe (m^2)
Biglide	5.95	3.05	1.67	16.45
Orthoglide	2.08	1.06	1.18	3.91

5. Conclusions

The design of two 2-DOF PKM dedicated to machining applications has been conducted in this paper, through a novel design method based on the analysis of VAF. The procedure applied is reminded here: first, boundaries on VAF and u-workspace shape and size have to be defined depending on the application to achieve. Secondly, the u-workspace orientation and position have to be found inside the C-workspace for the largest ratio between u-workspace and C-workspace. Then the u-workspace is grown in the found configuration until boundaries are met.

In the case studied here, the u-workspace is square and the boundaries on VAF are 1/3 and 3. The orientation and position of the u-workspace have also been chosen to avoid singularities inside the C-workspace and to achieve best compactness. The machine size of each resulting design is used as a comparative criterion and the 2-DOF Orthoglide appeared to have smaller dimensions than the Biglide.

Further comparisons between the mechanisms studied in this paper could deal with the way VAF vary inside the u-workspace.

References

Angeles, J. (1997), *Fundamentals of robotic mechanical systems*, Springer-Verlag.

Chablat, D., Wenger, P. and Angeles, J. (2000), Conception isotropique d'une morphologie parallèle: application à l'usinage, *3rd Int. Conf. on Integrated Design and Manufacturing in Mechanical Engineering*, Montreal, Canada.

Chablat, D., Wenger, P. (1999), On the characterization of the regions of feasible trajectories in the workspace of parallel manipulators, *10th World Congress on the Theory of Machines and Mechanisms*, Oulu, Finland.

Gosselin, C., Angeles, J. (1990), Singularity analysis of closed-loop kinematic chains, *Int. Conf. on Robotics and Automation*, Cincinnati, USA.

Kim, J., Park, C., Kim, J. and Park, F.C. (1997), "Performance Analysis of Parallel Manipulator Architectures for CNC Machining Applications", *IMECE Symposium on Machine Tools,* Dallas, USA.

Majou, F., Wenger, P. and Chablat, D. (2002), Design of a 3 axis parallel machine tool for high speed machining : the Orthoglide, *4th Conf. on IDMME*, Clermont Ferrand, France.

Merlet, J-P. (1997), *Les robots parallèles*, Paris, Hermes.

Wenger, P., Chablat, D. (2000), Kinematic analysis of a new parallel machine tool: the Orthoglide, *Advances in Robot Kinematic*, Portoroz, Slovenia.

Wenger, P., Gosselin, C. and Chablat, D., (2001), A comparative study of parallel kinematic architectures for machining applications, *2nd Workshop on Computational Kinematics,* Seoul, South Korea.

Yoshikawa, T. (1985), Manipulability of Robot Mechanisms, *The Int. Journal of Robotics Research*, no. 2, vol. 4, pp.3-9.

Zlatanov, D. S., Fenton, R. G., Benhabib, B. (1994), Analysis of the instantaneous kinematics and singular configurations of hybrid-chain manipulators, *23rd ASME Mechanisms Conference*, Minneapolis, USA.

KINEMATICALLY EQUIVALENT SPATIAL MECHANISMS WITH REVOLUTE PAIR AT INPUT AND PRISMATIC PAIR AT OUTPUT

Eduard Peisach

Technical University Chemnitz, Faculty of Mechanical and Process Engineering

D-09107, Chemnitz, Germany

eduard.peisach@mb1.tu-chemnitz.de

Abstract In the paper is shown that the spatial four-bar mechanism of structure RSSP has the property of kinematic equivalence. It is established that one and the same position function of output link can be reproduced not by some one mechanism RSSP, but by any mechanism belonging to some one-parametrical family of mechanisms of the same structure. The relations between constant parameters of the mechanisms forming the family are derived. Among all such mechanisms the mechanism with the minimum possible value of the maximal pressure angle is isolated.

Keywords: Kinematically equivalent mechanisms, spatial linkages, pressure angle

1. Introduction

Two (or more) linkages of the same structure, but with incongruous values of all or some constant parameters of its kinematic diagram, we shall name as *kinematically equivalent* mechanisms if they reproduce one and the same position function of the output link.

If there exist kinematically equivalent mechanisms among the linkages of the given structure, we shall say that the mechanisms of this structure have the property of *kinematic equivalence.*

As known, the linkages of kind R...R have the mentioned property. At such mechanisms the input and output links form the rotation pairs with the frame, that is the angular displacement at input will be transformed by the mechanism to angular displacement at output. The position function of such mechanisms remains the same, if all their linear constant parameters are multiplied by one and the same scale factor, and the angular constant parameters are left unchanged.

Some linkages of kind R...R have not only one mentioned above (trivial) case of kinematic equivalence. For example, the RRRR linkage Bennett keeps the same input-output relation by corresponding change of two constant angular parameters. I am grateful to Professor Kurt

J. Lenarčič and F. Thomas (eds.), Advances in Robot Kinematics, 329–336.

Wohlhart who has paid my attention to this case of kinematic equivalence.

This paper deals with the nontrivial case of kinematic equivalence revealed in spatial linkages of kind R...P (angular displacement at input is transformed by the mechanism to linear displacement at output). At such mechanisms the simple proportional change of the link sizes does not result to obtaining kinematically equivalent mechanisms, since the position function of the output link changes in this connection. Meanwhile, and for mechanisms of kind R...P it is possible to specify such (nonlinear) transformation of the constant parameters, in which the position function of the output link remains invariable.

In this paper it is shown that the spatial mechanism of structure RSSP (fig. 1) possesses the property of kinematic equivalence. It is established that any RSSP mechanism with arbitrary constant parameters belongs to some one-parametrical family of mechanisms of the same structure, which are kinematically equivalent with respect to the given mechanism. The relations between the constant parameters of the mechanisms forming the family are derived. Among all such mechanisms the mechanism with the minimum possible value of the maximal pressure angle is isolated.

The property of kinematic equivalence established for RSSP mechanisms is not solely of theoretical interest. It can be efficiently used in synthesizing such mechanisms.

2. Position function of RSSP mechanism

The following notations are used for a spatial mechanism of RSSP structure (fig. 1): 1, 2, 3, 4 are the ordinal numbers of links (1 is the fixed frame; 2 is the input link; 4 is the output link); $Oxyz$ is the stationaty coordinate system; DX is the stationaty axis running along the line of the hinge centre C motion; h_1, θ_1, l_{12}, h_2, h_3 are the constant parameters of the mechanism; α is the angular coordinate of link 2; X is the linear coordinate of link 4; $X{=}X(\alpha)$ is the position function of the output link.

The constant parameters should satisfy the following conditions:

$$h_1 \geq 0,\ \ l_{12} \geq 0,\ \ h_2 > 0,\ \ h_3 > 0,\ \ \cos\theta_1 \neq 0. \tag{1}$$

The position function $X{=}X(\alpha)$ is expressed by the formula:

$$X(\alpha) = B(\alpha) + M\sqrt{D(\alpha)}\,, \tag{2}$$

where M =+1 or −1 indicates one of the two possible assembly modes of the RSSP mechanism;

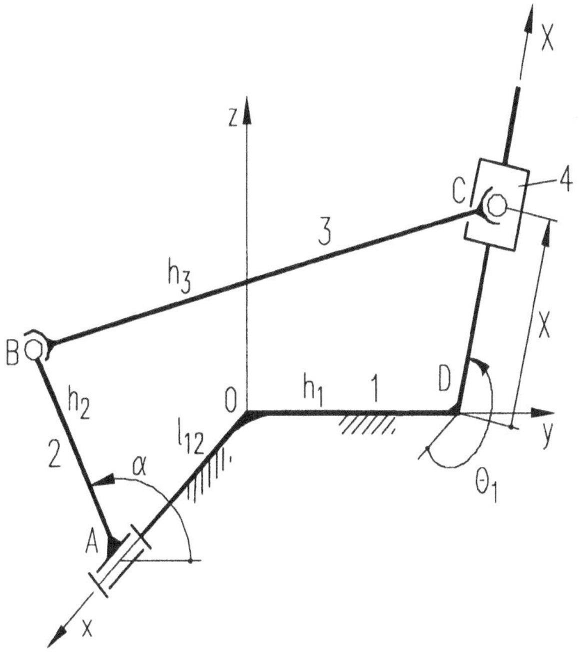

Figure 1. Mechanism RSSP

$$B(\alpha) = l_{12}\cos\theta - h_2 \sin\theta \sin\alpha \ ,$$
$$D(\alpha) = B(\alpha)^2 + 2h_1 h_2 \cos\alpha - l_{12}^2 - h_1^2 - h_2^2 + h_3^2 \ . \qquad (3)$$

The position function $X{=}X(\alpha)$ exists at those values of the argument α, at which is satisfied the following condition:

$$D(\alpha) \geq 0 \ . \qquad (4)$$

3. Family of kinematically equivalent RSSP mechanisms

The formula (2) for position function $X{=}X(\alpha)$ of RSSP mechanism can be written down in such kind:

$$X = p_1 - p_2 \sin\alpha + M\sqrt{(p_1 - p_2 \sin\alpha)^2 + 2p_3 \cos\alpha - p_4} \ , \qquad (5)$$

where the following notations are accepted:

$$p_1 = l_{12}\cos\theta_1,\quad p_2 = h_2\sin\theta_1,\quad p_3 = h_1 h_2,\quad p_4 = l_{12}^2 + h_1^2 + h_2^2 - h_3^2. \tag{6}$$

From here it may be seen, that the position function $X=X(\alpha)$ depends on four parameters p_1, p_2, p_3, p_4, while the RSSP mechanism has five constant parameters: h_1, θ_1, l_{12}, h_2, h_3. Thus, all mechanisms, at which five constant parameters h_1, θ_1, l_{12}, h_2, h_3 satisfy to four relations (6) (provided that the parameters p_1, p_2, p_3, p_4 are given), reproduce one and the same position function (5). Let's consider that all such mechanisms form some *family* of RSSP mechanisms. According to the definition entered in item 1, all mechanisms included in the given family are kinematically equivalent.

Since the number of relations (6) is smaller by one than the number of the parameters, the mentioned family of mechanisms is one-parametrical. As the parameter of this family, any of the constant parameters h_1, θ_1, l_{12}, h_2, h_3 can be taken, for instance, h_2.

Quantities p_1, p_2, p_3, p_4 determining a family of kinematically equivalent RSSP mechanisms we shall name as *the basic constants* of the family.

Using the relations (6), we express the constant parameters h_1, θ_1, l_{12}, h_3 of the mechanisms belonging to the mentioned family through the basic constants p_1, p_2, p_3, p_4 in function of the family parameter h_2:

$$h_1 = \frac{p_3}{h_2},\quad l_{12} = \frac{|p_1|h_2}{\sqrt{h_2^2 - p_2^2}},\quad h_3 = \sqrt{\frac{p_1^2 h_2^2}{h_2^2 - p_2^2} + \frac{p_3^2}{h_2^2} + h_2^2 - p_4},$$
$$\sin\theta_1 = \frac{p_2}{h_2},\quad \cos\theta_1 = \frac{\mu}{h_2}\sqrt{h_2^2 - p_2^2},\quad \text{where}\ \mu = \begin{cases} \operatorname{sign}(p_1), & \text{if}\ p_1 \neq 0, \\ \pm 1, & \text{if}\ p_1 = 0 \end{cases} \tag{7}$$

The basic constants p_1, p_2, p_4 can have any values, and the constant p_3 should satisfy to the condition: $p_3 \geq 0$.

As to the family parameter h_2, it can be chosen arbitrary, but in view of the following inequalities:

$$h_2 > |p_2|,$$
$$f_1(h_2) = h_2^6 + (p_1^2 - p_2^2 - p_4)h_2^4 + (p_3^2 + p_2^2 p_4)h_2^2 - p_2^2 p_3^2 > 0. \tag{8}$$

4. Synthesis of a mechanism with the optimal pressure angle

All mechanisms belonging to a family of kinematically equivalent RSSP mechanisms reproduce one and the same position function $X=X(\alpha)$ of the output link. But they differ in other geometrical characteristics. The latter include, for instance, the angle $\vartheta = \vartheta(\alpha)$ between line BC and axis DX (fig. 1), called *the pressure angle* (as known, the pressure angle ϑ characterizes the quality of the transfer of motion and forces from the input link to the output link). This specific feature of kinematically equivalent mechanisms can be efficiently used in their synthesizing.

Let it be required to determine the constant parameters of the mechanism on the basis of the major condition of synthesis: the position function must differ slightly only (on any norm) from the given function on a given segment of the input angle variation. Besides, the additional condition should be taken into account: the maximal value ϑ_{max} of the pressure angle ϑ on the given segment should not exceed the given allowable value.

If the synthesized mechanism has the property of kinematic equivalence, there is an opportunity of the separate account of the major and additional synthesis conditions. Let us illustrate this by considering the RSSP mechanism.

The task of synthesis is solved in two stages. At the first stage the major condition is taken into account only and the values of four basic constants p_1, p_2, p_3, p_4 appearing in the expression (5) for the position function $X=X(\alpha)$ are defined. Thereby, some family of kinematically equivalent RSSP mechanisms is determined. The concrete values of five constant mechanism parameters h_1, θ_1, l_{12}, h_2, h_3 are not found for the present, since it is necessary to choose previously a value of the free parameter h_2.

At the second stage of synthesis the additional condition is taken into account. It is formulated here in such form: among the mechanisms of the obtained family it is required to find a mechanism with the minimum possible value of the maximum pressure angle ϑ_{max}. The search of such mechanism reduces to determination of the optimal value $(h_2)_{opt}$ of the independent parameter h_2, at which the condition $\vartheta_{max} \to \min$ is satisfied.

At the final stage of computation the mechanism parameters are determined on the basis of the formulas (7) at $h_2=(h_2)_{opt}$.

Let us consider the second stage of synthesis in more detail. The pressure angle ϑ is determined by the formula:

$$\vartheta(\alpha) = \arccos\left(\sqrt{D(\alpha)}/h_3\right) \tag{9}$$

On the basis (9) it is obtained:

$$\vartheta_{max} = \arccos\left(\sqrt{D_{min}}/h_3\right), \tag{10}$$

where D_{min} is the least value of the function $D(\alpha)$ on the given segment of argument α variation. Note, that $D_{min} > 0$ always, since the condition $D(\alpha) > 0$ at all values α belonging to the given segment is necessarily taken into account at the first stage of synthesis.

According to the formulas (3) and (6), the function $D(\alpha)$ can be submitted in such form:

$$D(\alpha) = (p_1 - p_2 \sin\alpha)^2 + 2p_3 \cos\alpha - p_4 . \tag{11}$$

From here it can be seen, that the function $D(\alpha)$ and the quantity D_{min} do not depend on the parameter h_2 of the found family of mechanisms. Hence, when the basic constants p_1, p_2, p_3, p_4 are known, the quantity D_{min} is some constant for the family.

In such case the angle ϑ_{max} is minimum at the point, where the function $h_3 = h_3(h_2)$ is minimum. Thus, for definition $(\vartheta_{max})_{min}$ it is necessary to find the optimum value $(h_2)_{opt}$ of the parameter h_2, for which $h_3 = (h_3)_{min}$.

Analyzing the function $h_3 = h_3(h_2)$ defined by (7) for the extremum, we get the equation of 4th degree:

$$f_2(u) = u^4 - 2p_2^2 u^3 + (p_2^4 - p_3^2 - p_1^2 p_2^2)u^2 + 2p_2^2 p_3^2 u - p_2^4 p_3^2 = 0, \tag{12}$$

where $u = h_2^2$. The sought value $h_2=(h_2)_{opt}$ is equal to $\sqrt{u_*}$, where u_* is one of real roots of the equation (12). It is necessary to choose such root, for which are satisfied the inequalities (8). If the number of such roots is more than one, of interest is only the root corresponding to the smaller value of h_3.

5. Numerical example

Let us assume that as a result of synthesis of the RSSP mechanism the following values of the basic constants are obtained: $p_1=10$, $p_2=-10$, $p_3=5$, $p_4=-200$. All mechanisms belonging to the found family of kinematically equivalent mechanisms reproduce one and the same position function $X=X(\alpha)$ (it is accepted here that $M=1$):

$$X = 10 + 10\sin\alpha + \sqrt{100(1+\sin\alpha)^2 + 10\cos\alpha + 200} . \tag{13}$$

The condition $D(\alpha) > 0$ is fulfilled for any values of the argument α. Further we find: $D_{min}=196.641$.

According to the inequalities (8), free parameter h_2 of the obtained family of mechanisms should satisfy to the condition: $h_2 > 10$. At the figures 2 and 3 are given the diagrams of dependence of parameters h_1, h_3, l_{12}, θ_1 and ϑ_{max} from h_2 for $10 < h_2 \le 20$.

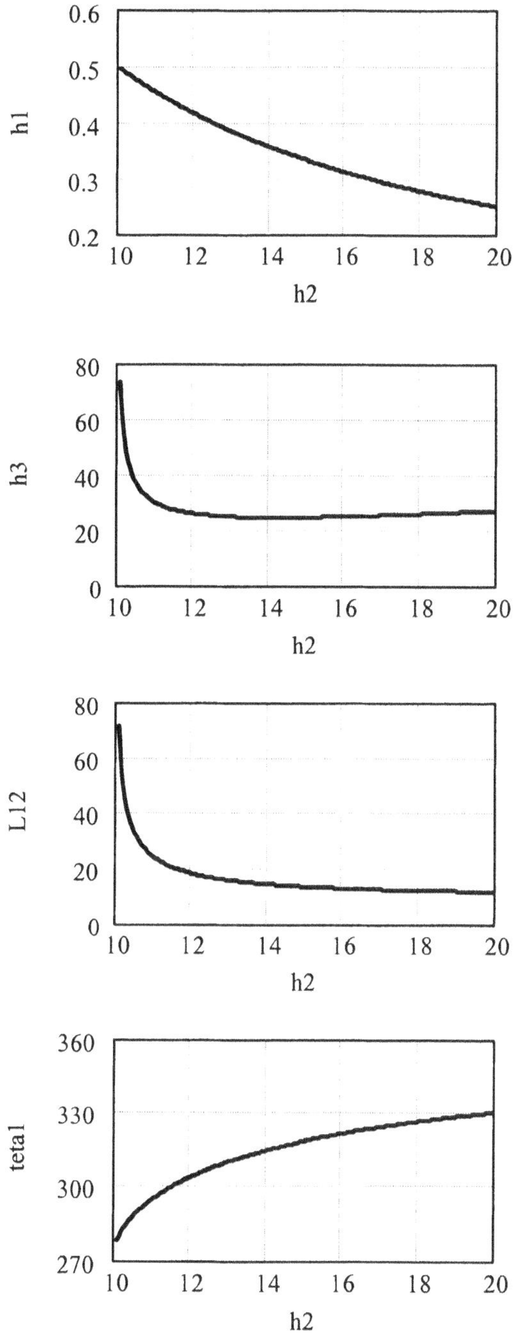

Figure 2. Parameters h_1, h_3, l_{12}, θ_1 as functions of h_2 (at the example)

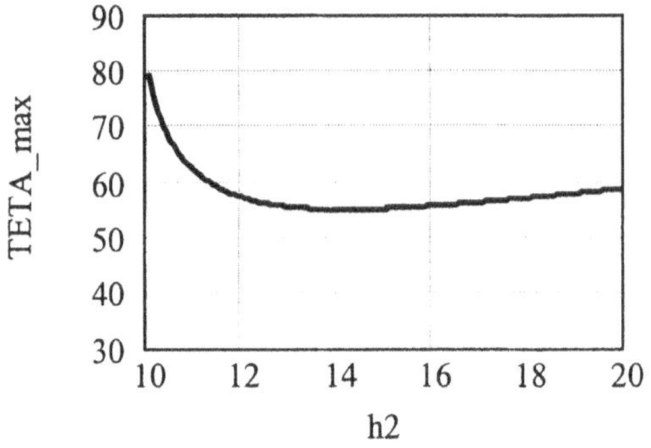

Figure 3. Angle ϑ_{max} as function of h_2 (at the example)

The equation (12) has only one root satisfying to the mentioned above conditions: $u = u_* = 200.0313$. From here: $h_2=(h_2)_{opt}=14.143$, $h_3=(h_3)_{min}=24.497$, $(\vartheta_{max})_{min}=55.081°$. Next, using formulas (7) we find values of other constant parameters of the mechanism: $h_1=0.3535$, $l_{12}=14.141$, $\theta_1=315.0045°$.

6. Conclusions

In the paper is shown, that the spatial linkage RSSP has the property of kinematic equivalence. The concept of the basic constants of one-parametrical family of kinematically equivalent RSSP mechanisms reproducing one and the same position function of a output link is entered. Among mechanisms of the family one mechanism with minimum possible value of the maximum pressure angle is selected. Stated results can be used in synthesis of the RSSP mechanism.

THE OPTIMAL SYNTHESIS OF PARALLEL MANIPULATORS FOR DESIRED WORK-SPACES

A.M. Hay and J.A. Snyman
Multidisciplinary Design Optimization Group, Department of Mechanical Engineering, University of Pretoria, Pretoria, South Africa, 0002
jan.snyman@eng.up.ac.za

Abstract The main objective of this paper is to find a fundamentally sound and robust numerical methodology for synthesizing parallel manipulators for desired workspaces. The constrained optimization formulation presented here is aimed at determining a manipulator design so that a prescribed workspace is fully enclosed, but that the workspace is also well-conditioned with respect to some performance index. The particular manipulator used to illustrate and evaluate the formulation is a simple planar 3-dof parallel manipulator. Solutions to the manipulator design problems are found in an efficient and convincing manner.

Keywords: parallel manipulator, workspace, optimization, condition number

1. Introduction

It is well known that parallel manipulators possess a number of advantages over traditional serial manipulators (Merlet, 2000). One disadvantage of parallel manipulators is that the particular architecture of the manipulator results in smaller workspaces than their serial counterparts. Parallel manipulators can also be difficult to design (Gosselin et al., 1995), since the relationships between design parameters and the workspace, and behavior of the manipulator throughout the workspace, are not intuitive by any means. Indeed, Merlet, 2000 argues that since the performance of parallel manipulators is so dependent on their dimensions, customization of these manipulators for each application is absolutely necessary. An effective way of performing this customization and of addressing the problems stated above is through the use of optimization techniques in the design process. Depending on the particular application, certain manipulator performance criteria may be of more importance than others.

The main objective of this paper is to find a fundamentally sound and robust numerical methodology for synthesizing parallel manipulators for a desired workspace. The constrained optimization formula-

J. Lenarčič and F. Thomas (eds.), Advances in Robot Kinematics, 337–346.

tion presented here is aimed at determining a manipulator design so that a prescribed workspace is fully enclosed, but that the workspace is also well-conditioned with respect to some performance index. The performance measure used here is the condition number of the manipulator Jacobian, although any of the other performance measures, such as the quality index or manipulator stiffness could also have been used. The optimization method used in performing the optimization is the Dynamic-Q method (Snyman and Hay, 2000a).

In the next section the proposed optimization formulation is described. This formulation is then applied to a planar 3-dof planar parallel manipulator.

2. The constrained optimization problem

2.1 Definition of the workspace

As described by Haug et al., 1994, *generalized* coordinates $\mathbf{q} = [q_1, q_2, \ldots, q_{nq}]^T \in R^{nq}$ are used to describe the motion of the manipulator. These generalized coordinates can be divided into *input* coordinates $\mathbf{v} = [v_1, v_2, \ldots, v_{nv}]^T \in R^{nv}$, used to control the manipulator, *output* coordinates $\mathbf{u} = [u_1, u_2, \ldots, u_{nu}]^T \in R^{nu}$, describing the functionality of the mechanism, and the remaining *intermediate* coordinates $\mathbf{w} = [w_1, w_2, \ldots, w_{nw}]^T \in R^{nw}$, where $nw = nq - nu - nv$. In the vicinity of an assembled configuration the input and output coordinates satisfy m independent kinematic constraint equations of the form

$$\mathbf{\Phi}(\mathbf{q}) = \mathbf{\Phi}(\mathbf{u}, \mathbf{v}, \mathbf{w}) = \mathbf{0} \tag{1}$$

where $\mathbf{\Phi} : R^{nq} \rightarrow R^m$ is a smooth function.

Limits imposed by the construction of the manipulator are often described by means of inequality constraints placed on the input, or sometimes the intermediate coordinates. These respectively take the forms

$$\mathbf{v}^{\min} \leq \mathbf{v} \leq \mathbf{v}^{\max} \tag{2}$$

$$\mathbf{w}^{\min} \leq \mathbf{w} \leq \mathbf{w}^{\max} \tag{3}$$

These constraints, together with the geometry of the manipulator, determine the size and shape of the workspace A_c of the manipulator. The workspace is the set of points that can be reached by the working point of the manipulator. The boundary ∂A_c of the workspace is defined as: (Snyman et al., 2000)

$$\begin{aligned}\partial A_c = \{&\mathbf{u} \in R^{nu} : \mathbf{u} \in A_c \text{ and } \exists \text{ an } \mathbf{s} \in R^{nu} \text{ such that for } \mathbf{u}' = \mathbf{u} + \lambda \mathbf{s},\\ &\lambda \in R \text{ arbitrarily small and either positive or negative,}\end{aligned} \tag{4}$$

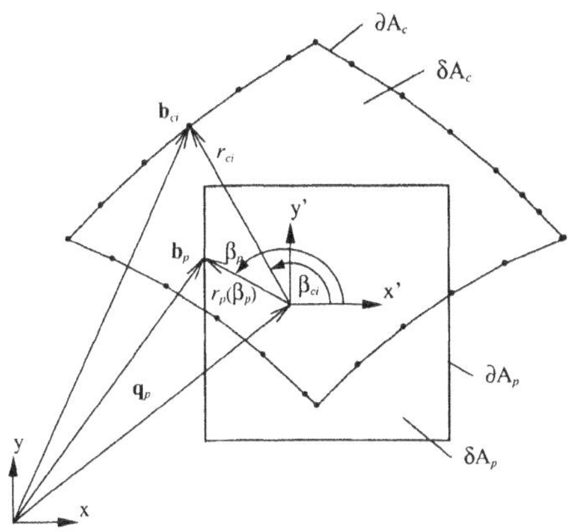

Figure 1. Prescribed and calculated workspaces

no $\mathbf{v}$ and $\mathbf{w}$ exist that satisfy $\mathbf{\Phi}(\mathbf{u}', \mathbf{v}, \mathbf{w}) = \mathbf{0}$; as well as inequalities (2) and (3)}

There exist a number of different methods for determining workspaces of manipulators. In this study, an optimization method that is based on the above definition of the workspace is used. The particular optimization method used is the chord method (Snyman and Hay, 2000b). The optimization method uses the LfopC algorithm (Snyman, 2000) to determine discrete points $\mathbf{b}_{ci}, i = 1, 2, \ldots, n_{bc}$ along the boundary ∂A_c of the workspace A_c, as shown in Fig. 1.

2.2 Formulation

The problem studied here is to determine a manipulator design vector $\mathbf{d}$ which has a corresponding workspace A_c that fully includes a prescribed workspace A_p, and such that the included A_p is well-conditioned. As indicated in Fig. 1, the prescribed workspace is defined by polar coordinates (r_p, β_p) centered on a local coordinate system $x' - y'$ at $\mathbf{q}_p$. The boundary of the workspace A_c associated with design $\mathbf{d}$ may be defined in a similar manner (Fig. 1). Here the boundary point $\mathbf{b}_{ci}$, generated by the chord method (Snyman and Hay, 2000b), corresponds to angle β_{ci} and ray length r_{ci}.

The part of workspace A_p not intersecting A_c is denoted δA_p, and the part of workspace A_c not intersecting A_p is denoted δA_c. The calculation of approximations to the areas δA_p and δA_c is performed using a simple numerical scheme. Areas δA_p and δA_c can be determined by summing the incremental contributions for sequential pairs of boundary points $(\mathbf{b}_{ci}, \mathbf{b}_{ci+1}), i = 1, \ldots, n_{bc} - 2$ where n_{bc} is the number of boundary points generated by the chord method. The closing boundary section is calculated by considering points $(\mathbf{b}_{c(n_{bc}-1)}, \mathbf{b}_{c1})$.

It is evident that for any prescribed workspace there is a very large number of manipulator designs which will result in a workspace that fully includes the prescribed one. Murray et al., 1997 and Merlet, 2000 have proposed different methods for determining the set of manipulator configurations, the workspaces of which include prescribed points, or line segments. The most suitable manipulator with respect to some performance criterion or criteria can then be selected from this set. The method proposed here differs from these methods firstly in that the entire workspace boundary is prescribed, and secondly in that the optimum solution is obtained directly in one step. Thus the constrained optimization formulation, proposed here, is aimed at optimizing manipulators with respect to some performance index, subject to the constraint that the workspace of the optimal manipulator should include the prescribed workspace.

The specific performance measure chosen here is the inverse of the condition number of the Jacobian matrix of the manipulator. The accuracy of control of the manipulator is dependent on the condition number, denoted here by κ (Gosselin and Angeles, 1991). Since this quantity tends to infinity as the manipulator approaches a singular position, maximizing the inverse condition number, κ^{-1}, will ensure that the manipulator remains far away from singular positions. For a parallel manipulator the inverse kinematics are easy to solve, resulting in an inverse transformation relating the input and output velocities. For this manipulator $m = n = nv = nu + nw$.

$$\mathbf{J}\dot{\mathbf{u}} = \dot{\mathbf{v}} \tag{5}$$

The condition number κ of the $n \times n$ Jacobian $\mathbf{J}$ is defined as

$$\kappa = ||\mathbf{J}||||\mathbf{J}^{-1}|| \tag{6}$$

where $||\cdot||$ denotes any norm of its matrix argument. The norm adopted here is the same as that used by Gosselin and Angeles, 1991, namely

$$||\mathbf{J}|| = \sqrt{\text{trace}(\mathbf{J}\mathbf{W}\mathbf{J}^T)} \tag{7}$$

where $\mathbf{W}$ is defined as n^{-1} multiplied by the $n \times n$ identity matrix. The lower the condition number, the better the behavior of the manipulator, with the lowest possible value of κ being unity. When the condition number is equal to 1 then the manipulator is in an isotropic configuration (Gosselin and Angeles, 1989). The *inverse* of the condition number thus lies between 0 and 1, and is used as the objective function because it is bounded and better conditioned than the condition number itself.

The optimization problem is thus:

$$\begin{aligned} &\max_{\mathbf{d}} \left\{ \min_{\mathbf{u} \in A_p} \kappa^{-1} (\mathbf{d}, \mathbf{u}) \right\} \\ &\text{subject to the inequality constraint} \\ &c(\mathbf{d}) \leq 0 \end{aligned} \tag{8}$$

where the inequality constraint function $c(\mathbf{d})$ is defined as follows. The displacement vector between the prescribed workspace boundary and calculated workspace boundary, measured along a ray emanating from $\mathbf{q}_p$ at angles β_{ci} is denoted by $r_i \mathbf{e}_i$, where $\mathbf{e}_i$ is a unit outward vector at angle β_{ci} and $r_i = r_p(\beta_{ci}) - r_{ci}$. If $r_j = \min_i \{r_i, i = 1, 2, \ldots, n_{bc}\}$ then set $r = r_j$. The constraint function is now defined as follows:

$$c(\mathbf{d}) = \begin{cases} \delta A_p & \text{if } \delta A_p > 0 \\ -r^2 & \text{if } \delta A_p = 0 \end{cases} \tag{9}$$

This is done to improve the topology of the inequality constraint which in turn improves the behavior of the optimization algorithm.

The solution to optimization problem (8) seeks to improve the *single worst point* with respect to chosen performance measure, κ^{-1}, within the *prescribed* workspace, A_p. This philosophy differs from that proposed by Gosselin and Angeles, 1991 and used by numerous other researchers where the *average* performance index over the *entire* workspace is optimized. Since in this case it is assumed that the manipulator's movements will be limited to the prescribed workspace it is only necessary to ensure good performance qualities within this workspace and thus it is better here to optimize the single worst value instead of the average.

One point which arises concerns the nested part of optimization problem (8) and the question of how to determine the smallest value of κ^{-1} over the set $\mathbf{u} \in A_p$. One approach that has been used (Tsai and Joshi, 2000), is to apply the Monte Carlo method to determine representative values over the entire set. Based on numerical experimentation for a planar 2-dof manipulator studied previously (Hay and Snyman, 2001), it was concluded that the single least value of κ^{-1} lies on the boundary ∂A_p of the prescribed workspace. The assumption is made that this behavior is similar for the particular 3-dof manipulator to be investigated here. The minimum value of the inverse condition number κ^{-1} can thus efficiently be approximated by calculating κ^{-1} at points $\mathbf{b}_{pi}$, $i = 1, \ldots, n_{bc}$ simultaneously to the determination of the boundary points $\mathbf{b}_{ci}$, $i = 1, \ldots, n_{bc}$. The overall minimum of the κ^{-1} values at these candidate points may then easily be determined. Based on the results presented later in this paper it appears that the above assumption is valid.

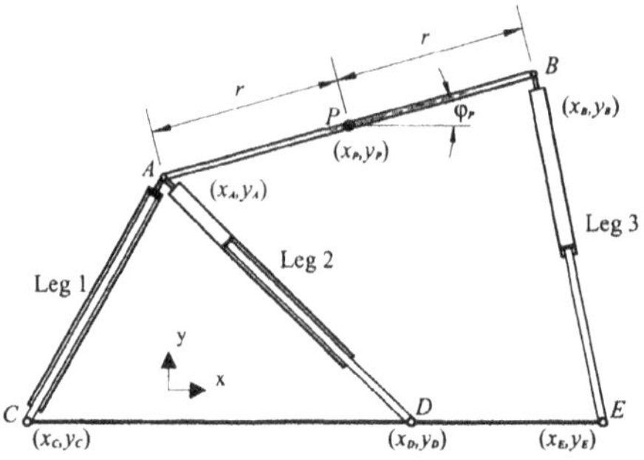

Figure 2. The 3-dof planar parallel manipulator

3. Three-dof planar parallel manipulator

The three-dof planar parallel manipulator taken from Haug et al., 1994 is shown in Fig. 2. The mechanism consists of a platform of length $2r$ connected to a base by three linear actuators, which control the three output degrees of freedom of the platform. The actuators have leg lengths l_1, l_2 and l_3 and are joined to the base and platform by means of revolute joints identified by the letters $A - E$. It will be assumed that $y_C = y_D = y_E$. The coordinates of point P, the working point of the platform, are (x_P, y_P) and the orientation of the platform is ϕ_P. With reference to the definitions given in sections 2.1 and 2.2, the actuator leg lengths are the input variables, i.e. $\mathbf{v} = [l_1, l_2, l_3]^T$. The global coordinates of the working point P form the output coordinates, i.e. $\mathbf{u} = [x_P, y_P]^T$. The rotation angle of the platform is the only intermediate coordinate, i.e. $\mathbf{w} = w = [\phi_P]$. The generalized coordinates for the Stewart platform are given by $\mathbf{q} = [\mathbf{u}^T, \mathbf{v}^T, \mathbf{w}^T]^T = [x_P, y_P, l_1, l_2, l_3, \phi_P]^T$.

It is assumed the maximum and minimum leg lengths are fixed at some lengths, i.e. the actuators have been chosen. The remaining five design variables for the problem are thus

$$\mathbf{d} = [x_C, y_C, x_D, x_E, r]^T \tag{10}$$

Certain points arise now due to the platform's orientation ability. First amongst these is the fact that the Jacobian of the manipulator contains entries related to both translational and rotational abilities of the platform. The condition number will thus inherently contain a mix of these terms. It is thus important to normalize the translational terms of the Jacobian matrix so that translational and rotational abilities are equally represented by the condition number. Pittens and Podhorodeski, 1993 and Stoughton and Arai, 1993 note this occurrence and suggest that the best approach is to normalize the translational terms of the Jacobian with respect to the platform radius r, a suggestion which is adopted here.

The second point arising is that some strategy needs to be implemented for dealing with the fact that the condition number needs to be minimized throughout the manipulator's three-dimensional (x, y, ϕ) workspace. Indeed, the prescribed workspace should also be 3-dimensional. It is envisaged that this point should be addressed by evaluating the problem at various angular "slices" through the workspace. This is the approach used by Boudreau and Gosselin, 2001 in the unconstrained case. As a first step in developing a more general method, the optimization procedure is restricted to a single angular slice through the workspace. In practical terms this means that, in this paper, the manipulator is optimized for a single fixed orientation ϕ_P of the moving platform.

Optimization problem (8) was solved using the Dynamic-Q algorithm (Snyman et al., 1994, Snyman and Hay, 2000a). In this method , the LfopC algorithm is used to solve successive spherical approximations of the optimization problem. During extensive testing of the method on a set of representative test problems, the solutions of the successive approximate subproblems $P[i], i = 1, 2, \ldots$ invariably converged towards local minima (Snyman and Hay, 2000a).

The method described above has been applied to the 3-dof planar parallel manipulator for three different prescribed workspaces with the platform fixed horizontally, i.e. $\phi_P = 0$. The prescribed workspaces are centered at $\mathbf{q}_p = [1, 1.5]^T$ and are labelled P1-P3 in Fig. 3a. Actuator limits were chosen as $l_i^{\text{min}} = \sqrt{2}$, $l_i^{\text{max}} = 2, i = 1, 2$ and $l_3^{\text{min}} = 1$, $l_3^{\text{max}} = \sqrt{2}$. Actuator leg lengths and initial design vector $\mathbf{d}_0 = [-1, 0, 1, 2, 1]^T$ were chosen to correspond to the manipulator studied by Haug et al., 1994. As well as the prescribed workspaces P1-P3, the workspace corresponding to the initial design vector and inverse condition number contours for this design are shown in Fig. 3a. For these runs the move limit used was $\triangle = 0.1$ and the chord length for calculating the workspace was 0.02. The optimization algorithm termination parameters on the stepsize and function value used were $\varepsilon_f = 10^{-6}$ and $\varepsilon_x = 10^{-4}$.

Numerical results for each of the optimization runs corresponding to the different prescribed workspaces are reported in Tb. 1. This table gives the number of gradient evaluations n_{ge} required for convergence, the initial function value $f(\mathbf{d}_0)$, converged objective function value f^*, inequality constraint c^* and components of the design vector $\mathbf{d}^*$. The resultant workspaces and inverse condition number contours correspond-

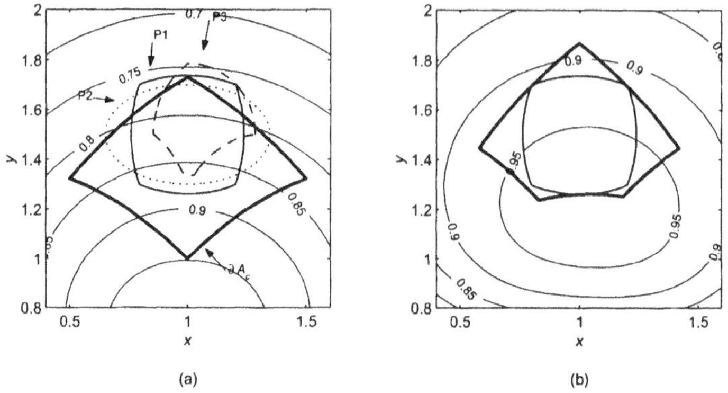

Figure 3. 3-dof planar parallel manipulator (a) Prescribed workspaces P1-P3, starting manipulator workspace $\partial A_c(\mathbf{d}_0)$ and κ^{-1} contours and (b) Prescribed workspace P1, manipulator workspace and κ^{-1} contours

ing to P1-P3 are shown in Fig. 3b, 4a and 4b respectively. Finally Fig. 5 gives the optimization histories for the various prescribed workspaces.

Table 1. 3-dof planar parallel manipulator constrained optimization solutions

	n_{ge}	$f(\mathbf{d}_0)$	f^*	c^*	x_C^*	y_C^*	x_D^*	x_E^*	r^*
P1	21	0.7585	0.9119	0.19E-5	-1.110	0.2632	1.277	1.967	0.9159
P2	17	0.7680	0.9273	0.39E-9	-1.153	0.3003	1.275	1.906	0.9401
P3	71	0.7467	0.9427	0.98E-5	-1.170	0.3365	1.463	1.946	0.9108

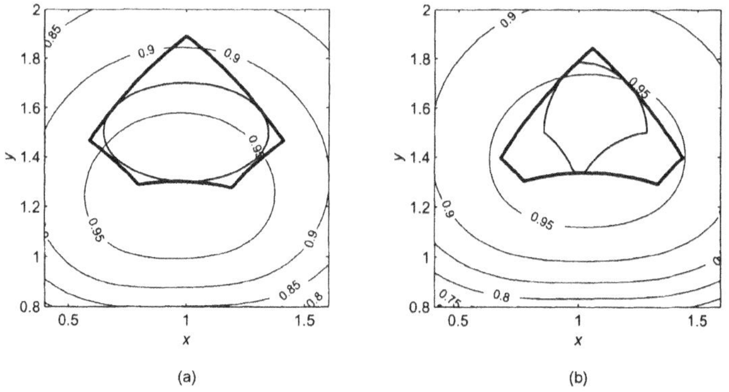

Figure 4. 3-dof planar parallel manipulator (a) Prescribed workspace P2, manipulator workspace and κ^{-1} contours and (b) Prescribed workspace P3, manipulator workspace and κ^{-1} contours

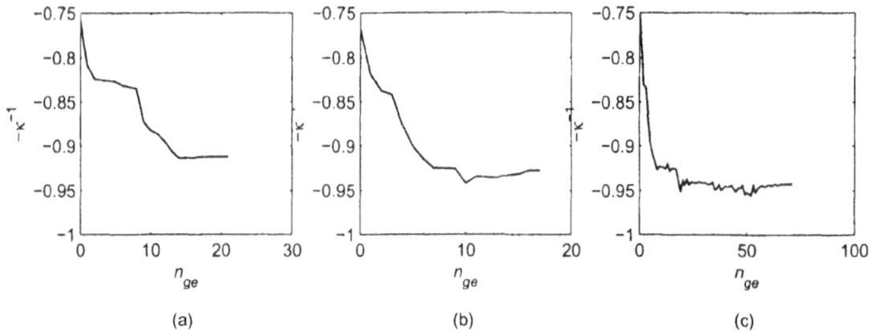

Figure 5. 3-dof planar parallel manipulator optimization histories for (a) P1, (b) P2 and (c) P3

4. Conclusion

For the 3-dof manipulators studied here the results obtained are extremely encouraging, with accurate and optimum solutions having been obtained with minimal computational effort. Although not reported here, the methodology has also been applied to simpler 2-dof planar manipulators with very good results. In each case the algorithm not only determines manipulator dimensions so that the prescribed workspace can be reached by the manipulator, but also places the calculated workspace so that the inverse condition number is as high as possible throughout the prescribed workspace. The proposed methodology produces convincing results, indicating it to be a stable and efficient method for designing planar parallel manipulators. Although the application of this numerical methodology for synthesizing parallel manipulators, which is believed to be fundamentally sound and robust, has been restricted to the simple 2-dof and 3-dof fixed orientation cases, it nevertheless led to a scheme that appears to be general. The Dynamic-Q optimization algorithm used in the synthesis methodology exhibits high efficiency in solving the associated optimization problem. Although the manipulators studied here are relatively simple, it is hoped that the principles developed here can be extended to similar, but more complicated spatial manipulator problems.

References

Boudreau, R., and Gosselin, C.M. (2001), La synthèse d'une plate-forme de Gough-Stewart pour un espace atteignable prescrit, *Mechanism and Machine Theory*, vol. 36, pp. 327-342.

Haug, E.J., Luh, C.M., Adkins, C.M., and Wang, J.Y. (1994), Numerical algorithms for mapping boundaries of manipulator workspaces, *Concurrent Engineering Tools for Dynamic Analysis and Optimization: IUTAM Fifth Summer School on Mechanics*, Aalborg, Denmark.

Gosselin, C.M., and Angeles, J. (1989), The optimum kinematic design of a spherical three-degree-of-freedom parallel manipulator, *ASME Journal of Mechanisms, Transmissions and Automation in Design*, vol. 113, pp. 220-226.

Gosselin, C.M., and Angeles, J. (1991), A global performance index for the kinematic optimization of robotic manipulators, *ASME Journal of Mechanical Design*, vol. 111, pp. 202-207.

Gosselin, C.M., Perrault, L., and Vaillancourt, C. (1995), Simulation and computer-aided kinematic design of three-degree-of-freedom spherical parallel manipulators, *Journal of Robotic Systems*, no. 12, vol. 12, pp. 857-869.

Hay, A.M., and Snyman, J.A. (2001), Methodologies for the optimal design of parallel manipulators, Submitted to *International Journal for Numerical Methods in Engineering*.

Merlet, J-P. (2000), *Parallel Robots*, Dordrecht, The Netherlands, Kluwer Academic Publishers.

Murray, A.P., Pierrot, F., Dauchez, P., and McCarthy, J.M. (1997), A planar quaternion approch to the kinematic synthesis of a parallel manipulator, *Robotica*, vol. 15, pp. 361-365.

Pittens, K.H., and Podhorodeski, R.P. (1993), A family of Stewart platforms with optimal dexterity, *Journal of Robotic Systems*, no. 4, vol. 10, pp. 463–479.

Snyman, J.A. (2000), The LFOPC leap-frog algorithm for constrained optimization, *Computers and Mathematics with Applications*, vol. 40, pp. 1085-1096.

Snyman, J.A., Du Plessis, L.J., and Duffy, J. (2000), An optimization approach to the determination of the boundaries of manipulator workspaces, *ASME Journal of Mechanical Design*, vol. 122, pp. 447-455.

Snyman, J.A., and Hay, A.M. (2000a), The Dynamic-Q method: An alternative to SQP?, *Proceedings of the International Workshop on Multidisciplinary Design Optimization*, Pretoria, South Africa, pp. 163-172.

Snyman, J.A., and Hay, A.M. (2000b), *Advances in Robot Kinematics*, Dordrecht, The Netherlands, Kluwer Academic Publishers.

Snyman, J.A., Roux, W.J., and Stander, N. (1994), A dynamic penalty function method for the solution of structural optimization problems, *Applied Mathematical Modelling*, vol. 18, pp. 435-460.

Stoughton, R.S., and Arai, T. (1993), A modified Stewart platform manipulator with improved dexterity, *IEEE Transactions on Robotics and Automation*, no. 2, vol. 9, pp. 166-173.

Tsai, L-W., and Joshi, S. (2000), Kinematics and optimization of a spatial 3-UPU parallel manipulator, *ASME Journal of Mechanical Design*, vol. 122, pp. 439-446.

CONNECTING ASSEMBLY MODES FOR WORKSPACE ENLARGEMENT

Jürgen Hesselbach, Manfred B. Helm and Sven Soetebier

Institute of Machine Tools and Production Technology (IWF)

Technical University of Braunschweig, Germany

m.helm@tu-bs.de

Abstract This paper introduces an approach for increasing the useable workspace by realizing a continuous motion between different assembly modes. The workspace envelope is defined as the set of all elements of all workspaces of the different assembly modes. For a planar PRRRP parallel manipulator with 2 DOF this definition is explained. The crossover poses from one assembly mode to another are given by different singular configurations. Using the planar manipulator and its workspace and singularities, different methods are developed how a continuous motion from one assembly mode to another, i.e. passing the singularities, can be realized. Finally a prototype of the PRRRP parallel manipulator is presented on which the described approach could be implemented. First experiments prove the principle.

Keywords: kinematics, workspace, singularities, control, handling

1. Introduction

Research on parallel mechanisms already brought up a large number of technical solutions for robot architectures since the first manipulators with kinematic loops have been proposed. Different theoretical and practical problems associated with those architectures have been addressed by many authors since then. Two fundamental problems which are discussed in relation with parallel manipulators are related to workspace and singularities.

A classification of singularities is done by GOSSELIN and ANGELES. The classification is based on the singularities of the Jacobian matrices (**J** for the inverse kinematics and **K** for the direct kinematics) and distinguishes between three types of singularities. Singularities of the first type occur at poses of the manipulator where the Jacobian matrix of the inverse kinematics becomes singular ($\det(\mathbf{J})=0$; $\det(\mathbf{K})\neq 0$). These poses of an manipulator are always on the boundaries of the manipulators workspace. The geometric explanation for those positions is at least one completely folded or extended leg, which leads to an undermobility of the manipulator. At least one actuator can fulfill an infinitesimal small motion in such a position without moving the

J. Lenarčič and F. Thomas (eds.), Advances in Robot Kinematics, 347–356.

endeffector. Singularities of the second type occur at poses of the manipulator where the Jacobian matrix of the direct kinematics becomes singular (det(**K**)=0; det(**J**)≠0). Geometric explanation for singularities of the second type can be developed using the Grassmann geometry. These poses of an manipulator can occur inside the manipulator workspace and therefore lead to additional control problems. In such poses for the endeffector an infinitesimal small motion without any actuated joint being moved is possible which leads to an overmobility of the manipulator. Singularities of the third type occur at poses where both Jacobian matrices become singular (det(**K**)=0; det(**J**)=0).

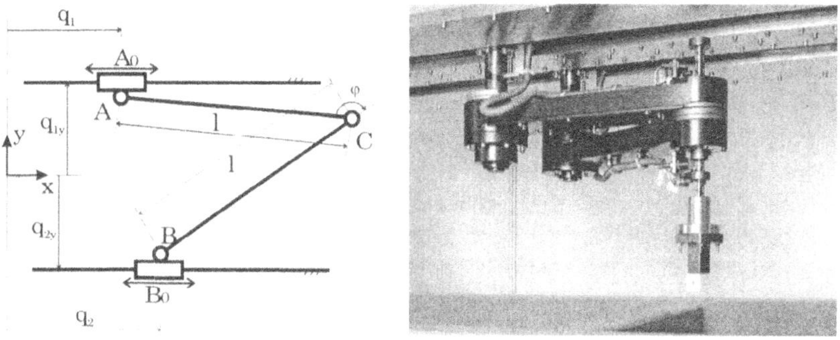

Figure 1. Kinematics and prototype of the planar PRRRP parallel manipulator

The workspace of a manipulator based on a parallel mechanism describes the limitations of its endeffector motions. They can be restricted by different factors. One of those factors are singularities. Several different types of workspaces are used to describe the possible motions of a manipulator. Despite of which type of workspace is referred to in literature there is always the workspace of one assembly mode described. The regarded assembly mode is always only one of the several possible solutions of the direct kinematics. This means that "all the points from a given workspace subset correspond to the same assembly mode of the parallel robot, i.e., the parallel manipulator can move continuously between any two points from a given workspace subset, without having to be disassembled" (BONEV). But generally a manipulator could move – if wanted - in any possible assembly mode given by the different real solutions of the direct kinematics.

2. Planar PRRRP Parallel Manipulator

In the following a planar PRRRP parallel manipulator is used to explain the presented approach. The kinematics of the manipulator is depicted in Fig. 1. The prismatic joints A_0 and B_0 are actuated. The Points A, B and C are revolute joints.

The endeffector is, together with a serial phi-z-axis, attached to the center of the passive joint C (Fig. 1, prototype). All actuations of the prototype are realized with rotational synchronous drives combined with gears and synchronous belts and synchronous belt pulleys. The drives for the phi-z-axis are mounted on the sliding carriages of the prismatic joints and the drives for the prismatic joints are mounted on the frame.

The kinematic equations are derived from Fig. 1. The equations are restricted to the planar translational motions for the reason that the serial phi-z-axis does not have any effect on the planar workspace or singular configurations. For a simple kinematic structure like a five bar mechanism they directly show the possible number of assembly modes and the singular configurations.

The inverse kinematics

$$q_1 = x + k_1 \cdot \sqrt{l^2 - (y - q_{1y})^2};\ k_1 = \pm 1 \tag{1}$$

$$q_2 = x + k_2 \cdot \sqrt{l^2 - (y - q_{2y})^2};\ k_2 = \pm 1 \tag{2}$$

has four solutions. All possible solutions are combined in Fig. 2.

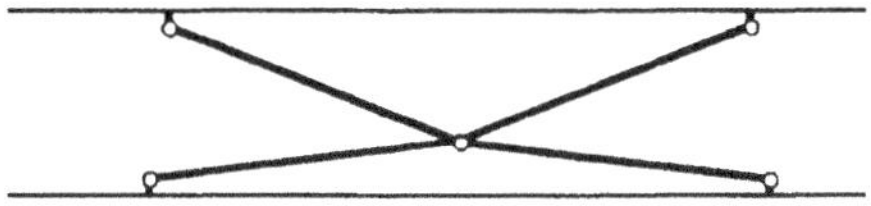

Figure 2. Solutions for the inverse kinematics

The inverse kinematics becomes singular if one arm stands perpendicular to the attached prismatic joint (singularities of type one: Fig. 3).

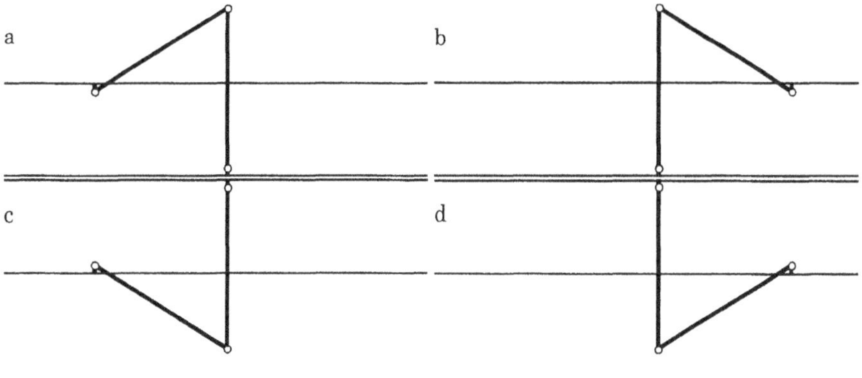

Figure 3. Singular configurations of the inverse kinematics

The direct kinematics

$$x = q_1 + \frac{A_x \cdot H_1 - A_y \cdot H_2}{2 \cdot H_1} \tag{3}$$

$$y = q_{1y} - a + \frac{A_x \cdot H_2 + A_y \cdot H_1}{2 \cdot H_1} \tag{4}$$

using $$A_x = q_2 - q_1\,;\ A_y = q_{2y} - q_{1y}\,;\ H_1 = A_x^2 + A_y^2 \qquad (5)(6)(7)$$

$$H_2 = k_3 \cdot \sqrt{H_1 \cdot (2 \cdot b)^2 - H_1^{\,2}}\,;\ k_3 = \pm 1 \tag{8}$$

has two solutions. All possible solutions are combined in Fig. 4.

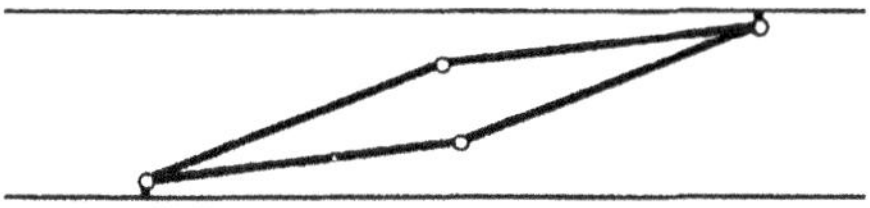

Figure 4. Solutions for the direct kinematics

The direct kinematics becomes singular if both arms stand parallel (collinear) to each other (singularities of type two: Fig. 5).

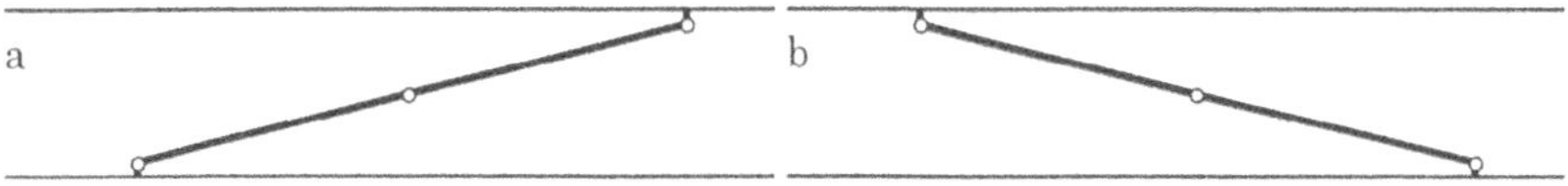

Figure 5. Singular configurations of the direct kinematics

Each of the solutions of the direct and the inverse kinematics result in different workspaces. Therefore for the planar PRRRP parallel manipulator six different workspaces exist. The four workspaces of the four solutions of the inverse kinematics are shown in Fig. 6.

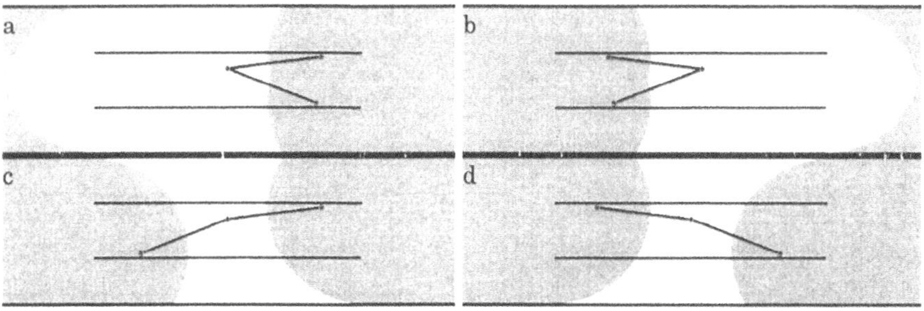

Figure 6. The four workspaces of the inverse kinematics

The remaining two workspaces of the direct kinematics are – for the discussed manipulator – equal to the first two (a and b) in Fig. 6.

The sketches of the manipulator show another predicate of singular configurations. They are crossover poses from one solution to another solution for the associated kinematic problem.

3. Crossover Poses

Assuming that a crossover pose for movement between different workspaces exists, a workspace envelope $\boldsymbol{W}$ can be defined. This workspace envelope is the set of all elements of all workspaces, one for each different assembly mode. Taking two different workspaces ($\boldsymbol{A}$ and $\boldsymbol{B}$; $\boldsymbol{W} = \boldsymbol{A} \cup \boldsymbol{B} \cup ...$) from one workspace envelope, three different relative positions (in general in a 6-DOF space) of the two workspaces can occur. In the first case $\boldsymbol{A}$ completely infoldes $\boldsymbol{B}$ or as a subcase $\boldsymbol{A}$ and $\boldsymbol{B}$ are equal. Here $\boldsymbol{B}$ is hence a subset of $\boldsymbol{A}$ and their intersection $\boldsymbol{I}$ is equal to (case one: $\boldsymbol{I}$=$\boldsymbol{B} \wedge \boldsymbol{I} \subseteq \boldsymbol{A}$). Secondly $\boldsymbol{A}$ and $\boldsymbol{B}$ have common poses and $\boldsymbol{I}$ is smaller than each of the sets (case two: $\boldsymbol{I} \subseteq \boldsymbol{B} \wedge \boldsymbol{I} \subseteq \boldsymbol{A}$). Third $\boldsymbol{A}$ and $\boldsymbol{B}$ do not have common poses. In consequence $\boldsymbol{I}$ is the null set (case three: $\boldsymbol{I}$=$\{\}$). Even if $\boldsymbol{B}$ is completely identical to $\boldsymbol{A}$ a crossover pose is not compellent, where the manipulator could pass from $\boldsymbol{A}$ to $\boldsymbol{B}$ and back. For a crossover pose it is not sufficient to have common poses, but for poses of that kind the articular coordinates have to be equal as well. Therefore an additional condition is needed to realize a manipulator that could change from one workspace to another.

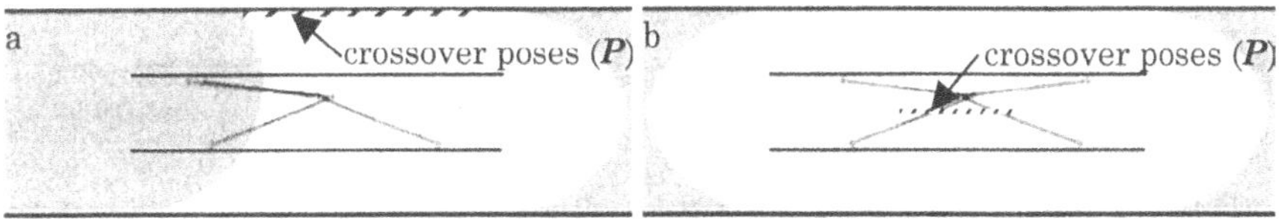

Figure 7. Crossover poses (the set $\boldsymbol{P}$)

As mentioned above, the crossover poses are always singular configurations. Hence an additional set is defined. $\boldsymbol{S}$ is the set of all singular configurations. Now the condition for the possibility of moving from $\boldsymbol{A}$ to $\boldsymbol{B}$ can be specified. Only if the set $\boldsymbol{P}$, the intersection of $\boldsymbol{I}$ and $\boldsymbol{S}$ ($\boldsymbol{P} = \boldsymbol{I} \cap \boldsymbol{S}$) is not the null set, crossover poses from $\boldsymbol{A}$ to $\boldsymbol{B}$ exist. $\boldsymbol{P}$ proves to be the set of all possible crossover poses from $\boldsymbol{A}$ to $\boldsymbol{B}$. Fig. 7 a shows the combination of the two workspaces of the inverse kinematics Fig. 6 b (here $\boldsymbol{A}$) and Fig. 6 d (here $\boldsymbol{B}$). The associated set of crossover poses $\boldsymbol{P}$ includes the singular configurations of the type one that are shown in Fig. 3 a. In Fig. 7 b the more interesting case is shown. The workspaces $\boldsymbol{A}$ and $\boldsymbol{B}$ are those of the direct kinematics (equal to a and b in Fig. 6). The associated set $\boldsymbol{P}$ of crossover poses includes all singular configurations of the second type (a and b in Fig. 5). In order to pass singular configurations, one main difference between the singularities of the type one and those of the type two becomes important. As long as the manipulator is controlled in articular coordinates, singularities of the type one do not need to be regarded at all. Therefore the crossing from one solution of the inverse kinematics to another one requires only that the manipulator is not controlled in generalized coordinates, but in articulated coordinates (Fig. 7 a).

On the other hand, singularities of the type two can not be treated as simple. If one actuated joint is moved in a singular configuration of the second type, there is no unequivocal answer to the question on which branch of that bifurcation the manipulator continues its move. Therefore a secure movement or positioning of the manipulator cannot be guaranteed after a singular configuration of the type two has been reached before. Even if appropriate sensors are implemented to detect the actual generalized coordinates the crossing of that singular configuration would still not be secure. In contrast to the crossing of singularities of the type one is only the crossing of singularities of the type two, i.e. a change of assembly modes, a workspace enlargement.

4. Passing Singularities of the Second Type

From this point on the term "singularity" always signifies a singular configuration of the second type. As said before, singularities have usually to be avoided because they result in a lack of distinction of the manipulator. But for moving from one workspace of the direct kinematics to another workspace of the direct kinematics a singularity has to be passed.

To make that possible special effort is necessary, concerning the control and the design of the manipulator. Four conditions have to be fulfilled to pass such a singularity. First of all, the passive and actuated joints have to allow motion on both sides of the singularity. This is an additional design condition, that could be easily fulfilled for the discussed planar PRRRP parallel manipulator (Fig. 11). It is a crucial condition when the approach is to be adapted to spatial mechanisms. Secondly, the control system has to allow an on-line change of the kinematic equations. The algebraic sign of the square roots in the kinematic equations have to be adapted after each crossover into another workspace. This can be realized by implementing the algebraic signs (k_1, k_2 and k_3 in Eq. 1, 2 and 8) that have to be changed as additional variable parameters in the control system. Thirdly, a sensor has to be added, which allows to recognize the active workspace in a way, that a known and secure reference position can be reached from any unknown pose. For the discussed manipulator a sensor in C (Fig. 1) is sufficient detecting the direction in which the angle φ (in Fig. 1) points. A secure and known reference pose can afterwards always be reached by moving the manipulator in articular coordinates. This as well becomes much more complicated when transferring the approach to spatial mechanisms. Fourthly and most important is an additional control algorithm that can under certain conditions depend on the applied design additions, has to be implemented. This last fact is discussed in the following chapter.

5. Securing the singularity

Two principally different approaches are possible to enable a mechanism to pass a singularity. Firstly, the singularity can be momentarily removed by changing the mechanism itself. Secondly a force can be applied to the parallel mechanism manipulator in the singularity pose to apply a preferred direction out of the singularity. Since the regarded singularities are no configuration space singularities their positions can be moved by changing the mechanisms actuated joints.

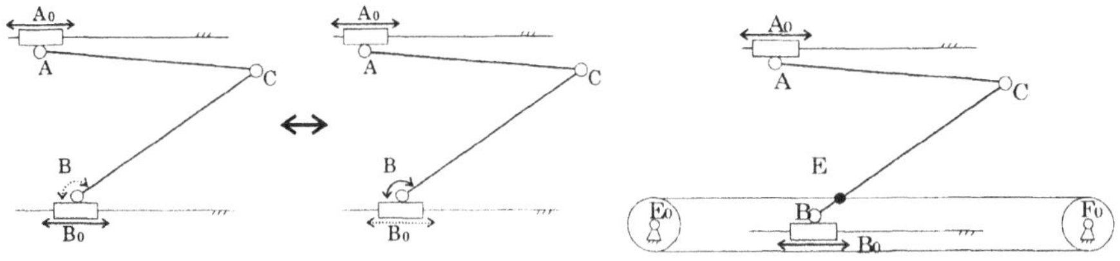

Figure 8. Momentary reconfiguration by moving the actuation and by extension of the mechanism

In Fig. 8 (left) this principle is explained. An additional actuator is attached to the joint B. This actuator can be very small and does not need to guarantee any accuracy. It has only to be activated when the singularity has been approached. In the same moment the actuator in B_0 is deactivated and the actuated joint B can move the manipulator across the no longer existing singularity. After the singularity has been passed the actuator in B is deactivated and the one in B_0 is reactivated. Subsequent to that procedure the manipulator can be moved in the new workspace.

A second change of the mechanism is to extend it with an additional elastic two-bar that can be braked. In Fig. 8 (right), a circulatory toothed belt is added as a two-bar. If in E_0 or F_0 a brake is applied to the synchronous pulleys and one of the actuators is disabled the mechanism can be moved through the removed singularity. If the belt provides sufficient elasticity this procedure could even be executed without disabling any of the actuators.

By not modifying the mechanism to remove the singularity the effort to realize a singularity passing can be significantly reduced. In Fig. 9 the mechanism is extended by a compression spring at the joint B. Using the spring in a thus implemented toggle lever a preferred direction out of the singularity is applied.

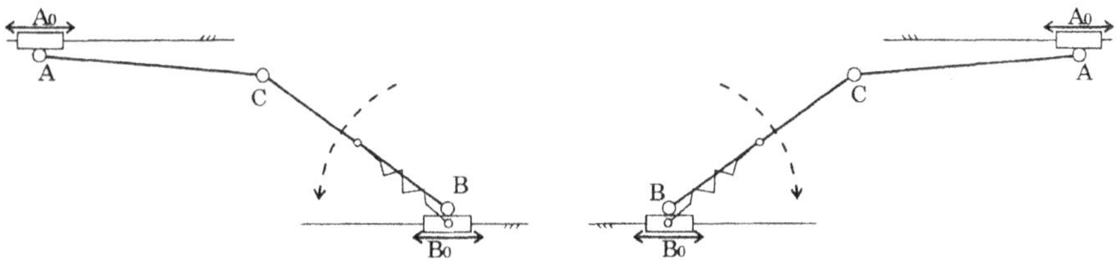

Figure 9. Compression spring for applying a preferred direction

By using the two different solutions of the inverse kinematics where singularities appear one single spring is sufficient to move the manipulator from one workspace of the direct kinematics to the other and back. Of course a torsion spring can be used as well. Only the directions of the force applied to the manipulator will change and thus the preferred direction out of the singularity.

Despite of its simplicity the solution in Fig. 9 bears one severe conflict of aims. To realize a fast singularity passing the spring has to be very stiff to guarantee the crossover even if the appearing dynamic forces are high due to the speed of the manipulator. On the other hand the spring has to be very soft to lower the forces that have to be compensated by the actuators during the normal motion inside one workspace.

To overcome this problem a fourth solution was developed. Without any changes of the design of the manipulator (except for additional sensors) the masses of the manipulator links can be used to cross a singularity. When the manipulator is moved into the singularity with one drive disabled the inertia of the links slides the manipulator through the singularity. The implementation of this approach to cross singularities is explained in the next chapter.

6. Implementation to the Prototype

The crossing of singularities using the inertia of the machine elements has been implemented to the manipulator control system (DSP-based) in eight steps. It is programmed as one function and can be called during any motion cycle.

In the first step, the manipulator is moved in articulated coordinates from its actual pose to a pose close to the singularity (e.g. Fig. 11). As the second step, one drive is disabled (idle, not braked). In step three, the still enabled drive moves the manipulator into the singularity. When the singularity is reached a minimum speed is obligatory. If the minimum speed could not be obtained, the friction forces in the joints are higher then the inertia forces of the masses moved and the singularity will not be crossed. Immediately when reaching the singularity the drive has to decelerate to prevent an oscillation about the singularity. If the active drive is decelerated too late an oscillation about the singularity can move the manipulator back into the initial assembly mode.

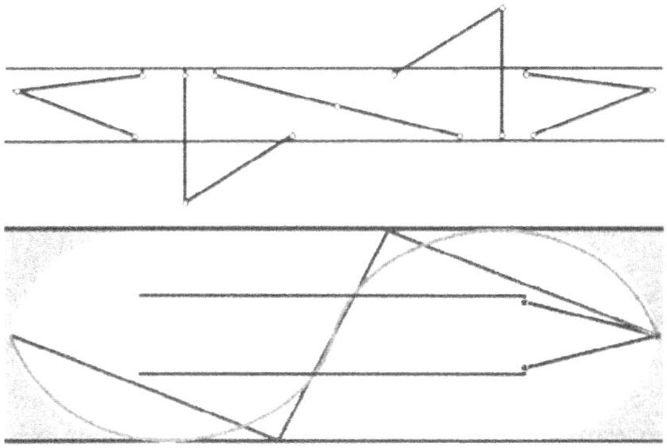

Figure 10. Path through the complete workspace envelope

In step four, the sensor that detects the assembly mode is checked whether the crossover was successful. As step five, the kinematic equations are adjusted in the control software. In the sixth step, the disabled drives are enabled. This is done by increasing the proportional parameter of the drive controller from zero to its optimal value with an exponential function. In step seven, the manipulator is moved in articulated coordinates to the target pose. In the last step, the crossover is reported successful, and the manipulator controller continues with its prior program. A sample path through the workspace is shown in Fig. 10.

Figure 11. Prototype on both sides of a singularity

First experiments with the prototype show that the method is very reliable. After the minimum speed and the deceleration parameters for the active drive had been determined by experiments, the crossover itself has not once been unsuccessful. The time that is required for the crossover is mainly determined by the power up time of the disabled drives. The main difference to moving in only one workspace is that not any path is possible. But this is no restriction for handling tasks.

7. Conclusions

The functioning of the presented approach to enlarge a parallel manipulator workspace by combining several workspaces to one workspace envelope and move from one workspace to another through singularities could be proved for a planar parallel 2-DOF manipulator.

The implementation to other mechanisms will be very promising. Especially the workspace in rotational generalized coordinates could be enlarged for most mechanisms by changing the assembly modes. For spatial manipulators another question has to be solved. How can a pose be passed where the Jacobian loses rank by more then one?

Further work is necessary in several other disciplines as well. Procedures have to be developed to optimize a mechanism considering the possibility of crossing singularities.

For the implementation of the presented procedure on control systems that can be used in industrial environments additional efforts will be necessary. Only if an industrial control system is able to provide this additional functionality a transfer to industry could be thought of.

References

Bonev, I. (2001), http://www.parallemic.org/Terminology/Kinematics.html

Dietmaier P. (1998). The Stewart-Gough platform of general geometry can have 40 real postures. *ARK 1998*: 7-16, Strobl, Austria

Hesselbach, J. (2001), High Dynamic Robot Systems for Assembly and Handling Applications. *7th International Conference on Production Engineering, Design and Control*, Alexandria, Egypt

Kock, S. (2001), *Parallelroboter mit Antriebsredundanz*, Düsseldorf, VDI Verlag

Merlet, J.-P. (2000), *Parallel Robots*, Dordrecht, Kluwer Academic Publishers

Merlet J-P.; Gosselin C.; Mouly N. (1996), Workspaces of planar parallel manipulators, *Mechanism and Machine Theory*, no. 33, vol. 2, pp. 7-20

Mohammadi Daniali, H.R.; Zsombor-Murray, P.J.; Angeles, J. (1995), Singularity Analysis of Planar Parallel Manipulators, *Mechanism and Machine Theory*, no. 30, vol. 5, pp. 665-678

Gosselin C. et Angeles J. (1990), Singularity analysis of closed-loop kinematic chains, *IEEE Trans. on Robotics and Automation*, no. 6, vol. 3, pp. 281-290

Zlatanov, D.; Bonev, I.; Gosselin, C. (2001): Constraint Singularities as Configuration Space Singularities, *http://www.parallemic.org*

THE ISOTROPIC CONDITIONS OF PARALLEL MANIPULATORS OF DELTA TOPOLOGY

Luc Baron
Département de génie mécanique, École Polytechnique
C.P. 6079, succ. CV, Montréal, Québec, Canada, H3C 3A7
Luc.Baron@polymtl.ca

Xiaoyu Wang
Département de génie mécanique, École Polytechnique
C.P. 6079, succ. CV, Montréal, Québec, Canada, H3C 3A7
xwang@meca.polymtl.ca

Guy Cloutier
Département de génie mécanique, École Polytechnique
C.P. 6079, succ. CV, Montréal, Québec, Canada, H3C 3A7
Guy.Cloutier@polymtl.ca

Abstract This paper proposes a necessary and sufficient number of 24 geometric parameters allowing to describe the design manifold of the Delta topological class, i.e., all geometries of manipulators having the same topology and mobility of the end-effector as the Delta parallel manipulator. Two isotropic conditions are then applied on this manifold in order to define the constraint manifold of isotropic designs, i.e., those having an isotropic Jacobian matrix. These constraints would facilitates the design of isotropic manipulators in this topological class.

Keywords: Parallel manipulator, geometrical synthesis, kinematic isotropy.

1. Introduction

As shown in Fig. 1, the Delta parallel manipulator (Clavel, 1985) allows 3D translations of the end-effector (EE) using three motorized revolute joints acting along three parallel legs. In general, parallel manipulators are particularly interesting because they possess complementary characteristics to serial manipulators. The former can be used in situations where the characteristics of the latter does not satisfy the application requirements. The design of parallel manipulators is a complex activity involving many technical and creative behaviors.

J. Lenarčič and F. Thomas (eds.), Advances in Robot Kinematics, 357–366.

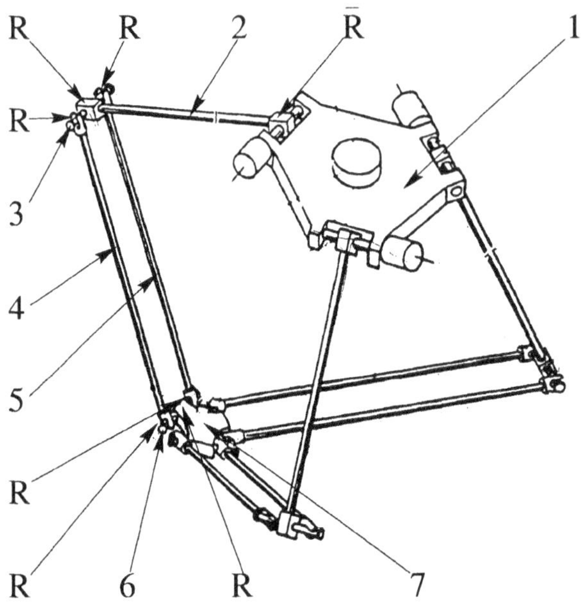

Figure 1. The Delta Parallel Manipulator

The designer must evaluate the capacity of each potential design to satisfy the application requirements. Among others, the *isotropy* (Salisbury and Craig, 1982)—i.e., the capacity of a manipulator to produce an equal motion of the EE in all directions from an equal motion of its actuators—is one important design criterion, because the manipulator exhibits, at this state, its best kinematic performances. Unfortunately, this property depends on both the manipulator postures and geometry. A manipulator is said isotropic if there exist at least one posture within the workspace that reach isotropy. Most of the today's industrial manipulators do not admit any isotropic postures, and hence, are not isotropic, because their geometries don't allow to reach such a property. In this paper, we propose a set parameters allowing to describe all the geometries of manipulators producing a 3D translational mobility of the EE and having the same topology as the Delta parallel manipulator. Isotropic constraints are then applied on this design manifold in order to define the constraint manifold describing the isotopic designs.

2. Problem Formulation

The kinematics of robotic manipulators can be described with the concept of *kinematic chain*, which carries both topological and geometrical informations. A kinematic chain is defined as a mechanical system

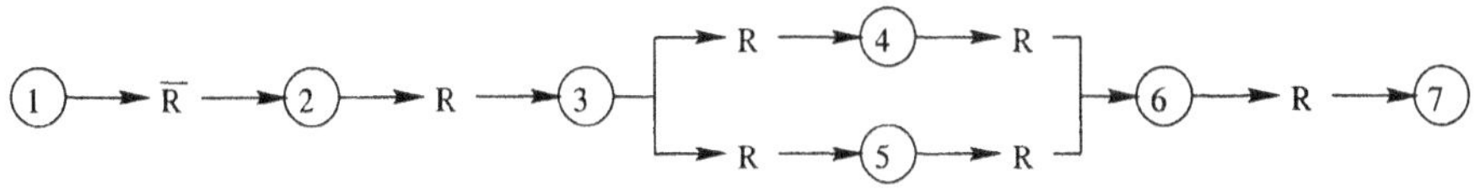

(a) Actual leg topology

(b) Simplified leg topology

Figure 2. Topology of Delta Parallel Manipulator

in which rigid bodies, called *links*, are coupled by *lower kinematic pairs*. There are six of such pairs, namely, revolute (R), prismatic (P), cylindrical (C), helicoidal (H), planar (E), and spherical (S). The *topology* refers to the layout of these pairs along the chain, while the *geometry* refers to the relative location of the pairs on each link. As shown in Fig. 2(a), the kinematic chain of the Delta parallel manipulator is composed of three serial chains in parallel of identical topology, i.e., an actuated R joint—denoted $\bar{R}$—connecting link 1 to link 2 (also called the *proximal* link), a passive R joint connecting link 2 to 3, two passive R joints connecting links 4 and 5 (also called the *distal* links) to link 6, and a passive R joint connecting link 6 to link 7, i.e., the EE. Since the EE is undergoing 3D translations, the motion of link 6 relative to link 3 of each leg can be regarded as the motion obtained with a circular prismatic pair (Hervé and Sparacino, 1985), denoted P_c, i.e., a circular path at constant orientation. The simplified leg topology is thus given as $\bar{R}RP_cR$ and shown in Fig. 2(b). The *geometrical synthesis* problem is to find values of a set of parameters describing the geometry of a manipulator that belong to a specific topological class. In this context, the following two definitions applies:

Definition 1 : Topological class

A topological class is the set of all mechanisms having the same topology and mobility of the EE disregarding links geometry.

Definition 2 : Minimal set of geometric parameters

A minimal set of geometric parameters must use the minimum number of parameters that allow to describe uniquely all the geometries of manipulators of a topological class.

Different sets of values of geometric parameters must describe different manipulators, and not only a displacement of frames. Consequently, a specific location of frame must be chosen without loss of generality. The next section proposes such a minimal set of geometric parameters.

3. Geometric Parameters

The geometry of the actuated joints on the base A can be described with the help of three arbitrarily located points $\{A_i\}_1^3$ together with three arbitrarily oriented unit vectors $\{\mathbf{e}_i\}_1^3$ along the joint axes and pointing toward the positive direction indicated by the right-hand rule. Without lost of generality, the following definition can be made

Definition 3 : Location of frame $\mathcal{A}$

The origin of frame $\mathcal{A}$ is located on the base A at the intersection point of $\mathbf{e}_1$ and the common perpendicular of $\mathbf{e}_1$ with $\mathbf{e}_2$, its x-axis aligned along $\mathbf{e}_1$ and its y-axis along the common perpendicular of $\mathbf{e}_1$ with $\mathbf{e}_2$, and directed from $\mathbf{e}_1$ toward $\mathbf{e}_2$.

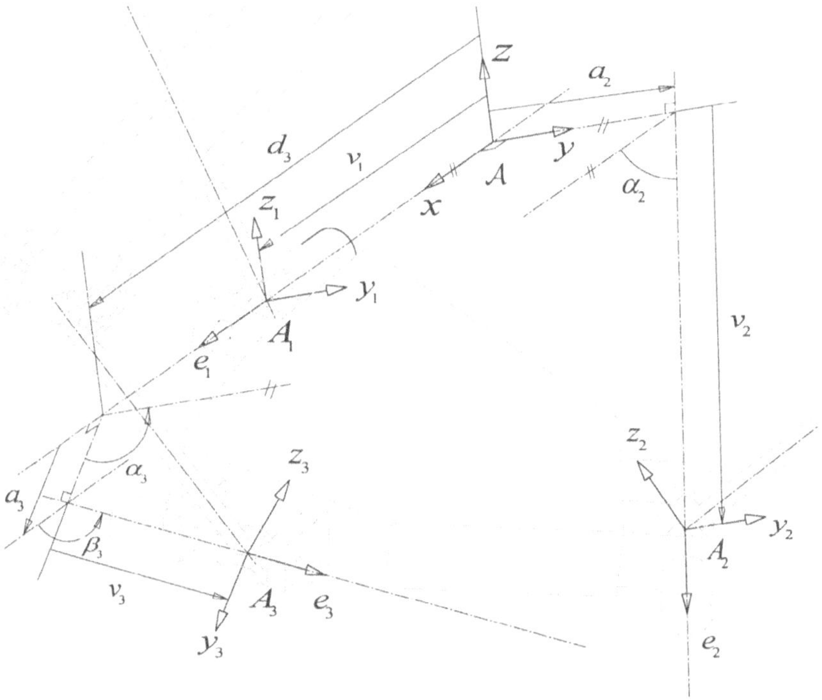

Figure 3. Geometric parameters of the base A

As shown in Fig. 3, the position vectors of points A_i in frame $\mathcal{A}$ are given as

$$\mathbf{a}_1 = v_1\mathbf{i}, \quad \mathbf{a}_2 = a_2\mathbf{j} + \mathbf{R}_y(\alpha_2)v_2\mathbf{i},$$
$$\mathbf{a}_3 = d_3\mathbf{i} + \mathbf{R}_x(\beta_3)(a_3\mathbf{j} + \mathbf{R}_y(\alpha_3)v_3\mathbf{i}) \tag{1}$$

while the orientation of the actuated joint axes in $\mathcal{A}$ are

$$\mathbf{e}_1 = \mathbf{i}, \quad \mathbf{e}_2 = \mathbf{R}_y(\alpha_2)\mathbf{i}, \quad \mathbf{e}_3 = \mathbf{R}_x(\beta_3)\mathbf{R}_y(\alpha_3)\mathbf{i}, \tag{2}$$

where v_1, a_2, α_2, v_2, d_3, β_3, a_3, α_3 and v_3 are the nine geometric parameters associated to the base, $\mathbf{R}_a(\beta)$ denotes the 3×3 rotation matrix of an angle β around axis a, while $\mathbf{i}$, $\mathbf{j}$, $\mathbf{k}$ are the unit vectors along x-, y- and z-axis, respectively.

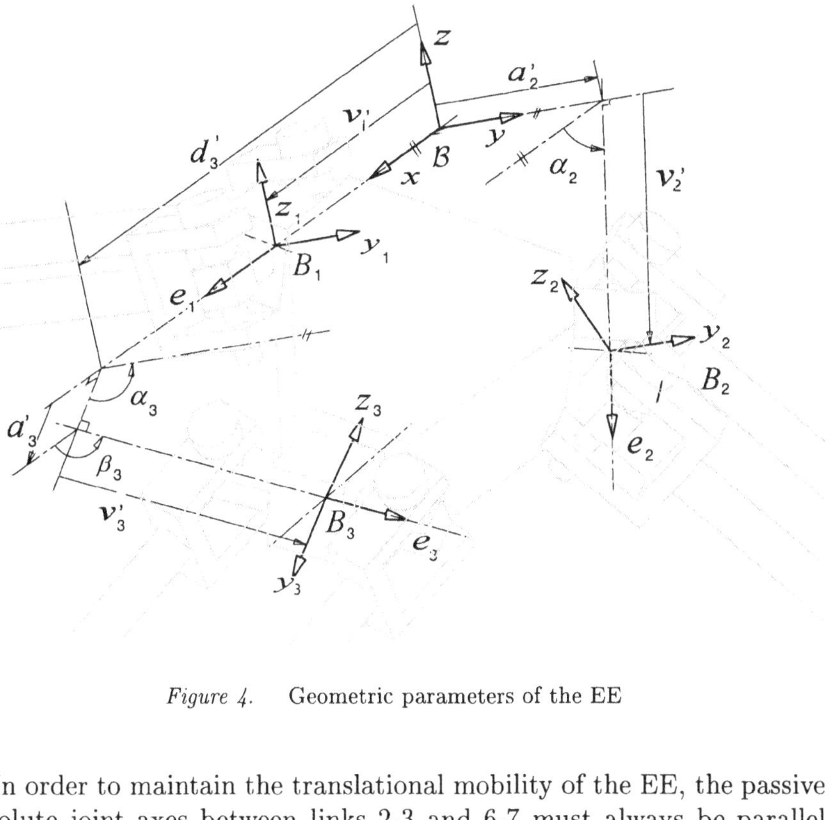

Figure 4. Geometric parameters of the EE

In order to maintain the translational mobility of the EE, the passive revolute joint axes between links 2-3 and 6-7 must always be parallel to the actuated joint axis $\mathbf{e}_i$ of the corresponding leg. Therefore, the rotation matrix describing in $\mathcal{A}$ the orientation of frame $\mathcal{B}$ attached to the EE, denoted B, is given as

$$\mathbf{R}_{\mathcal{B}}^{\mathcal{A}} = \mathbf{1}_{3\times3}. \tag{3}$$

In this context, the geometry of the EE can be described with the help of three arbitrarily located points $\{B_i\}_1^3$ together with the same unit vectors $\{\mathbf{e}_i\}_1^3$ along the passive revolute joint axes connecting links 6 to the EE. Without lost of generality, the following definition applies, i.e.,

Definition 4 : Location of frame $\mathcal{B}$

The origin of frame $\mathcal{B}$ is located on the EE, or body B, at the intersection point of $\mathbf{e}_1$ and the common perpendicular of $\mathbf{e}_1$ with $\mathbf{e}_2$, its x-axis aligned along $\mathbf{e}_1$ and its y-axis along the common perpendicular of $\mathbf{e}_1$ with $\mathbf{e}_2$, and directed from $\mathbf{e}_1$ toward $\mathbf{e}_2$.

As shown in Fig. 4, the position vectors of points B_i in frame $\mathcal{B}$ are given as

$$\mathbf{b}_1 = v_1'\mathbf{i}, \quad \mathbf{b}_2 = a_2'\mathbf{j} + \mathbf{R}_y(\alpha_2)v_2'\mathbf{i},$$
$$\mathbf{b}_3 = d_3'\mathbf{i} + \mathbf{R}_x(\beta_3)(a_3'\mathbf{j} + \mathbf{R}_y(\alpha_3)v_3'\mathbf{i}) \tag{4}$$

where v_1', a_2', v_2', d_3', a_3' and v_3' are the six geometric parameters associated to the EE.

The position of the end-point C_i of the proximal links (links 2 in Fig. 1) relative to points A_i are given as

$$\mathbf{c}_1 = \mathbf{R}_x(\theta_1)c_1\mathbf{j}, \quad \mathbf{c}_2 = \mathbf{R}_y(\alpha_2)\mathbf{R}_x(\theta_2)c_2\mathbf{j},$$
$$\mathbf{c}_3 = \mathbf{R}_x(\beta_3)\mathbf{R}_y(\alpha_3)\mathbf{R}_x(\theta_3)c_3\mathbf{j} \tag{5}$$

where c_1, c_2 and c_3 are the three geometric parameters associated with the proximal links; and θ_1, θ_2 and θ_3 the actuated joint positions. The length of the distal links d_i (links 4 and 5 in Fig. 1) are computed to close the kinematic loop of each leg i, when the EE, or frame $\mathcal{B}$, is arbitrarily located at $\mathbf{p}^0 = [x^0 \; y^0 \; z^0]^T$ for an arbitrarily actuated joint positions $\boldsymbol{\theta}^0 = [\theta_1^0 \; \theta_2^0 \; \theta_3^0]^T$, both called the home position of the manipulator. In summary, the minimal set of geometric parameters is composed of 24 parameters, i.e.: v_1, a_2, α_2, v_2, d_3, β_3, a_3, α_3, v_3, v_1', a_2', v_2', d_3', a_3', v_3', c_1, c_2, c_3, x^0, y^0, z^0, θ_1^0, θ_2^0 and θ_3^0, which define the design manifold of parallel manipulators of Delta topology.

4. Kinematic Model

Each leg of the parallel manipulator defines a kinematic loop passing through the origin of frames $\mathcal{A}$ and $\mathcal{B}$, and points A_i, C_i and B_i. As shown in Fig. 5, the closure of each kinematic loop can be expressed in the form

$$\mathbf{a}_i + \mathbf{c}_i + \mathbf{d}_i = \mathbf{p} + \mathbf{b}_i, \quad i = 1, 2, 3 \tag{6}$$

where $\mathbf{d}_i$ is the position vector, along the distal links, of point B_i relative to C_i in $\mathcal{A}$ and $\mathbf{p}$ the position vector of the origin of $\mathcal{B}$ in $\mathcal{A}$. At the home position, i.e., when $\{\theta_i\}_1^3 = \{\theta_i^0\}_1^3$, eq.(6) reduces to

$$\mathbf{a}_i + \mathbf{m}_i = \mathbf{p}^0 + \mathbf{b}_i, \tag{7}$$

where $\mathbf{p}^0$ is the home position of $\mathcal{B}$ in $\mathcal{A}$, and $\mathbf{m}_i$ is the home position of $\mathbf{c}_i + \mathbf{d}_i$, i.e., $\mathbf{m}_i = \mathbf{c}_i^0 + \mathbf{d}_i^0$. In general, the position of $\mathcal{B}$ in $\mathcal{A}$ can alternatively be written as

$$\mathbf{p} \equiv \mathbf{p}^0 + \boldsymbol{\Delta} \tag{8}$$

where $\boldsymbol{\Delta}$ is a displacement from $\mathbf{p}^0$. Substituting eqs.(8) and (7) into (6) yields

$$\mathbf{d}_i = \boldsymbol{\Delta} + \mathbf{m}_i - \mathbf{c}_i, \tag{9}$$

where $\mathbf{d}_i$ and $\mathbf{c}_i$ are the only two time-varying quantities, while the others are constant for a specific manipulator geometry.

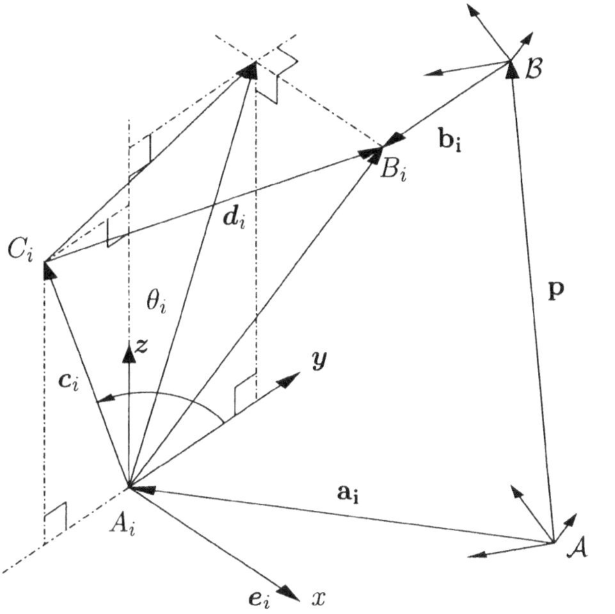

Figure 5. Kinematic loop

For trajectory planning, it is necessary to solve the inverse kinematic problem, i.e., the computation the joint variables $\{\theta_i\}_1^3$ from a desired EE position $\boldsymbol{\Delta}$. Upon taking the norm of both sides of eq.(9) to eliminate the unknown vector $\mathbf{d}_i$, we obtain

$$d_i^2 = \boldsymbol{\Delta}^T\boldsymbol{\Delta} + 2\boldsymbol{\Delta}^T\mathbf{m}_i + \mathbf{m}_i^T\mathbf{m}_i - 2(\boldsymbol{\Delta} + \mathbf{m}_i)^T\mathbf{c}_i + c_i^2 \tag{10}$$

where the unknown joint angle θ_i is contained in vector $\mathbf{c}_i$.

5. Jacobian Matrix

The Jacobian matrix is computed from the velocity relationship of the two end-points of the distal links, i.e., points C_i and B_i. Since the EE is undergoing translational motions only, all points of B undergo the same velocity as the origin of $\mathcal{B}$, and hence, the velocity of B_i in $\mathcal{A}$ is directly given as $\dot{\mathbf{p}}$. From eq.(5), the velocity of C_i in $\mathcal{A}$ can easily be written as

$$\dot{\mathbf{c}}_i = \frac{\partial \mathbf{c}_i}{\partial \theta_i}\dot{\theta}_i = \mathbf{e}_i \times \mathbf{c}_i\dot{\theta}_i \tag{11}$$

Since the distal link is rigid, the projection of the velocities of points C_i and B_i along the distal link must be equal, and hence, we have

$$\mathbf{d}_i^T\dot{\mathbf{p}} = \mathbf{d}_i^T\dot{\mathbf{c}}_i \tag{12}$$

which can be rewritten in a matrix form as

$$\mathbf{A}\dot{\mathbf{p}} = \mathbf{B}\dot{\boldsymbol{\theta}}, \tag{13}$$

where $\mathbf{A}$ and $\mathbf{B}$ are, respectively, the parallel and serial Jacobian matrices of the manipulator defined such as

$$\mathbf{A} \equiv \begin{bmatrix} \mathbf{d}_1^T \\ \mathbf{d}_2^T \\ \mathbf{d}_3^T \end{bmatrix}, \quad \mathbf{B} \equiv \begin{bmatrix} v_1 & 0 & 0 \\ 0 & v_2 & 0 \\ 0 & 0 & v_3 \end{bmatrix}, \tag{14}$$

$$v_i \equiv \mathbf{d}_i^T\mathbf{e}_i \times \mathbf{c}_i, \quad \dot{\boldsymbol{\theta}} \equiv [\dot{\theta}_1 \; \dot{\theta}_2 \; \dot{\theta}_3]^T \tag{15}$$

with $\dot{\theta}_i$ being the actuated-joint velocity. From eq.(13), it is clear that the numerical conditioning of the two Jacobian matrices of the manipulator depend on both the manipulator geometry and posture.

6. Isotropic Conditions

In order to design isotropic parallel manipulators, the Jacobian matrix, namely $\mathbf{J}$, must be isotropic at one or more posture(s) within the manipulator workspace, i.e.,

$$\dot{\mathbf{p}} = \mathbf{J}\dot{\boldsymbol{\theta}}, \quad \mathbf{J} \equiv \mathbf{A}^{-1}\mathbf{B} \tag{16}$$

For the sake of foregoing algebraic development, we rather use the inverse of $\mathbf{J}$, namely $\mathbf{J}^{-1}$, in order to define the isotropic constraints, i.e.,

$$\mathbf{J}^{-1} = \mathbf{B}^{-1}\mathbf{A} = \sigma\mathbf{Q}, \tag{17}$$

where $\sigma > 0$ and $\mathbf{Q}$ is orthogonal. Upon substituting eqs.(14) and (15) into (17) yields

$$\mathbf{J}^{-1} = \begin{bmatrix} \mathbf{d}_1^T/v_1 \\ \mathbf{d}_2^T/v_2 \\ \mathbf{d}_3^T/v_3 \end{bmatrix} = \sigma\mathbf{Q} \tag{18}$$

which is equivalent to the following two isotropic conditions, i.e.,

Condition 1 : Orthogonality of the distal links
The distal links axes must be orthogonal to each other, i.e.,

$$\mathbf{d}_i^T \mathbf{d}_j = 0, \qquad \forall\ i, j = 1, 2, 3 \text{ and } i \neq j. \tag{19}$$

Condition 2 : Identical effective proximal link length
The projection of the proximal link along the normal to the distal link parallelogram must be identical for all leg, i.e.,

$$c_i \cos\gamma_i = 1/\sigma, \qquad \forall\ i = 1, 2, 3 \tag{20}$$

where $c_i \equiv \|\mathbf{c}_i\|$ *and* γ_i *is the angle between* $\mathbf{c}_i$ *and* $\mathbf{e}_i \times \mathbf{d}_i$.

From eq.(18), condition 1 is obvious, while condition 2 is readily obtained as

$$\|\mathbf{d}_i^T / v_i\|_2 = d_i / v_i = 1/c_i \cos\gamma_i = \sigma, \tag{21}$$

where $v_i = c_i d_i \cos\gamma_i$ is the volume sustained by vectors $\mathbf{d}_i$, $\mathbf{e}_i$ and $\mathbf{c}_i$. Any parallel manipulator of Delta topology is therefore said isotopic if there exist at least one posture where the two isotropic conditions are simultaneously satisfied. For example, the original Delta parallel manipulator is isotropic at the posture shown in Fig. 6.

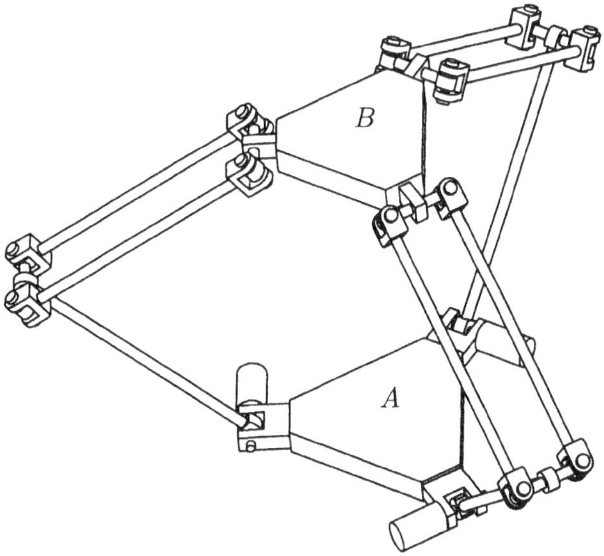

Figure 6. The Delta parallel manipulator at an isotropic posture

Another example is the new isotropic parallel manipulator of Delta topology shown in Fig. 7. This manipulator can produce the same translational mobility of the EE with three motorized revolute joints fixed on the base (not shown on the figure).

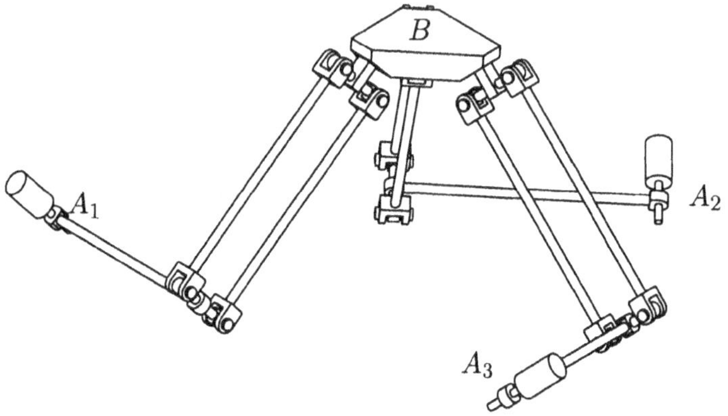

Figure 7. A new parallel manipulator of Delta topology at an isotropic posture

7. Conclusion

A minimal set of 24 geometric parameters has been proposed to describe the design manifold of parallel manipulators of the Delta topological class, i.e., manipulators having the same topology and mobility of the EE as the Delta parallel manipulator. Two isotropic conditions have been proposed to constraint this manifold to the isotropic designs.

8. Acknowledgements

The authors acknowledge the financial support of NSERC (National Sciences and Engineering Research Council of Canada) under grants OGPIN-203618 and RGPIN-138478, and FCAR (Fond Concerté d'aide à la Recherche of Québec) under grants NC-66861 and ER-3618.

References

Clavel, R. (1985), *Device For Displacing and Positioning an Element in Space*, International patent, No. WO 87/03528.

Salisbury, J.-K. and Craig, J.-J. (1982), Articulated Hands: Force Control and Kinematic Issues, *The Int. J. of Robotics Research*, vol. 1, no. 1, pp. 4–17.

Hervé, J.M. and Sparacino, (1985), Star, a new concept in robotics, *Advances in Robot Kinematics*, Ferrare, pp. 176–183.

Kinematic Analysis

J. Lenarčič and F. Thomas (eds.), Advances in Robot Kinematics, 367.

AN ACCURATE ALGORITHM FOR THE REAL-TIME SOLUTION OF THE DIRECT KINEMATICS OF 6-3 STEWART PLATFORM MANIPULATORS

R. Vertechy
DIEM, University of Bologna
Viale Risorgimento, 2, 40136 Bologna, Italy

G. R. Dunlop
Department of Mechanical Engineering, University of Canterbury
PB4800 Christchurch, New Zealand
email: r.dunlop@canterbury.ac.nz

V. Parenti-Castelli
DIEM, University of Bologna
Viale Risorgimento, 2, 40136 Bologna, Italy
email: vincenzo.parenticastelli@mail.ing.unibo.it

Abstract The direct kinematics of Stewart platform manipulators involves many solutions. However only the actual one is of practical interest for control purposes. This paper presents a simple and very fast procedure for estimating the actual configuration of 6-3 Stewart platform manipulators in real-time. The method minimizes the influence of measurement errors on the estimated pose of the end-effector. This is achieved through the use of three additional rotary sensors along with a proved simplifying hypothesis. By the addition of six extra rotary sensors, the method could by applied to 6-6 Stewart platform manipulators of general geometry. The effectiveness of the approach has been shown through Monte Carlo simulations and its performances have also been compared with other two recent algorithms proposed in the literature.

Keywords: real-time direct kinematics, Stewart Platform, extra sensors, errors.

1. Introduction

The Stewart Platform Manipulator (SPM) was introduced almost 40 years ago by Gough and Whitehall, 1962, and Stewart, 1965. The device features a mobile plate (platform) connected to a fixed plate (base) via six

J. Lenarčič and F. Thomas (eds.), Advances in Robot Kinematics, 369–378.

kinematic chains (legs). Each chain is connected to the base and to the platform, respectively, by means of universal and spherical pairs. In order to provide the platform with six degrees of freedom with respect to the base, the legs have adjustable length controlled by prismatic actuators. Different SPM are possible according to the leg attachment point positions on the base and on the platform. The 6-3 SPM, which is of interest in this paper, has six and three attachment points located on the base and on the platform respectively.

A topic of practical interest is the direct kinematic problem (DKP) that is finding the SPM configuration when the lengths of the six actuated legs are given. The problem involves the resolution of a system of non-linear implicit equations in the parameters describing the pose (position and orientation) of the platform and it is a challenging task. In fact several solutions are possible. However, only the actual configuration of the manipulator is of practical interest for control purposes.

As a matter of fact, Griffis and Duffy, 1989, Nanua et al., 1990, and Innocenti and Parenti-Castelli, 1990, solved in analytical form the DKP of the 6-3 SPM showing that sixteen solutions are possible. This was accomplished by finding the roots of a polynomial equation in one unknown. This approach is as robust as the root-solver is. Indeed, one-dimensional numerical methods, for instance Newton-Raphson, invariably fail near singularities of the manipulator, because the root being tracked makes the polynomial pathologically ill-conditioned. Moreover, once all the possible configurations are found, the actual configuration of the manipulator must be determined somehow by the use of additional information.

Even with the best current computer technology, polynomial-based approaches take considerable time to determine the actual posture of the 6-3 SPM and are therefore unsuited to real-time applications. For real-time applications, iterative methods (mostly Newton-Raphson) have been proposed to solve the system of non-linear equations that describe the DKP. These methods have several drawbacks: (1) they fail near a singularity, (2) at this time there is no way to determine the exact size of the convergence domain (of the numerical scheme) that is needed to ensure that the final pose of the moving platform will be the actual one, and finally (3) after a fixed number of iterations the error bound is not guaranteed.

Closed-form solutions of the DKP can be also obtained by suitably placing extra-sensors (Podhorodeski, 1992; Merlet, 1993) in addition to the six transducers measuring the leg lengths. Extra sensors allow reduction of the computational complexity as well as make it possible to obtain a one-to-one relation between active joint variables and the actual platform configuration. Because of the hardware cost and the need to calibrate and operate a complex sensory system, the number and the location of extra-sensors have to be carefully selected in order to find one solution (the actual one) of the DKP as well as to reduce the sensitivity to both computation and transducer accuracy. Innocenti, 1998, Parenti-Castelli and Di Gregorio, 1996, Han et al., 1996, have shown that, except for some particular configurations, the DKP of a SPM can be solved by the addition of only one extra-sensor, but these schemes are highly sensitive to the measurement noise and also to round-off errors. In order to minimize this sensitivity, a different algorithm, based on a number of extra sensors, has been proposed by Baron and Angeles, 2000. In fact, dealing with a more general manipulator, namely the 6-6 SPM (six attachment points both on the base and on the platform), they devised a procedure that is robust to measurement errors and that produces estimates whose accuracy, generally high, depends on the number and location of the transducers.

In this paper, a novel method is presented for searching the actual pose of a 6-3 SPM by using three additional angular sensors. It is noteworthy that this scheme will not provide the actual solution of the DKP, but rather only estimate thereof.

The proposed procedure has high accuracy and low sensitivity to noise in sensor readouts. Moreover, it is computationally more efficient than the previous algorithms. Examples are reported that show the efficiency and accuracy of the presented procedure.

2. Manipulator and Geometric Formulation of its Actual Configuration

From a kinematic point of view, the 6-3 SPM can be represented as in Fig. 1. Without loss of generality, we have assumed that it comprises a semi-regular hexagonal base (i.e. a base with sides alternatively of the same length) and an equilateral platform whose vertices are B_i ($i = 1,\ldots, 6$) and P_j ($j = 1, 3, 5$) respectively. In the same figure, the segments connecting points B_i and P_j correspond to the actuated manipulator legs whose length q_i is therefore variable. Legs coalesce pairwise at points P_j. In order to describe the manipulator pose we have attached the *O-XYZ* frame over the base and the *C-nsr* frame over the platform (*C* is the centroid of P_j). Therefore, the position and orientation of the platform can be expressed, respectively, by the vector $c = C - O$ and by the rotational matrix $\mathbf{R}$ describing the orientation of the *C-nsr* frame with respect to the *O-XYZ* frame.

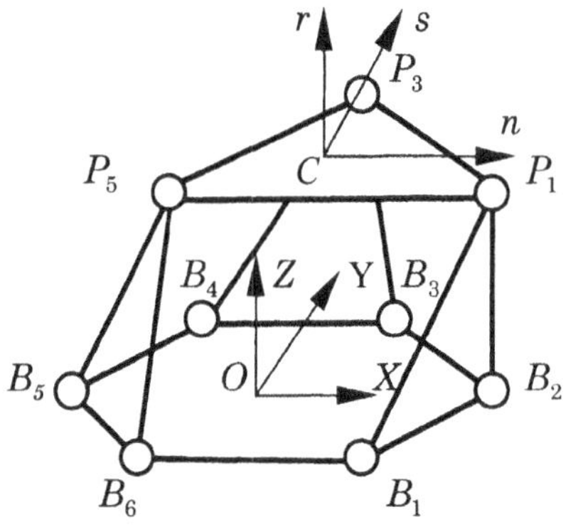

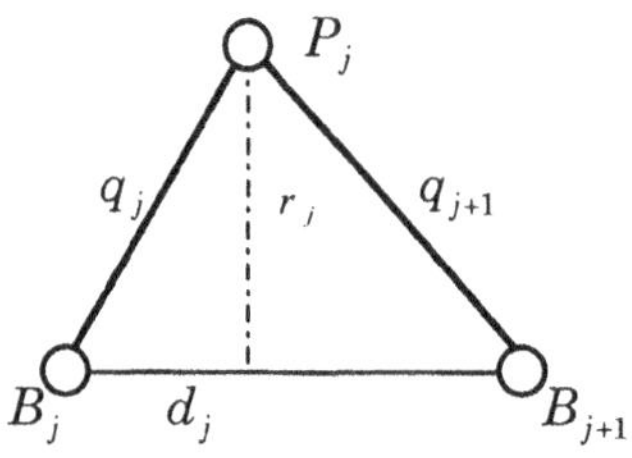

Figure 1. 6-3 Stewart platform *Figure 2.* Triangle $B_jP_jB_{j+1}$

Since the pose of a rigid body is determined once the position of three non-collinear points is identified, knowing the location of the three vertices P_j allows determining the platform pose (vector $\mathbf{c}$ and matrix $\mathbf{R}$) as follows:

$$\mathbf{c} = \frac{1}{3}\sum_{j=1,3,5} \mathbf{P}_j \quad , \quad \mathbf{R} = [\hat{\mathbf{n}} \;\; \hat{\mathbf{s}} \;\; \hat{\mathbf{r}}] \tag{1.a}$$

where $\mathbf{P}_j = P_j - O$ and

$$\hat{\mathbf{n}} = \frac{\mathbf{P}_1 - \mathbf{P}_5}{\|\mathbf{P}_1 - \mathbf{P}_5\|}, \; \hat{\mathbf{s}} = \frac{(\mathbf{P}_3 - \mathbf{P}_5) - \hat{\mathbf{n}}[(\mathbf{P}_3 - \mathbf{P}_5)\cdot\hat{\mathbf{n}}]}{\|(\mathbf{P}_3 - \mathbf{P}_5) - \hat{\mathbf{n}}[(\mathbf{P}_3 - \mathbf{P}_5)\cdot\hat{\mathbf{n}}]\|}, \; \hat{\mathbf{r}} = \hat{\mathbf{n}} \times \hat{\mathbf{s}}. \tag{1.b}$$

It is worth noticing that for a given set of leg lengths the 6-3 SPM becomes a structure. If the platform is momentarily removed from the legs at P_j, each of these points can describe a circle whose center and radius depend on the given set of leg lengths. As shown in Fig. 2 for j = 1, 3, 5, the location of each vertex P_j can be parameterised by an angle ϑ_j and two distances d_j and r_j. The angle ϑ_j measures the slope of the triangle $B_jP_jB_{j+1}$ with respect to the base plate, while d_j and r_j can be easily expressed as functions of the leg lengths q_j and q_{j+1}. In fact, considering the triangle $B_jP_jB_{j+1}$, we have:

$$d_j = \frac{q_j^2 - q_{j+1}^2 + |B_jB_{j+1}|^2}{2|B_jB_{j+1}|} \quad \text{and} \quad r_j = \sqrt{q_j^2 - d_j^2}\,. \tag{2}$$

Introducing $\boldsymbol{q} = [q_1 \quad q_2 \quad q_3 \quad q_4 \quad q_5 \quad q_6]^T$ and $\boldsymbol{\vartheta} = [\vartheta_1 \quad \vartheta_3 \quad \vartheta_5]^T$, we can write:

$$\boldsymbol{c} = \boldsymbol{c}(\boldsymbol{q}, \boldsymbol{\vartheta}) \text{ and } \mathbf{R} = \mathbf{R}(\boldsymbol{q}, \boldsymbol{\vartheta}). \tag{3}$$

If we measure both $\boldsymbol{q}$ and $\boldsymbol{\vartheta}$ and if we suppose the absence of measurement noise, the resolution of the DKP is straightforward. However, in real applications, the transducer readouts, $\boldsymbol{q}^m$ and $\boldsymbol{\vartheta}^m$, are always affected by errors and therefore the measured pose, i.e.

$$\boldsymbol{c}^m = \boldsymbol{c}(\boldsymbol{q}^m, \boldsymbol{\vartheta}^m) \text{ and } \mathbf{R}^m = \mathbf{R}(\boldsymbol{q}^m, \boldsymbol{\vartheta}^m) \tag{4}$$

will be highly inaccurate.

3. Alternative Formulation of the Direct Kinematic Problem

The pose of the mobile plate can be described by the generalized coordinate vector $\boldsymbol{X} = [\boldsymbol{c} \quad \boldsymbol{a}]$, where $\boldsymbol{a}$ is a vector containing the three orientation parameters. Thus, the platform vertices can be expressed as $\boldsymbol{P}_j = \boldsymbol{P}_j(\boldsymbol{X})$.

Due to noise, the measured points P_j^m will define a triangle that differs from the platform. Then, the DKP can be thought as the problem of finding the estimate $\boldsymbol{X}^E$ that minimizes the functional:

$$\Pi(\boldsymbol{X}) = \sum_{j=1,3,5} \left[P_j(\boldsymbol{q}^m, \boldsymbol{\vartheta}^m) - P_j(\boldsymbol{X})\right]^2. \tag{5}$$

Finding the minimum of a function in six variables is computationally very expensive and requires iterative schemes. However, according to the following consideration, the number of variables can be reduced. In fact, even using low precision sensors it is possible to accurately determine the plane in which the platform lies (the accuracy is greater if the triangles $B_jP_jB_{j+1}$ are nearly perpendicular to the platform plate). Since the pose of a rigid body is completely determined, in a plane, by three parameters, then the function to be minimized can be expressed as dependent on only three variables which describe the pose of the platform with respect to the following appropriate reference frame. As shown in Fig. 3 we place an origin at the measured point P_5^m. Then, we set the first axis $\hat{\boldsymbol{u}}$ along the segment $P_5^m P_1^m$ with the same direction. The second axis $\hat{\boldsymbol{v}}$ through P_5^m is orthogonal to the first and its direction lies on the same side as P_3^m.

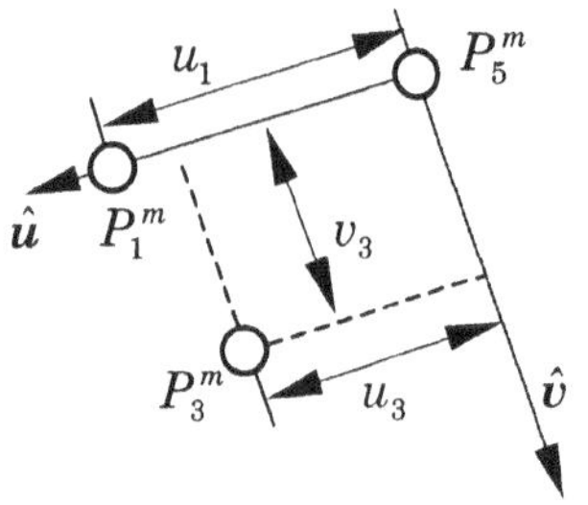

Figure 3. Reference frame

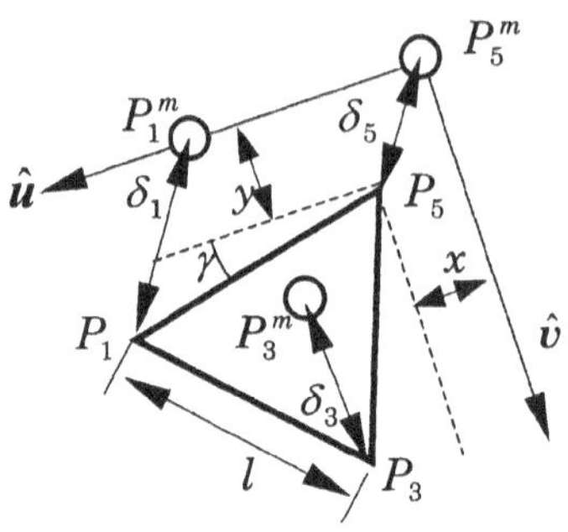

Figure 4. Platform pose

Therefore, we have the following unit vectors:

$$\hat{\boldsymbol{u}} = \frac{\boldsymbol{P}_1^m - \boldsymbol{P}_5^m}{u_1} \text{ and } \hat{\boldsymbol{v}} = \frac{\left(\boldsymbol{P}_3^m - \boldsymbol{P}_5^m\right) - u_3\hat{\boldsymbol{u}}}{v_3} \tag{6.a}$$

where u_1, u_3 and v_3 are the constants:

$$u_1 = \left\|\boldsymbol{P}_1^m - \boldsymbol{P}_5^m\right\|,\ u_3 = \left(\boldsymbol{P}_3^m - \boldsymbol{P}_5^m\right)\cdot\hat{\boldsymbol{u}},\ v_3 = \left\|\left(\boldsymbol{P}_3^m - \boldsymbol{P}_5^m\right) - u_3\hat{\boldsymbol{u}}\right\|. \tag{6.b}$$

Here, referring to the new frame, u_1 and u_3 are, respectively, the components of P_1^m and P_3^m along $\hat{\boldsymbol{u}}$, while v_3 is the component of P_3^m along $\hat{\boldsymbol{v}}$. Then, we express the three vertices of the platform, namely P_j, in the new reference frame as $\boldsymbol{P}^*_j$ (note that each P_j is related to P_j^m):

$$\boldsymbol{P}^*_5 = \begin{bmatrix} x \\ y \end{bmatrix},\ \boldsymbol{P}^*_1 = \begin{bmatrix} x + l\cos\gamma \\ y + l\sin\gamma \end{bmatrix} \text{ and } \boldsymbol{P}^*_3 = \begin{bmatrix} x + l\cos(\gamma + \pi/3) \\ y + l\sin(\gamma + \pi/3) \end{bmatrix} \tag{7}$$

where x and y are the coordinates of P_5 in the new frame, γ is the angle between the segment of the equilateral triangle (i.e. P_5P_1) and the axis $\hat{\boldsymbol{u}}$, while l is the length of the edge of the mobile plate. Thus, posing $\boldsymbol{x} = [x \quad y \quad \gamma]^T$ and

$$\boldsymbol{P}^{*m}_5 = \begin{bmatrix} 0 \\ 0 \end{bmatrix},\ \boldsymbol{P}^{*m}_1 = \begin{bmatrix} u_1 \\ 0 \end{bmatrix},\ \boldsymbol{P}^{*m}_3 = \begin{bmatrix} u_3 \\ v_3 \end{bmatrix} \tag{8}$$

the function to be minimized is:

$$\Pi(\boldsymbol{x}) = \sum_{j=1,3,5} \left|\boldsymbol{P}^{*m}_j - \boldsymbol{P}^*_j\right|^2 \tag{9}$$

which, as said before, is a function in three variables. Even considering low precision sensors, at least in the range of interest, the values of γ will be small. Therefore, after having plugged Eq. 7 and Eq. 8 into Eq. 9, we can replace, in Eq. 9, the trigonometric functions $\sin(\gamma)$ and $\cos(\gamma)$ with their Taylor expansions and then, discarding the terms whose order is greater than the second, the functional Π can be rewritten as a functional $\tilde{\Pi}$ quadratic in the variables $\boldsymbol{x}$:

$$\tilde{\Pi}(\boldsymbol{x}) = Ax^2 + By^2 + C\gamma^2 + Dx\gamma + Ey\gamma + Fx + Gy + H\gamma + I \tag{10}$$

where A, B, C, D, E, F, G, H, I are constant.

Since $\tilde{\Pi}(\boldsymbol{x})$ is a multivariable function, its minima are reached when the gradient $\nabla\tilde{\Pi} = \boldsymbol{0}$. Considering the Taylor series expansion for $\nabla\tilde{\Pi}(\boldsymbol{x})$ and recalling that $\tilde{\Pi}(\boldsymbol{x})$ is quadratic in the variables, we have:

$$\nabla\tilde{\Pi}(\boldsymbol{x}) = \nabla\tilde{\Pi}(\boldsymbol{x}_0) + \tilde{\mathbf{H}}(\boldsymbol{x} - \boldsymbol{x}_0) \tag{11}$$

where $\tilde{\mathbf{H}}$, here a constant matrix, is the Hessian of $\tilde{\Pi}$.

Forcing $\nabla\tilde{\Pi} = \boldsymbol{0}$, the minimum $\boldsymbol{x}_{\min}$ can be found as:

$$\boldsymbol{x}_{\min} = \boldsymbol{x}_0 - \tilde{\mathbf{H}}^{-1}\nabla\tilde{\Pi}(\boldsymbol{x}_0). \tag{12}$$

Notice that $\boldsymbol{x}_0$ is a near guess for the minimum. Since the Hessian matrix, in this case, is quite simple, we notice that we can evaluate its inverse, via symbolic calculations. Therefore, we have:

$$\tilde{\mathbf{H}}^{-1} = \frac{\mathrm{Adj}(\tilde{\mathbf{H}})}{\det(\tilde{\mathbf{H}})} \tag{13}$$

where Adj is the adjoint matrix of $\tilde{\mathbf{H}}$.

Once we know $\boldsymbol{x}_{\min}$, considering only the first order terms of the Taylor expansions of the trigonometric functions $\sin(\gamma)$ and $\cos(\gamma)$, the correspondent position of the platform vertices $\tilde{\boldsymbol{P}}_j$ in the aforementioned plane can be written as:

$$\tilde{\boldsymbol{P}}_5 = \begin{bmatrix} x_{\min} \\ y_{\min} \end{bmatrix}, \; \tilde{\boldsymbol{P}}_1 = \begin{bmatrix} x + l \\ y_{\min} + l\gamma_{\min} \end{bmatrix} \text{ and } \tilde{\boldsymbol{P}}_3 = \begin{bmatrix} x_{\min} + \left(1 - \sqrt{3}\gamma_{\min}\right)l/2 \\ y_{\min} + \left(\sqrt{3} + \gamma_{\min}\right)l/2 \end{bmatrix} \tag{14}$$

while regarding the estimated vector $\boldsymbol{P}_j^E$ we have:

$$\boldsymbol{P}_j^E = \boldsymbol{P}_5^m + \widetilde{P}_{j_u}\hat{\boldsymbol{u}} + \widetilde{P}_{j_v}\hat{\boldsymbol{v}} \tag{15}$$

where $\tilde{P}_{j_u}$ and $\tilde{P}_{j_v}$ are the components of $\widetilde{\boldsymbol{P}}_j$ in the reference system centered in P_5^m (see Fig. 4) with respect to $\hat{\boldsymbol{u}}$ and $\hat{\boldsymbol{v}}$ axes respectively.

From Eq. 1, it is then straightforward determining the posture of the EE. It is noteworthy that the procedures do not provide the actual solution of the DKP, but rather only estimate thereof.

4. Performance Evaluation

The manipulator we have chosen for the simulations has an equilateral platform whose vertices lie on a circle whose radius is $R_p = 450mm$, while the base is a semi-regular hexagon, whose vertices are grouped by pairs, which are uniformly spaced along a circle whose radius is $R_B = 1000mm$. Regarding the base, the offset angles between B_2 and B_3, B_4 and B_5, B_6 and B_1 have been set to $\pi/24$ *rad* (the angles are measured from the centre of the base circle).

In order to prove the effectiveness of our approach (*method 1*), its performances have been compared to the ones regarding two other different methods found in the literature. The first (*method 2*) has been derived through our application to the 6-3 SPM of the method proposed for the 6-6 SPM by Baron and Angeles, 2000. The second (*method 3*) is the algorithm suggested by Han et al., 1996. The three procedures have been implemented in Matlab programs with the aim of evaluating their performance. According to Baron and Angeles, 2000, we have considered two indices: the numerical efficiency of the procedures and the accuracy of the estimated pose while using simulated data.

4.1 Numerical Efficiency

Our method (*1*) involves 89 multiplications/divisions, 73 additions/subtractions, 6 trigonometric evaluations and 5 root extractions; therefore, a total of 168 *flops* and 5 transcendental operations. Notice that these values are strictly constant.

Conversely, the algorithm derived from the one proposed by Baron and Angeles (*2*) requires 105 multiplications/divisions, 93 additions/ subtractions, 6 trigonometric evaluations, 3 root extractions and, with regards to Baron and Angeles, 2000, an average of 177 flops for the polar decomposition algorithm that this method requires. Therefore, the mean value for the number of floating point operations is 381 *flops* and 3 transcendental operations.

Finally, the method (*3*) proposed by Han et al., 1996, requires 138 multiplications/divisions, 129 additions/subtractions, 2 trigonometric evaluations and 15 root extractions; therefore, a total of 269 *flops* and 15 transcendental operations.

4.2 Estimation Accuracy

According to Baron and Angeles, 2000, the estimation accuracy is measured by the position, δ, and orientation, ε, errors between the estimated and the actual pose, i.e.

$$\delta = \left\| \mathbf{c} - \mathbf{c}^{E} \right\| \text{ and } \varepsilon = \left\| \text{vect}\left(\left(\mathbf{R}^{E} \right)^{T} \mathbf{R} \right) \right\| \tag{16}$$

where vect$(\cdot)$ is the axial vector of its 3 x 3 matrix argument. The estimates are computed from data simulated through the addition of uniformly distributed noise between 0 and 0.02 *mm/rad* to the values of the transducer variables that correspond, in an ideal case, to the actual configuration. This latter is known since it coincides with the desired trajectory.

The data reported in Tab. 1 show that, when considering noise in the measurements, the method proposed by Han (*method 3*) is inaccurate and not robust. Moreover, it is possible to show that some configurations exist in which the use of only one extra sensor does not allow Han's method to solve uniquely the DKP. Conversely, our approach (*method 1*) and the one derived by Baron and Angeles (*method 2*) behave in the same way. In particular they show better accuracy and do not fail at any configuration.

Table 1. *Accuracy results*

	method 1	*method 2*	*method 3*
mean(δ) [*mm*]	5.40	5.39	8.12
mean(ε) [*rad*]	2.14 e-2	2.15 e-2	3.13 e-2

5. Conclusions

In this paper the determination of the actual pose of the 6-3 SPM with equilateral platform is addressed by using three extra rotary sensors in addition to the transducers included in the leg hardware equipment. The method can be easily extended to the 6-3 SPM with general geometry. Moreover, if we add a total of six extra rotary sensors, the method can be applied to the 6-6 SPM of general geometry. In fact, in this latter case, in order to measure the position of three hip-attachment points over the platform we have to add, respectively to three legs, two rotary sensors at each leg universal joint. The method we have proposed is fast and it does not require any iteration. Moreover it is robust to measurement errors and produces estimates with high accuracy. Comparison with other methods shows that our approach is computationally faster and beside, at least, equal in terms of sensitivity to measurement errors. The algorithm is therefore well suited when dealing with real-time applications for control purpose.

Acknowledgments

The funds of the Italian MURST and CNR are gratefully acknowledged.

References

Baron, L., and Angeles, J., (2000), The Kinematic Decoupling of Parallel Manipulators Under Joint-Sensor Data, in *IEEE Transactions on Robotics and Automation*, no. 6, vol. 16, pp. 644-651.

Gough, V.E., and Whitehall, S.G., (1962), Universal Tire Test Machine, *Proc. 9th Int. Technical Congress F.I.S.I.T.A*, vol. 117, pp. 117-135.

Griffis, M., and Duffy, J., (1989), A forward displacement analysis of a class of Stewart platform, *Journal of Robotics systems*, vol. 6, pp. 703-720.

Han, H., Chung, W., Youm, Y., (1996), New Resolution Scheme of the Forward Kinematics of Parallel Manipulators Using Extra Sensors, in *Journal of Mechanical Design*, vol. 118, pp. 214-219.

Innocenti, C., and Parenti-Castelli, V., (1990), Direct Position Analysis of the Stewart Platform Mechanism, *Mechanism and Machine Theory*, no. 6, vol. 25, pp. 611-621.

Innocenti, C., (1998), Closed-Form Determination of the Location of a Rigid Body by Seven in Parallel Linear Transducers, *Journal of Mechanical Design*, vol. 120, pp. 293-298.

Merlet, J.-P., 1993, Closed-form resolution of the direct kinematics of planar parallel manipulators using extra sensors data, *Proc of the IEEE Int. Conf. on Robotics and Automation*, Atlanta, pp. 200-204.

Nanua, P., Waldron, K.J., and Murty, V., (1990), Direct Solution of a Stewart Platform, *IEEE Transactions on Robotics and Automation*, no. 4, vol. 6, pp. 438-443.

Parenti-Castelli, V., and Di Gregorio, R., (1996), A Real-Time Computation Scheme for the Direct Position Analysis of the 6-3 Stewart Platform, *27th International Symposium on Industrial Robots, ISIR*, Milan, pp. 581-585.

Podhorodeski, R.P., (1992), Three branch hybrid-chain manipulators: Structure, displacement, uncertainty and redundancy related concerns, *Proc. of the 3rd ARK*, Ferrara, pp. 150-156.

Stewart, D., (1965), A Platform with Six Degree of Freedom, *Proc. of the Institution of Mechanical Engineers*, vol. 180 (Part 1, 15), pp. 371-386.

3-LEGGED SPATIAL 4-BAR PLATFORM KINEMATICS

P.J. Zsombor-Murray
McGill University
Department of Mechanical Engineering
Centre for Intelligent Machines
817 Sherbrooke St. W.
Montréal (Québec) Canada, H3A 2K6
paul@cim.mcgill.ca

J. Cervantes-Sanchez
Visiting Professor
McGill University
Centre for Intelligent Machines
3480 University St.
Montréal (Québec) Canada, H3A 2K6
jecer@cim.mcgill.ca

Abstract An R-S-S-R spatial four-bar mechanism provides a revolute actuated one dof platform attached to the S-S coupler with a third C-S leg. All three spherical joints are on the platform. This mechanism may have eight real, distinct assembly modes but the solution reduces to a univariate quartic equation which arises from the coplanar intersection of a circle and an ellipse. This input-output analysis is the direct kinematics(DKP) of three legged, six degree of freedom parallel platforms with 3-2-1 basal actuation and up to eight assemblies.

Keywords: mechanism, manipulator, parallel, kinematics, direct, geometry

1. Introduction

[Sandor and Zhuang, 1984] studied one degree of freedom(dof) platform mechanisms with two R-S and one C-S legs. C, R and S stand for cylindrical, revolute and spherical joint. There was no systematic input-output analysis, nor was any cited. A literature search revealed none. However this mechanism is ideal to show formulation of constraint equations in the projective space, with homogeneous coordinates of points, planes and lines. Then it is shown how to derive the quartic univariate, via dialytic elimination [Salmon, 1885], to neatly structure the symbolic computation of the five coefficients. The eight possible assembly modes

J. Lenarčič and F. Thomas (eds.), Advances in Robot Kinematics, 379–386.

of this one dof platform mechanism arise from the intersection between a coplanar circle and ellipse combined with the two solutions afforded by the kernel R-S-S-R spatial four bar mechanism. Consider a typical layout as shown in Fig. 1.

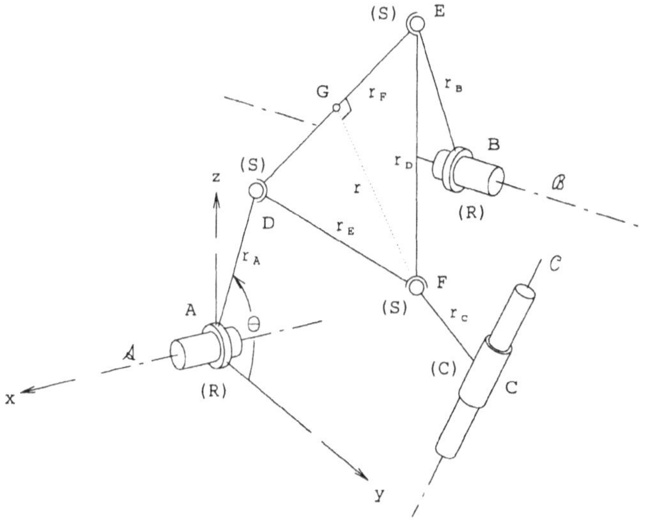

Figure 1. R-S-S-R Spatial 4-Bar with a Third C-S Leg

The actuated R-joint is centered on an origin at A. The joint axis is coincident with the fixed frame(FF) coordinate x-axis, along line $\mathcal{A}$. θ is the given input angle, measured from plane $z = 0$ to the input link AD where AD is perpendicular to $\mathcal{A}$ and terminates on a coupler S-joint at D. Another S-joint is centered on E. Lengths of the input link $AD = r_A$ and the coupler $DE = r_F$ are given as is that of the output link $BE = r_B$. The free output R-joint, axis on given line $\mathcal{B}$, is centred on given point B. Output link BE rotates in a plane perpendicular to $\mathcal{B}$. First one finds the positions of E as a function of θ. The third S-joint is centred on point F, located by two given sides r_E and r_D of the platform S-joint triangle. A secondary output link CF connects the free C-joint on given axis line $\mathcal{C}$. Length $CF = r_C$ is also given and CF is taken as perpendicular to $\mathcal{C}$. Locating point F, hence point C, completes the computational procedure. Point G, at the foot of the perpendicular from F to the coupler line DE, is the centre of a circle produced by revolving F on axis DE. The plane of this circle cuts line $\mathcal{C}$ at the centre of an ellipse formed by the plane section on a cylinder of revolution traced by F when free to be moved by the C-joint.

2. Position of Point E

First the input-output equation of the R-S-S-R spatial four-bar linkage is solved. Recall that the given input angle is θ and the actuated revolute axis is on $\mathcal{A}\{1:0:0:0:0:0\}$ where the origin is $A\{1:0:0:0\}$. The plane of revolution of input link AD is $\Pi_{AD}\{0:1:0:0\}$. These specifications are given as homogeneous radial line, point and plane coordinates. Now the coordinates of point D can be witten immediately.

$$D\{1:0:r_A\cos\theta:r_A\sin\theta\}\equiv\{d_0:d_1:d_2:d_3\}$$

The analysis that follows may seem roundabout to readers unfamiliar with the projective space but such an approach leads to tight, efficient code in the end. The procedure can be abbreviated as

$$E=((K_{DE}-K_{BE})\cap\Pi_{BE}\{B\cap\overline{\mathcal{B}})\cap K_{DE}$$

A sphere K_{DE}, centred on D with radius r_F describes a constraining surface on the output link's spherical joint at $E\{e_0:e_1:e_2:e_3\}\in K_{DE}$. This produces a sphere equation which will be solved as a univariate in the x-coordinate, e_1, of the point E.

$$\begin{aligned}K_{DE}: d_0^2e_1^2-2d_0d_1e_0e_1+d_0^2e_2^2-2d_0d_2e_0e_2+d_0^2e_3^2-2d_0d_3e_0e_3\\+(d_1^2+d_2^2+d_3^2-d_0^2r_F^2)e_0^2=0\end{aligned}\quad(1)$$

Similarly, sphere K_{BE} centred on B with radius r_B provides a second constraint on $E\in K_{BE}$.

$$\begin{aligned}K_{BE}: b_0^2e_1^2-2b_0b_1e_0e_1+b_0^2e_2^2-2b_0b_2e_0e_2+b_0^2e_3^2-2b_0b_3e_0e_3\\+(b_1^2+b_2^2+b_3^2-b_0^2r_B^2)e_0^2=0\end{aligned}\quad(2)$$

The equation of plane Π_E on circle $k_E=K_{DE}\cap K_{BE}$ is produced by the difference $\Pi_E=b_0^2K_{DE}-d_0^2K_{BE}=0$.

$$\begin{aligned}\Pi_E: 2(d_0^2b_0b_1-b_0^2d_0d_1)e_0e_1+2(d_0^2b_0b_2-b_0^2d_0d_2)e_0e_2+2(d_0^2b_0b_3\\-b_0^2d_0d_3)e_0e_3+[b_0^2(d_1^2+d_2^2+d_3^2-d_0^2r_F^2)-d_0^2(b_1^2+b_2^2+b_3^2-b_0^2r_B^2)]e_0^2=0\end{aligned}\quad(3)$$

or $E_1e_1+E_2e_2+E_2e_3+E_0e_0=0$ which yields plane coordinates

$$\Pi_E\{E_0:E_1:E_2:E_3\}$$

Now a line $\mathcal{E}=\Pi_E\cap\Pi_{BE}$ is defined to produce the point $E=\mathcal{E}\cap K_{DE}$. Note that plane $\Pi_{BE}=B\cap\overline{\mathcal{B}}$ is on R-joint centre B and is perpendicular to axis $\mathcal{B}$ hence $E\in\Pi_{BE}$. $\overline{\mathcal{B}}$, on ω the plane at infinity, is a line which closes all planes perpendicular to line $\mathcal{B}$. If the radial coordinates

$$\mathcal{B}\{b_{01}:b_{02}:b_{03}:b_{23}:b_{31}:b_{12}\}$$

are given then the axial coordinates

$$\overline{\mathcal{B}}\{\overline{B}_{01}:\overline{B}_{02}:\overline{B}_{03}:\overline{B}_{23}:\overline{B}_{31}:\overline{B}_{12}\}\equiv\{b_{01}:b_{02}:b_{03}:0:0:0\}$$

are available by inspection so as to produce the plane coordinates

$$\Pi_{BE}\{B_0 : B_1 : B_2 : B_3\}$$

as the following intersection.

$$B_i = \sum_{j=0}^{3} b_j \overline{B}_{ij}, \quad B_0 = b_1 b_{01} + b_2 b_{02} + b_3 b_{03},$$
$$B_1 = -b_0 b_{01}, \quad B_2 = -b_0 b_{02}, \quad B_3 = -b_0 b_{03}$$

The axial line coordinates $\mathcal{E}\{E_{01} : E_{02} : E_{03} : E_{23} : E_{31} : E_{12}\}$ are

$$E_{01} = B_0 E_1 - E_0 B_1, \quad E_{02} = B_0 E_2 - E_0 B_2, \quad E_{03} = B_0 E_3 - E_0 B_3,$$
$$E_{23} = B_2 E_3 - E_2 B_3, \quad E_{31} = B_3 E_1 - E_3 B_1, \quad E_{12} = B_1 E_2 - E_1 B_2$$

Before computing the coordinates of the output link S-joint centre point $E = \mathcal{E} \cap K_{DE}$ recall the four doubly dependent equations which state $E \in \mathcal{E}$, *i.e.*,

$$\sum_{j=1}^{3} e_j E_{ij} = 0$$

for all $j = 0, 1, 2, 3.$

$$\begin{array}{ccccc} 0 & +E_{01}e_1 & +E_{02}e_2 & +E_{03}e_3 & = 0 \\ -E_{01}e_0 & +0 & +E_{12}e_2 & -E_{31}e_3 & = 0 \\ -E_{02}e_0 & -E_{12}e_1 & +0 & +E_{23}e_3 & = 0 \\ -E_{03}e_0 & +E_{31}e_1 & -E_{23}e_2 & +0 & = 0 \end{array}$$

The last two lines of the array above, substituted into Eq. 1, produce the univariate quadratic in e_1 below. Note the following simplifications which occur due to choice of coordinate frame and to dehomogenization of the original sphere equation. By setting

$$d_0 = 1, \quad d_1 = 0, \quad e_0 = 1, \quad w = 1, \quad x = e_1, \quad y = e_2, \quad z = e_3$$

Eq. 1 becomes

$$K_{DE} : e_1^2 + e_2^2 - 2d_2 e_2 + e_3^2 - 2d_3 e_3 + (d_2^2 + d_3^2 - r_F^2) = 0$$

After substitution for e_2 and e_3 and collecting coefficients of e_1, the result is Eq. 4. With e_1 in hand the last two equations of the four equation array $E \in \mathcal{E}$ generate the remaining two required coordinates of E.

$$\begin{aligned}(E_{23}^2 + E_{31}^2 + E_{12}^2)e_1^2 - 2[E_{03}E_{31} - E_{02}E_{12} + (d_2 E_{31} + d_3 E_{12})E_{23}]e_1 \\ +[E_{02}^2 + E_{03}^2 + 2(d_2 E_{03} - d_3 E_{02})E_{23} + d_2^2 + d_3^2 - r_F^2] = 0 \qquad (4)\end{aligned}$$

3. Position of Points F and C

3.1 Three Constraint Surfaces on F

Consider $F\{f_0 : f_1 : f_2 : f_3\} = K_{DF} \cap K_{EF} \cap K_{CF}$. K_{DF} and K_{EF} are the spheres centred on D and E and with radii r_E and r_D, respectively. K_{CF} is the cylinder of revolution on axis C with radius r_C. The derivation of the cylinder point equation will be explained later.

$$
\begin{aligned}
&(f_1-d_1)^2+(f_2-d_2)^2+(f_3-d_3)^2-r_E^2=0\\
&(f_1-e_1)^2+(f_2-e_2)^2+(f_3-e_3)^2-r_D^2=0\\
&(c_{23}+c_{02}f_3-c_{03}f_2)^2+(c_{31}+c_{03}f_1-c_{01}f_3)^2\\
&+(c_{12}+c_{01}f_2-c_{02}f_1)^2-r_C^2(c_{01}^2+c_{02}^2+c_{03}^2)=0
\end{aligned}
\tag{5}
$$

These may be reduced to standard quadric form after replacing the second sphere with the plane on the circle of intersection $K_{DF} \cap K_{EF}$.

$$
\begin{aligned}
a_{17}f_1+a_{18}f_2+a_{19}f_3+a_{1A}&=0\\
a_{21}f_1^2+a_{22}f_2^2+a_{23}f_3^2+a_{27}f_1+a_{28}f_2+a_{29}f_3+a_{2A}&=0\\
a_{31}f_1^2+a_{32}f_2^2+a_{33}f_3^2+a_{34}f_1f_2+a_{35}f_2f_3+a_{36}f_3f_1&\\
+a_{37}f_1+a_{38}f_2+a_{39}f_3+a_{3A}&=0
\end{aligned}
\tag{6}
$$

Careful rearrangement reduces the numerical computation, of the five univariate quatric coefficients from their general symbolic form, to a mere 200 FLOPS, to allow construction of feasible real time, microcontroller based algorithms.

$$
\begin{aligned}
a_{17}=2(e_1-d_1),\ a_{18}=2(e_2-d_2),\ a_{19}=a(e_3-d_3)\\
a_{1A}=(e_1-d_1)^2+(e_2-d_2)^2+(e_2-d_2)^2-r_D^2+r_E^2\\
a_{21}=a_{22}=a_{23}=1,\ a_{27}=-2d_1,\ a_{28}=-2d_2,\ a_{29}=-2d_3,\ a_{2A}=d_1^2+d_2^2+d_3^2-r_E^2\\
a_{31}=c_{02}^2+c_{03}^2,\ a_{32}=c_{03}^2+c_{01}^2,\ a_{33}=c_{01}^2+c_{02}^2\\
a_{34}=-2c_{01}c_{02},\ a_{35}=-2c_{02}c_{03},\ a_{36}=-2c_{03}c_{01}\\
a_{37}=2(c_{31}c_{03}-c_{12}c_{02}),\ a_{38}=2(c_{12}c_{01}-c_{23}c_{03}),\ a_{39}=2(c_{23}c_{02}-c_{31}c_{01})\\
a_{3A}=c_{23}^2+c_{31}^2+c_{12}^2-r_C^2(c_{01}^2+c_{02}^2+c_{03}^2)
\end{aligned}
$$

Eliminating f_3 from the last two equations with the first produces the next level of compression in coefficients b_{ij}; two conics in f_1 and f_2.

$$
\begin{aligned}
b_{11}f_1^2+b_{12}f_2^2+b_{13}f_1f_2+b_{14}f_1+b_{15}f_2+b_{16}=0\\
b_{21}f_1^2+b_{22}f_2^2+b_{23}f_1f_2+b_{24}f_1+b_{25}f_2+b_{26}=0
\end{aligned}
$$

where

$$
\begin{aligned}
b_{11}=a_{17}^2a_{23}-a_{19}^2a_{21},\ b_{12}=a_{18}^2a_{23}-a_{19}^2a_{22},\ b_{13}=2a_{17}a_{18}a_{23}\\
b_{14}=a_{17}(2a_{1A}a_{23}-a_{19}a_{29})+a_{19}^2a_{27},\ b_{15}=a_{18}(2a_{1A}a_{23}-a_{19}a_{29})+a_{19}^2a_{28}\\
b_{16}=a_{19}(a_{19}a_{2A}-a_{1A}a_{29})\\
b_{21}=a_{17}^2a_{33}+a_{19}^2a_{31},\ b_{22}=a_{18}(a_{18}a_{33}-a_{19}a_{35})+a_{19}^2a_{32}\\
b_{23}=a_{17}(2a_{18}a_{33}-a_{19}a_{35})+a_{19}^2a_{34}\\
b_{24}=a_{17}[2a_{1A}a_{33}-a_{19}(a_{36}+a_{39})]+a_{19}^2a_{37}\\
b_{25}=a_{19}[a_{19}a_{38}-a_{1A}a_{35}-a_{18}(a_{36}+a_{39})]+2a_{18}a_{1A}a_{33}\\
b_{26}=a_{19}[a_{19}a_{3A}-a_{1A}(a_{36}+a_{39})]a_{1A}^2a_{33}
\end{aligned}
$$

Two quadratics in f_2 emerge when rearrangment absorbs the retained variable f_1 into the coefficients.

$$
\begin{aligned}
b_{12}f_2^2+(b_{13}f_1+b_{15})f_2+(b_{11}f_1^2+b_{14}f_1+b_{16})=0\\
b_{22}f_2^2+(b_{23}f_1+b_{25})f_2+(b_{21}f_1^2+b_{24}f_1+b_{26})=0
\end{aligned}
$$

Another coefficient compression stage produces

$$
P_0f_2^2+P_1f_2+P_2=0,\ \ Q_0f_2^2+Q_1f_2+Q_2=0
$$

Using 0th order Bezout multipliers Q_0 and P_0 then 1st order multipliers $Q_0 f_2 + Q_1$ and $P_0 f_2 + P_1$ yields

$$(P_0Q_1 - Q_0P_1)f_2 + (P_0Q_2 - Q_0P_2) = 0$$
$$(P_0Q_2 - Q_0P_2)f_2 + (P_1Q_2 - Q_1P_2) = 0 \qquad (7)$$

Therefore f_2 is eliminated with the following 2×2 determinant.

$$\begin{vmatrix} (P_0Q_1 - Q_0P_1) & (P_0Q_2 - Q_0P_2) \\ (P_0Q_2 - Q_0P_2) & (P_1Q_2 - Q_1P_2) \end{vmatrix} = 0$$

So the quartic univariate in f_1, which can be solved reliably and explicitly [Zsombor-Murray, 2001], is

$$Af_1^4 + Bf_1^3 + Cf_1^2 + Df_1 + E = 0$$

where

$$A = k_2k_3 - k_4^2,\ m_1 = k_6 + k_7,\ B = k_1k_3 + k_2m_1 - 2k_4k_5,\ m_2 = k_9 + k_{10}$$
$$C = k_1m_1 + k_2m_2 - 2k_4k_8 - k_5^2,\ D = k_1m_2 + k_2k_{11} - 2k_5k_8,\ E = k_1k_{11} - k_8^2$$

such that

$$k_1 = b_{22}b_{15} - b_{12}b_{25},\ k_2 = b_{22}b_{13} - b_{12}b_{23},\ k_3 = b_{23}b_{11} - b_{13}b_{21},\ k_4 = b_{22}b_{11} - b_{12}b_{21}$$
$$k_5 = b_{22}b_{14} - b_{12}b_{24},\ k_6 = b_{25}b_{11} - b_{15}b_{21},\ k_7 = b_{23}b_{14} - b_{13}b_{24},\ k_8 = b_{22}b_{16} - b_{12}b_{26}$$
$$k_9 = b_{25}b_{14} - b_{15}b_{24},\ k_{10} = b_{23}b_{16} - b_{13}b_{26},\ k_{11} = b_{25}b_{16} - b_{15}b_{26}$$

Linear backsubstitution to find unique corresponding values of f_2 and f_3 proceeds using either of Eq. 7 and the linear equation in the set Eq. 6 and the procedure to find point F is complete. All that remains is to intersect line $\mathcal{C}$ with a perpendicular plane Π_C, on F, the centre of the C-joint revolute C.

3.2 Position of Point C

Plane Π_C must have coordinates $\Pi_C\{C_0' : c_{01} : c_{02} : c_{03}\}$ because it is normal to cylinder axis $\mathcal{C}$. Since $F \in \Pi_C$ the plane equation can be written.

$$C_0' f_0 + c_{01}f_1 + c_{02}f_2 + c_{03}f_3 = 0$$

Therefore

$$\Pi_C\{-(c_{01}f_1 + c_{02}f_2 + c_{03}f_3) : c_{01}f_0 : c_{02}f_0 : c_{03}f_0\} \equiv \{C_0 : C_1 : C_2 : C_3\}$$

Using

$$C = \mathcal{C} \cap \Pi_C,\ c_i = \sum_{j=0}^{3} c_{ij}C_j$$

produces

$$c_0 = c_{01}^2 f_0 + c_{02}^2 f_0 + c_{03}^2 f_0$$
$$c_1 = c_{01}(c_{01}f_1 + c_{02}f_2 + c_{03}f_3) + c_{02}c_{12}f_0 - c_{03}c_{31}f_0$$
$$c_2 = c_{02}(c_{01}f_1 + c_{02}f_2 + c_{03}f_3) - c_{01}c_{12}f_0 + c_{03}c_{23}f_0$$
$$c_3 = c_{03}(c_{01}f_1 + c_{02}f_2 + c_{03}f_3) + c_{01}c_{31}f_0 - c_{02}c_{23}f_0 \qquad (8)$$

3.3 Line Complex and Cylinder Equation

The Sperical Tangent Line Complex. The set of all lines $\mathcal{T}\{t_{01} : t_{02} : t_{03} : t_{23} : t_{31} : t_{12}\}$ tangent to a given sphere, radius R, centred on point $M\{m_0 : m_1 : m_2 : m_3\}$ can be deduced by equating the square of the moment magnitude of a typical tangent line's direction numbers, about the point M, obtained in two different ways. First, the square of the difference between the typical tangent line's moment $t_{23} : t_{31} : t_{12}$ and the moment of a line on M which has the direction numbers $t_{01} : t_{02} : t_{03}$ of the typical tangent line. This can be expressed as

$$\left(\begin{bmatrix} t_{23} \\ t_{31} \\ t_{12} \end{bmatrix} - \frac{1}{m_0}\begin{bmatrix} m_1 \\ m_2 \\ m_3 \end{bmatrix} \times \begin{bmatrix} t_{01} \\ t_{02} \\ t_{03} \end{bmatrix}\right)^2 = \left(\begin{bmatrix} t_{23} - (m_2 t_{03} - m_3 t_{02})/m_0 \\ t_{31} - (m_3 t_{01} - m_1 t_{03})/m_0 \\ t_{12} - (m_1 t_{02} - m_2 t_{01})/m_0 \end{bmatrix}\right)^2$$

Then the square of the moment of the typical line's direction numbers about M is formed directly as

$$\left(R\begin{bmatrix} t_{01} \\ t_{02} \\ t_{03} \end{bmatrix}\right)^2$$

Therefore the desired equation of the line complex is the difference multiplied by m_0^2.

$$\begin{aligned} &[m_0 t_{23} - (m_2 t_{03} - m_3 t_{02})]^2 + [m_0 t_{31} - (m_3 t_{01} - m_1 t_{03})]^2 \\ &+[m_0 t_{12} - (m_1 t_{02} - m_2 t_{01})]^2 - (m_0 R)^2 (t_{01}^2 + t_{02}^2 + t_{03}^2) = 0 \end{aligned} \tag{9}$$

Cylinder of Revolution. Beginning with Eq. 9 consider the cylinder of revolution of radius r_C and whose axis is line $\mathcal{C}$. First the line complex is converted to the surface equation in terms of typical ruling lines

$$\mathcal{P}\{p_{01} : p_{02} : p_{03} : p_{23} : p_{31} : p_{12}\}$$

by noting that $t_{0j} = c_{0j}$ and moments $m_i t_{0j} - m_j t_{0i} = c_{ij}$ and making these substitutions to yield

$$(p_{23} - c_{23})^2 + (p_{31} - c_{31})^2 + (p_{12} - c_{12})^2 - r_C^2(c_{01}^2 + c_{02}^2 + c_{03}^2) = 0$$

This becomes a one parameter set of lines when combined with the Plücker condition, in this case a linear equation.

$$c_{01}p_{23} + c_{02}p_{31} + c_{03}p_{12} = 0$$

The point equation for F is produced by using two elements of the expansion $F \in \mathcal{P}$ which gives the degenerate plane Π_P defined by a line and a point on it.

$$P_i = \sum_{j=0}^{3} P_{ij} f_j = 0$$

The cylinder equation, Eq. 5, is produced upon substituting radial for axial line coordinates.

$$\begin{aligned} &(c_{23}f_0 + c_{02}f_3 - c_{03}f_2)^2 + (c_{31}f_0 + c_{03}f_1 - c_{01}f_3)^2 \\ &+(c_{12}f_0 + c_{01}f_2 - c_{02}f_1)^2 - r_C(c_{01}^2 + c_{02}^2 + c_{03}^2) = 0 \end{aligned}$$

In this way the implicit form can be written from its geometric definition; no coordinate transformation or trigonometric functions.

4. Conclusion

This platform mechanism's input-output equations establish the position of points D, E and F on the platform S-joint centres. However they also engender the DKP solution for a certain subclass of three-legged, six dof spatial manipulator, *viz.*, 3-2-1 basally actuated three-legged robots. One leg has three actuated joints which position D. Therefore the points E and F move on concentric spheres of respective radii r_F and r_E. Regardless of its architecture the doubly actuated second leg BE, point E must in general lie on a circle or line. Therefore solution of a quadratic equation is sufficient to find the position of E. If the single actuator of leg CF confines F to a sphere then a second, decoupled quadratic equation establishes the position of F. The four possible platform poses are avaiable as a product of quadratic factors. If, however, the third leg's single actuator confines F on a cylinder of revolution then one arrives at the eight pose solution containing quadratic and quartic factors. Topics of continuing interest in this vein include the extraction of a general octic in convenient symbolic form, if possible, for all basally actuated 2-2-2 six dof three-legged platforms where S-joints move on circles. A related type [Zsombor-Murray and Shum, 2001] has already been solved but the three circle class DKP [O'Brien and Wen, 2001], which is of great economic importance, remains elusive.

5. Acknowledgements

This research is supported by grants provided by NSERC 139964, Natural Sciences and Engineering Research Council and FCAR ER73234, Fonds pour la Formation des Chercheurs et l'Aide à la Recherche.

References

O'Brien, J.F. and Wen, J.T. (2001), On Kinematic Instability of Parallel Robots, *Proc. CK2001 2nd Wkshp. on Comp. Kin.*, Seoul, Korea, pp. 279–290. (See references [16,17,18] therein.)

Salmon, G. (1885), *Lessons Introductory to the Modern Higher Algebra*, 4th Ed., Hodges, Figgis and Co., Dublin, pp. 76–90.

Sandor, G.N. and Xirong Zhuang (1984), On the Elimination of the Branching Problem in the Synthesis of Spatial Motion Generators with Spheric Joints. Part 1: Theory and Part 2: Application, *ASME J. Mech. Trans. and Autom. Des.*, v.106, pp. 312–326.

Zsombor-Murray, P.J. (2001), Explicit Solution of Cubic and Quartic Univariate Polynomials, *CIM Int. Rpt. 01-04-03*, 5 pp.

Zsombor-Murray, P.J. and Shum, J.F.C (2001), The Three-Points-on-Three-Lines Problem, *Proc. 18th Cdn. Conf. of Appl. Mech.*, St. John's, Canada, pp. 361–362.

DYNAMIC MODELING OF THE ORTHOGLIDE

Sylvain Guégan and Wisama Khalil
Institut de Recherche en Communications et Cybernétique de Nantes (IRCCyN)
BP 92101, 1 rue de la Noë, 44321 Nantes Cedex 03, France
wisama.khalil@irccyn.ec-nantes.fr

Abstract This paper presents a novel solution for the inverse and forward dynamic models of the Orthoglide, a 3-DOF parallel robot that moves in the Cartesian space with fixed orientation. The models are computed in terms of the Cartesian dynamic model of the legs and of the Newton-Euler equation of the platform. The base inertial parameters of the robot, which constitute the minimum number of inertial parameters, are determined. The numbers of operations to compute the inverse and direct dynamic models are given.

Keywords: Parallel robots, inverse dynamic model, forward dynamic model, base parameters, operational space, inertial parameters, computation cost.

1. Introduction

The dynamic modeling of parallel manipulators presents an inherent complexity due to their closed-loop structure and kinematic constraints. The inverse dynamic model is important for high performance control algorithms, and the forward dynamic model is required for their simulation. The dynamics of parallel robots, are classically obtained by computing the dynamic model of an equivalent tree structure, then the Lagrange multipliers are used to obtain the complete closed-loop dynamics [Nguyen and Pooran, 1989, Ait-Ahmed, 1993, Bhattacharya et al., 1997, Liu et al., 2000]. The principle of virtual work has been used in [Codourey and Burdet, 1997, Tsai, 2000]. The Newton-Euler formulation has been applied by Reboulet and Berthomieu, 1991, Gosselin, 1993, Dasgupta and Mruthyunjaya, 1998, Dasgupta and Choudhury, 1999, and Ji, 1993. In [Khalil and Guegan, 2002] we have proposed a Newton-Euler modeling method for the 6 DOF Gough-Stewart parallel robot. The models are obtained in terms of the Cartesian inertia and forces of the legs as perceived at the points connecting the platform and the legs. This paper proposes to apply our method on the Orthoglide [Wenger and Chablat, 2000], figure 1. We also treat the problem of the determination of the base inertial parameters and the computational complexity of the inverse dynamic model for on-line application.

J. Lenarčič and F. Thomas (eds.), Advances in Robot Kinematics, 387–396.

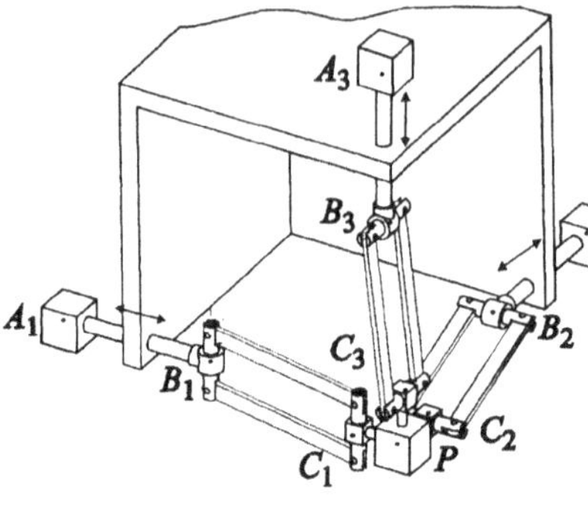

Figure 1. Orthoglide kinematic architecture

2. Kinematic modeling of the Orthoglide

The Orthoglide has three parallel PRPaR identical legs (where P, R and Pa stand for Prismatic, Revolute and Parallelogram joint, respectively). Each leg is composed of six passive revolute joints and 1 active prismatic joint, figures 1 and 2. It has a complex structure with two spatial closed loops and three planar closed loops.

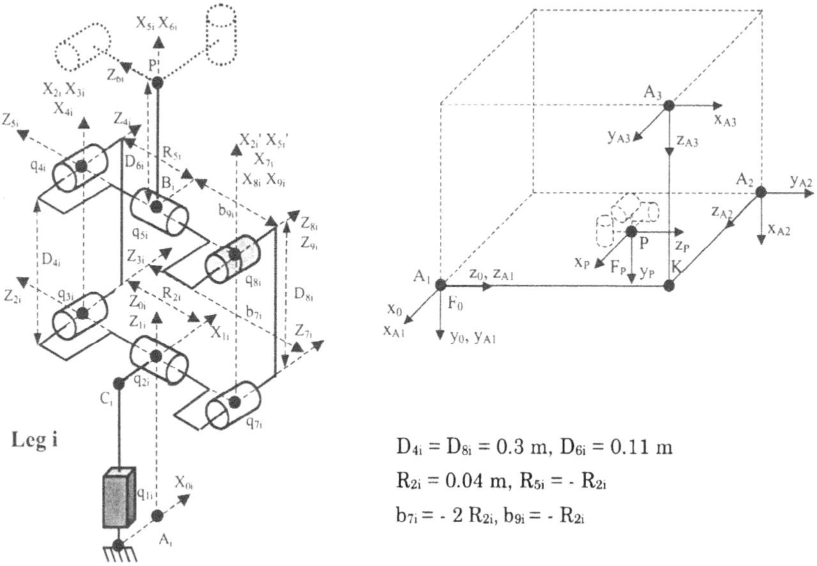

Figure 2. Link frames of one leg

We define frame F_0 fixed with the base and frame F_P fixed with the mobile platform. Their origins are A_1 and P respectively. Their axes (x_0, y_0, z_0) and (x_P, y_P, z_P) are parallel. The base frames of the legs are defined by the frames F_{A1}, F_{A2} and F_{A3}, whose origins are the points A_1, A_2 and A_3 respectively. Their z_{Ai} axes are along the prismatic joint axes. Figure 2 shows these different frames. The notations of Khalil and Kleinfinger, 1986, are used to describe the geometry of the system.

The following notations are used:

(X_p, Y_p, Z_p) Coordinates of point P with respect to frame F_0;

(X_{Ai}, Y_{Ai}, Z_{Ai}) Coordinates of point A_i with respect to frame F_0;

$\mathbf{L}$ (3×1) vector of the motorized joint variables: $\mathbf{L} = [q_{11} \quad q_{12} \quad q_{13}]^T$;

$\dot{\mathbf{L}}$ (3×1) vector of the motorized joint velocities: $\dot{\mathbf{L}} = [\dot{q}_{11} \quad \dot{q}_{12} \quad \dot{q}_{13}]^T$;

${}^0\mathbf{V}_p$ (3×1) vector of the linear velocity of the origin of the platform;

${}^0\dot{\mathbf{V}}_p$ (3×1) vector of the linear acceleration of the origin of the platform.

The following geometric and kinematic models can be obtained [Wenger and Chablat, 2000, Merlet, 2000, Khalil and Dombre, 2002]:

i) The inverse geometric model of the robot:

$$\mathbf{L} = \begin{bmatrix} q_{11} \\ q_{12} \\ q_{13} \end{bmatrix} = \begin{bmatrix} Z_P - \cos(q_{21})\cos(q_{31})D_{41} - D_{61} \\ X_P - X_{A2} - \cos(q_{22})\cos(q_{32})D_{42} - D_{62} \\ Y_P - Y_{A3} - \cos(q_{23})\cos(q_{33})D_{43} - D_{63} \end{bmatrix} \quad (1)$$

With: $q_{31} = \sin^{-1}\left(\frac{-Y_P}{D_{41}}\right)$, $q_{21} = -\left(\sin^{-1}\left(\frac{-X_P}{\cos(q_{31})D_{41}}\right) + \frac{\pi}{2}\right)$, $q_{32} = \sin^{-1}\left(\frac{-Z_P + Z_{A2}}{D_{42}}\right)$

$q_{22} = -\left(\sin^{-1}\left(\frac{-Y_P + Y_{A2}}{\cos(q_{32})D_{42}}\right) + \frac{\pi}{2}\right)$, $q_{33} = \sin^{-1}\left(\frac{-X_P + X_{A3}}{D_{43}}\right)$, $q_{23} = -\left(\sin^{-1}\left(\frac{-Z_P + Z_{A3}}{\cos(q_{33})D_{43}}\right) + \frac{\pi}{2}\right)$

ii) The inverse kinematic model of the robot:

$$\dot{\mathbf{L}} = {}^0\mathbf{J}_p^{-1}\, {}^0\mathbf{V}_p \quad (2)$$

Where ${}^0\mathbf{J}_P^{-1}$ is the inverse Jacobian matrix of the robot:

$${}^0\mathbf{J}_p^{-1} = \begin{bmatrix} -\frac{1}{T_{21}} & \frac{T_{31}}{S_{21}} & 1 \\ 1 & -\frac{1}{T_{22}} & \frac{T_{32}}{S_{22}} \\ \frac{T_{33}}{S_{23}} & 1 & -\frac{1}{T_{23}} \end{bmatrix} \quad (3)$$

With: ${}^0\mathbf{J}_p^{-1}$ always regular, $T_{2i} = \tan(q_{2i})$, $S_{3i} = \sin(q_{3i})$, $T_{3i} = \tan(q_{3i})$.

iii) The inverse kinematic model of a leg which gives the joint velocities $(\dot{q}_{1i}, \dot{q}_{2i}, \dot{q}_{3i})$ as functions of the velocity of point P:

$$\dot{\mathbf{q}}_i = {}^0\mathbf{J}_i^{-1}\,{}^0\mathbf{V}_p \tag{4}$$

Where ${}^0\mathbf{J}_i^{-1}$ is the inverse Jacobian matrix of the leg:

$${}^0\mathbf{J}_1 = \begin{bmatrix} 0 & -D_{41}C_{31}S_{21} & -D_{41}S_{31}C_{21} \\ 0 & 0 & -D_{41}C_{31} \\ 1 & -D_{41}C_{31}C_{21} & D_{41}S_{31}S_{21} \end{bmatrix}, \quad {}^0\mathbf{J}_2 = \begin{bmatrix} 1 & -D_{42}C_{32}C_{22} & D_{42}S_{32}S_{22} \\ 0 & -D_{42}C_{32}S_{22} & -D_{42}S_{32}C_{22} \\ 0 & 0 & -D_{42}C_{32} \end{bmatrix}$$

$${}^0\mathbf{J}_3 = \begin{bmatrix} 0 & 0 & -D_{43}C_{33} \\ 1 & -D_{43}C_{33}C_{23} & D_{43}S_{33}S_{23} \\ 0 & -D_{43}C_{33}S_{23} & -D_{43}S_{33}C_{23} \end{bmatrix} \tag{5}$$

The inverse of these matrices can be obtained explicitly. The velocities of the other joints of the leg can be obtained as functions of $\dot{\mathbf{q}}_i$.

iv) The second order inverse kinematic model of the leg:

$$\ddot{\mathbf{q}}_i = {}^0\mathbf{J}_i^{-1}\left({}^0\dot{\mathbf{V}}_p - {}^0\dot{\mathbf{J}}_i\dot{\mathbf{q}}_i\right) \tag{6}$$

3. Inverse dynamic model

The inverse dynamic model gives the motorized forces, Γ, in terms of the desired trajectory of the mobile platform ${}^0\mathbf{P}_p, {}^0\mathbf{V}_p, {}^0\dot{\mathbf{V}}_p$.

The dynamic model is computed in two steps. First we calculate, the reaction forces of the platform on the legs at point P, which is denoted by $\mathbf{f}_i$, then the Newton-Euler equation of the platform is applied to obtain Γ. We note that the reaction moments at point P are zero.

The general form of the inverse dynamic model of a leg i, is written as (see Appendix):

$$\Gamma_i = \mathbf{H}_i(\mathbf{q}_i, \dot{\mathbf{q}}_i, \ddot{\mathbf{q}}_i) + {}^0\mathbf{J}_i^T\,{}^0\mathbf{f}_i \tag{7}$$

Where:

$\mathbf{H}_i$ is the inverse dynamic model of leg i, when terminal point is free.

Γ_i is composed of the independent torques/forces of the joints of the leg i, where Γ_{1i} and Γ_{2i} are zero:

$$\Gamma_i = [\Gamma_{1i} \quad \Gamma_{2i} \quad \Gamma_{3i}]^T = [\Gamma_{1i} \quad 0 \quad 0]^T \tag{8}$$

Using equation (7) the reaction forces $\mathbf{f}_i$ can be written as:

$${}^0\mathbf{f}_i = -{}^0\mathbf{J}_i^{-T}\mathbf{H}_i(\mathbf{q}_i, \dot{\mathbf{q}}_i, \ddot{\mathbf{q}}_i) + {}^0\mathbf{J}_i^{-T}\Gamma_i = -\mathbf{H}_{xi}(\mathbf{q}_i, \dot{\mathbf{q}}_i, \ddot{\mathbf{q}}_i) + {}^0\mathbf{J}_i^{-T}\Gamma_i \tag{9}$$

The vector $\mathbf{H}_{Xi}(\mathbf{q}_i, \dot{\mathbf{q}}_i, \ddot{\mathbf{q}}_i)$, is the transformation of the vector $\mathbf{H}_i(\mathbf{q}_i, \dot{\mathbf{q}}_i, \ddot{\mathbf{q}}_i)$ into the position Cartesian space at point P_i [Khatib, 1987, Lily and Orin, 1990].

Using equations (3) and (5), we obtain:

$$ {}^{0}\mathbf{f}_i = -\mathbf{H}_{xi}(\mathbf{q}_i,\dot{\mathbf{q}}_i,\ddot{\mathbf{q}}_i) + {}^{0}\mathbf{J}^{-T}_{p\,[:,i]}\Gamma_i \tag{10} $$

Where ${}^{0}\mathbf{J}^{-T}_{p\,[:,i]}$ represents the ith column of the inverse transpose Jacobian matrix of the robot.

The Newton-Euler equation around the origin of the platform is written as (no rotation):

$$ {}^{0}\mathbf{F}_P = {}^{0}\dot{\mathbf{V}}_P\, M_p - M_p\,\mathbf{g} \tag{11} $$

With:

${}^{0}g$ acceleration of gravity, referred to frame F_0;

M_p Mass of the platform;

${}^{0}\mathbf{F}_p$ Total external forces on the platform.

Since ${}^{0}\dot{\mathbf{V}}_p$ is known, the vector ${}^{0}\mathbf{F}_p$ can be computed using equation (11).

The reaction forces of the legs on the platform are given as:

$$ {}^{0}\mathbf{F}_p = \sum_{i=1}^{3} {}^{0}\mathbf{f}_i \tag{12} $$

Substituting equation (10) into equation (12), we obtain:

$$ \Gamma_{robot} = {}^{0}\mathbf{J}_p^{T}\left({}^{0}\mathbf{F}_p + \sum_{i=1}^{3}\left[\mathbf{H}_{xi}(\mathbf{q}_i,\dot{\mathbf{q}}_i,\ddot{\mathbf{q}}_i)\right]\right) \tag{13} $$

Thus we can obtain:

$$ \mathbf{\Gamma}_{robot} = {}^{0}\mathbf{J}_p^{T}\mathbf{H}_{robot} \tag{14} $$

With:

$$ \mathbf{\Gamma}_{robot} = \begin{bmatrix}\Gamma_{11} & \Gamma_{12} & \Gamma_{13}\end{bmatrix}^T $$

$$ \mathbf{H}_{robot} = {}^{0}\mathbf{F}_p + \sum_{i=1}^{3}\left[\mathbf{H}_{xi}(\mathbf{q}_i,\dot{\mathbf{q}}_i,\ddot{\mathbf{q}}_i)\right] \tag{15} $$

From equation (13) we deduce that the dynamics of leg i is equivalent to a resistant force $\mathbf{H}_{xi}$ on the mobile platform.

Equation (14) represents the inverse dynamic model. Beside the Jacobian matrices, we need to determine the inverse dynamic model of the legs, $\mathbf{H}_i(\mathbf{q}_i,\dot{\mathbf{q}}_i,\ddot{\mathbf{q}}_i)$. Different methods can be used to calculate these vectors [Khalil and Dombre, 2002, Luh et al., 1979, Megahed and Renaud, 1982].

To reduce the computational cost, the Newton-Euler method using customized symbolic method and the base inertial parameters could be used [Khalil and Kleinfinger, 1987].

4. Forward dynamic model

The direct dynamic model of the robot gives the platform acceleration as a function of the state variables of the robot (position and velocity of the platform) and the input forces ${}^{0}\dot{V}_p = f({}^{0}P_p, {}^{0}V_p, \Gamma)$.

It can be obtained using the same manner as the inverse dynamic model, but we have to replace $\mathbf{H}_i$ of equation (7) by:

$$\mathbf{H}_i(q_i, \dot{q}_i, \ddot{q}_i) = \mathbf{A}_i \ddot{q}_i + \mathbf{h}_i(q_i, \dot{q}_i) \tag{16}$$

Where: $\mathbf{A}_i$ is the (3x3) inertia matrix of leg i and $\mathbf{h}_i$ is the (3x1) vector of Coriolis, centrifugal and gravity forces.

Using (6), (9) and (16) we obtain:

$${}^{0}\mathbf{f}_i = -\mathbf{A}_{xi}\, {}^{0}\dot{\mathbf{V}}_p + \mathbf{A}_{xi}\, {}^{0}\dot{\mathbf{J}}_i\, \dot{q}_i - \mathbf{h}_{xi}(q_i, \dot{q}_i) + {}^{0}\mathbf{J}_{p\,[:,i]}^{-T} \Gamma_i \tag{17}$$

Where $\mathbf{A}_{xi} = {}^{0}\mathbf{J}_i^{-T} \mathbf{A}_i\, {}^{0}\mathbf{J}_i^{-1}$ is the 3x3 Cartesian inertia matrix of leg i, $\mathbf{h}_{xi} = {}^{0}\mathbf{J}_i^{-T} \mathbf{h}_i$ is the 3x1 Coriolis, centrifuge and gravity forces of leg i referred to the Cartesian space [Khatib, 1987].

Substituting (17) into the Newton-Euler equation (12), we obtain:

$$\left(\sum_{i=1}^{3}[\mathbf{A}_{xi}] + \mathbf{I}_3 M_P\right) {}^{0}\dot{\mathbf{V}}_p + \sum_{i=1}^{3}\left[\mathbf{h}_{xi}(q_i, \dot{q}_i) - \mathbf{A}_{xi}\, {}^{0}\dot{\mathbf{J}}_i\, \dot{q}_i\right] - M_P\, \mathbf{g} = {}^{0}\mathbf{J}_P^{-T} \Gamma_{robot} \tag{18}$$

Consequently, the platform acceleration can be determined from:

$${}^{0}\dot{\mathbf{V}}_p = \mathbf{A}_{robot}^{-1}\left({}^{0}\mathbf{J}_P^{-T} \Gamma_{robot} - \mathbf{h}_{robot}\right) \tag{19}$$

With:

$$\mathbf{A}_{robot} = \sum_{i=1}^{3}[\mathbf{A}_{xi}] + \mathbf{I}_3 M_p \tag{20}$$

$$\mathbf{h}_{robot} = \sum_{i=1}^{3}\left[\mathbf{h}_{xi}(q_i, \dot{q}_i) - \mathbf{A}_{xi}\, {}^{0}\dot{\mathbf{J}}_i\, \dot{q}_i\right] - M_P\, \mathbf{g} \tag{21}$$

Finally, the joint accelerations $\ddot{q}_{1i}$, $\ddot{q}_{2i}$, $\ddot{q}_{3i}$ (i = 1, 2, 3) of the legs could be computed, by equation (6).

$\mathbf{A}_{robot}$ is the (3x3) inertia matrix of the robot. It is composed of the Cartesian matrix of the leg and the mass of the platform.

We note that the Coriolis, centrifugal and gravity forces of each leg are transformed into the (3x1) Cartesian force vector $\mathbf{h}_{xi}$ at point P.

5. Base inertial parameters of the robot

The standard inertial parameters of link j are collected in the (11×1) vector:

$$\chi_j = \left[XX_j \;\; XY_j \;\; XZ_j \;\; YY_j \;\; YZ_j \;\; ZZ_j \;\; MX_j \;\; MY_j \;\; MZ_j \;\; M_j \;\; Ma_j\right]^T \tag{22}$$

Where:

- XX_j, ..., ZZ_j are the elements of the inertia matrix around the origin of frame j;
- MX_j, MY_j, MZ_j define the first moments of link j;
- M_j is the mass of link j;
- Ma_j is the mass of motor j referred to the joint side, it is defined for the motorized joints.

The base inertial parameters represent the minimum number of parameters from which the dynamic model can be calculated. The dynamic model complexity is reduced when computed by the base inertial parameters. Besides, they constitute the only identifiable parameters [Gautier and Khalil, 1990]. They can be deduced from the standard parameters, by eliminating the dynamic parameters that have no effect on the dynamic model and by grouping some others. To determine them, we use a numerical method, which is based on the QR decomposition [Gautier, 1990]. First we determine the base parameters of each leg, then we determine the effect of connecting the platform.

There are 12 base parameters for legs 1 and 2. They are given by (i=1, 2): M_{1Ri}, ZZ_{2Ri}, MX_{2i}, MY_{2Ri}, XX_{3Ri}, XY_{3Ri}, XZ_{3Ri}, YZ_{3Ri}, ZZ_{3Ri}, MX_{3Ri}, MY_{3Ri}, MX_{4i}. The index R indicates that some parameters are grouped on that one:

$$M_{1Ri} = M_{1i} + Ma_{1i} + M_{2i} + M_{3i} + M_{4i} + M_{5i} + M_{6i} + M_{7i}$$
$$ZZ_{2Ri} = ZZ_{2i} + YY_{3i} + YY_{4i} + 0.09\, M_{4i} + 0.09\, M_{5i} + 0.09\, M_{6i} + YY_{7i}$$
$$MY_{2Ri} = MY_{2i} + MZ_{3i} + MZ_{4i} + MZ_{7i}$$
$$XX_{3Ri} = XX_{3i} - YY_{3i} + XX_{7i} - YY_{7i} - 0.09\, M_{4i} - 0.09\, M_{5i} - 0.09\, M_{6i}$$
$$XY_{3Ri} = XY_{3i} + XY_{7i}$$
$$XZ_{3Ri} = XZ_{3i} - 0.3\, MZ_{4i} + XZ_{7i}$$
$$YZ_{3Ri} = YZ_{3i} + YZ_{7i}$$
$$ZZ_{3Ri} = ZZ_{3i} + ZZ_{7i} + 0.09\, M_{4i} + 0.09\, M_{5i} + 0.09\, M_{6i}$$
$$MX_{3Ri} = MX_{3i} + MX_{7i} + 0.3\, M_{4i} + 0.3\, M_{5i} + 0.3\, M_{6i}$$
$$MY_{3Ri} = MY_{3i} + MY_{7i}$$

Since, the prismatic joint of leg 3 is along gravity, there are 13 base parameters for leg 3, the motor inertia Ma_{13} is not grouped with M_{1R3}. Taking into account the three legs and the platform, we obtain that the parameters MX_{3Ri} (i = 1, 2, 3) are grouped to the mass of the platform. The 35 base inertial parameters of the Orthoglide are given in table 1.

The number of standard inertial parameters is 223, while the number of the base inertial parameters is 35. The number of operations to determine the complete inverse and direct dynamic models of the Orthoglide, using customized symbolic method and the base inertial parameters [Khalil and Creusot, 1997], are 307 additions / 308 multiplications and 342 additions / 481 multiplications respectively.

Table 1. Base parameters of the Orthoglide

M'_{1R1}	ZZ'_{2R1}	MX_{21}	MY_{2R1}	XX'_{3R1}	XY_{3R1}	XZ_{3R1}	YZ_{3R1}	ZZ'_{3R1}	MY_{3R1}	MX_{41}	
M'_{1R2}	ZZ'_{2R2}	MX_{22}	MY_{2R2}	XX'_{3R2}	XY_{3R2}	XZ_{3R2}	YZ_{3R2}	ZZ'_{3R2}	MY_{3R2}	MX_{42}	
M'_{1R3}	Ma_{13}	ZZ'_{2R3}	MX_{23}	MY_{2R3}	XX'_{3R3}	XY_{3R3}	XZ_{3R3}	YZ_{3R3}	ZZ'_{3R3}	MY_{3R3}	MX_{43}
M_{RP}											

$$M_{RP} = M_P + \frac{10}{3} MX_{3R1} + \frac{10}{3} MX_{3R2} + \frac{10}{3} MX_{3R3}$$

$$\left.\begin{array}{l} M'_{1Ri} = M_{1Ri} - \frac{10}{3} MX_{3Ri}\,,\; ZZ'_{2Ri} = ZZ_{2Ri} - 0.3\, MX_{3Ri} \\ XX'_{3Ri} = XX_{3Ri} + 0.3\, MX_{3Ri}\,,\; ZZ'_{3Ri} = ZZ_{3Ri} - 0.3\, MX_{3Ri} \end{array}\right\} \quad \text{for } i = 1, 2, 3$$

If we consider the fact that the links are symmetrical, then we can define the set of 20 essential parameters, given in table 2:

Table 2. Essential inertial parameters of the Orthoglide

M'_{1R1}	ZZ'_{2R1}	MY_{2R1}	XX'_{3R1}	ZZ'_{3R1}	MX_{41}	M'_{1R2}	ZZ'_{2R2}	MY_{2R2}	XX'_{3R2}	ZZ'_{3R2}
MX_{42}	M'_{1R3}	Ia_{13}	ZZ'_{2R3}	MY_{2R3}	XX'_{3R3}	ZZ'_{3R3}	MX_{4R3}	M_P		

The number of operations to determine the inverse and direct dynamic models, using the essential inertial parameters, are 211 additions / 239 multiplications and 234 additions / 406 multiplications respectively.

6. Conclusion

In this paper, we highlight an interesting physical interpretation of the final form of the inverse and forward dynamic models of the Orthoglide robot. The approach is straightforward and could be applied to other fully parallel manipulators. These models are computed in terms of the Cartesian leg dynamic forces and inertia as perceived at the points connecting the mobile platform with the legs. Consequently, the computational complexity of this model can make use of the techniques that are developed many years ago for the serial robots. To reduce the computation cost, the base inertial parameters of the robot have been determined. The number of operations given can be computed on-line using one processor. Its calculation could be easily distributed on three processors (each leg model on a processor) to reduce more computation time.

References

Ait-Ahmed, M. (1993), Contribution à la modélisation géométrique et dynamique des robots parallèles, *Thèse de doctorant*, LAAS, Toulouse.

Bhattacharya, S., Hatwal, H., and Ghosh, A. (1997), An on-line estimation scheme for generalized Stewart platform type parallel manipulators, *Mechanism and Machine Theory*, Vol. 32(1), pp.79-89.

Codourey, A., and Burdet, E. (1997), A body oriented method for finding a linear form of the dynamic equatiojns of fully parallel robot, *In IEEE Int. Conf. on Robotics and Automation*, Albuquerque, New Mexico, U.S., pp.1612-1618.

Dasgupta, B., and Choudhury, P. (1999), A general strategy based on the Newton-Euler approach for the dynamic formulation of parallel manipulators, *Mechanism and Machine Theory*, Vol. 34(6), pp.801-824.

Dasgupta, B., and Mruthyunjaya T.S. (1998), A Newton-Euler formulation for the inverse dynamics of the Stewart platform manipulator, *Mechanism and Machine Theory*, Vol. 33(8), pp.1135-1152.

Gautier, M., and Khalil, W. (1990), Direct calculation of minimum set of inertial parameters of serial robots, *IEEE Trans. on Robotics and Automation*, Vol6(3), pp.368-373.

Gautier, M. (1990), Numerical calculation of the base inertial parameters, *IEEE Int. Conf. on Robotics and Automation*, Cincinnati, U.S., pp.1020-1025.

Gosselin, C. M. (1993), Parallel computationnal algorithms for the kinematics and dynamics of parallel manipulators, *IEEE Int. Conf. on Robotics and Automation*, New York, U.S., Vol.(1), pp.883-889.

Ji, Z. (1993), Study of the effect of leg inertia in Stewart platform, *In IEEE Int. Conf. on Robotics and Automation*, Atlanta, U.S., pp.121-126.

Khalil, W., and Dombre, E. (2002), Modeling, identification and control of robots, *Hermes Penton*, London-Paris.

Khalil, W., and Creusot, D. (1997), SYMORO+: a system for the symbolic modelling of robots, *Robotica*, Vol. 15, pp.153-161.

Khalil, W., and Guegan, S. (2002), A novel solution for the dynamic models of Gough-Stewart manipulators, *IEEE Int. Conf. on Robotics and Automation*, Washington, U.S., to appear.

Khalil, W., and Kleinfinger J.-F. (1987), Minimum operations and minimum parameters of the dynamic model of tree structure robots, *IEEE J. of Robotics and Automation*, Vol. RA-3(6), pp.517-526.

Khalil, W., and Kleinfinger, J.-F. (1986), A new geometric notation for open and closed-loop robots, *Proc. IEEE Conf. on Robotics and Automation*, San Francisco, US, pp.1174-1180.

Khatib, O. (1987), A unified approach for motion and force control of robot manipulators: the operational space formulation, *IEEE J. of Robotics and Automation*, Vol. RA-3(1), pp.43-53.

Lilly, K.W., and Orin D.E. (1990), Efficient O(N) computation of the operational space inertia matrix, *Proc. IEEE Int. Conf. on Robotics and Automation*, Cincinnati, US, pp.1014-1019.

Liu, M-J., Li, C-X., and Li, C-N. (2000), Dynamics analysis of the Gough-Stewart platform manipulator, *IEEE Transaction on Robotics and Automation*, Vol. 16(1), pp.94-98.

Luh, J.Y.S, Walker, M.W., and Paul, R.C.P. (1979), On-line computational scheme for mechanical manipulators, *Proc. 2nd IFAC/IFIP Symp. on*

Information Control Problems in Manufacturing Technology, Stuttgart, pp.165-172.
Megahed, S., and Renaud, M. (1982), Minimization of the computation time necessary for the dynamic control, *Proc. 12th Int. Symp. on Industrial Robots,* Paris, France, pp.469-478.
Merlet, J.-P. (2000), Parallel robots, *Kluwer Academic Publ., Dordrecht,* The Netherland.
Nguyen, C.C., and Pooran, F.J. (1989), Dynamic analysis of a 6 d.o.f. CKCM robot end-effector for dual-arm telerobot systems, *Robotics and Autonomous Systems,* Vol. 5, pp.377-394.
Reboulet, C., and Berthomieu, T. (1991), Dynamic models of a six degree of freedom parallel manipulators, *ICAR,* Pise, Italy, pp.1153-1157.
Tsai, L-W (2000), Solving the inverse dynamics of a Stewart-Gough manipulator by the principle of virtual work, *Journal of Mechanical design,* Vol.122, pp.3-9.
Wenger, P., and Chablat D. (2000), Kinematic analysis of a new parallel machine tool : the Orthoglide, *Advances in Robot Kinematic,* Kluwer Academic Publishers, pp.305-314.

Appendix: Dynamic model of a leg

Each leg has a planar parallelogram closed loop. The inverse dynamic model of the equivalent tree structure is obtained by cutting the revolute joint q_{8i} (i = 1, 2, 3), figure 2:

$$\Gamma_{tr_i} = \mathbf{H}_{tr_i}(\mathbf{q}_i, \dot{\mathbf{q}}_i, \ddot{\mathbf{q}}_i) \tag{23}$$

Leg i is isolated from the platform, so we can consider the variables q_{1i}, q_{2i} and q_{3i} to be independent. In the complete model of the robot, only q_{1i} is active and the torques Γ_{2i} and Γ_{3i} are zero.

Let the vector $\mathbf{q}_{a_i}$ is composed of the independent joints and the vector $\mathbf{q}_{p_i}$ is composed of the passive joints of leg i:

$$\mathbf{q}_{a_i} = [q_{1i} \quad q_{2i} \quad q_{3i}]^T \qquad \mathbf{q}_{p_i} = [q_{4i} \quad q_{5i} \quad q_{7i}]^T \tag{24}$$

The constraint equations of the loop are:

$$q_{4i} = -q_{3i}, \quad q_{5i} = -q_{2i} - \frac{\pi}{2}, \quad q_{7i} = q_{3i}, \quad q_{8i} = -q_{3i} \tag{25}$$

The dynamic model of the leg is obtained from Γ_{tr_i} and the constraint equations by [Khalil and Dombre, 2002]:

$$\Gamma_i = [\Gamma_{1i} \quad \Gamma_{2i} \quad \Gamma_{3i}]^T = \mathbf{G}_i^T \Gamma_{tr_i} = \mathbf{H}_i(\mathbf{q}_i, \dot{\mathbf{q}}_i, \ddot{\mathbf{q}}_i) = \mathbf{A}_i(\ddot{\mathbf{q}}_i) + \mathbf{H}_i(\mathbf{q}_i, \dot{\mathbf{q}}_i) \tag{26}$$

With: $\mathbf{G}_i = \dfrac{\partial \mathbf{q}_{tr_i}}{\partial \mathbf{q}_{a_i}}$ and $\mathbf{q}_{tr_i} = \left[\mathbf{q}_{a_i}^T \quad \mathbf{q}_{p_i}^T\right]^T$, thus:

$$\mathbf{G}_i^T = \begin{bmatrix} 1 & 0 & 0 & 0 & 0 & 0 \\ 0 & 1 & 0 & 0 & -1 & 0 \\ 0 & 0 & 1 & -1 & 0 & 1 \end{bmatrix} \tag{27}$$

The dynamic model of the tree structure is obtained by any classical method. We used recursive symbolic Newton-Euler method with the base or essential parameters.

WORKSPACE ANALYSIS OF THE ORTHOGLIDE USING INTERVAL ANALYSIS

D. Chablat, Ph. Wenger

Institut de Recherche en Communications et Cybernétique de Nantes, 1, rue de la Noë, 44321 Nantes, France

Damien.Chablat@irccyn.ec-nantes.fr

J. Merlet

INRIA Sophia-Antipolis, 2004 Route des Lucioles, 06902 Sophia Antipolis, France

Merlet@sophia.inria.fr

Abstract This paper addresses the workspace analysis of the orthoglide, a 3-DOF parallel mechanism designed for machining applications. This machine features three fixed parallel linear joints which are mounted orthogonally and a mobile platform which moves in the Cartesian x-y-z space with fixed orientation. The workspace analysis is conducted on the bases of prescribed kinetostatic performances. The interesting features of the orthoglide are a regular Cartesian workspace shape, uniform performances in all directions and good compactness. Interval analysis based methods for computing the dextrous workspace and the largest cube enclosed in this workspace are presented.

Keywords: Optimal design, Parallel Mechanism, Machining, Workspace, Interval Analysis, Transmission factor.

1. Introduction

Parallel kinematic machines (PKM) are commonly claimed to offer several advantages over their serial counterparts, like high structural rigidity, high dynamic capacities and high accuracy (Treib and Zirn, 1998; Wenger et al., 1999). Thus, PKM are interesting alternative designs for high-speed machining applications.

This is why parallel kinematic machine-tools attract the interest of more and more researchers and companies. Since the first prototype presented in 1994 during the IMTS in Chicago by Gidding&Lewis (the Variax), many other prototypes have appeared.

However, the existing PKM suffer from two major drawbacks, namely, a complex Cartesian workspace and highly non linear input/output rela-

J. Lenarčič and F. Thomas (eds.), Advances in Robot Kinematics, 397–406.

tions. For most PKM, the Jacobian matrix which relates the joint rates to the output velocities is not constant and not isotropic. Consequently, the performances (e.g. maximum speeds, forces accuracy and rigidity) vary considerably for different points in the Cartesian workspace and for different directions at one given point. This is a serious drawback for machining applications (Kim et al., 1997; Treib and Zirn, 1998; Wenger et al., 2001). To be of interest for machining applications, a PKM should preserve good workspace properties, that is, regular shape and acceptable kinetostatic performances throughout. In milling applications, the machining conditions must remain constant along the whole tool path (Rehsteiner et al., 1999; Rehsteiner et al., 1999). In many research papers, this criterion is not taking into account in the algorithmic methods used for the optimization of the workspace volume (Luh et al., 1996; Merlet, 1999; Ottaviano, (1991)).

The orthoglide is a 3-axis PKM with the advantages a classical serial PPP machine tool but not its drawbacks. It is an optimized version of the Delta mechanism defined by Clavel, (1990). Most industrial 3-axis machine-tool have a serial PPP kinematic architecture with orthogonal linear joint axes along the x, y and z directions. Thus, the motion of the tool in any of these directions is linearly related to the motion of one of the three actuated axes. Also, the performances are constant in the most part of the Cartesian workspace, which is a parallelepiped. The main drawback is inherent to the serial arrangement of the links, namely, poor dynamic performances.

The orthoglide is a PKM with three fixed linear joints mounted orthogonally. The mobile platform is connected to the linear joints by three articulated parallelograms and moves in the Cartesian x-y-z space with fixed orientation. Its workspace shape is close to a cube whose sides are parallel to the planes xy, yz and xz respectively. The interval analysis is conducted on the basis of prescribed bounded velocity and force transmission factors. Interval analysis based method is used to compute dextrous workspace as well as the largest cube enclosed in this workspace (Merlet, 2000).

2. Description of the Orthoglide

Most existing PKM can be classified into two main families. The PKM of the first family have fixed foot points and variable length struts and are generally called "hexapods". The second family of PKM has been more recently investigated and have variable foot points and fixed length struts. PKMs of the second family are more interesting because the actuators are fixed and thus the moving masses are lower than in the hexapods and tripods.

The orthoglide studied in this article is a 3-axis translational parallel kinematic machine and is belongs to the second family. Figure 1 shows the general kinematic architecture of the orthoglide. The orthoglide has three parallel $PRPaR$ identical chains (where P, R and Pa stands for Prismatic, Revolute and Parallelogram joint, respectively). The actuated joints are the three orthogonal linear joints. These joints can be actuated by means of linear motors or by conventional rotary motors with ball screws. The output body is connected to the linear joints through a set of three parallelograms of equal lengths $L = B_iC_i$, so that it can move only in translation. The first linear joint axis is parallel to the x-axis, the second one is parallel to the y-axis and the third one is parallel to the z-axis. In figure 1, the base points A_1, A_2 and A_3 are fixed on the ith linear axis such that $A_1A_2 = A_1A_3 = A_2A_3$, B_i is at the intersection of the first revolute axis $\mathbf{i}_i$ and the second revolute axis $\mathbf{j}_i$ of the ith parallelogram, and C_i is at the intersection of the last two revolute joints of the ith parallelogram. When each B_iC_i is aligned with the linear joint axis A_iB_i , the orthoglide is in an isotropic configuration and the tool center point P is located at the intersection of the three linear joint axes. In this configuration, the base points A_1, A_2 and A_3 are equally distant from P. The symmetric design and the simplicity of the kinematic chains (all joints have only one degree of freedom, fig. 2) should contribute to lower the manufacturing cost of the orthoglide. The orthoglide is free of singularities and self-collisions. The workspace has a regular, quasi-cubic shape. The input/output equations are simple and the velocity transmission factors are equal to one along the x, y and z direction at the isotropic configuration, like in a serial PPP machine (Wenger and Chablat, 2001).

3. Kinematic Equations and Singularity Analysis

We recall briefly here the kinematics of the Orthoglide (See Wenger and Chablat, 2001 for more details).

3.1 Kinematic Equations

Let θ_i and β_i denote the joint angles of the parallelogram about the axes $\mathbf{i}_i$ and $\mathbf{j}_i$, respectively (fig. 2). Let ρ_1, ρ_2, ρ_3 denote the linear joint variables, $\rho_i = A_iB_i$ and L denote the length of the tree legs, B_iC_i. The position vector $\mathbf{p}$ of the tool center point P is defined in a reference frame (O, x, y, z) centered at the intersection of the three linear joint axes (note that the reference frame has been translated in Fig. 1 for more legibility).

Let $\dot{\boldsymbol{\rho}}$ be referred to as the vector of actuated joint rates and $\dot{\mathbf{p}}$ as the velocity vector of point P:

$$\dot{\boldsymbol{\rho}} = [\dot{\rho}_1\ \dot{\rho}_2\ \dot{\rho}_3]^T, \quad \dot{\mathbf{p}} = [\dot{x}\ \dot{y}\ \dot{z}]^T$$

$\dot{\mathbf{p}}$ can be written in three different ways by traversing the three chains $A_iB_iC_iP$:

$$\dot{\mathbf{p}} = \mathbf{n}_1\dot{\rho}_1 + (\dot{\theta}_1\mathbf{i}_1 + \dot{\beta}_1\mathbf{j}_1) \times (\mathbf{c}_1 - \mathbf{b}_1) \tag{1a}$$
$$\dot{\mathbf{p}} = \mathbf{n}_2\dot{\rho}_1 + (\dot{\theta}_2\mathbf{i}_2 + \dot{\beta}_2\mathbf{j}_2) \times (\mathbf{c}_2 - \mathbf{b}_2) \tag{1b}$$
$$\dot{\mathbf{p}} = \mathbf{n}_3\dot{\rho}_3 + (\dot{\theta}_3\mathbf{i}_3 + \dot{\beta}_3\mathbf{j}_3) \times (\mathbf{c}_3 - \mathbf{b}_3) \tag{1c}$$

where $\mathbf{b}_i$ and $\mathbf{c}_i$ are the position vectors of the points B_i and C_i, respectively, and $\mathbf{n}_i$ is the direction vector of the linear joints, for $i = 1, 2, 3$.

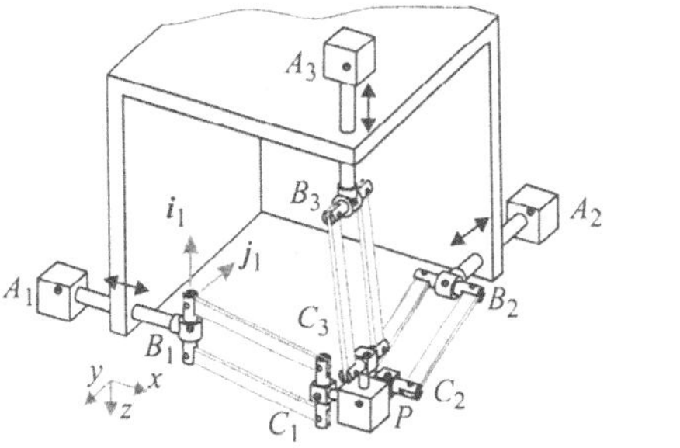

Figure 1. Orthoglide kinematic architecture

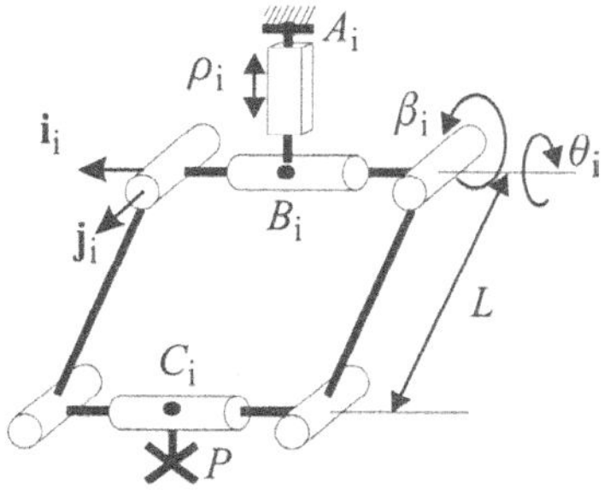

Figure 2. Leg kinematics

3.2 Singular configurations

We want to eliminate the two idle joint rates $\dot{\theta}_i$ and $\dot{\beta}_i$ from Eqs. (1a–c), which we do upon dot-multiplying Eqs. (1a–c) by $\mathbf{c}_i - \mathbf{b}_i$:

$$(\mathbf{c}_1 - \mathbf{b}_1)^T\dot{\mathbf{p}} = (\mathbf{c}_1 - \mathbf{b}_1)^T\mathbf{n}_1\dot{\rho}_1 \tag{2a}$$
$$(\mathbf{c}_2 - \mathbf{b}_2)^T\dot{\mathbf{p}} = (\mathbf{c}_2 - \mathbf{b}_2)^T\mathbf{n}_2\dot{\rho}_2 \tag{2b}$$
$$(\mathbf{c}_3 - \mathbf{b}_3)^T\dot{\mathbf{p}} = (\mathbf{c}_3 - \mathbf{b}_3)^T\mathbf{n}_3\dot{\rho}_3 \tag{2c}$$

Equations (2a–c) can now be cast in vector form, namely

$$\mathbf{A}\dot{\mathbf{p}} = \mathbf{B}\dot{\boldsymbol{\rho}}$$

where $\mathbf{A}$ and $\mathbf{B}$ are the parallel and serial Jacobian matrices, respectively:

$$\mathbf{A} = \begin{bmatrix} (\mathbf{c}_1 - \mathbf{b}_1)^T \\ (\mathbf{c}_2 - \mathbf{b}_2)^T \\ (\mathbf{c}_3 - \mathbf{b}_3)^T \end{bmatrix} \quad \text{and} \quad \mathbf{B} = \begin{bmatrix} \eta_1 & 0 & 0 \\ 0 & \eta_2 & 0 \\ 0 & 0 & \eta_3 \end{bmatrix} \tag{3}$$

with $\eta_i = (\mathbf{c}_i - \mathbf{b}_i)^T \mathbf{n}_i$ for $i = 1, 2, 3$.

Parallel singularities (Chablat and Wenger, 1998) occur when the determinant of the matrix $\mathbf{A}$ vanishes, *i.e.* when $det(\mathbf{A}) = 0$. In such configurations, it is possible to move locally the mobile platform whereas the actuated joints are locked. These singularities are particularly undesirable because the structure cannot resist any force. Eq. (3a) shows that the parallel singularities occur when:

$$(\mathbf{c}_1 - \mathbf{b}_1) = \alpha(\mathbf{c}_2 - \mathbf{b}_2) + \lambda(\mathbf{c}_3 - \mathbf{b}_3)$$

that is when the points B_1, C_1, B_2, C_2, B_3 and C_3 are coplanar. A particular case occurs when the links B_iC_i are parallel:

$$(\mathbf{c}_1-\mathbf{b}_1)//(\mathbf{c}_2-\mathbf{b}_2) \quad \text{and} \quad (\mathbf{c}_2-\mathbf{b}_2)//(\mathbf{c}_3-\mathbf{b}_3) \quad \text{and} \quad (\mathbf{c}_3-\mathbf{b}_3)//(\mathbf{c}_1-\mathbf{b}_1)$$

Serial singularities arise when the serial Jacobian matrix $\mathbf{B}$ is no longer invertible *i.e.* when $det(\mathbf{B}) = 0$. At a serial singularity a direction exists along which any cartesian velocity cannot be produced. Eq. (3b) shows that $\det(\mathbf{B}) = 0$ when for one leg i, $(\mathbf{b}_i - \mathbf{a}_i) \perp (\mathbf{c}_i - \mathbf{b}_i)$.

When $\mathbf{A}$ and $\mathbf{B}$ are not singular, we obtain the relations,

$$\dot{\mathbf{p}} = \mathbf{J}\dot{\rho} \text{ with } \mathbf{J} = \mathbf{A}^{-1}\mathbf{B} \tag{4}$$

3.3 Velocity transmission factors

For joint rates belonging to a unit ball, namely, $||\dot{\rho}|| \leq 1$, the Cartesian velocities belong to an ellipsoid such that:

$$\dot{p}^T(\mathbf{J}\mathbf{J}^T)\dot{p} \leq 1$$

The eigenvectors of matrix $\mathbf{J}\mathbf{J}^T$ define the direction of its principal axes of this ellipsoid and the square roots ψ_1, ψ_2 and ψ_3 of the eigenvalues σ_1, σ_2 and σ_3 of $\mathbf{J}\mathbf{J}^T$, *i.e.* the lengths of the aforementioned principal axes are the velocity transmission factors in the directions. To limit the variations of this factor in the Cartesian workspace, we set

$$\psi_{min} \leq \psi_i \leq \psi_{max} \tag{5}$$

throughout the workspace. To simplify the problem, we set $\psi_{min} = 1/\psi_{max}$ where the value of ψ_{max} depends on the performance requirements.

4. Determination of the dextrous workspace

The dextrous workspace $\mathcal{W}$ is here defined as the loci of the points for which all the eigenvalues of the matrix $\mathbf{J}\mathbf{J}^T$, *i.e.* the velocity transmission factors, lie within a predefined range $[\sigma_{min}, \sigma_{max}]$. These eigenvalues are determined by solving the third degree characteristic polynomial of the matrix which is defined only for the points within the intersection $\mathcal{I}$ of the three cylinders defined by

$$x^2 + y^2 \leq L \quad x^2 + z^2 \leq L \quad y^2 + z^2 \leq L \tag{6}$$

To solve numerically the above equations, the length of the legs is normalized, *i.e.* we set $L = 1$.

Our purpose in this section is to determine an approximation of $\mathcal{W}$ as a set of 3D Cartesian boxes for any point of which we are sure that the constraints on the eigenvalues are satisfied. The width of all the boxes in the list will be greater than a given threshold and the value of this threshold will define the quality of the approximation.

4.1 Box verification

A basic tool of the algorithm is a module $\mathcal{M}(B)$ that takes as input a box B belonging to $\mathcal{I}$ and whose output is:

- either that for any point in the box the eigenvalues lie in the range $[\sigma_{min}, \sigma_{max}]$
- or that for any point in the box one of the eigenvalues is either lower than σ_{min} or larger than σ_{max}
- or that the two previous conditions does not hold for all the points of the box *i.e.* that for some points the eigenvalues lie in the range $[\sigma_{min}, \sigma_{max}]$ while this is not true for some other points

The first step of this module consists in considering an arbitrary point of the box (e.g. its center) and to compute the eigenvalues at this point: either all of them lie in the range $[\sigma_{min}, \sigma_{max}]$ or at least one of them lie outside this range.

In the first case if we are able to check that there is no point in B such that the eigenvalue at this point may be equal to σ_{min} or σ_{max}, then we may guarantee that for any point of B the eigenvalues will be in the range $[\sigma_{min}, \sigma_{max}]$. Indeed assume that at a given point of B the lowest eigenvalue is lower than σ_{min}: this implies that somewhere along the line joining this point to the center of the box the lowest eigenvalue will be exactly σ_{min}. To perform this check we substitute in the polynomial λ successively by σ_{min} and σ_{max} to get a polynomial in x, y, z only. We have now to verify if there is at least one value for these three variables that cancel the polynomial, being understood that these values have to define a point belonging to B: this is done by using an interval analysis algorithm from the `ALIAS` library (Merlet, 2000).

Assume now that at the center of the box the largest eigenvalue is greater than σ_{max}. If we are to determine that there is no point of B such that one of the eigenvalue is equal to σ_{max}, then we may guarantee that for any point of B the largest eigenvalue will always be greater than σ_{max}. This check is performed by using the same method than in the previous case. Hence the $\mathcal{M}$ module will return:

- 1: if for all points of B the eigenvalues lie in $[\sigma_{min}, \sigma_{max}]$ (hence B is in the dextrous workspace)
- -1: if for all points of B either the largest eigenvalue is always greater than σ_{max} or the lowest eigenvalue is lower than σ_{min} (hence B is outside the dextrous workspace).
- 0: in the other cases *i.e.* parts of B may be either outside or inside the dextrous workspace

4.2 Algorithm for the determination of the dextrous workspace

The principle of the algorithm is pretty simple: we will maintain a list $\mathcal{L}$ of boxes, indexed by i, which is is initialized with the box [-1,1], [-1,1], [-1,1]. A minimal width ϵ for the ranges in a box is defined and the operator $\mathcal{W}(B_i)$ will return the largest width of the ranges in B_i. An error index $\mathcal{E}$ will be computed as the total volume of the boxes that are not in the approximation but may contain points that are inside the dextrous workspace. We then apply $\mathcal{M}(B_i)$:

- if $\mathcal{M}(B_i)$=1: we store B_i as part of the dextrous workspace and consider the next box in $\mathcal{L}$
- if $\mathcal{M}(B_i)$=-1: we consider the next box in $\mathcal{L}$
- if $\mathcal{M}(B_i)$=0:
  - if $\mathcal{W}(B_i) \geq \epsilon$: we create 2 new boxes from B_i by bisecting the range of B_i with the largest width. The two new boxes are put at the end of $\mathcal{L}$
  - otherwise we add the volume of B_i to $\mathcal{E}$

The algorithm stops when all the boxes in $\mathcal{L}$ have been processed. Note that this basic algorithm has to be modified in order to consider only boxes that belongs to $\mathcal{I}$ but this can be done using the same principle. The algorithm returns a description of the dextrous workspace as a list of boxes and the comparison between the volume of the approximation and $\mathcal{E}$ allows to determine the quality of the approximation. Note that

$\mathcal{E}$ is very conservative as part of this volume consists in points that do not belong to $\mathcal{I}$ or to the dextrous workspace.

4.3 Implementation and results

The previous algorithms has been implemented in Maple with a system call to a C++ program that implements the $\mathcal{M}$ module. For an ϵ of 0.05 we found in about 5 hours that the volume of the dextrous workspace is 1.468 with an error bound of [0,0.48] with $\sigma_{min} = 0.25$ and $\sigma_{max} = 4$, *i.e.* $\psi_{min} = 1/2$ and $\psi max = 2$ (Fig. 3).

5. Determination of the largest cube enclosed in the dextrous workspace

For usual machine tools, the Cartesian workspace is generally given as a function of the size of a right-angled parallelepiped. Due to the symmetrical architecture of the orthoglide, the Cartesian workspace has a fairly regular shape in which it is possible to include a cube whose sides are parallel to the planes xy, yz and xz respectively (Fig. 4).

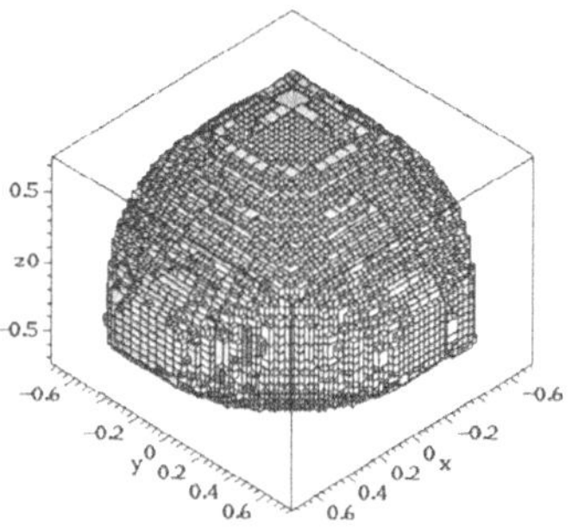

Figure 3. Dextrous workspace of the Orthoglide mechanism

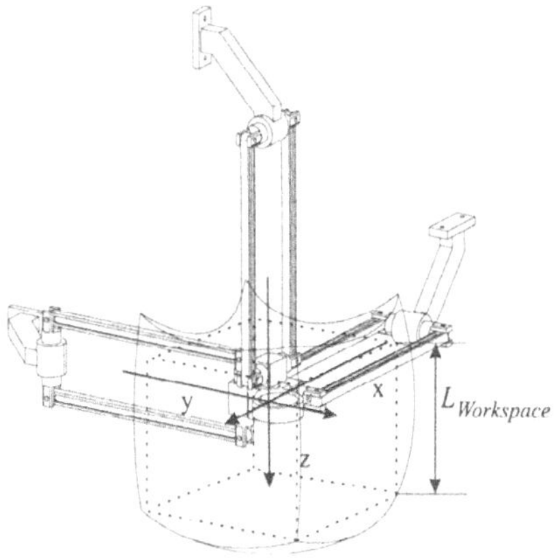

Figure 4. Cartesian workspace and isotropic configuration of the Orthoglide mechanism

We will now describe a method for determining a cube that is enclosed in this workspace whose edge length is $2R$ such that there is no other cube enclosed in the workspace with an edge length of $2(R + \alpha)$, where α is an accuracy threshold fixed in advance.

The first step is to determine the largest cube enclosed in the workspace with a center located at (0,0,0). This is done by using the $\mathcal{M}$ module on the box $[-A - k\alpha, A + k\alpha], [-A - k\alpha, A + k\alpha], [-A - k\alpha, A + k\alpha]$ where k is an integer initialized to 1 and A a number initialized to 0.

Each time the $\mathcal{M}$ module returns 1 (which means that the cube with edge length $2(A + k\alpha)$ is enclosed in the dextrous workspace) we set A to $A + k\alpha$ and we double the value of k. When this module returns -1 if k is greater than 1 we reset k to 1 and restart the process, otherwise the process stops and we have determined that the cube with edge length $2A$ is enclosed in the workspace while the cube with edge length $2(A + \alpha)$ is not: thus A is an initial value for R.

We use then the same algorithm than for the determination of the workspace with the following modifications:

- for each box B_i in $\mathcal{L}$ we test if at the 8 corners of the cube centered at the center of the box and with edge length $2(R + \alpha)$ the eigenvalues satisfies the constraints:
  - if this not the case let u be the maximal half-width of the ranges of B_i and assume that $u_1 = R - u > 0$. If for at least for one corner of the cube $\mathcal{C}$ with center of B_i and edge length u_1 the eigenvalues do not satisfy the constraint, then the center of the largest cube cannot be located in B_i: indeed any cube with a center in B_i and edge length $2R$ will include all the corners of $\mathcal{C}$ and thus contain at least a point for which the eigenvalues do not satisfy the constraints.
  - if at the corners the eigenvalues are all valid then we search for the largest cube centered at the center of the box using the same method than for the determination of the largest cube centered at (0,0,0). If we find a cube with a larger edge length than $2R$, then the value of R is updated.
- the boxes are bisected only if the largest width of their ranges is greater than 2α: indeed even if the largest cube has a center located in such box, then the maximal edge length will be at most $2(R + \alpha)$.

Using this algorithm with a value of 0.001 for α we found out that the largest cube has its center located at $(0.085938, 0.085938, 0.085938)$ and that its edge length was $L_{Workspace} = 0.643950$ while we may guarantee that there is no cube with edge length larger than 0.643952 enclosed in the workspace

6. Conclusions

The dextrous Cartesian workspace and the largest cube enclosed in this workspace is computed using interval analysis based method for the Orthoglide. Unlike most existing PKMs, the dextrous workspace is fairly regular and the performances are homogeneous in it. Thus, the entire workspace is really available for tool paths. The bounds of velocity and force transmission factors used in this paper are given as an example because it depends on the performance requirements for machining applications. A 1:3-scale prototype of this mechanism is under construction in our laboratory with these bound contraints.

References

Treib, T. and Zirn, O. (1998), "Similarity laws of serial and parallel manipulators for machine tools", Proc. Int. Seminar on Improving Machine Tool Performance, pp. 125–131, Vol. 1.

Wenger, P. Gosselin, C. and Maille. B. (1999), "A Comparative Study of Serial and Parallel Mechanism Topologies for Machine Tools", Proc. PKM'99, Milano, pp. 23–32.

Kim J. , Park C., Kim J. and Park F.C., (1997), "Performance Analysis of Parallel Manipulator Architectures for CNC Machining Applications", Proc. IMECE Symp. On Machine Tools, Dallas.

Wenger, P. Gosselin, C. and Chablat, D. (2001), "A Comparative Study of Parallel Kinematic Architectures for Machining Applications", Proc. Workshop on Computational Kinematics', Seoul, Korea, pp. 249–258.

Rehsteiner, F., Neugebauer, R. Spiewak, S. and Wieland, F., (1999), "Putting Parallel Kinematics Machines (PKM) to Productive Work", Annals of the CIRP, Vol. 48:1, pp. 345–350.

Tlusty, J., Ziegert, J, and Ridgeway, S., (1999), "Fundamental Comparison of the Use of Serial and Parallel Kinematics for Machine Tools", Annals of the CIRP, Vol. 48:1, pp. 351–356.

Luh C-M., Adkins F. A., Haug E. J. and Qui C. C., (1996), "Working Capability Analysis of Stewart platforms", Transactions of ASME, pp. 220–227.

Merlet J-P., (1999), "Determination of 6D Workspace of Gough-Type Parallel Manipulator and Comparison between Different Geometries", The Int. Journal of Robotic Research, Vol. 19, No. 9, pp. 902–916.

Golub, G. H. and Van Loan, C. F., (1989), Matrix Computations, The John Hopkins University Press, Baltimore.

Salisbury J-K. and Craig J-J., (1982), "Articulated Hands: Force Control and Kinematic Issues", The Int. J. Robotics Res., Vol. 1, No. 1, pp. 4–17.

Angeles J., (1997), Fundamentals of Robotic Mechanical Systems, Springer-Verlag.

Wenger, P., and Chablat, D., (2000), "Kinematic Analysis of a new Parallel Machine Tool: the Orthoglide", in Lenarčič, J. and Stanišić, M.M. (editors), *Advances in Robot Kinematic*, Kluwer Academic Publishers, June, pp. 305–314.

Chablat D. and Wenger P., (1998), "Working Modes and Aspects in Fully-Parallel Manipulator", IEEE Int. Conf. On Robotics and Automation, pp. 1964–1969.

Merlet J-P., (2000), "ALIAS: an interval analysis based library for solving and analyzing system of equations", SEA, June. Automation, pp. 1964–1969.

Clavel, R., (1990), "Device for the Movement and Positioning of an Element in Space", US Patent No. 4,976,582, December 11.

Ottaviano E., Ceccarelli M., (2001), "Optimal Design of CAPAMAN (Cassino Parallel Manipulator) with Prescribed Workspace", 2nd Workshop on Computational Kinematics CK 2001, Seoul. pp.35-44.

A DIVIDE-AND-CONQUER METHOD FOR INVERSE KINEMATICS OF HYPER-REDUNDANT MANIPULATORS

Yunfeng Wang, Gregory S. Chirikjian
Department of Mechanical Engineering
Johns Hopkins University
Baltimore, MD 21218
{yfwang,gregc}@jhu.edu

Abstract Hyper-redundant manipulators have a large number of redundant degrees of freedom. They have been recognized as a means to improve manipulator performance in complex and unstructured environments. However, the high degree of redundancy also causes difficulty in inverse kinematics calculations. Motivated by the cascading of workspace densities, this paper develops a "divide-and-conquer" method for inverse kinematics using the workspace density generated by a partial differential equation. This method does not involve a high dimensional Jacobian matrix and offers high accuracy. Its computational complexity is only $\mathcal{O}(log_2 P)$ for a manipulator with P modules. Numerical simulations are performed to demonstrate this method.

Keywords: Inverse Kinematics, Hyper-Redundant Manipulator, Workspace Density

1. Introduction

Hyper-redundant manipulators, which are also called snake-like, serpentine or highly-articulated manipulators, possess conformational freedom far superior to that of conventional manipulators. Substantial research has been done on this type of manipulator, and many different prototypes have been constructed [Hannan and Walker, 2001; Hirose, 1993; Koganezawa et al., 2001]. To avoid complicated control schemes, binary-actuated manipulators are proposed in [Chirikjian, 1994]. They adopt a variable geometry truss structure. Figure 1 shows a spatial binary hyper-redundant manipulator built in the second author's laboratory [Ebert-Uphoff, 1997]. It contains 6 modules each which has 6 actuators and can reach around 7×10^{10} states.

The high degree of redundancy of hyper-redundant manipulators usually results in an infinite number of inverse kinematic solutions. Various approaches have been proposed to select an optimal one under some criteria from among these infinite number of solutions. For a manipulator

J. Lenarčič and F. Thomas (eds.), Advances in Robot Kinematics, 407–414.

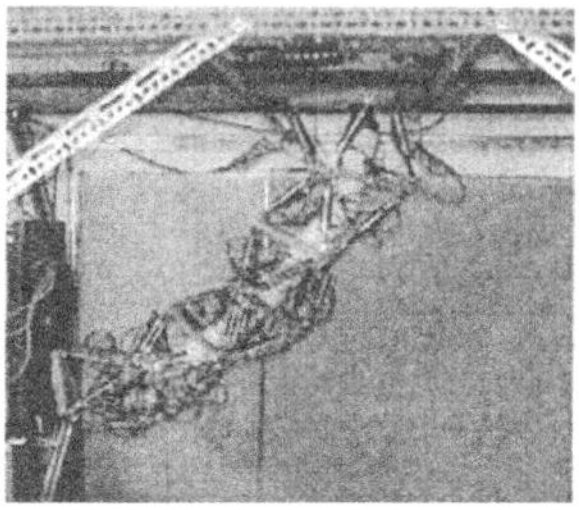

Figure 1. A binary manipulator with 6 modules [Ebert-Uphoff, 1997]

with only a few extra degrees of freedom (DOF) the problem is usually solved by strict optimization of a cost function [Deo and Walker, 1997; Krzysztof, 2000; Wunderlich et al., 1996]. Those kinds of optimization procedures become very inefficient in the case of highly articulated manipulators because the parameter optimization is performed in a very high dimensional space. A backbone curve method was developed in [Chirikjian and Burdick, 1994; Zanganeh and Angeles, 1995], in which the shape of the manipulator is prescribed by a backbone curve. The authors of [Ebert-Uphoff and Chirikjian, 1996] used the workspace density function in a breadth-first search for solving the inverse kinematics problem.

In this paper, we propose a "divide-and-conquer" method for solving inverse kinematics of hyper-redundant manipulators. The general idea is to divide the manipulator into virtual segments and find the optimal manipulator configuration by cascading the corresponding workspace densities and singling out the case that results in the maximum workspace density over the target spot. Using this method, the computational complexity is only $\mathcal{O}(log_2 P)$ for a manipulator with P modules. Also, the high dimensional Jacobian matrix is avoided.

2. The Divide-and-Conquer Method

2.1 Workspace Density

One major input for the proposed divide-and-conquer method is the workspace density of a manipulator with different modules. Workspace density describes the density of reachable points/frames in any portion of the workspace [Ebert-Uphoff, 1997], where the workspace is discretized into small blocks and the density of points/frames is the number of reachable points/frames per unit workspace volume. Workspace density is a probabilistic measure of accuracy over the workspace. Higher density at a point means the manipulator can reach that point more accurately. When the concept of workspace density is used for continuously-actuated manipulators, the range of motion of each degree of freedom (DOF) will be discretized first. For example, a manipulator with n DOF each sampled at K values results in a discretized workspace with K^n positions and orientations.

We generate the workspace density by solving a diffusion-type equation [Wang, 2001], which describes the evolution of the workspace density function depending on manipulator length and kinematic properties. This diffusion-type equation is a partial differential equation defined on the motion group and can be easily solved using the techniques of non-commutative harmonic analysis as developed in [Chirikjian and Kyatkin, 2000]. The computational complexity of this workspace generation algorithm is independent of the number of DOF/modules.

2.2 The Scheme for Using Workspace Density

Some notational conventions are clarified first. As shown in Figure 2, g_k denotes the homogenous transformation that relates the distal end of the kth segment to its own base, where $k \in \{1, ..., P\}$. The transformation of the distal end of the kth segment relative to the base of the whole manipulator is then

$$g^{(k)} = g_1 \circ g_2 \circ \cdots \circ g_k$$

and the transformation that relates the distal end of the manipulator to the distal end of the kth segment is

$$(g^{(k)})^{-1} \circ g^{(P)} = g_{k+1} \circ g_{k+2} \circ \cdots \circ g_P.$$

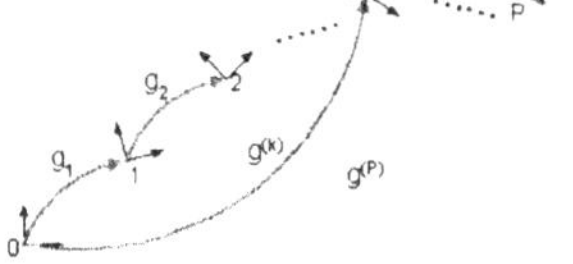

Figure 2. The notation of transformations

Consider a manipulator with P independent modules and denote the workspace density function for the ith module as $\rho_{i-1,i}(g)$. With this notation, $\rho_{0,P}(g)$ will be the workspace density function for the whole manipulator.

It was shown in [Chirikjian and Ebert-Uphoff, 1998] that the workspace density function for two concatenated modules is the convolution

$$\rho_{i-1,i+1}(g) \;=\; (\rho_{i-1,i} * f_{i,i+1})(g) \qquad (1)$$

and the workspace density function for the whole manipulator is the multiple convolution

$$\rho_{0,P}(g) = (\rho_{0,1} * \rho_{1,2} * \cdots * \rho_{P-1,P})(g).$$

Using the fact revealed in (1), we virtually divide the manipulator into segments. Let the lower segment serve as a transport device for the workspace density of the upper segment. We then fix the configuration of lower segment by choosing the one that satisfies the given criteria.

2.3 Realization Procedure

One way to realize the above scheme is to start from the middle module and proceed concurrently by dividing the manipulator in half as depicted in Figure 3. This is an example of the very general divide-and-conquer method often employed in computer science. In Figure 3, the shaded areas denote the workspace density and g_{des} is the target spot. We first divide the whole manipulator into half and find the transformation h such that the density $\rho_{p/2}(h) \cdot \rho_{p/2}(h^{-1} \circ g_{des})$ is maximized. Note that "$\cdot$" is a scalar multiplication. Suppose h_{opt} is such a transformation. Then we divide each segment into half again as shown in Figure 3b. We concurrently search for the transformation r that maximizes the density $\rho_{p/4}(r) \cdot \rho_{p/4}(r^{-1} \circ h_{opt})$ and s that maximizes the density $\rho_{p/4}(s) \cdot \rho_{p/4}(s^{-1} \circ g_{des})$. This dividing and searching procedure is performed until the segments consist of only one module. After we find the optimal states of all the modules, we can configure the manipulator. It is easy to notice that this is a fast computation with computational complexity of $\mathcal{O}(log_2 P)$. When P is not a power of 2, we first fix the modules from base one using the Ebert-Uphoff method [Chirikjian and Kyatkin, 2000; Ebert-Uphoff and Chirikjian, 1996] until the number of the rest modules are the number of power of 2.

For this divide-and-conquer method, two factors may affect the target-reaching accuracy. One is the error of workspace density for a manipulator with a few modules. The other is that the predicted state of the manipulator is used to decide the next step. To increase the accuracy, two strategies can be applied as follows.

- For a discretely-actuated or a discretized continuously-actuated manipulator with a few modules, the workspace is sparse. Using the workspace density to describe the workspace has relatively

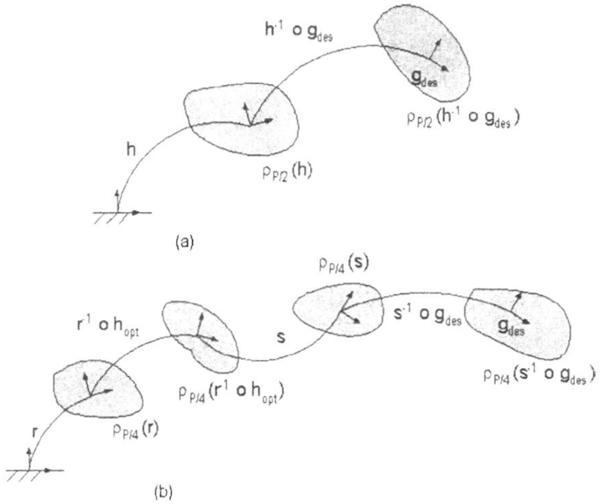

Figure 3. The divide-and-conquer method

large error. Hence, when the segments consist of a few modules, we directly apply brute force instead of using the information of workspace density. In this sense, the transformations decided by the divide-conquer method can be viewed as intermediate targets.

- Instead of using the previous intermediate target as the base state for the next intermediate target, we use the state decided by brute force. Hence, the current states of the manipulator's configuration are incorporated in the method.

We now use the example shown in Figure 4 to restate the above strategies concretely. In this example, the divide-and-conquer method is applied until each segment consist of 2 modules. $h1_{opt}, h2_{opt}, \cdots$ are the transformations of each segment decided by the divide-and-conquer method. Hence, $A, B \cdots$ will serve as intermediate targets. From the base of the manipulator, we start to use brute force to decide the transformation of each module for all the segments. b_1 and b_2 are the resulting transformation for the first segment. When we apply brute force for the second segment, we use the current state A' as its base instead of the predicted state A.

3. Numerical Simulations

Consider a discretely-actuated serial manipulator. Each module can reach 16 different states as shown in Figure 5. The workspace and the

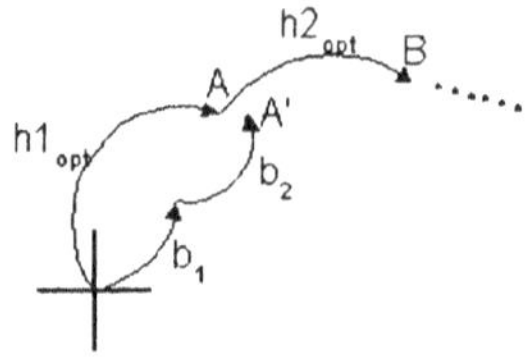

Figure 4. The strategies for increasing accuracy

workspace density of this manipulator with 4 modules are illustrated in Figures 6 (a) and (b) respectively. They are generated by brute force and the diffusion-type equation [Wang, 2001].

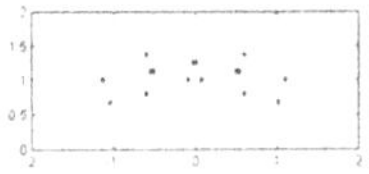

Figure 5. The workspace of a manipulator with one module

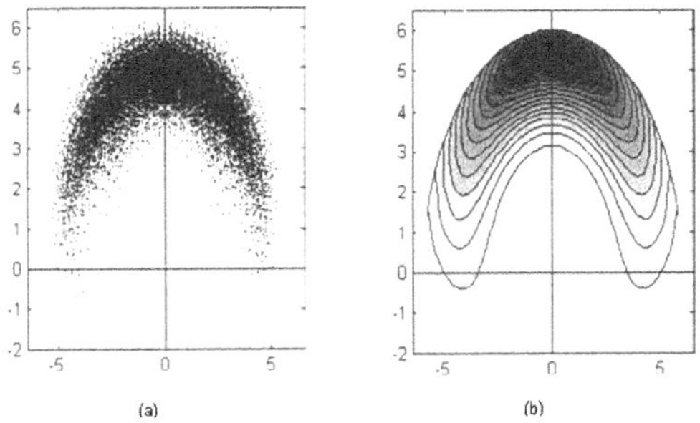

Figure 6. The workspace (a) and workspace density (b) of a 4-module manipulator

This manipulator with 16 modules is used to simulate the divide-and-conquer method. Brute force is applied when each segment consists of 4 modules. Targets located in different areas of the workspace are tested. Figure 7 shows the resulting configuration of the manipulator for the targets located in areas of high workspace density. Figures 8 give the resulting configuration of the manipulator for the targets located in areas of low workspace density. In these figures, the contour plot on the background of the figure indicates the workspace density of this manipulator with 16 modules. The cross denotes the position of the fixed end of the manipulator. Segmented lines display the corresponding configurations of the manipulator, where each segment represents a module. Circles indicate intermediate targets and end-effector targets. These simulations show that our method can make the manipulator reach targets very accurately.

The above simulations are implemented using Matlab on a Pentium III 64MB RAM PC. It takes less than 0.6 seconds to compute the inverse kinematics solution for each target. Note that the computation time for generating the workspace densities is not counted since we store the workspace density of this manipulator with different numbers of modules in advance.

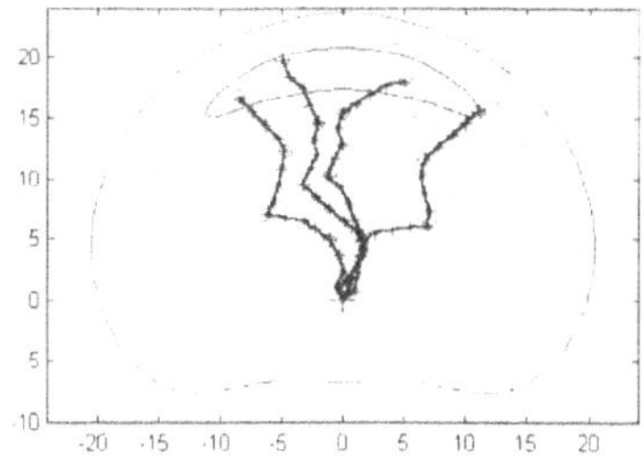

Figure 7. Targets located in areas of high workspace density

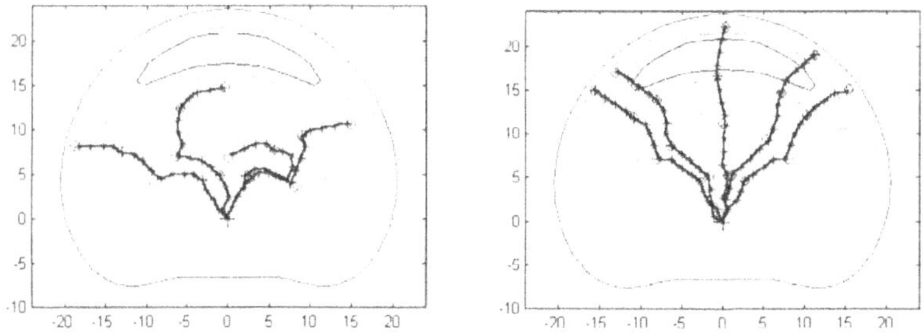

Figure 8. Targets located in areas of low workspace density

4. Conclusion

Inverse kinematics is an important issue for applications of hyper-redundant manipulators to be realized. This paper presented a divide-and-conquer method by dividing the manipulator into virtual segments and cascading the corresponding workspace densities. Using this method, the inverse kinematics of a hyper-redundant manipulator can be solved simply and efficiently with a high target-reaching accuracy as demenstrated by numerical simulations. This method has computational complexity of $\mathcal{O}(log_2 P)$ for a manipulator with P modules.

5. Acknowledgements

The authors would like to acknowledge the support of the National Science Foundation for support under grant IIS-0098382.

References

Chirikjian, G.S., and Kyatkin, A.B. (2000) *Engineering Applications of Noncommutative Harmonic Analysis*, CRC Press.

Chirikjian, G.S., and Ebert-Uphoff, I. (1998), Numerical convolution on the Euclidean group with applications to workspace generation, *IEEE Trans. on Robotics and Automation*, vol. 14, no. 1, pp. 123–136.

Chirikjian, G.S. (1994), A binary Paradigm for Robotic Manipulators, *IEEE International Conference on Robotics and Automation*, pp. 3063–3069.

Chirikjian, G.S., and Burdick, J.W. (1994), A hyper-redundant manipulator, *IEEE Robotics and Automation Magazine*, pp. 22–29.

Deo, A.S., and Walker, I.D. (1997), Minimum effort inverse kinematics for redundant manipulators, *IEEE Trans. on Robotics and Automation*, vol. 13, no. 5, pp. 767–775.

Ebert-Uphoff, I. (1997), *On the development of discretely-actuated hybrid-serial-parallel manipulators*, Ph.D. Dissertation, Johns Hopkins University.

Ebert-Uphoff, I., and Chirikjian, G.S. (1996), Inverse kinematics of discretely actuated hyper-redundant manipulators using workspace densities, *IEEE Int. Conf. on Robotics and Automation*, pp. 139–145.

Hannan, M.W., and Walker, I.D. (2001), The elephant trunk manipulator, design and implementation, *2001 IEEE/ASME Int. Conf. on Advanced Intelligent Mechatronics*, Como, Italy, pp. 14–19.

Hirose, S. (1993), *Biologically Inspired Robots*, Oxford, Oxford University Press.

Koganezawa, K., and Kinoshita, T. (2001), Hyper redundant manipulator using compound three-bar linkages, *2001 IEEE/ASME Int. Conf. on Advanced Intelligent Mechatronics*, Como, Italy, pp. 8–13.

Krzysztof, T. (2000), Hyperbolic normal forms for manipulator kinematics, *IEEE Trans. on Robotics and Automation*, vol. 16, no. 2, pp. 196–201.

Wang, Y.F. (2001), *Applications of Diffusion Processes in Robotics, Optical Communications and Polymer Science*, Ph.D. Dissertation, Johns Hopkins University.

Wunderlich, J. and Boncelet, Ch. (1996), Local optimization of redundant manipulator kinematics within constrained workspaces, *IEEE Int. Conf. on Robotics and Automation*, pp. 127–138.

Zanganeh, K.R., and Angeles, J. (1995), The inverse kinematics of hyper-redundant manipulators using splines, *IEEE Int. Conf. on Robotics and Automation*, pp. 2797–2802.

INTERVAL PROPAGATION FOR SOLVING PARALLEL SPHERICAL MECHANISMS

Enric Celaya

Institut de Robòtica i Informàtica Industrial

Llorens i Artigas, 4-6, 08028 Barcelona

celaya@iri.upc.es

Abstract An interval propagation method for spherical kinematic loops is used in a branch and prune algorithm to solve the direct kinematics of parallel spherical mechanisms. The algorithm finds all solutions with a desired resolution. The use of specific properties of the rotation equations involved allows the method to be more efficient than more general algorithms for this problem.

Keywords: Interval methods, interval propagation, parallel spherical mechanisms.

1. Introduction

The direct kinematics of parallel manipulators has proven to be very complex. In the spatial domain, for example, it is well known that the general Stewart-Gough platform can have as many as 40 different solutions for a given position of the actuators (Dietmaier, 1998). If we stay into the domain of spherical mechanisms, we still find that the parallel platform studied in Gosselin et al., 1994 is described by an 8-degree polynomial and has 8 possible solutions. An alternative to the traditional algebraic approaches to find the solutions of the direct kinematics of parallel manipulators consists in using interval methods to progressively reduce the length of the interval containing a solution through a branch and prune process, until the desired precision is reached. Typical problems of branch and prune interval methods are the sensitivity to the size of the initial intervals, the availability of an efficient pruning mechanism, and the decision about how much pruning must be done before branching takes place.

Here we present an interval method for spherical mechanisms that is well suited for this problem, since it uses an efficient algorithm for determining the set of all allowed values for a variable in a spherical kinematic loop. Besides this, a low cost interval propagation algorithm that provides the set of values for one variable compatible with a given interval of another one with no overestimation is used for pruning. Using enough precision in the computations, the method grants to find all the

J. Lenarčič and F. Thomas (eds.), Advances in Robot Kinematics, 415–422.

solutions of the mechanism. Interval methods are also well suited for underconstrained problems whose solution presents one or more degrees of freedom, so that the set of allowed values for a variable is not a finite number of isolated points, but one or more intervals of values. In this case, the algorithm will return a set of boxes (of the size of the working resolution) containing the solution.

This paper is structured as follows: first, the procedure to find the allowed values for variables in rotation equations is described in Section 2, while the algorithm for interval propagation is described in Section 3. In Section 4, the description of the branch and prune algorithm for the solution of the direct kinematics of parallel spherical manipulators is presented, and an application example is given in Section 5. The main conclusions of the paper are resumed in Section 6.

2. Solution sets for variables in spherical kinematic loops

A spherical kinematic loop yields a rotation equation of the form:

$$\prod_{i=1}^{n} \mathbf{Rx}(\theta_i)\,\mathbf{Rz}(\alpha_i) = I, \tag{1}$$

where Rx and Rz are rotations around the x and z axis, respectively, θ_i are the rotation variables, and α_i are constant angular parameters that can be assumed to be in the interval $[0, \pi]$. The rotation equation written in this form has an intuitive geometric interpretation, according to which α_i are the sides of a spherical polygon, and θ_i its exterior angles. For $n > 3$, equation (1) has, in general, an infinite number of solutions, that is, each variable θ_i can take an infinite number of values. In Celaya and Torras, 1994, a way to describe the solution of a rotation equation is proposed, that consists in determining the set of allowed values for each variable. Such a description provides a kind of information that is not directly available when the solution is expressed in algebraic form, as for example, the enclosing box (or set of boxes) that contains all the solution vectors $(\theta_1, ..., \theta_n)$. This description permits obtaining any solution vector by sequentially fixing the value of a variable within its solution set, then finding the set of allowed values for the next variable in the resulting equation, fixing it to a value within this set, and so on.

The computation of the set of allowed values for a variable θ_s is simple provided all parameters α_j, $(j{\neq}s)$ are $\leq\pi/2$, a condition that can always be fulfilled, since the equation can be rewritten by appropriately substituting sides of length $\alpha_i{>}\pi/2$ by sides of length, $(\pi{-}\alpha_i)$, what corresponds to consider a supplementary polygon whose relation with

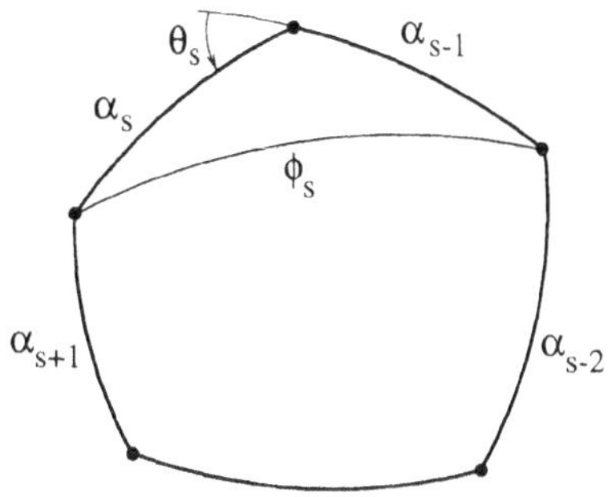

Figure 1. Solving for θ_s in a spherical polygon.

the original one is straightforward (Chiang, 1988). Denoting by ϕ_s the arc delimited by the extreme points of sides α_s and α_{s-1} (Fig. 1), we have, according to the cosine law of spherical trigonometry:

$$\cos\phi_s = \cos\alpha_{s-1}\cos\alpha_s - \sin\alpha_{s-1}\sin\alpha_s\cos\theta_s, \tag{2}$$

On the other hand, ϕ_s is also the arc delimited by the extreme points of the chain of sides α_j, ($j{\neq}s$,s-1), which is a function of θ_j, ($j{\neq}s$-1,s, s+1). Assuming that $\alpha_j{\leq}\pi/2$, $\forall j{\neq}s$, it can be shown that $\cos\phi_s$ can only take values in the interval $[\cos\sigma, \cos\delta]$, where $\sigma{=}\min(\pi, \sum_{i\neq s,s-1}\alpha_i)$, and $\delta{=}\max(0, \alpha_M{-}\sum_{i\neq M,s,s-1}\alpha_i)$, with $\alpha_M{=}\max_{i\neq s,s-1}(\alpha_i)$ (see Celaya, 1992 or Celaya and Torras, 1994 for details). Then, using (2), we obtain that the allowed values for cos θ_s are those that satisfy:

$$\frac{\cos\alpha_{s-1}\cos\alpha_s - \cos\delta}{\sin\alpha_{s-1}\sin\alpha_s} \leq \cos\theta_s \leq \frac{\cos\alpha_{s-1}\cos\alpha_s - \cos\sigma}{\sin\alpha_{s-1}\sin\alpha_s} \tag{3}$$

Taking into account that the arccos function is double-valued, the set of allowed values for θ_s, according to (3), will consist of two symmetric intervals, except when the left hand side of (3) is ≤ -1 or the right-hand side is ≥ 1 (or both), in which cases the two intervals join in a single one.

A function denoted by $SolveSpherical(eq, \theta)$ will be used in the remaining of this paper to provide the set of solution values of variable θ in the rotation equation eq as explained above.

3. Interval propagation between variables in spherical kinematic loops

To obtain the set of values $S_i[\theta_j{=}x_j]$ of variable θ_i that are compatible with the value x_j of θ_j, we can rewrite the rotation equation (1) substituting θ_j for this value and find the set of allowed values for θ_i in the resulting equation using the *SolveSpherical*() function. But, if what we are interested in is the set of values of θ_i that are compatible with an *interval of values* I_j of θ_j, then we will use the propagation algorithm described next.

Clearly, the set $S_i[\theta_j \in I_j]$ of values of θ_i that are compatible with $\theta_j \in I_j$, is a subset of $S_i = SolveSpherical(eq, \theta_i)$. Since $S_i[\theta_j \in I_j]$ is defined as the set of all those values of θ_i that are compatible with any value of $\theta_j \in I_j$, it must include, in particular, all the values compatible with the extrema of the interval. Thus, if $I_j = [a, b]$, then $(S_i[\theta_j = a] \cup S_i[\theta_j = b]) \subset S_i[\theta_j \in I_j]$. Therefore, all we need is to check the compatibility in the difference set

$$D_i = S_i - (S_i[\theta_j = a] \cup S_i[\theta_j = b]).$$

Note that the set D_i is composed of at most six non-intersecting intervals, since each of the S_i consists in, at most, two separate intervals. Due to the symmetry properties of the functions involved, it is possible to show (see Celaya and Torras, 1994) that each of the intervals composing D_i is either completely included in, or completely excluded from $S_i[\theta_j \in I_j]$. As a consequence, it is sufficient to check the compatibility of a single value into each interval forming D_i. This is done by choosing the central point x_k of each interval (six in the worst case), substituting it in the equation, and computing the intersection $S_j[\theta_i = x_k] \cap I_j$. If this intersection is $\neq \emptyset$, then the whole interval must be included in $S_i[\theta_j \in I_j]$, otherwise it must be excluded. A function denoted by $Prop(eq, \theta_j, I_j, \theta_i)$ implementing this propagation algorithm will be used in the remaining of this paper to compute the set $S_i[\theta_j \in I_j]$.

4. Direct kinematics of parallel spherical mechanisms

Solving the direct kinematics of a spherical mechanism consists in computing the orientation of the output link from the values of those variables of the mechanism's equations that are taken as input variables. For sure, the orientation of the output link can be determined once the values of all variables in the mechanism's equations are known, though, in general, this orientation can be directly expressed in terms of just a subset of the variables that can be considered as the *output variables* of the mechanism. So, we can formulate the problem of solving the direct kinematics of a mechanism as that of finding the values that the output variables can take when the input variables are fixed to some values.

Consider a parallel spherical mechanism with two independent kinematic loops, i.e., two rotation equations of the form (1). Let k be the number of variables shared between the two equations, excluding the input variables (since they remain fixed during the solution process).

Obviously, if k=0, each equation can be solved independently, and the problem does not really correspond to a parallel mechanism. In this paper, we will restrict our discussion to the cases of k=1 and k=2.

4.1 Two equations sharing one variable

If k=1, the set of solution values for the shared variable θ_s can be obtained by computing the set of values allowed by each equation for this variable and intersecting them. Then, for each solution value of θ_s, solution values for all the output variables can be found by solving the corresponding equation with the fixed value of θ_s.

4.2 Two equations sharing two variables

In the case of k=2, to obtain the solution values for the shared variables θ_s and θ_t, it is not sufficient to find the sets of values allowed for them by both equations: it is also necessary that the values of the two variables are simultaneously compatible with each other in the two equations. To grant this, the propagation algorithm can be used in an iterative process: Once the set of values for θ_s allowed by the two equations is obtained, it is propagated to θ_t through both equations, and the resulting sets are intersected. The so obtained set of values for θ_t is then propagated back to θ_s through both equations and intersected with the initial set for θ_s to get a smaller set for it. This process is iterated until the convergence to a stable set for θ_s is reached.

If the process converged to an isolated value, this is a solution for θ_s, and the corresponding values of θ_t and of all output variables can be obtained as in the case of k=1, after fixing θ_s to the solution value. If the process converged to a set that is not an isolated value, it must be split in two subsets, and the process must be recursively repeated in each subset until all of them either vanish or converge to isolated values (or to subsets of a size smaller than the working resolution ϵ). The resulting algorithm, called $SolveParallelSpherical()$, is shown in Fig. 2. It receives 6 arguments, the two rotation equations, the shared variables θ_s and θ_t, and their initial sets of values I_s and I_t into which solution values are looked for. The algorithm returns the set of all solution values for the shared variable θ_s contained in I_s.

In order to improve efficiency, the loop for the convergence to a stable set of values for θ_s is stopped as soon as the reduction obtained in a single iteration is less than 50%, in which case, the set is immediately split in two halves. Since, when one half of the set does not contain a solution, it usually vanishes after the first iterations, this strategy makes that most iterations result in a reduction of the set in, at less, one half,

```
SolveParallelSpherical(eqA, eqB, θs, Is, θt, It)

As = SolveSpherical(eqA, θs)
Bs = SolveSpherical(eqB, θs)
Is = Is ∩ As ∩ Bs
L0 = ∞
Ls = length(Is)

DO WHILE (Ls > ε) AND (Ls < L0/2)
        At = Prop(eqA, θs, Is, θt)
        Bt = Prop(eqB, θs, Is, θt)
        It = It ∩ At ∩ Bt
        As = Prop(eqA, θt, It, θs)
        Bs = Prop(eqB, θt, It, θs)
        Is = Is ∩ As ∩ Bs
        L0 = Ls
        Ls = length(Is)
END DO

IF (Ls < ε) THEN RETURN(Is)
ELSE
        Split Is into I1s, I2s
        S1 = SolveParallelSpherical(eqA, eqB, θs, I1s, θt, It)
        S2 = SolveParallelSpherical(eqA, eqB, θs, I2s, θt, It)
        RETURN(S1 ∪ S2)
END
```

Figure 2. Recursive branch and prune algorithm to find the solution values of variable θ_s for two rotation equations sharing variables θ_s and θ_t.

which implies exponential convergence. Other improvements have been implemented to skip the execution of propagation steps through one of the equations when the resulting interval reduction is not large enough.

5. Example

We consider the spherical parallel manipulator analized in Gosselin et al., 1994. It consists of two platforms connected by three identical legs, each formed by two links joined by a revolute pair. The schematics of the general mechanism, assuming some symmetries in the parameters, is shown in Fig. 3.

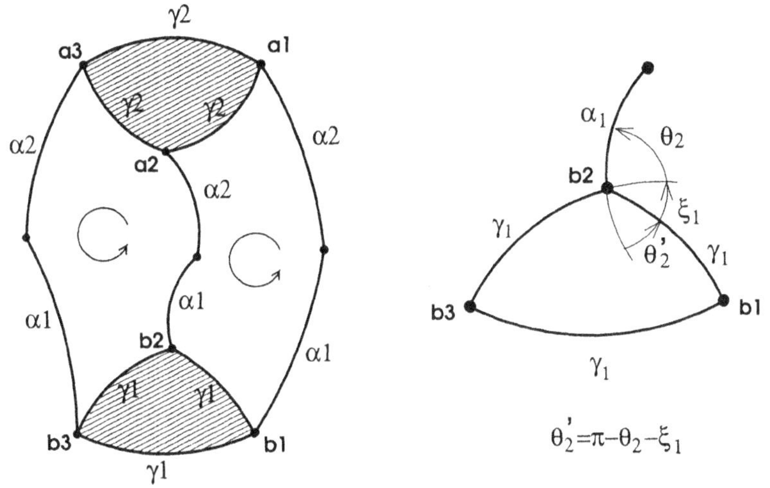

Figure 3. Parallel spherical manipulator and relation between the shared variables.

Two independent rotation equations are obtained from cycles involving legs 1,2 and 2,3, respectively:

$$\begin{aligned}&\mathbf{Rx}(\theta_1)\,\mathbf{Rz}(\alpha_1)\,\mathbf{Rx}(\mu_1)\,\mathbf{Rz}(\alpha_2)\,\mathbf{Rx}(\rho_1)\,\mathbf{Rz}(\gamma_2)\\ &\qquad \mathbf{Rx}(\rho_2')\,\mathbf{Rz}(\alpha_2)\,\mathbf{Rx}(-\mu_2)\,\mathbf{Rz}(\alpha_1)\,\mathbf{Rx}(\theta_2')\,\mathbf{Rz}(\gamma_1) = I,\end{aligned}$$

$$\begin{aligned}&\mathbf{Rx}(\theta_2)\,\mathbf{Rz}(\alpha_1)\,\mathbf{Rx}(\mu_2)\,\mathbf{Rz}(\alpha_2)\,\mathbf{Rx}(\rho_2)\,\mathbf{Rz}(\gamma_2)\\ &\qquad \mathbf{Rx}(\rho_3')\,\mathbf{Rz}(\alpha_2)\,\mathbf{Rx}(-\mu_3)\,\mathbf{Rz}(\alpha_1)\,\mathbf{Rx}(\theta_3')\,\mathbf{Rz}(\gamma_1) = I,\end{aligned}$$

where the inputs are the angles $\theta_1, \theta_2, \theta_3$ of the joints b_i of the base link, that we define as shown in Fig. 3. Angles θ_i' are related to the input angles θ_i in the form: $\theta_i' = \pi\text{-}\theta_i\text{-}\xi_1$, where ξ_1 is the exterior angle of the spherical equilateral triangle of sides γ_i, which, according to the cosine law, is given by $\cos\xi_1 = (\cos^2\gamma_1 - \cos\gamma_1)/\sin^2\gamma_1$. Similarly, for the angles ρ_i of the end effector joints a_i, we have $\rho_i' = \pi\text{-}\rho_i\text{-}\xi_2$, with $\cos\xi_2 = (\cos^2\gamma_2 - \cos\gamma_2)/\sin^2\gamma_2$.

In summary, once the input variables θ_i are given fixed values, we have two rotation equations sharing two variables, μ_2 and ρ_2, that can be solved with the *SolveParallelSpherical()* algorithm.

Taking the same parameter values used in Gosselin et al., 1994, $\gamma_1{=}\pi/7$, $\gamma_2{=}\pi/2$, $\alpha_1{=}1.12$, $\alpha_2{=}1.32$, and solving the mechanism for the input values $\theta_i{=}-\pi/3$, the algorithm returns the eight solutions shown in Tb. 2, whose correctness can be checked by substituting these values into the rotation equations. In this case, to find the 8 solutions the algorithm performs the minimal number of interval partitions: 7, and the mean reduction rate at each iteration of the pruning process is above 75%.

Table 1. Solution values for $\theta_1 = \theta_2 = \theta_3 = -\pi/3$.

	1	2	3	4	5	6	7	8
ρ_2	3.71014	4.14384	.01532	5.45702	6.03943	1.55548	1.81455	2.39696
μ_2	3.74975	3.10884	.96645	5.27620	4.75242	5.89215	2.10617	1.58239
ρ_1	2.39695	1.81454	.01532	4.14384	3.71013	1.55548	5.45703	6.03944
μ_1	1.58239	2.10617	.96645	3.10884	3.74975	5.89215	5.27620	4.75242
ρ_3	6.03943	5.45702	.01532	1.81455	2.39696	1.55548	4.14384	3.71013
μ_3	4.75242	5.27620	.96645	2.10618	1.58239	5.89215	3.10884	3.74975

6. Conclusions

Interval propagation methods are a good alternative to numerical approaches when closed form algebraic solutions are not available. The method presented here is specific for spherical mechanisms, thus it may take advantage of the availability of the interval propagation algorithm to reduce the solution interval, what makes the algorithm better suited for this problem than more general interval methods. The algorithm is rather insensitive to the selection of the initial intervals, since it starts by reducing the sets of values for the variables to those allowed by the two equations, which in combination with the propagation method, tends to quickly isolate the different solutions.

7. Acknowledgements

This research was financed by the CICYT, project TIC2000-0696.

References

Celaya, E. (1992), "Geometric Reasoning for the Determination of the Position of Objects Linked by Spatial Relationships", Ph.D. dissertation, Dept. LSI, Universitat Politècnica de Catalunya.

Celaya, E. and Torras, C. (1994), "Solving Multiloop Linkages with Limited-Range Joints", *Mechanism and Machine Theory*, vol. 29, No. 3, pp. 373-391.

Chiang, C.H. (1988), *Kinematics of spherical mechanisms.* Cambridge University Press. Cambridge.

Dietmaier, P. (1998), "The Stewart-Gough platform of general geometry can have 40 real postures" in ARK, Strobl, pp. 7-16.

Gosselin, C.M., Sefrioui, J., and Richard, M.J. (1994), "On the Direct Kinematics of Spherical Three-Degree-of-Freedom Parallel Manipulators of General Architecture" *ASME Journal of Mechanical Design*, Vol. 116, No. 2, pp. 594–598.

KINEMATIC ANALYSIS OF SPATIAL 3-DOF PARALLELEPIPED MECHANISMS

Yangnian Wu
Département de Génie Mécanique, Université Laval
Québec, Québec, Canada, G1K 7P4
wu@gmc.ulaval.ca

Clément M. Gosselin
Département de Génie Mécanique, Université Laval
Québec, Québec, Canada, G1K 7P4
gosselin@gmc.ulaval.ca

Abstract In this paper, the mechanical architecture and kinematic analysis of a novel spatial 3-DOF parallel mechanism referred to as a parallelepiped mechanism are first presented. Two types of actuation schemes of the mechanism are considered. The inverse and direct kinematic problems of these two instances of the mechanism are solved. The expressions of the singularity loci and their graphical representation are obtained. Finally, geometrical algorithms and discretization methods are used for the determination of the workspace for the two cases of the mechanism. The spatial 3-DOF parallelepiped mechanisms are used as legs to synthesize 6-DOF dynamically balanced parallel mechanisms.

Keywords: 3-DOF parallelepiped mechanism, inverse kinematics, singularity loci, workspace, geometrical algorithm, discretization

1. Introduction

Parallel mechanisms have received considerable attention from both researchers and manufacturers over the past 30 years because of their potential for high stiffness, low inertia and high speed capability. Several kinds of spatial 3-DOF parallel manipulators have been presented for different applications in the literature (Clavel, 1988, Lee and Shah, 1988, Gosselin and Angeles, 1989, Di Gregorio and Parenti-Castelli, 1998, Siciliano, 1999, Tsai, 1999, Tsai and Joshi, 2000). In order to design light mechanisms with high stiffness, deformable trusses have also been contemplated (Reinholtz and Gokhale, 1987, Arun et al., 1990, Salerno and Reinholtz, 1994, Hertz and Hughes, 1994). Deformable trusses are akin to parallel mechanisms. However, they are generally built by assembling

J. Lenarčič and F. Thomas (eds.), Advances in Robot Kinematics, 423–432.

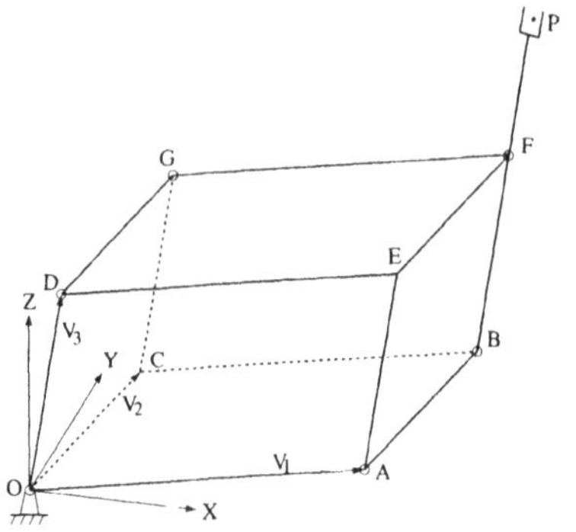

Figure 1. Schematic representation of a 3-DOF parallelepiped mechanism

deformable 1-dof modules in series, thereby augmenting the complexity and degrading the stiffness.

Furthermore, dynamically balanced (or reactionless) multi-degree-of-freedom mechanisms have been of interest to many researchers for potential applications such as space robots or telescope mirror mechanisms (see for instance Ricard and Gosselin, 2000, Foucault and Gosselin, 2002). Foucault and Gosselin have presented a planar 3-DOF reactionless parallel mechanism using the legs of parallelograms. In order to synthesize a 6-DOF reactionless manipulator mechanism, a novel 3-DOF parallel mechanism referred to as a parallelepiped mechanism is first presented in this paper. This mechanism may be considered as a "truly parallel" deformable truss, i.e., a multi-dof truss-type mechanism in which all actuators are mounted in parallel. After describing the design techniques, the kinematic analysis including the inverse and direct kinematics as well as the determination of the singularity loci and the workspace of the mechanism will be solved.

2. Description of the 3-DOF Parallelepiped Mechanisms

A schematic representation of a spatial 3-DOF parallelepiped mechanism is shown in Fig. 1. Point P is the end-effector and is located on link BFP. The mechanism consists of a parallelepiped with 6 faces. The parallelepiped can be deformed with the faces remaining planar and with opposite faces remaining parallel, while providing three degrees of freedom, and can therefore be used to position point P in space. The three edges attached to the ground and denoted by V_1, V_2 and V_3 respectively are actuated using fixed revolute actuators. The following groups of links have equal length, namely: $OA = CB = GF = DE, OC =$

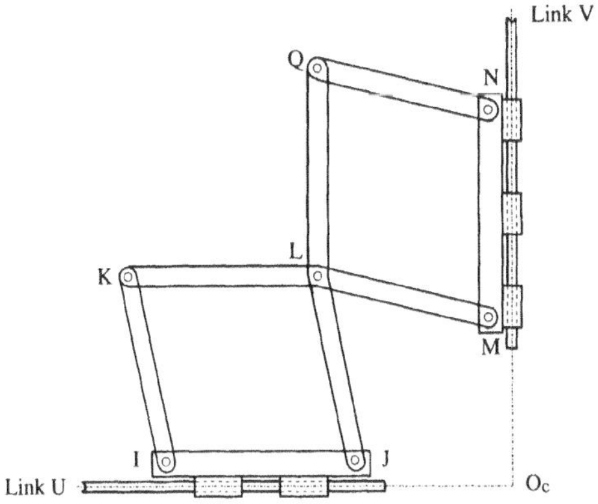

Figure 2. An offset planar hinge

$AB = EF = DG, OD = AE = BF = CG$. All the opposite links remain parallel throughout the motion, i.e., $OA \parallel CB \parallel GF \parallel DE, OC \parallel AB \parallel EF \parallel DG$ and $OD \parallel AE \parallel BF \parallel CG$.

Several feasible designs of the parallelepiped mechanism have been investigated. A simple implementation is to use an offset planar hinge (Hamlin and Sanderson, 1994) at each corner of the 6 faces of the mechanism. The offset planar hinge (Fig. 2) is a Watt I type mechanism with two straight brackets (IK, NQ), two identical bent brackets (KLM, JLQ), and two leaves (IJ, MN). Each leaf can rotate around the corresponding link (or pin) of the parallelepiped. Moreover, $IJ = KL = LQ = MN$, $IK = JL = ML = NQ$, and the offsets of the leaf pivots relative to their leaf axes are the same. By suitable design of the parameters of the hinge mechanism, the intersection point O_c of the two leaf axes becomes the certer of rotation for the links U and V while moving. Hence, the three adjacent links can rotate about the vertex center point at each vertex of the parallelepiped as a spherical joint while the faces of the parallelepiped remain planar.

The fixed coordinate axes of the inertial frame are denoted by $OXYZ$ and are attached to the base. Two types of actuation schemes of the mechanism designated as Case I and Case II respectively (Fig. 3 and Fig. 4) are considered in the present work. In Case I, V_1 and V_2 rotate around the Z axis and V_3 around the X axis, while in Case II, V_1, V_2 and V_3 rotate around the Y, Z and X axes respectively.

3. Inverse Kinematics

In parallelepiped mechanisms (Fig. 1), let the Cartesian coordinates of P be (x, y, z) expressed in the fixed frame. Moreover, let l_1, l_2 and l_3

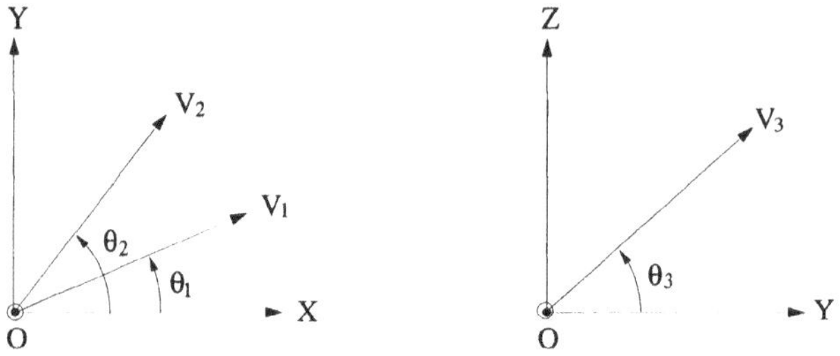

Figure 3. The first type of actuation of the parallelepiped mechanism (Case I).

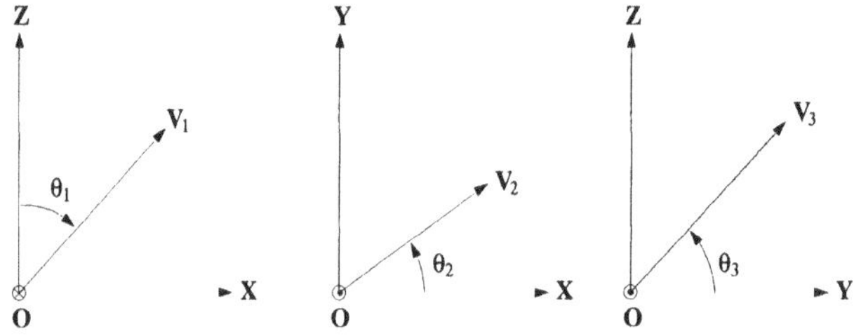

Figure 4. The second type of actuation of the parallelepiped mechanism (Case II).

be the lengths of the three edges of the parallelepiped mechanism and let l_e be the length of the extended part of link BF attached to the end-effector.

3.1 Case I

The position of the end-effector can be written as

$$\mathbf{p} = \begin{bmatrix} l_1 \cos\theta_1 + l_2 \cos\theta_2 \\ l_1 \sin\theta_1 + l_2 \sin\theta_2 + (l_3 + l_e)\cos\theta_3 \\ (l_3 + l_e)\sin\theta_3 \end{bmatrix} = \begin{bmatrix} x \\ y \\ z \end{bmatrix} \tag{1}$$

From the above equation, two solutions for θ_3 — obtained from the z component — and four solutions for θ_1 and θ_2 are obtained.

For the velocity analysis, one can write

$$\dot{\mathbf{p}} = \mathbf{J}\dot{\boldsymbol{\theta}} \tag{2}$$

where $\mathbf{J}$ is the Jacobian matrix, which can be written as follows

$$\mathbf{J} = \begin{bmatrix} -l_1 \sin\theta_1 & -l_2 \sin\theta_2 & 0 \\ l_1 \cos\theta_1 & l_2 \cos\theta_2 & -(l_3 + l_e)\sin\theta_3 \\ 0 & 0 & (l_3 + l_e)\cos\theta_3 \end{bmatrix} \tag{3}$$

Hence the actuated joint velocities can be obtained by solving the linear system of equations given in Eq. 2.

3.2 Case II

For Case II, one has

$$\mathbf{p} = \begin{bmatrix} l_1 \sin\theta_1 + l_2 \cos\theta_2 \\ l_2 \sin\theta_2 + (l_3 + l_e)\cos\theta_3 \\ l_1 \cos\theta_1 + (l_3 + l_e)\sin\theta_3 \end{bmatrix} = \begin{bmatrix} x \\ y \\ z \end{bmatrix} \tag{4}$$

The Jacobian matrix is written as

$$\mathbf{J} \;=\; \begin{bmatrix} l_1 \cos\theta_1 & -l_2 \sin\theta_2 & 0 \\ 0 & l_2 \cos\theta_2 & -(l_3 + l_e)\sin\theta_3 \\ -l_1 \sin\theta_1 & 0 & (l_3 + l_e)\cos\theta_3 \end{bmatrix} \tag{5}$$

From Eq. 4, the actuated joint coordinates can be obtained by solving a univariate 8th-order polynomial equation. Consequently, the inverse kinematics leads to a maximum of eight real solutions in this case. This result is readily obtained by applying elimination between the three equations.

4. Direct Kinematics

For given actuated joint coordinates (θ_1, θ_2 and θ_3), the Cartesian coordinates can be calculated from Eq. 1 and 4 for Case I and Case II respectively. The solution is unique and straightforward.

5. Singularity Analysis

The singularity loci can be obtained using the expression of the determinant of the Jacobian matrix, a quantity which is known to vanish in the singular configurations (Gosselin and Angeles, 1990), i.e.,

$$det(\mathbf{J}) = 0 \tag{6}$$

Hence, for Case I, substituting Eq. 3 into Eq. 6 leads to

$$\cos\theta_3 \sin(\theta_2 - \theta_1) = 0 \tag{7}$$

i.e., $\theta_3 = \pm\frac{\pi}{2}$, or $\theta_2 = \theta_1$ or $\theta_2 = \pi + \theta_1$.

For simplicity and generality, the joint limits and mechanical interferences of the mechanism are ignored for the determination of the singularity loci and workspace in this paper. A mechanism with the lengths of $100mm$ for l_1 and $50mm$ for l_2, l_3 and l_e respectively is taken as an example.

The condition ($\theta_3 = \pm\frac{\pi}{2}$) means that the first zones of singularity in Cartesian space are two annular zones parallel to the xy plane — each of which is enclosed by two concentric circles with the outer and inner radii of $(l_1 + l_2)$ and $(l_1 - l_2)$ respectively.

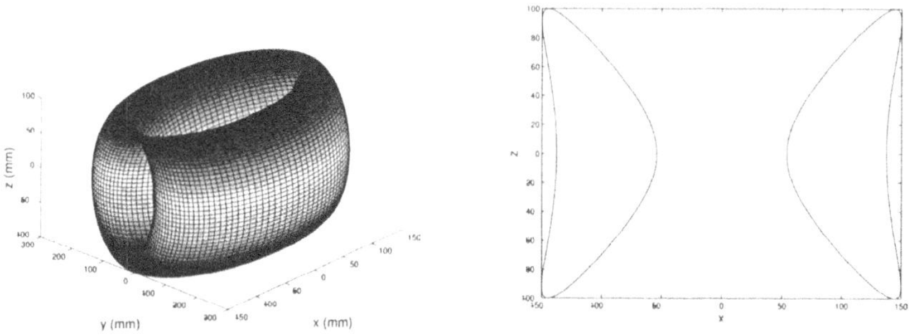

(a) Singularity surface

(b) y = 40 mm

Figure 5. Singularity surface and its section for Case I ($\theta_2 = \theta_1$)

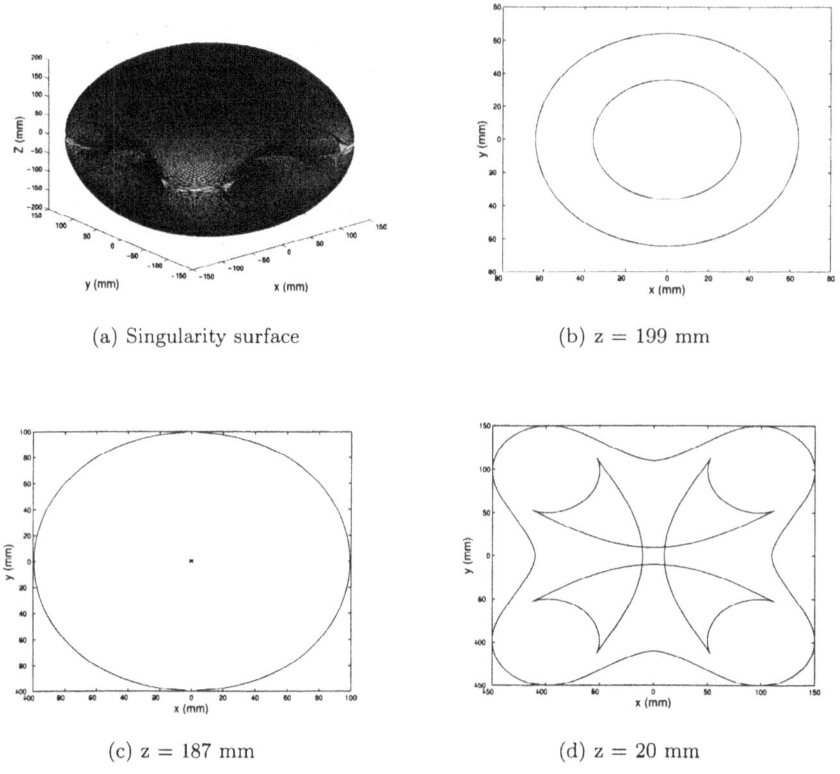

(a) Singularity surface

(b) z = 199 mm

(c) z = 187 mm

(d) z = 20 mm

Figure 6. Singularity surface and its sections for different values of z (Case II)

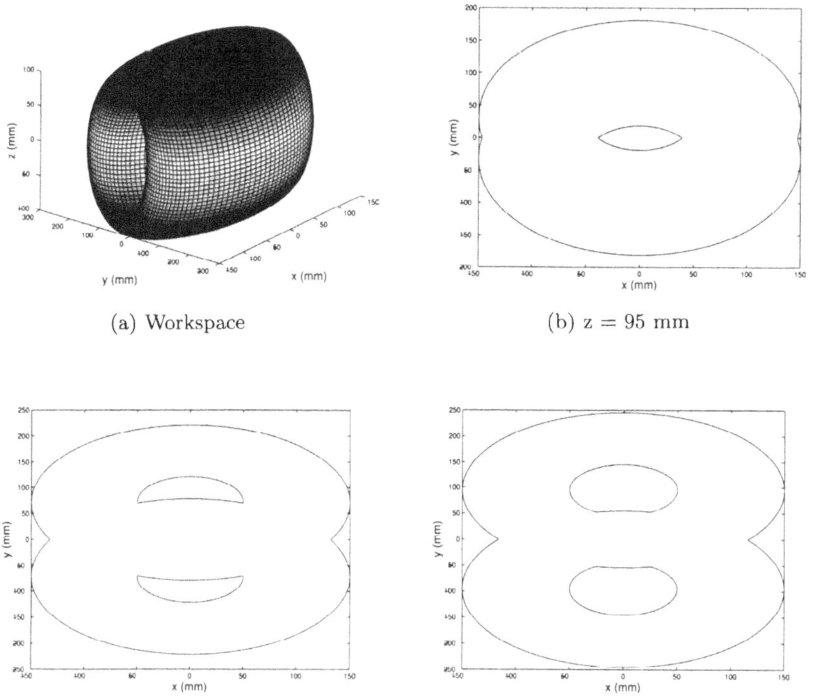

(a) Workspace

(b) z = 95 mm

(c) z = 70 mm

(d) z = 30 mm

Figure 7. Workspace and its boundaries for different values of z (Case I)

The latter conditions express that link OA and OC are colinear but in the same and opposite direction respectively, the vertex B moves consequently on circles in the xy plane, centered at the origin but with radii of $(l_1 + l_2)$ and $(l_1 - l_2)$ respectively. Hence, the corresponding singularity surfaces are formed by a circle — parallel to the yz plane and with a radius of $(l_3 + l_e)$ — the center of which traces circles of radius $(l_1 + l_2)$ and $(l_1 - l_2)$ respectively. In order to better demonstrate the singularity surface, different sections are also obtained. The singularity surface associated with the condition $(\theta_2 = \theta_1)$ and a section with $y = 40mm$ for the example mechanism are shown in Fig. 5.

For Case II, substituting Eq. 5 into Eq. 6 leads to

$$\cos\theta_1 \cos\theta_2 \cos\theta_3 - \sin\theta_1 \sin\theta_2 \sin\theta_3 = 0 \tag{8}$$

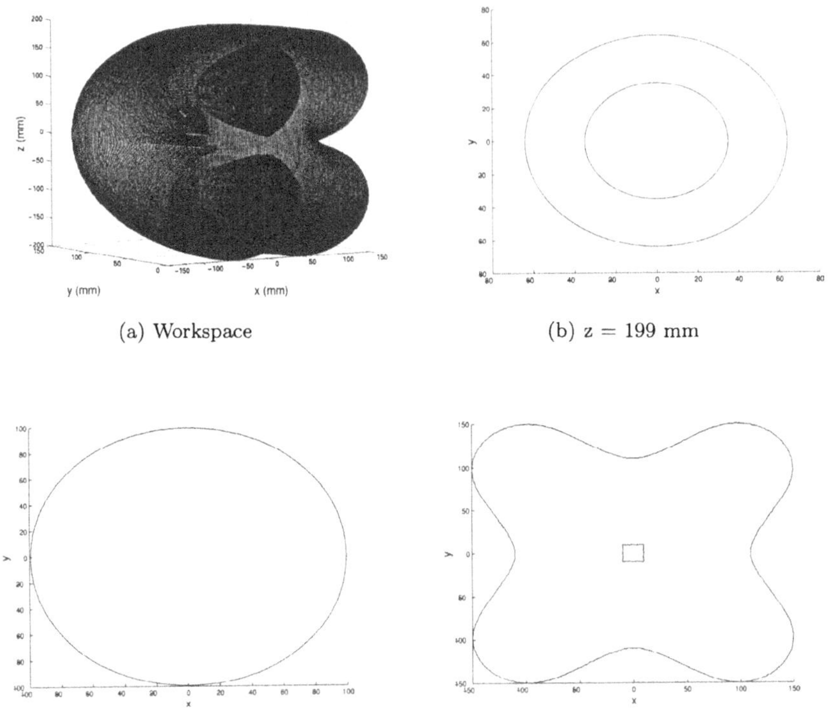

(a) Workspace (b) z = 199 mm

(c) z = 187 mm (d) z = 20 mm

Figure 8. Workspace and its boundaries for different values of z (Case II)

Eq. 8 corresponds a surface in joint space. From Eq. 4 and Eq. 8, the singularity surface and its sections for different values of z in Cartesian space can be also obtained as demonstrated in Fig. 6.

6. Workspace

It is of primary importance to develop some efficient tools for the determination of the workspace of parallel mechanisms (Gosselin, 1990). For Case I, it is found by inspection that the three kinds of singularity surfaces corresponding to the three conditions presented above define the top and bottom, outer and inner boundaries of the workspace, respectively. Since the section of the singularity surface for a given z is defined by two circles (Fig. 5), the section of the workspace on planes parallel to the xy plane can be obtained by the intersection of regions bounded by circles. The workspace and its boundaries for the example mechanism are finally shown in Fig. 7. For Case II, it is very difficult to determine the workspace by a geometrical method. A discretization algorithm is therefore used for this case. The philosophy of such an algorithm consists roughly in discretizing the three-dimensional space, solving the inverse kinematic problem at each point, and searching the boundaries of the workspace. Finally, the workspace and its boundary for the example mechanism are demonstrated in Fig. 8 (because of symmetry, only half of the workspace is shown in Fig. 8a). The workspace includes the space between the outer and inner boundaries as well as two lines from point $(0, 0, 14)$ to point $(0, 0, 187)$ and from point $(0, 0, -14)$ to point $(0, 0, -187)$ respectively.

By comparing Figs. 6 – 8, it has been found that for Case I, singularities only occur at the boundaries of the Cartesian workspace, while for Case II, they occur not only at the boundaries but inside the workspace as well.

Nevertheless, the workspace of the parallelepiped mechanism will decrease if the joint limits and mechanical interferences of the mechanism are considered. However, a workspace optimization with these constraints can be undertaken to meet the requirements of practical applications.

7. Conclusion

The mechanical architecture and kinematic analysis of a novel 3-DOF parallel mechanism referred to as a parallelepiped mechanism are first presented in this paper. Two types of actuation schemes of the mechanism are considered. The inverse and direct kinematic problems of two cases of the mechanisms have been solved. The expressions of the singularity loci and their graphical representations have been obtained. A geometrical algorithm and a discretization method are respectively used for the determination of the workspace for the two cases of the mechanism. It has been found that for Case I, singularities only occur at the boundaries of the Cartesian workspace, while for Case II they do not only occur at the boundaries but inside the workspace as well. All the results will be provided as guides for the future design of the mechanisms and the spatial 3-DOF parallelepiped mechanisms will be used as legs to synthesize 6-DOF dynamically balanced parallel mechanisms.

8. Acknowledgements

The authors would like to acknowledge the financial support of the Natural Sciences and Engineering Research Council of Canada (NSERC) as well as the financial support of the Canada Research Chair Program.

References

Arun, V., Reinholtz ,C.F., and Watson, L.T., (1990), Enumeration and analysis of variable geometry truss manipulators, *Proceedings of the 1990 ASME Design Technical Conference, DE-vol. 26*, Chicago, USA.

Clavel, R., (1988) DELTA, a fast robot with parallel geometry, *18th International Symposium on industrial robots*, Lausanne.

Di Gregorio, R., and Parenti-Castelli, V., (1998), A translational 3-DOF parallel manipulator, *Advances in Robot Kinematics: Analysis and Control*, Kluwer Academic Publishers.

Foucault, S., and Gosselin, C.M., (2002), On the development of a planar 3-DOF reactionless paralle mechanism, *Proceedings of the 2002 ASME Design Engineering Technical Conferences*, Montréal, Canada.

Gosselin, C.M., (1990), Determination of the workspace of 6-DOF parallel manipulators, *ASME Journal of Mechanical Design*, vol. 112, no. 9, pp. 331–336.

Gosselin, C.M., and Angeles, J., (1989), The optimum kinematic design of a spherical three-degree-of-freedom parallel manipulator, *ASME Journal of Mechanisms, Transmissions, and Automation in Design*, vol. 111, no. 2, pp. 202–207.

Gosselin, C.M., and Angeles, J., (1990), Singularity analysis of closed-loop kinematic chains, *IEEE Transactions on Robotics and Automation*, vol. 6, no. 11, pp. 281–290.

Hamlin, G.J., and Sanderson, A.C. (1994), A novel concentric multilink spherical joint with parallel robotics applications, *Proceedings of the 1994 IEEE International Conference on Robotics and Automation*, San Diego, USA.

Hertz, R.B., Hughes, P.C., (1994), Kinematic analysis of a class of spatial parallel manipulators with three degrees of freedom, *Proceedings of the 1994 ASME Design Technical Conference, DE-Vol. 72*, Minneapolis, USA.

Lee, K.M., and Shah, D.K., (1988), Kinematic analysis of a three-degree-of-freedom in-parallel actuated manipulator, *IEEE Transactions on Robotics and Automation*, vol. 4, no. 3, pp. 354–360.

Reinholtz, C.F., and Gokhale, D., (1987), Design and analysis of variable geometry truss robots, *Proceedings of the 9th Applied Mechanisms Conference*, New Orleans, USA.

Ricard, R. and Gosselin, C.M., (2000), On the development of reactionless parallel manipulators, *Proceedings of the 2000 ASME Design Engineering Technical Conferences*, Baltimore, USA.

Salerno, R.J., Reinholtz, C.F., (1994), A modular, long-reach, truss-type manipulator for waste storage tank remediation, *Proceedings of the 1994 ASME Design Technical Conference, DE-vol. 72*, Minneapolis, USA.

Siciliano, B., (1999), The tricept robot: inverse kinematics, manipulability analysis and closed-loop direct kinematics algorithm, *Robotica*, vol. 17, pp. 437–445.

Tsai, L.W., (1999), *Robot analysis: the mechanics of serial and parallel manipulators*, New York, John Wiley & Sons.

Tsai, L.W., and Joshi, S., (2000), Kinematics and optimization of a spatial 3-UPU parallel manipulator, *ASME Journal of Mechanical Design*, vol. 122, no. 4, pp. 439–446.

KINEMATICS OF A PLANAR ROBOT WITH ROLLING CONTACT JOINTS

Curtis L. Collins
Department Mechanical and Aerospace Engineering
University of California, Irvine
clc@engr.ucr.edu

Abstract A rolling contact joint is formed when two adjacent links roll without slip over each other. When the rolling profiles are circular, the joint can be physically realized by a pair of revolute joints coupled by gears or cables. In this paper, we investigate the forward and inverse kinematics of a planar three degree-of-freedom serial chain with rolling contact joints that have equal circular profiles. We show analytically and geometrically that there are 16 inverse kinematic solutions.

Keywords: Serial Manipulator, Rolling Contact Joint, Inverse Kinematics

1. Introduction

Articulated mechanisms used in robots utilize combinations of revolute and prismatic joints to achieve a desired motion workspace. It is commonly recognized that actuators such as DC motors, when attached to distal links of a serial chain, add undesirable weight and inertia to the system and generally degrade the dynamic performance of the device (Tsai, 1999). Because of this, it is common to find serial robots with two or more actuators mounted to a fixed base. Power from the actuators is transmitted through linkages or flexible transmission elements such as cables at the expense of coupling the joint motions and increasing the kinematic complexity of the device.

Base mounted actuation is especially important in robot finger designs where compactness and dexterity are desired. Early finger designs utilized cables to actuate the finger joints with motors mounted on the wrist or forearm as in, for example, the Salisbury/JPL hand (Mason and Salisbury, 1985). This introduced coupling between the joint motions which increased the kinematic complexity of the design. As an alternative to cables, linkages have also been used to transmit power from base/wrist mounted actuators (see for example Lovchik and Diftler, 1999). Like cables, they increase kinematic complexity by coupling the joint motions.

Other robot finger designs utilize kinematic coupling to reduce the number of actuators required to perform fundamental motions. For ex-

J. Lenarčič and F. Thomas (eds.), Advances in Robot Kinematics, 433–440.

ample, Crisman, Kanojia, and Zeid, 1996 find the combination of link lengths, pulley diameters, and cable routing that produce a desired motion using only one motor. The concept of motion design using cable coupled joints has also been generalized for custom applications by Krovi, Ananthasuresh, and Kumar, 1997.

The first application of coupled joints for a multi-dof manipulator was implemented at JPL for a telerobotic micro-surgery project (Schenker, Das, and Ohm, 1995). In their design, it was recognized that two revolute joints, when coupled through cables, allowed for the decoupled actuation of the distal joints of the manipulator serial chain through base mounted actuators. This double joint design is a special case of what we call a *rolling contact joint.*

In this paper we investigate the kinematics of a three degree-of-freedom planar serial chain with rolling contact joints. We define a rolling contact joint as a system of constraints that allow a link to roll without slip over another link. A serial chain composed of n rolling contact joints has n degrees of freedom. Cables running through a rolling contact joint are not influenced by the motion of the joint and can be used to independently actuate distal joints in the chain. This property makes them will suited for use in multiple serial chain architectures such robot hands and parallel manipulators (Collins and Long, 1995) where decoupled actuation through base mounted actuators is desired.

2. Forward Kinematics

The kinematic diagram of our three degree-of-freedom chain with rolling contact joints is shown in Figure 1. We assume that the joints constrain the links to roll without slipping, and that the contact profiles are circular. Figure 2 shows how we can use cables to actuate two adjacent joints independently. The contact profiles can be thought of as the pitch circles of meshed gears, for example.

The joint configuration shown in Figure 1 is considered the zero or initial configuration. We have attached local reference frames at the center of each rolling contact profile. We label the links sequentially from 0 to 3, and use the following subscript convention to identify variables associated with different links. The first subscript refers to the link the variables is defined on, while the second refers to the adjacent link that forms the rolling contact joint. For example, let the radius of the profile of link zero be denoted r_{01} and the radius of the adjacent profile on link 1 be denoted r_{10}.

Now consider when one link moves with respect to the other.We seek the relationship between the angle defined by the motion of the point of

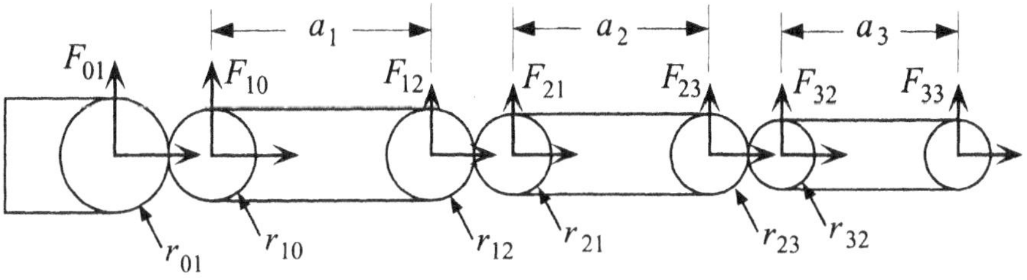

Figure 1. A four link three joint chain in its initial configuration

contact and the relative angle between the links. For example, as the link 1 rolls without slipping on link 0, the point of contact moves. We denote the angle of the motion of the point of contact by θ_{01} on link 0 and θ_{10} on link 1. The fundamental constraint is that the arc length defined by the moving point of contact is the same for both links. This is expressed as

$$r_{01}\theta_{01} = r_{10}\theta_{10}. \tag{1}$$

Upon closer examination of the angles, we find that

$$\theta_1 = \theta_{01} + \theta_{10}. \tag{2}$$

where θ_1 is the orientation of link 1 with respect to link 0.

For the three joint chain shown in Figure 1, we can write the transformation from the fixed to the moving frame as:

$$^{F}\mathbf{D}_M = {}^{01}\mathbf{D}_{12}{}^{12}\mathbf{D}_{23}{}^{23}\mathbf{D}_{33}, \tag{3}$$

where

$$\begin{aligned} {}^{01}\mathbf{D}_{12} &= \mathbf{R}(\theta_{01})\mathbf{X}(r_1)\mathbf{R}(\theta_{10})\mathbf{X}(a_1) \\ {}^{12}\mathbf{D}_{23} &= \mathbf{R}(\theta_{12})\mathbf{X}(r_2)\mathbf{R}(\theta_{21})\mathbf{X}(a_2) \\ {}^{23}\mathbf{D}_{33} &= \mathbf{R}(\theta_{23})\mathbf{X}(r_3)\mathbf{R}(\theta_{32})\mathbf{X}(a_3). \end{aligned} \tag{4}$$

$\mathbf{D}$ denotes the total displacement matrix between the frames, $\mathbf{R}()$ denotes a pure rotation in the plane, and $\mathbf{X}()$ denotes a pure translation along the local x-axis. a_1, a_2, and a_3 denote the link lengths, and $r_1 = r_{01} + r_{10}$, $r_2 = r_{12} + r_{21}$ and $r_3 = r_{23} + r_{32}$. Expanding, we obtain

$$^{F}\mathbf{D}_M = \begin{bmatrix} \cos(\theta_{01,\ldots,32}) & -\sin(\theta_{01,\ldots,32}) & x \\ \sin(\theta_{01,\ldots,32}) & \cos(\theta_{01,\ldots,32}) & y \\ 0 & 0 & 1 \end{bmatrix}, \tag{5}$$

where $\theta_{01,\ldots,ij}$ represents a summation of joint angles from θ_{01} to θ_{ij} going up the chain, and x and y are:

$$\begin{aligned} x &= a_1\cos(\theta_{01,10}) + a_2\cos(\theta_{01,\ldots,21}) + a_3\cos(\theta_{01,\ldots,32}) \\ &\quad + r_1\cos(\theta_{01}) + r_2\cos(\theta_{01,\ldots,12}) + r_3\cos(\theta_{01,\ldots,23}), \end{aligned} \tag{6}$$

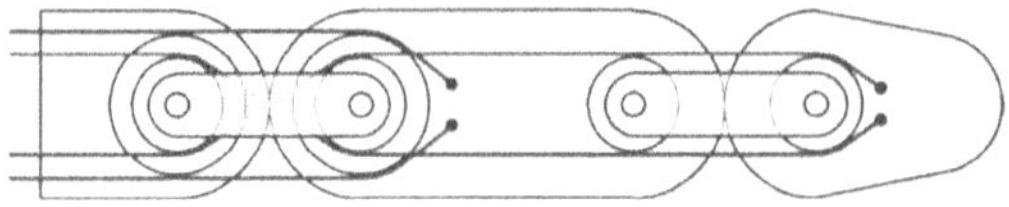

Figure 2. Example of how cables can be used to actuate two joints independently.

and

$$\begin{aligned} y &= a_1 \sin(\theta_{01,10}) + a_2 \sin(\theta_{01,\ldots,21}) + a_3 \sin(\theta_{01,\ldots,32}) \\ &\quad + r_1 \sin(\theta_{01}) + r_2 \sin(\theta_{01,\ldots,12}) + r_3 \sin(\theta_{01,\ldots,23}). \end{aligned} \tag{7}$$

Note that these equations are equivalent, in form, to the kinematics of a 6R planar serial chain. We can introduce the coupling constraints due to rolling contact,

$$\begin{aligned} r_{ij}\theta_{ij} &= r_{ji}\theta_{ji} \\ \theta_j &= \theta_{ij} + \theta_{ji}, \end{aligned} \tag{8}$$

where $i = 0, 1, 2$ and $j = i + 1$, in order to rewrite the equations in terms of any three independent joint variables.

3. Inverse Kinematics

The number of inverse kinematics solutions for mechanisms with circular rolling contact joints is dependent on the contact profile radii. For rational contact ratios, it is known that the relative motion of a single joint is algebraic (see Sandor, 1964, or Hunt, 1990). In this section we derive the inverse kinematics equations for the three joint chain specialized with 1:1 contact ratios.

Assuming that the full position and orientation of the moving frame are known, equations (6) and (7) can be rewritten as

$$\begin{aligned} p &= r_1 \cos(\theta_{01}) + a_1 \cos(\theta_{01} + \theta_{10}) + r_2 \cos(\theta_{01} + \theta_{10} + \theta_{12}) + \\ &\quad a_2 \cos(\theta_{01} + \theta_{10} + \theta_{12} + \theta_{21}) + r_3 \cos(\theta_{01} + \theta_{10} + \theta_{12} + \theta_{21} + \theta_{23}) \end{aligned} \tag{9}$$

and

$$\begin{aligned} q &= r_1 \sin(\theta_{01}) + a_1 \sin(\theta_{01} + \theta_{10}) + r_2 \sin(\theta_{01} + \theta_{10} + \theta_{12}) + \\ &\quad a_2 \sin(\theta_{01} + \theta_{10} + \theta_{12} + \theta_{21}) + r_3 \sin(\theta_{01} + \theta_{10} + \theta_{12} + \theta_{21} + \theta_{23}) \end{aligned} \tag{10}$$

where

$$p = x - a_3 \cos(\theta_e), \tag{11}$$

$$q = y - a_3 \sin(\theta_e), \tag{12}$$

and

$$\theta_e = \theta_{01} + \theta_{10} + \theta_{12} + \theta_{21} + \theta_{23} + \theta_{32}. \tag{13}$$

We reduce the number of unknowns in equations (9) and (10) by using the contact angle relations (8) and our knowledge of the moving frame orientation. Assuming we have 1:1 contact ratios, we find the expression for θ_{23} in terms of θ_{01}, θ_{12}, and θ_e has the form

$$\theta_{23} = \frac{\theta_e}{2} - \theta_{01} - \theta_{12}, \tag{14}$$

where θ_e is the known orientation of the moving frame. Equations (9) and (10) can then be written as

$$\begin{aligned} p &= r_1 \cos(\theta_{01}) + a_1 \cos(2\theta_{01}) + r_2 \cos(2\theta_{01} + \theta_{12}) \\ &\quad + a_2 \cos(2\theta_{01} + 2\theta_{12}) + r_3 \cos(\theta_{01} + \theta_{12} + \tfrac{\theta_e}{2}), \end{aligned} \tag{15}$$

and

$$\begin{aligned} q &= r_1 \sin(\theta_{01}) + a_1 \sin(2\theta_{01}) + r_2 \sin(2\theta_{01} + \theta_{12}) \\ &\quad + a_2 \sin(2\theta_{01} + 2\theta_{12}) + r_3 \sin(\theta_{01} + \theta_{12} + \tfrac{\theta_e}{2}). \end{aligned} \tag{16}$$

We now have a nonlinear system of two equations in two unknowns. We have tried a number of different techniques to eliminate different variables from Equations (15) and (16), and have found the following procedure provides the most consistent and efficient results in *Mathematica*.

1 Expand equations (15) and (16) using the trigonometric multiple angle relations. After the expansion, the equations will contain sums and products of the sine and cosine of the two angles θ_{01} and θ_{12}.

2 Convert the sine and cosine terms into parameters by applying the substitution $\sin(\theta_{01}) = s_1$, $\cos(\theta_{01}) = c_1$, $\sin(\theta_{12}) = s_2$, and $\cos(\theta_{12}) = c_2$. Eliminate a pair (c_1, s_1) or (c_2, s_2) using the original equations plus the standard circular constraints $s_1^2 + c_1^2 = 1$ and $s_2^2 + c_2^2 = 1$. The result is a single equation in say (c_2, s_2).

3 Apply the half-tangent substitution $c_2 = (1 - u^2)/(1 + u)^2$ and $s_2 = 2u/(1 + u^2)$ where $u = \tan(\theta_{12}/2)$ and expand to find a univariate polynomial in the unknown u.

4 The roots of the polynomial in u lead to the solutions for θ_{12}. Corresponding solutions for θ_{01} can be found using equations (15) and (16), and for θ_{23} using equation (14).

The result is an 8th order polynomial that leads to 16 solutions for the three joint case. The 16 solutions can be found by first solving for 8 solutions corresponding to the angle $0 \leq \theta_e \leq 2\pi$, and then solving for the solutions corresponding to the angle $\theta_e + 2\pi$. An example set of 16 solutions is shown in Table 1 and Figure 3.

Table 1. Example of 16 inverse kinematics solutions for $r_1 = r_2 = r_3 = 0.5$, $a_1 = a_2 = 1.5$, $a_3 = 1.0$, $\theta_e = 0°$, $x = 2.0$, and $y = 0.5$.

Solution	θ_{12}	θ_{01}	θ_{23}
1	-2.59119	3.02852	2.70426
2	-2.18377	-0.05456	2.23833
3	-2.09440	2.61799	-0.52360
4	-1.85243	-0.28592	5.27994
5	-1.59286	1.13197	0.46090
6	-1.41647	0.63065	3.92741
7	-1.31367	-2.00067	6.45593
8	-1.06300	-2.44269	3.50568
9	1.06300	2.77750	-3.84050
10	1.31367	-0.17275	2.00067
11	1.41647	2.35577	-0.63065
12	1.59286	-0.46090	-1.13197
13	1.85243	1.00325	0.28592
14	2.09440	0.52360	-2.61799
15	2.18377	-2.23833	0.05456
16	2.59119	-2.70426	3.25467

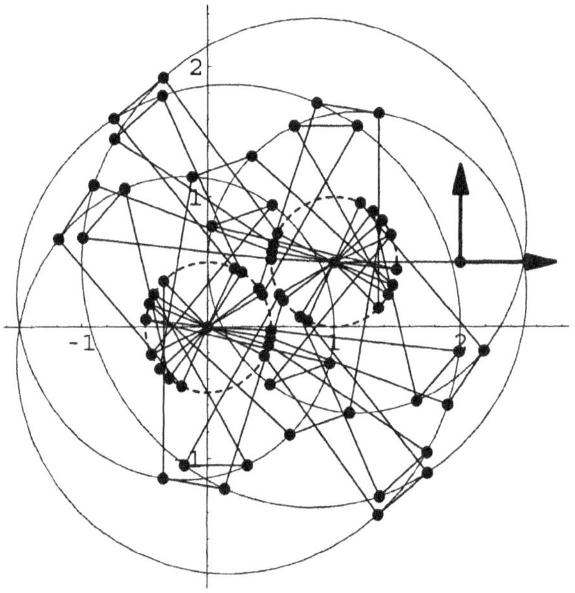

Figure 3. Graphical depiction of the 16 solutions shown in Table 1. A 6R kinematic skeleton is used to demonstrate the robot configurations.

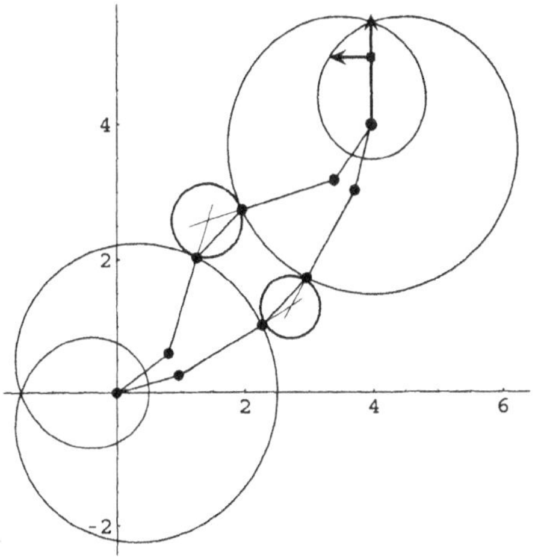

Figure 4. Example showing a set of two inverse kinematics solutions. The circles are tangent to the cycloids, and the cord between the tangent points has length r_2.

4. Discussion

The inverse kinematics solutions have a unique geometric interpretation depicted in Figure 4. In this case, there are only two real solutions. Geometrically, once the end frame is fixed, so is the last link. The solutions are found by noting that the origin of frame F_{12} traces out a cycloid as θ_1 varies, while the origin of frame F_{21} traces out another cycloid as θ_3 varies. Note that for the $3R$ chain, these curves are circles whose intersection defines the possible location of the second joint axis in the chain. In our case, the possible solutions for θ_2 can be found by noting that the distance between the frames must be r_2, and the normals to the cycloids must form an isosceles triangle with the line segment between the two origins. This is depicted in Figure 4. Each cycloid has degree 4 and circularity 2 (Hunt, 1990), and can have up to $4 \cdot 4 - 2(2 \cdot 2) = 8$ real solutions. Furthermore, since the second joint also has circularity 2, there are up to two inverse kinematics solutions near each of the 8 intersections of the cycloids, leading to at most 16 solutions. Solutions of equations (9) and (10) for other rational ratios and are currently under investigation.

5. Conclusions

The forward and inverse kinematics equations for a three degree-of-freedom planar serial chain with rolling contact joints is derived. The solution of the inverse kinematics problem is complicated by the fact that only rational contact ratios lead to a finite number of solutions. For the 1:1 case, the solution to the inverse kinematics equations can be reduced to the solution of an 8th order polynomial. However, 16 solutions exist since each joint has a range of motion of 4π. It is shown that the geometry of the solutions are related to the intersection of two cycloids. Example parameters that lead to 16 solutions are provided. While rolling contact joints increase the kinematic complexity of the mechanism, advantages such as decoupled motions, compactness, increased range of motion, foldability, potentially low friction, and local transmission ratio combine to provide a useful component for new robot finger and manipulator designs.

6. Acknowledgements

This work was supported by a grant from the UC Regents Faculty Fellowship Program. The author would like to thank Prof. J. Michael McCarthy for his feedback on various stages of the research, and Paul Stovall for his design advice and wire EDM fabrication support.

References

Collins, C.L., and Long, G.L., 1995. The Singularity Analysis of an In-Parallel Hand Controller for Force-Reflected Teleoperation, *IEEE Transactions on Robotics and Automation*, 11(5):661-669.

Crisman, J.D., Kanojia, C., and Zeid, I., 1996. Graspar: a flexible easily controllable robotic hand, *IEEE Robotics & Automation Magazine*, June, 1996, pp. 32–38.

Hunt, K. H.,1990. Kinematic Geometry of Mechanisms, Oxford University Press, New York.

Krovi, V., Ananthasuresh, G. K., and Kumar, V., 1997. Synthesis of coupled multi-link serial chain mechanisms, *Proceedings of the 5th National Applied Mechanisms & Robotics Conference*, Cincinnati, Ohio, October 12–15, 1997.

Lovchik, C.S., and Diftler, M.A., 1999. The Robonaut hand: a dexterous robot hand for space, *Proceedings of the 1999 IEEE International Conference on Robotics and Automation*, Detroit, MI, May 1999, pp. 907–912.

Mason, M.T., and Salisbury, J.K., 1985. *Robot Hands and the Mechanics of Manipulation*, MIT Press.

Sandor G. N., 1964. On the existence of a cycloidal Burmester theory in planar kinematics, *Transactions of the ASME Journal of Applied Mechanics* Vol 86(E), December, pp. 694–699.

Schenker, P.S., Das, H., and T. R. Ohm, 1995. A new robot for high dexterity microsurgery, *1995 Conference on Computer Vision, Virtual Reality and Robotics in Medicine*, April 3–6, Nice France.

Tsai, L.W., 1999. *Robot Analysis: The Mechanics of Serial and Parallel Manipulators*, John Wiley & Sons, New York.

Synthesis of Mechanisms and Motion

DUAL QUATERNION SYNTHESIS OF CONSTRAINED ROBOTS

Alba Perez
Robotics and Automation Laboratory, University of California, Irvine
Irvine, CA 92697
maperez@uci.edu

J. Michael McCarthy
Robotics and Automation Laboratory, University of California, Irvine
Irvine, CA 92697
jmmccart@uci.edu

Bruce Bennett
Department of Mathematics, University of California, Irvine
Irvine, CA 92697
bbennett@uci.edu

Abstract This paper presents a synthesis methodology for robots that have less than six degrees of freedom, termed constrained robots. The goal is to determine the physical parameters of the chain that fit its workspace to a given set of spatial positions. Our formulation uses the dual quaternion form of the kinematics equations of the constrained robot. Here we develop the theory and formulate the synthesis equations for the spatial RPR robot. Their solution ensures that the three dimensional workspace of this robot contains a given set of four spatial positions.

1. Introduction

The focus of this paper is on the design of robots that have less than six degrees of freedom that we call *constrained robots.* These robots have workspaces of dimension two through five in the Lie group of spatial displacements. Fitting a constrained robot's workspace to a given set of task positions can be viewed as similar to fitting a surface to a set of points in space. This theory is an extension of the kinematic synthesis of linkages (McCarthy, 2000b).

The geometric design of a robots generally focuses on systems with six or more degrees of freedom. In this case, the workspace is sized to enclose the task space, and the challenge is to ensure that the system

J. Lenarčič and F. Thomas (eds.), Advances in Robot Kinematics, 443–452.

has desired differential properties at specified task positions, Kumar and Waldron, 1981. Chedmail, 1998 and Gosselin, 1998 present optimization techniques for the design serial and parallel robotic systems, respectively, that provide desired properties of the workspace.

In constrast, constrained robots have lower dimensional workspaces and the challenge is to locate and shape it such that it passes through the specified task positions.

1.1 Linkage Synthesis

Spatial linkage synthesis uses the geometric properties of a serial chain to formulate algebraic equations that must be satisfied at each of a discrete set of positions in the workspace (Suh and Radcliffe, 1978). This yields algebraic equations that are solved to determine the dimensions of the chain, also see McCarthy, 2000a. See, for example, the synthesis of spatial RR chains (Tsai and Roth, 1973, Perez, 2000), spatial CC chains (Chen, 1969, Huang, 2000) and SS chains (Innocenti, 1994, Liao, 1998).

Recently, new methods have been developed that use the kinematics equations of the robot to form the design equations. In particular, Mavroidis and Lee used the kinematics equations of the spatial RR and RRR robots to formulate its design equations. Larochelle, 2000 uses a similar approach with planar quaternions to define an approximate synthesis for planar robots. This method introduces the joint parameters of the chain at each of the goal positions as additional variables in the design equations (Mavroidis, 2001, Lee, 2002). The advantage is that it can be systematically applied to a broad range of robotic systems.

1.2 Overview

In this paper, we follow Mavrodis' basic ideas, however, we use successive screw displacements (Gupta, 1986, Tsai, 1999) formulated in terms of dual quaternions to represent the kinematics equations of the robot. Dual quaternions were introduced to linkage analysis by Yang and Freudenstein, 1964. They form an eight dimensional Clifford algebra that contains a subset, known as unit dual quaternions, which is isomorphic to the group of spatial displacements (McCarthy, 1990). Also see Angeles, 1998.

There are two advantages in this formulation. The first is that successive screw displacements provide a convenient formulation for the kinematics equations in terms of the joint axes directly. Secondly, it reduces the number of equations needed for each goal position.

2. The Kinematics Equations

The kinematics equations of the robot equate the 4×4 homogeneous transformation $[D]$ between the end-effector and base frame. to the sequence of local coordinate transformations along the chain (Craig 1989),

$$[D] = [G][Z(\theta_1, d_1)][X(\alpha_{12}, a_{12})][Z(\theta_2, d_2)] \dots \\ [X(\alpha_{n-1,n}, a_{n-1,n})][Z(\theta_n, d_n)][H]. \tag{1}$$

The parameters (θ, d) define the movement at each joint and (α, a) are the length and twist of each link, collectively know as the Denavit-Hartenberg parameters. The transformation $[G]$ defines the position of the base of the chain relative to the fixed frame, and $[H]$ locates the tool relative to the last link frame.

2.1 Successive Screw Displacements

These kinematics equations can be transformed into successive screw displacements choosing a reference position $[D_0]$. We then compute $[D_{0i}] = [D_i][D_0]^{-1}$, that is

$$\begin{aligned}[D_{0i}] &= [D_i][D_0]^{-1} \\ &= ([G][Z(\theta_{1i}, d_{1i})] \dots [Z(\theta_{ni}, d_{ni})][H])([G][Z(\theta_{10}, d_{10})] \dots [Z(\theta_{n0}, d_{n0})][H])^{-1}.\end{aligned} \tag{2}$$

This can be viewed as

$$[D_{0i}] = [T(\Delta\theta_1, \mathsf{S}_1)] \dots [T(\Delta\theta_n, \mathsf{S}_n)], \tag{3}$$

where

$$\begin{aligned}[T(\Delta\theta_1, \mathsf{S}_1)] &= [G][Z(\theta_{1i}, d_{1i})][Z(\theta_{10}, d_{10})]^{-1}[G]^{-1}, \\ [T(\Delta\theta_2, \mathsf{S}_2)] &= ([G][Z(\theta_{10}, d_{10})][X(\alpha_{12}, a_{12})][Z(\theta_{2i}, d_{2i})]) \\ &\qquad ([G][Z(\theta_{10}, d_{10})][X(\alpha_{12}, a_{12})][Z(\theta_{20}, d_{20})])^{-1}, \\ &\vdots \\ [T(\Delta\theta_n, \mathsf{S}_n)] &= ([G][[Z(\theta_{10}, d_{10})] \dots)[Z(\theta_n, d_n)][Z(\theta_{n0}, d_{n0})]^{-1}([G][[Z(\theta_{10}, d_{10})] \dots)^{-1}.\end{aligned} \tag{4}$$

The displacements $[T(\Delta\theta_i, \mathsf{S}_i)]$ are the relative transformations along the joint axes of the robot from the reference configuration. Notice that by expressing them in this way, the initial transformation $[G]$ is absorbed in the first joint axis and the final transformation $[H]$ does not appear in the expression.

2.2 Dual Quaternions

The workspace of the robot can also be expressed by using the Clifford algebra of the *dual quaternions.* A spatial displacement can be represented as a dual quaternion,

$$\hat{Q}(\hat{\theta}) = \sin(\frac{\hat{\theta}}{2})\mathsf{S} + \cos(\frac{\hat{\theta}}{2}), \tag{5}$$

where $\mathsf{S} = \mathbf{s} + \epsilon\mathbf{s}^0$, with $\epsilon^2 = 0$, is the screw axis of the transformation. The dual numbers $\cos(\frac{\hat{\theta}}{2}) = \cos\frac{\theta}{2} + \epsilon(-\frac{d}{2}\sin\frac{\theta}{2})$ and $\sin(\frac{\hat{\theta}}{2}) = \sin\frac{\theta}{2} + \epsilon(\frac{d}{2}\cos\frac{\theta}{2})$ contain the information about the rotation and the displacement along the screw axis. The components of the dual quaternions are easily computed from the homogeneous matrix transformation.

The spatial displacements can be represented as the set of points $\mathbf{Z} = (\mathbf{Z}, \mathbf{Z}^0)$ in $\mathbf{R}^8$ which are subject to two constraints: $\mathbf{Z} \cdot \mathbf{Z} = 1$ and $\mathbf{Z} \cdot \mathbf{Z}^0 = 0$. The workspace lies on a six-dimensional submanifold of $\mathbf{R}^8$.

The dual quaternion form for the kinematics equations of the robot are obtained by transforming eq.(3) into

$$\hat{D}^i = \hat{S}_1(\Delta\hat{\theta}_1^i)\dots\hat{S}_n(\Delta\hat{\theta}_n^i), \tag{6}$$

where $\hat{D}^i$ is the dual quaternion for $[D_{0i}]$ and $\hat{S}_i$ is the dual quaternion for $[T(\Delta\theta_i, \mathsf{S}_i)]$.

This approach yields the kinematics equations as successive screw transformations from the reference position. It is a useful formulation from the synthesis point of view because the components of each axis appears explicitly in the base frame coordinates.

3. Synthesis of Constrained Robots

Let $[T(\theta_1, \dots, \theta_k)]$ be the kinematics equations of a serial robot, and let a discrete approximation of the desired workspace be given in the form of n goal transformations $[D_i], i = 0, \dots, n-1$. The synthesis problem consists of solving the n matrix equations

$$[T(\theta_{1,i}, \dots, \theta_{k,i})] = [D_i], \quad \imath = 0, \dots, n-1. \tag{7}$$

We now transform these equations to successive screw displacements in dual quaternion form. The result is $n-1$ goal positions $\hat{D}^i, i = 1, \dots, n-1$ and the kinematics equations $\hat{Q}(\hat{\theta}_1, \dots, \hat{\theta}_k)$ to obtain the $n-1$ equations

$$\hat{Q}_i(\hat{\theta}_1^i, \dots, \hat{\theta}_k^i) = \hat{D}^i, \quad i = 1, \dots, n-1 \tag{8}$$

For each of the $n-1$ positions we have eight component equations. However, due to the structure of the dual quaternions, only six of them are independent.

For a robot chain represented by a series of j revolute joints, each of the joints has an axis defined by six Plucker coordinates, which yields $6j$ unknowns. The j joint variables take different values at each of the $n-1$ positions, which add $j(n-1)$ unknowns. This yields $6j+j(n-1)$ unknowns. For each joint axis, there are two constraints associated with its Plucker coordinates. For each of the $n-1$ goal positions we obtain eight equations, which can be reduced to six. Thus, we have $2j+6(n-1)$ equations. Equating the number of unknowns to the number equations, we obtain

$$6j + j(n-1) = 6(n-1) + 2j. \tag{9}$$

Solving for n we obtain $n = \frac{3j+6}{6-j}$. We have that $2R$, $3R$, $4R$ and $5R$ spatial chains require 3, 5, 9, 21 positions, respectively.This analysis has been extended to include other types of joints and topologies. In general, the number of orientations we can reach is limited by the fact that rotations operate independently in spatial displacements. To compute complete spatial positions, we need to check whether the orientations are limited. Assume that our robot consists of l revolute joints and k prismatic joints. The number of spherical positions we can reach is

$$3l + l(n_R - 1) = 3(n_R - 1) + l, \tag{10}$$

while the accounting for both orientation and translation is

$$6(l+k) + (l+k)(n-1) = 6(n-1) + 2l + k. \tag{11}$$

From the rotation equation, $n_R = \frac{3+l}{3-l}$, and notice that this coincides with the results for spherical robots. If the number of complete positions is restricted by n_R we will reach only that many arbitrary orientations, and the rest will be just translational components of dual quaternions in which rotations will have to be bounded to the given workspace.

Notice also that here we assume that the axes of the rotational and translational joints are not related, but it is easy to adapt the formula to particular cases in which the joints are constrained.This result is used to determine the number of positions we need to define in the synthesis process in order to obtain a finite number of solutions.

4. Solving the Design Equations

The design equations for constrained robots contain joints variables and axis variables. The goal is to eliminate the joint variables and solve for the axis variables, which define the physical dimensions of the robot. The elimination of the joint parameters is called "implicitization" of the parametric equations, which we apply to the kinematics equations for each position.

The methodology consists of solving linearly for two of the revolute parameters in the rotational part of the dual quaternion. The result is substituted in the moment part of the dual quaternion and combined with geometric constraints. This yields equations that can be solved for the remaining translational and rotational parameters. The following derivation for the spatial RPR robot has been applied to spatial RR, CC and RRR robots as well.

5. Synthesis of a Spatial RPR Robot

The *RPR* robot is a three-degree of freedom robot. The fixed axis G allows rotation of angle θ about it and it is followed by a translation d along an arbitrary direction P and finally a rotation of angle ϕ about an arbitrary axis W, see Figure 1.

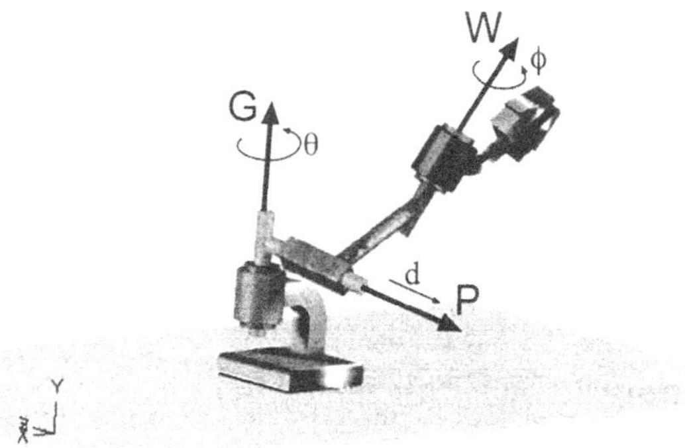

Figure 1. The spatial RPR robot

The dual quaternion representation for the relative displacements of the chain is given by

$$\hat{Q}_{RPR} = \hat{G}(\theta, 0)\hat{P}(0, d)\hat{W}(\phi, 0), \tag{12}$$

When applying the dual quaternion product we obtain the expression $\hat{Q}_{RPR} = Q^0 + \mathsf{Q}$, where the point is

$$Q^0 = \cos\frac{\theta}{2}\cos\frac{\phi}{2} - \epsilon\frac{d}{2}\sin\frac{\theta}{2}\cos\frac{\phi}{2}\mathsf{G}\cdot\mathsf{P} - \epsilon\frac{d}{2}\cos\frac{\theta}{2}\sin\frac{\phi}{2}\mathsf{P}\cdot\mathsf{W} -$$
$$\sin\frac{\theta}{2}\sin\frac{\phi}{2}\mathsf{G}\cdot\mathsf{W} - \epsilon\frac{d}{2}\sin\frac{\theta}{2}\sin\frac{\phi}{2}(\mathsf{G}\times\mathsf{P})\cdot\mathsf{W}, \tag{13}$$

and the vector

$$\mathsf{Q} = \sin\frac{\theta}{2}\cos\frac{\phi}{2}\mathsf{G} + \cos\frac{\theta}{2}\sin\frac{\phi}{2}\mathsf{W} + \epsilon\frac{d}{2}\cos\frac{\theta}{2}\cos\frac{\phi}{2}\mathsf{P} - \epsilon\frac{d}{2}\sin\frac{\theta}{2}\sin\frac{\phi}{2}(\mathsf{G}\cdot\mathsf{P})\mathsf{W} + \epsilon\frac{d}{2}\sin\frac{\theta}{2}\cos\frac{\phi}{2}(\mathsf{G}\times\mathsf{P}) + \epsilon\frac{d}{2}\cos\frac{\theta}{2}\sin\frac{\phi}{2}(\mathsf{P}\times\mathsf{W}) + \sin\frac{\theta}{2}\sin\frac{\phi}{2}(\mathsf{G}\times\mathsf{W}) + \epsilon\frac{d}{2}\sin\frac{\theta}{2}\sin\frac{\phi}{2}(\mathsf{G}\times\mathsf{P})\times\mathsf{W}. \tag{14}$$

The expansion of this equations componentwise leads to a set of equations in the components of the fixed rotation axis, $\mathsf{G} = (g_x, g_y, g_z) + \epsilon(g_x^0, g_y^0, g_z^0)$, the moving prismatic axis $\mathbf{P} = (p_x, p_y, p_z)$, and the moving rotation axis $\mathsf{W} = (w_x, w_y, w_z) + \epsilon(w_x^0, w_y^0, w_z^0)$.

The number of positions needed to obtain finite number of solutions is as follows: we have $15 + 3(n-1)$ unknowns (for the prismatic joint P only the direction matters) and $5+6(n-1)$ equations. Therefore we can define up to $n = 4 + \frac{1}{3}$ positions. The fractional value of n means that we can define 4 full positions plus two out of the six parameters that define a fifth position. Another option is to specify some extra relation for the joint axes. For instance, if we specify that the slider P must be perpendicular to the fixed rotation axis G we are adding one constraint and the counting gives finite number of solutions for $n = 4$ positions. This is the case that we use in the example.

To create the design equations we equate the expanded eq.(14) to the goal dual quaternion $\hat{D}$, that is,

$$\hat{Q}_{RPR}(\theta, d, \phi) - \hat{D} = \vec{0}, \tag{15}$$

to obtain one of the sets of design equations. To eliminate the joint parameters we consider first the direction equations, which can be solved linearly for θ and ϕ,

$$\begin{bmatrix} w_x & g_x & g_y w_z - g_z w_y & 0 \\ w_y & g_y & g_z w_x - g_x w_z & 0 \\ w_z & g_z & g_x w_y - g_y w_x & 0 \\ 0 & 0 & -(g_x w_x + g_y w_y + g_z w_z) & 1 \end{bmatrix} \begin{Bmatrix} \cos\frac{\theta}{2}\sin\frac{\phi}{2} \\ \sin\frac{\theta}{2}\cos\frac{\phi}{2} \\ \sin\frac{\theta}{2}\sin\frac{\phi}{2} \\ \cos\frac{\theta}{2}\cos\frac{\phi}{2} \end{Bmatrix} = \begin{Bmatrix} p_x \\ p_y \\ p_z \\ p_w \end{Bmatrix} \tag{16}$$

The solution exists when the matrix is invertible (the determinant is zero only for the degenerate case when both directions are parallel, hence creating a planar rotation) and when the directions $\mathbf{g}$, $\mathbf{w}$ make the system solvable, which we can assume will be given by the solution of the design equations. The obtained values are substituted in the four moment equations, to obtain four equations which are linear in the parameter d. We eliminate linearly the joint variable d . We obtain three implicit equations per goal position. These 9 equations together with 6 constraints,

$$\begin{aligned} g_x^2 + g_y^2 + g_z^2 = 1, \quad & g_x g_x^0 + g_y g_y^0 + g_z g_z^0 = 0 \\ w_x^2 + w_y^2 + w_z^2 = 1, \quad & w_x w_x^0 + w_y w_y^0 + w_z w_z^0 = 0 \\ p_x^2 + p_y^2 + p_z^2 = 1, \quad & g_x p_x + g_y p_y + g_z p_z = 0 \end{aligned} \tag{17}$$

allows us to solve for the 15 unknowns corresponding to the three joint axes.

We present an example in which we want to design the RPR robot to reach the following four positions:

Table 1. The goal positions

Position	*Axis*	*Rotation*	*Translation*
position 1	$(1.0, 0.0, 0.0) + \epsilon(0.0, 0.0, 0.0)$	0°	0
position 2	$(0.52, 0.59, 0.61) + \epsilon(-0.97, 1.35, -0.47)$	120°	-0.32
position 3	$(0.54, 0.56, 0.63) + \epsilon(-0.10, 0.84, -0.66)$	44°	-0.05
position 4	$(-0.33, -0.63, -0.70) + \epsilon(0.00, 1.30, -1.18)$	39°	0.01

One of the obtained solutions is presented, as the joint axes in the first position,in Table(2). Figure 2 shows the robot while reaching each of the positions.

Table 2. The joint axes

Joint Axis	*Direction*	*Moment*
G	$(0.394, 0.593, 0.701)$	$(-1.379, -0.074, 0.839)$
W	$(0.594, 0.518, 0.615)$	$(0.388, 0.976, -1.199)$
P	$(-0.897, 0.085, 0.432)$	$(-0.629, -0.946, -1.119)$

6. Conclusions

This paper introduces a new formulation for the kinematic synthesis of constrained serial robots. These robots have less than six degrees of freedom. The standard kinematics equations of the chain are transformed into successive screw displacements and then expressed using dual quaternions. The result is an explicit set of axis parameters that define the robot and a set of joint parameters that can be eliminated

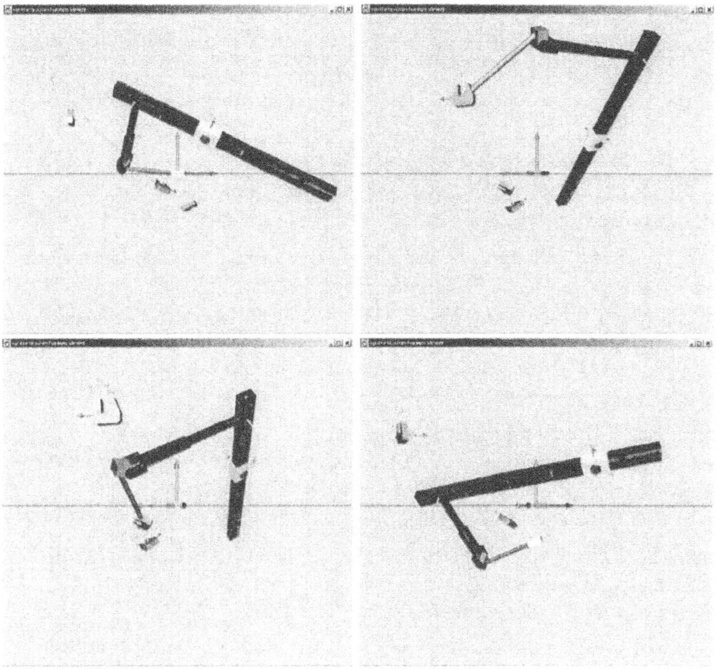

Figure 2. The spatial RPR robot reaching the four goal positions

algebraically. The structure of these equations provides a convenient strategy for this elimination, which we demonstrate with the design of a spatial RPR robot.

7. Acknowledgements

The authors gratefully acknowledge the support of the National Science Foundation and discussions with Curtis Collins and Haijun Su.

References

Angeles, J., 1998, "The application on dual algebra to kinematic analysis," *Computational Methods in Mechanical Systems, NATO ASI Series* (ed. J. Angeles and E. Zakhariev) Springer, Berlin.

Bottema, O., and Roth, B., 1979, *Theoretical Kinematics*, North Holland. (reprinted Dover Publications 1990).

Chedmail, P., 1998, "Optimization in Multi-DOF Mechanisms," *Computational Methods in Mechanical Systems, NATO ASI Series* (ed. J. Angeles and E. Zakhariev) Springer, Berlin.

Chen, P., and Roth, B., 1969, Design Equations for the Finitely and Infinitesimally Separated Position Synthesis of Binary Links and Combined Link Chains, *ASME J. Eng. Ind.* 91(1):209219.

Cox, D., Little, J. and O'Shea, D., 1998, *Using Algebraic Geometry*, Springer, New York.

Craig, J., 1986, *Introduction to Robotics*, Addison-Wesley.

Gosselin, C. M., 1998, "On the design of efficient parallel mechanisms," *Computational Methods in Mechanical Systems, NATO ASI Series* (ed. J. Angeles and E. Zakhariev) Springer, Berlin.

Gupta, K.C., 1986, "Kinematic Analysis of Manipulators Using Zero Reference Position Description", *Int. J. Robot. Res.*, 5(2):5-13

Huang, C., and Chang, Y.-J., 2000, "Polynomial Solution to the Five-Position Synthesis of Spatial CC Dyads via Dialytic Elimination," *Proceedings of the ASME Design Technical Conferences*, September 1013, 2000, Baltimore MD, Paper Number DETC2000/MECH-14102.

Innocenti, C., 1994, "Polynomial Solution of the Spatial Burmester Problem." *Mechanism Synthesis and Analysis, ASME DE* vol. 70.

Kumar, A., and Waldron, K. J., 1981, "The Workspaces of a Mechanical Manipulator," *ASME Journal of Mechanical Design*, vol. 103, pp 665-672.

Larochelle, P., 2000, "Approximate motion synthesis via parametric constraint manifold fitting," *Advances in Robot Kinematics* (eds. J. Lenarcic and M. M. Stanisic) Kluwer Acad. Publ., Dordrecht.

Mavroidis, C., Lee, E. and Alan, M. 2001, "A New Polynomial Solution to the Geometric Design Problem of Spatial R-R Robot Manipulators Using the Denavit and Hartenberg Parameters," *ASME J. of Mechanical Design* 123(2):58-67.

Lee, E., and Mavrodis, D., 2002 (in press), "Solving the Geometric Design Problem of Spatial 3R Robot Manipulators Using Polynomial Homotopy Continuation," *ASME J. of Mechanical Design.*

Liao, Q. and McCarthy, J.M., 1998, "On the seven position synthesis of a 5-SS platform linkage", *ASME Journal of Mechanical Design.*

McCarthy, J. M., 1990, *An Introduction to Theoretical Kinematics*, MIT Press.

McCarthy, J. M., 2000a, *Geometric Design of Linkages*, Springer, New York.

McCarthy, J. M., 2000b, "Mechanisms Synthesis Theory and the Design of Robots," *Proceedings of the 2000 IEEE International Conference on Robotics and Automation*, April 2428 2000, San Francisco, CA.

Perez, A., and McCarthy, J. M., 2000, "Dimensional Synthesis of Spatial RR Robots," Advances in Robot Kinematics, Lenarcic, J., ed., Piran-Portoroz, Slovenia, June 2630, 2000.

Suh, C.H. and Radclie, C.W.,1978, *Kinematics and mechanisms design.* John Wiley & Sons, 1978.

Tsai, L.W., 1999, *Robot Analysis.* John Wiley and Sons, New York.

Tsai, L.W. and Roth, B., 1973, "A Note on the Design of Revolute-Revolute Cranks". *Mechanisms and Machine Theory*, Vol.8, pp 23-31.

Yang, A.T., and Freudenstein, F., 1964, "Application of Dual-Number Quaternion Algebra to the Analysis of Spatial Mechanisms", *ASME Journal of Applied Mechanics*, June 1964, pp.300-308.

TYPE SYNTHESIS OF LINEAR TRANSLATIONAL PARALLEL MANIPULATORS

Xianwen Kong
Département de Génie Mécanique, Université Laval
Québec, Québec, Canada, G1K 7P4
xwkong@gmc.ulaval.ca

Clément M. Gosselin
Département de Génie Mécanique, Université Laval
Québec, Québec, Canada, G1K 7P4
gosselin@gmc.ulaval.ca

Abstract Linear TPMs (3-leg 3-DOF translational parallel manipulators) refer to TPMs whose forward displacement analysis can be performed by solving a set of linear equations. A method is proposed for the type synthesis of linear TPMs. The proposed method is based on the geometric interpretation of the forward displacement analysis and the fact that the intersection of three planes can be obtained by solving a set of linear equations. Then, all linear TPMs are obtained. Finally, all I-O (input-output) decoupled linear TPMs are revealed. An I-O decoupled linear TPM refers to a linear TPM for which three translations of the moving platform along three orthogonal directions can be controlled independently by three actuators.

Keywords: Translational parallel manipulator, I-O decoupled manipulator, Type synthesis, Forward kinematics

1. Introduction

A TPM (3-leg 3-DOF translational parallel manipulator) is composed of a moving platform and a base connected by three legs in parallel (Fig. 1). Under the action of the total constraints of its three legs, the moving platform can only translate with respect to the base.

TPMs have a wide range of applications such as assembly and machining. Several types of TPMs have been proposed, e.g., in (Appleberry, 1992; Clavel, 1990; Di Gregorio and Parenti-Castelli, 1998; Hervé and Sparacino, 1991; Hervé, 1995; Tsai, 1999a; Tsai, 1999b; Zhao and Huang, 2000). Several systematic approaches have also been proposed for the

J. Lenarčič and F. Thomas (eds.), Advances in Robot Kinematics, 453–462.

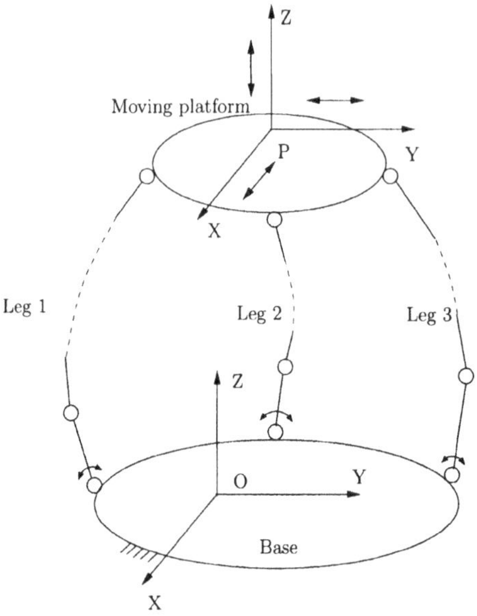

Figure 1. A TPM.

type synthesis of TPMs, such as the method based on the displacement group theory (Hervé and Sparacino, 1991; Hervé, 1995) and the methods based on screw algebra or screw theory (Frisoli, Checcacci, Salsedo and Bergamasco, 2000; Kong and Gosselin, 2001a). A comprehensive list of TPMs is also given in (Kong and Gosselin, 2001a).

In fact, previous works on the systematic type synthesis of TPMs (Hervé and Sparacino, 1991; Frisoli, Checcacci, Salsedo and Bergamasco, 2000; Kong and Gosselin, 2001a) deal mainly with the systematic type synthesis of TPKCs (translational parallel kinematic chains). As a consequence, the selection of actuators of TPMs has not been dealt with systematically.

Recently, (Kong and Gosselin, 2001b; Kong and Gosselin, 2001c) revealed the condition for all the translational degrees of freedom of the C (cylindrical) joints of the 3-CRR translational parallel kinematic chain — which is implicitly proposed in (Hervé and Sparacino, 1991) — to be actuated and thus proposed a 3-CRR TPM with linear actuators. It has been revealed that the forward displacement analysis of the 3-CRR TPM can be performed by solving a set of linear equations.

This paper deals with the type synthesis of linear TPMs. A linear TPM refers to a TPM whose forward displacement analysis can be performed by solving a set of linear equations. A method is proposed for the type synthesis of linear TPMs in section 2. By inspecting the list

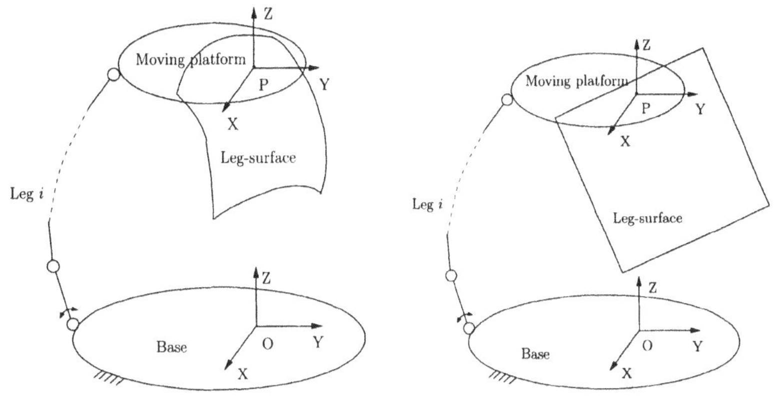

(a) The general case.

(b) The planar case.

Figure 2. Leg-surfaces of legs for TPMs.

of legs for TPKCs, all legs for linear TPMs are determined in section 3. Then, all linear TPMs are obtained in section 4. All I-O (input-output) decoupled linear TPM are revealed in section 5. Finally, conclusions are drawn.

2. An approach to the type synthesis of linear TPMs

When the actuator of a given leg of a TPM is locked, the moving platform will translate along a surface with its orientation unchanged under the action of the total constraints on the moving platform by all of the legs of the TPM. For brevity, the above surface is referred to as the leg-surface (Fig. 2) of the leg. Thus, the forward displacement analysis of the TPM can be described geometrically as follows: it consists in finding the intersection of three leg-surfaces. The higher the degree of the leg-surfaces, the more complicated the forward displacement analysis of the TPM. The simplest cases are TPMs with three planar leg-surfaces. As the intersection of three planes can be obtained by solving a set of linear equations, a TPM is a linear TPM if all of its three leg-surfaces are planes. Thus, type synthesis of linear TPMs is reduced to the type synthesis of legs for TPMs whose leg-surfaces are planes (Fig. 2(b)), which are referred to as legs for linear TPMs in the following.

The type synthesis of linear TPMs can be performed in two steps. The first is to perform the type synthesis of legs for linear TPMs, the second is to perform the type synthesis of linear TPMs.

3. Type synthesis of legs for linear TPMs

In the type synthesis of legs for linear TPMs, we make the assumption that the first joint, i.e., the joint located on the base, is actuated while the other joints are unactuated. The type of a leg is denoted by a chain of characters denoting the joint types starting from the base to the moving platform. For simplicity reasons, R and $\bar{R}$ are both used to denote R (revolute) joints. R joints within one leg denoted by the same letter have parallel axes while R joints within one leg denoted by different letters do not.

As the leg-surface of a leg for linear TPMs is a plane, legs for linear TPMs should have the following characteristics.

In a leg for linear TPMs, the unactuated joints should be a combination of two P joints, a combination of two R joints with parallel axes and one P joint whose axis is perpendicular to the axes of the R joints with parallel axes, or a combination of three R joints with parallel axes.

Based on the characteristics of legs for linear TPMs, legs for linear TPMs can be obtained by inspection of the legs for TPKCs, which are listed in (Hervé and Sparacino, 1991; Frisoli, Checcacci, Salsedo and Bergamasco, 2000; Kong and Gosselin, 2001a).

All the legs for linear TPMs have been obtained. Table 1 lists the basic types of legs for linear TPMs. A basic type of legs for linear TPMs is those legs composed of R and P (prismatic) joints only. Other types of legs are the variations of the basic types. They can be obtained by (1) substituting a combination of one R joint and one P joint along parallel axes with a C joint, (2) substituting a combination of two adjacent R joints with unparallel axes with a U (universal) joint, (3) substituting an actuated P joint with a Π joint, i.e., a parallelogram, whose plane is not parallel to the axes of the two unactuated P joints or not perpendicular to the axes of the unactuated R joints within the same leg, (4) substituting an unactuated P joint with a Π joint whose plane is parallel to the axes of the two unactuated P joints or perpendicular to the axes of the unactuated R joints within the same leg, (5) substituting the actuated R joint with a H (helical) joint whose axis is parallel to the axis of the actuated R joint, (6) substituting the only unactuated R joint whose axis is parallel to the axis of the actuated R joint with a H joint whose axis is parallel to the axis of the actuated R joint, and/or (7) substituting a pair of unactuated R joints with parallel axes, whose axes are not parallel to the axes of any other R joint in the same leg, with a pair of H joints of the same pitch whose axes are parallel to the axes of the pair of unactuated R joints.

Table 1. Legs for linear TPMs.

No	Type	Geometric condition	Normal of planar leg-surface
1	PPP	The axes of the three P joints are not parallel to a plane	Perpendicular to the axes of the two unactuated P joints
2 3 4	PPRR PRPR PRRP	The axis of the second P joint is perpendicular to the axes of the R joints	Parallel to the axes of R joints
5	PRRR	The axis of the P joint is not perpendicular to the axes of the R joints	Parallel to the axes of R joints
6 7 8	RPPR RPRP RRPP	The axes of the R joints are not perpendicular to the axes of all the P joints	Perpendicular to the axes of the two unactuated P joints
9 10 11	RRPR̄R̄ RRR̄PR̄ RRR̄R̄P	The axis of the P joint is perpendicular to the axes of the R̄ joints	Parallel to the axes of R̄ joints
12	RRR̄R̄R̄	-	Parallel to the axes of R̄ joints
13 14 15	RPR̄R̄R RR̄PR̄R RR̄R̄PR	The axis of the P joint is perpendicular to the axes of the R̄ joints	Parallel to the axes of R̄ joints
16	RR̄R̄R̄R	-	Parallel to the axes of R̄ joints

For a better understanding of the legs for linear TPMs, their basic types are classified into the following four classes.

1) Legs for linear TPMs with an actuated P joint and two unactuated P joints. This class of legs includes the PPP (Fig. 3(a)) leg.

2) Legs for linear TPMs with an actuated P joint and a combination of two unactuated R joints with parallel axes and one unactuated P joint whose axis is perpendicular to the axes of the unactuated R joints with parallel axes, or a combination of three unactuated R joints with parallel axes. This class of legs includes the PPRR, PRPR, PRRP and PRRR (Fig. 3(b)) legs.

3) Legs for linear TPMs with an actuated R joint and two unactuated P joints. This class of legs includes the RPPR, RPRP (Fig. 3(c)) and RRPP legs.

4) Legs for linear TPMs with an actuated R joint and a combination of two unactuated R joints with parallel axes and one un-

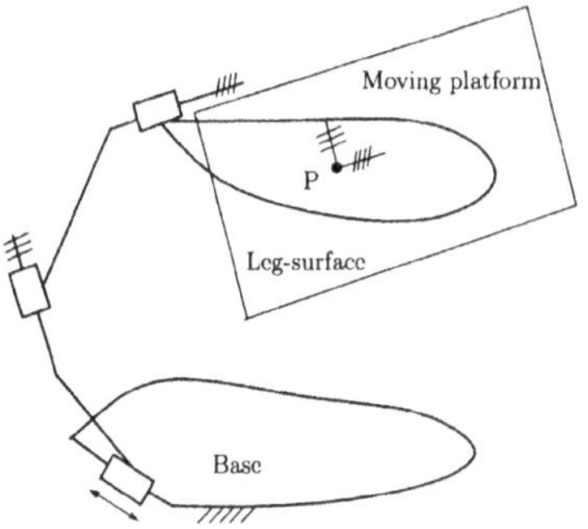

(a) The PPP leg.

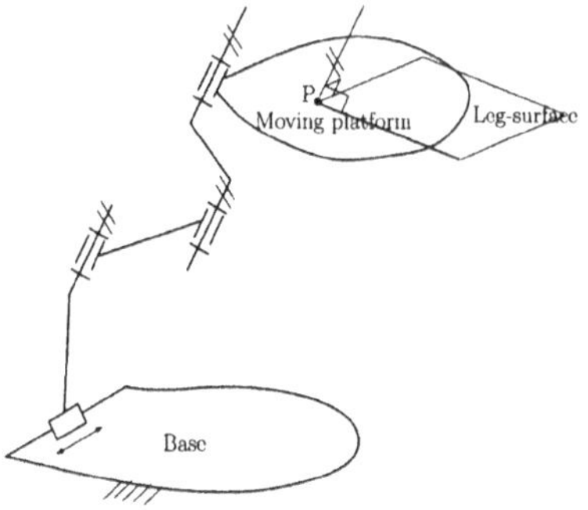

(b) The PRRR leg.

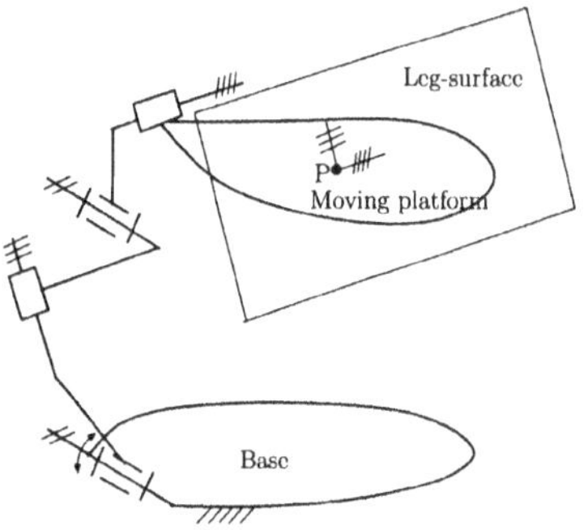

(c) The RPRP leg.

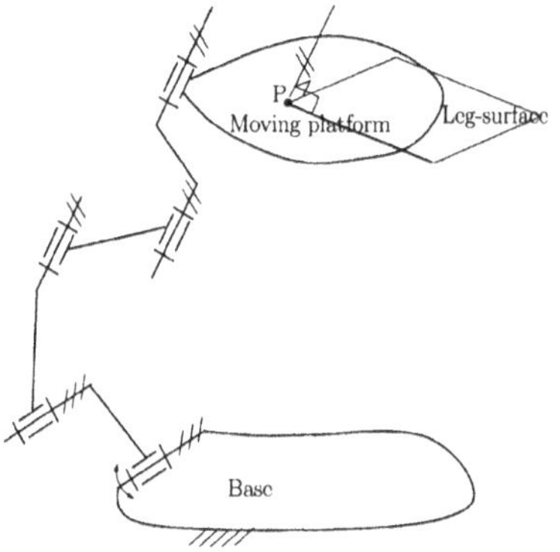

(d) The RR$\bar{R}\bar{R}\bar{R}$ leg.

Figure 3. Some legs for linear TPMs.

actuated P joint whose axis is perpendicular to the axes of the unactuated R joints with parallel axes, or a combination of three unactuated R joints with parallel axes. This class of legs includes the RRP$\bar{R}\bar{R}$, RR$\bar{R}$P$\bar{R}$, RR$\bar{R}\bar{R}$P, RR$\bar{R}\bar{R}\bar{R}$ (Fig. 3(d)), RP$\bar{R}\bar{R}$R, R$\bar{R}$P$\bar{R}$R, R$\bar{R}\bar{R}$PR, and R$\bar{R}\bar{R}\bar{R}$R legs.

It is noted that in a leg for linear TPMs with an actuated R or H joint, there must be one and only one unactuated R or H joint whose axis is parallel to the axis of the actuated R or H joint. Once the actuated R or H joint is locked, the unactuated R or H joint with parallel axis is also locked by the total constraints on the moving platform of all the legs in a linear TPM. Otherwise, the orientation of the moving platform will change. In the HPPR, HPRP, HRPP, RHPP, RPHP and RPPH legs for linear TPMs, the axes of the H and R joints are not necessarily perpendicular to the axes of the two unactuated P joints. In the HHPP, HPHP and HPPH legs for linear TPMs, the axes of the two H joints are also not necessarily perpendicular to the axes of the two unactuated P joints if the pitches of the two H joints are different.

It is also noted that in the legs for TPKCs corresponding to the legs for linear TPMs with a combination of two unactuated R joints with parallel axes and one unactuated P joint whose axis is perpendicular to the axes of the unactuated R joints with parallel axes, the axis of the above P joint is not necessarily perpendicular to the above mentioned R joints.

4. Type synthesis of linear TPMs

Taking a combination of three legs for linear TPMs, a linear TPM can be obtained. A linear TPM should satisfy the following two conditions.

- All the rotational degrees of freedom of the moving platform are eliminated by the total constraints on the moving platform of its three legs.
- In a general configuration, the three planar leg-surfaces have one and only one point in common.

The first condition has been dealt with in (Hervé and Sparacino, 1991; Frisoli, Checcacci, Salsedo and Bergamasco, 2000; Kong and Gosselin, 2001a) and is easy to satisfy. The second condition is actually the condition which guarantees that the selected actuators do work.

It is noted that for three planes having one or more points in common, the condition for the number of common points to be 1 is that the normals of these planes are not parallel to a plane. Detailed conditions can thus be obtained for linear TPMs.

To perform the type synthesis of linear TPMs, the normal of the planar leg-surface for each leg is first found as listed in column 4 of Table 1. Then all the types of linear TPMs can be obtained. Those with three legs of the same type are listed in Table 2.

5. I-O decoupled linear TPMs

An I-O decoupled linear TPM refers to a linear TPM for which three translations along three orthogonal directions can be controlled independently by three actuators. In this case, the forward displacement analysis of linear TPMs can be further simplified.

Recalling the geometric interpretation of the forward displacement analysis of TPMs, the forward displacement analysis of linear TPMs

Table 2. Linear TPMs.

No	Type	Geometric condition
1	3-PPP	Three lines, each being perpendicular to the axes of the last two P joints of one leg, are not parallel to a plane
2	3-PPRR	The axes of all the R joints are not parallel to a plane
3	3-PRPR	
4	3-PRRP	
5	3-PRRR	
6	3-RPPR	-
7	3-RPRP	-
8	3-RRPP	Three lines, each being perpendicular to the axes of the last two P joints of one leg, are not parallel to a plane
9	3-RRP$\bar{R}\bar{R}$	The axes of all the $\bar{R}$ joints on are not parallel to a plane
10	3-RR$\bar{R}$P$\bar{R}$	
11	3-RR$\bar{R}\bar{R}$P	
12	3-RR$\bar{R}\bar{R}\bar{R}$	
13	3-RP$\bar{R}\bar{R}$R	-
14	3-R$\bar{R}$P$\bar{R}$R	
15	3-R$\bar{R}\bar{R}$PR	
16	3-R$\bar{R}\bar{R}\bar{R}$R	

consists in finding the intersection of three planar leg-surfaces. When the three planar leg-surfaces are perpendicular (see planes PABC, PADE and PCFE in Fig. 4), the translation of the moving platform along the intersection of any two of the three planar leg-surfaces is controlled by the position of the other planar leg-surface. For example, the translation of the moving platform along PC, i.e., the intersection of planes PABC and PCFE, is controlled by the position of plane PADE. We can thus conclude that for a TPM to be I-O decoupled, its three planar leg-surfaces should be perpendicular.

Among the linear TPMs, the 3-PPP, 3-PPRR, 3-PRPR, 3-PRRP, 3-PRRR, 3-RRPP, 3-RRP$\bar{R}\bar{R}$, 3-RR$\bar{R}$P$\bar{R}$, 3-RR$\bar{R}\bar{R}$P and 3-RR$\bar{R}\bar{R}\bar{R}$ linear TPMs are TPMs with constant-orientation planar leg-surfaces. When the three planar leg-surfaces for these TPMs are perpendicular, these linear TPMs are I-O decoupled. I-O decoupled linear TPMs are thus obtained and listed in Table 3.

Figure 5 shows the CAD model of the 3-CRR linear TPM. The 3-CRR TPM is a specific case with the minimum number of links of the

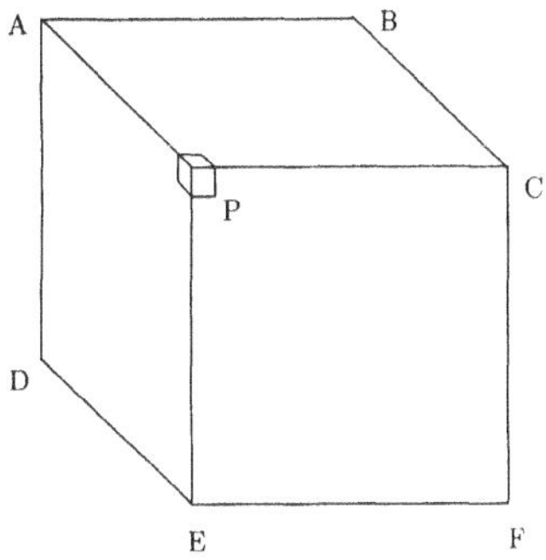

Figure 4. The intersection of three othorgonal planes.

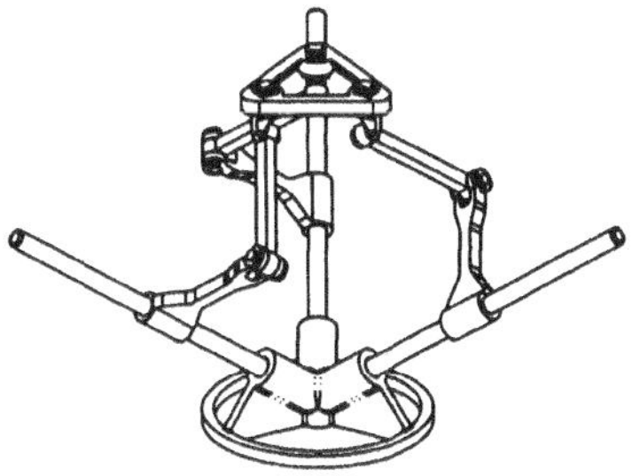

Figure 5. The CAD Model of a 3-CRR TPM, from (Kong and Gosselin, 2001c).

Table 3. I-O decoupled linear TPMs.

No	Type	Geometric condition
1	3-PPP	Three lines, each being perpendicular to the axes of the last two P joints of one leg, are perpendicular
2	3-PPRR	The axes of the R joints in any one leg are perpendicular to the axes of R joints in other legs
3	3-PRPR	
4	3-PRRP	
5	3-PRRR	
6	3-RRPP	Three lines, each being perpendicular to the axes of the last two P joints of one leg, are perpendicular
7	3-RRP$\bar{R}\bar{R}$	The axes of the $\bar{R}$ joints in any one leg are perpendicular to the axes of $\bar{R}$ joints in other legs
8	3-RR$\bar{R}$P$\bar{R}$	
9	3-RR$\bar{R}\bar{R}$P	
10	3-RR$\bar{R}\bar{R}\bar{R}$	

3-PRRR I-O decoupled linear TPM. It has been further pointed out in (Kong and Gosselin, 2001c) that the 3-CRR TPM is globally isotropic.

6. Conclusions

An approach has been proposed for the type synthesis of linear TPMs. Legs for linear TPMs as well as linear TPMs have been obtained. I-O decoupled linear TPMs have also been revealed.

The results of this paper should be of great interest in the development of fast TPMs and parallel kinematic machines with high-performance.

7. Acknowledgments

The authors would like to acknowledge the financial support of the Natural Sciences and Engineering Research Council of Canada (NSERC). Clément M. Gosselin is holding a Canada Research Chair and would like to acknowledge the financial support of the Chair program. The authors also thank Pierre-Luc Richard and Thierry Laliberté for building the CAD model of the isotropic 3-CRR TPM.

References

Appleberry, W. T. (1992), Anti-rotation positioning mechanism. *United States Patent*, No. 5156062.

Clavel, R. (1990), Device for the movement and positioning of an element in space. *United States Patent*, No. 4976582.

Di Gregorio, R., and Parenti-Castelli V. (1998), A translational 3-DOF parallel manipulator. *Advances in Robot Kinematics: Analysis and Control*, (J. Lenarčič and M. L. Husty, eds.), Kluwer Academic Publishers, pp. 49-58.

Frisoli, A., Checcacci, D., Salsedo, F. , and Bergamasco, M. (2000), Synthesis by screw algebra of translating in-parallel actuated mechanisms, *Advances in Robot Kinematics*, (J. Lenarčič and M. M. Stanišić, eds.), Kluwer Academic Publishers, pp. 433-440.

Hervé, J. M., and Sparacino F. (1991), Structural synthesis of parallel robots generating spatial translation. *Proceedings of the fifth International Conference on Advanced Robotics*, Pisa, Italy, Vol. 1, 808-813.

Hervé, J. M. (1995), Design of parallel manipulators via the displacement group. *Proceedings of the Ninth World Congress on the Theory of Machines and Mechanisms*, Milan, Italy, Vol. 3, 2079-2082.

Kong, X., and Gosselin, C. M. (2001a), Generation of parallel manipulators with three translational degrees of freedom based on screw theory. *Proceedings of 2001 CCToMM Symposium on Mechanisms, Machines and Mechatronics*, Saint-Hubert (Montreal).

Kong, X., and Gosselin, C. M. (2001b), Analytical manipulators with three translational degrees of freedom. *United States Patent application* (Submitted).

Kong, X., and Gosselin, C. M. (2001c), Kinematics and singularity analysis of a 3-CRR 3-DOF translational parallel manipulator. *International Journal of Robotics Research* (To appear).

Tsai, L. W. (1999a), *Robot Analysis: the mechanics of serial and parallel manipulators*, John Willey & Sons Inc..

Tsai, L. W. (1999b), The enumeration of a class of three-DOF parallel manipulators. *Proceedings of the tenth World Congress on the Theory of Machines and Mechanisms*, Oulu. Vol. 3, 1121-1126.

Zhao, T. S., and Huang, Z. (2000), A novel three DOF translational platform mechanism and its kinematics. *Proceedings of 2000 ASME Design Engineering Technical Conferences*, Baltimore. DETC2000/MECH-14101.

STIFFNESS SYNTHESIS OF A VARIABLE GEOMETRY PLANAR ROBOT

Nabil Simaan and Moshe Shoham

Robotics Laboratory, Department of Mechanical Engineering

Technion - Israel Institute of Technology

Haifa 32000

ISRAEL

http://robotics.technion.ac.il

nabil@tx.technion.ac.il

shoham@tx.technion.ac.il

Abstract This paper addresses the problem of task-based stiffness synthesis of a variable geometry three DOF (Degrees Of Freedom) planar robot. The synthesis considers the case where the robot has a limited number of free geometric parameters and constant actuator stiffness coefficients. This defines twenty problems of stiffness synthesis, in which, three parameters of the stiffness matrix are controlled according to task requirements. These problems are modeled as systems of polynomials in the free geometric parameters of the robot's base platform. Using Gröbner bases, the solubility of these polynomial systems is characterized. It is shown that arbitrary desired values of the Cartesian stiffness elements (k_{xx} and k_{yy}) are unattainable when only the geometry of the base platform is variable. An example of synthesizing three stiffness elements of the planar robot is solved and shown to have at most 48 solutions in the complex plane. In a numerical case study, sixteen real solutions are obtained, of which only eight are non-singular.

Keywords: Parallel robot, Re-configurable, Stiffness Synthesis, Gröbner bases.

1. Introduction

Robots are used to perform various tasks involving complex manipulations and interactions with their environment. Consequently, there are inevitable compromises when using a fixed-geometry robot for some tasks. To overcome this problem, the use of variable geometry robots is considered. In particular we concentrate here on variable geometry parallel robots. These robots can change the geometry of their base/moving platforms to accommodate the required characteristics, e.g. stiffness, specific to each task.

Researchers used various methods to enhance parallel robots' capabilities for better fitting task requirements in terms of stiffness, singularity avoidance, and inclusion of specific paths in the workspace.

J. Lenarčič and F. Thomas (eds.), Advances in Robot Kinematics, 463–472.

Actuation redundancy was used by Yi and Freeman (1993), Kim et al. (2000) for stiffness modulation. Kinematic redundancy of the robot for a given task was used by Merlet, et al. (2000) for path inclusion in the workspace and singularity avoidance. Stiffness/compliance synthesis algorithms were presented by Huang and Schimmels, (1998), and Roberts, (1999) for a system of springs supporting a rigid body.

Works directly addressing variable geometry parallel robots are limited in number. Zhiming and Song (1998) investigated the design aspects of modular Stewart-Gough platforms with workspace and joint limits considerations. Zhiming and Zhenqun (1999) presented an algorithm for identifying the parameters of the joint locations on the base in a modular Stewart-Gough platform. Merlet, (1997, 2000) presented a design algorithm for achieving a constant-orientation workspace of Stewart-Gough robots, which can be adapted for workspace modification of variable geometry robots♠.

In the present investigation, a case study of stiffness synthesis for a point in a given path of a planar 3 DOF robot with a variable geometry of its base platform is presented. The aim of the synthesis is to obtain a specific stiffness in a given position/orientation of the robot's moving platform. Under a simplifying assumption that the stiffness coefficients of the redundant actuators that change the base geometry are considerably larger than the coefficients of the other actuators, this work may be viewed as an algorithm for changing the geometry of the base platform of a variable geometry 3 DOF planar robot for obtaining a required stiffness in a point along a path specified by the given task.

2. Variable Geometry Planar Robot

The planar robot of Fig. 1 has an equilateral triangular moving platform connected to a circular base by three kinematic chains composed of a slider on the circular base, a revolute joint, a prismatic joint, and a revolute joint on its moving platform. The sliders on the circular base control the geometry of the base platform and the prismatic actuators manipulate the moving platform. This introduces a kinematic redundancy in this three DOF planar robot. The objective of this paper is to determine the geometry of

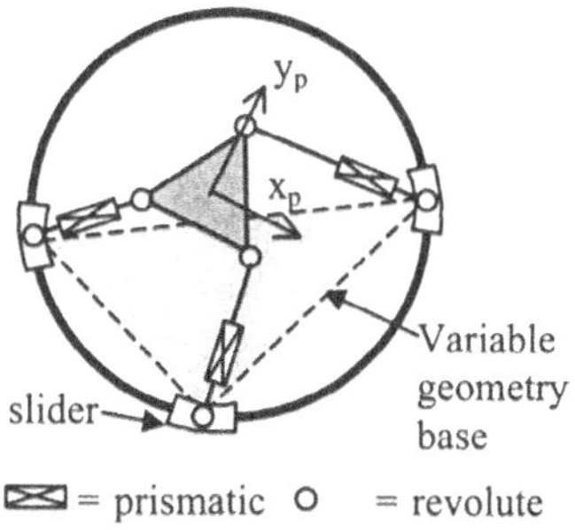

Figure 1. Planar Robot with variable geometry base platform

♠ The authors acknowledge the valuable discussions with J.-P. Merlet on this subject the base platform by computing the locations of these three redundant sliders for achieving a desired stiffness goal in a point of a path defined by the task.

3. Stiffness Polynomial Formulation

In this section, the stiffness matrix of the robot of Fig. 1 is formulated as a function of the positions of the sliders on the circular base and is described in a Platform–attached Coordinate System (PCS). The location and orientation of the moving platform are given by the task while the orientations of the prismatic actuators are given by the stiffness synthesis solution and are easily transformed to desired positions of the redundant sliders on the circular base.

The transformation of the desired stiffness from World Coordinate System (WCS) to PCS is given by, Tsai (1999):

$$ {}^{P}\mathbf{K}_{des} = \mathbf{A}^{T\,W}\mathbf{K}_{des}\mathbf{A} \qquad \mathbf{A} = \begin{bmatrix} \mathbf{R} & \mathbf{0} \\ \mathbf{0} & \mathbf{R} \end{bmatrix} \tag{1} $$

where $\mathbf{R}$ is the 3x3 rotation matrix from the PCS to WCS, ${}^{P}\mathbf{K}_{des}$ and ${}^{W}\mathbf{K}_{des}$ are the desired stiffness matrix in PCS and WCS, respectively. Hereafter, all vectors and matrices are represented in PCS.

The only controllable geometric variables by the sliders' locations are the unit vectors of the prismatic actuators' axes, ($\hat{\mathbf{l}}_i$, i=1,2,3), Figs. 1-2.

$$ \hat{\mathbf{l}}_i = a_i\,\hat{\mathbf{e}}_{i1} + b_i\,\hat{\mathbf{e}}_{i2} \quad i = 1,2,3 \tag{2} $$

where the symbol ^ indicates a unit vector, $\hat{\mathbf{e}}_{i1}$ and $\hat{\mathbf{e}}_{i2}$ are unit vectors indicated in Fig. 2, $\hat{\mathbf{l}}_i$ is a unit vector along the ith prismatic actuator, and a_i, b_i are the projections of $\hat{\mathbf{l}}_i$ on $\hat{\mathbf{e}}_{i1}$ and $\hat{\mathbf{e}}_{i2}$. In order for $\hat{\mathbf{l}}_i$ to be a unit vector, a_i and b_i (i=1,2,3) must obey:

$$ a_i^2 + b_i^2 + 2a_i b_i(\hat{\mathbf{e}}_{i1} \cdot \hat{\mathbf{e}}_{i2}) - 1 = 0 \tag{3} $$

For an equilateral platform, Eq. (3) is an ellipse in a_i-b_i plane:

$$ a_i^2 + b_i^2 + a_i b_i - 1 = 0 \qquad i = 1,2,3 \tag{4} $$

A simplifying assumption is made that the sliders on the circular platform have a mechanical means to lock rigidly on the circular base once the desired geometry of the base is obtained or that the stiffness coefficients of the sliders are considerably larger than the stiffness coefficients of the prismatic actuators. With this simplifying assumption, the stiffness matrix depends only on the stiffness coefficients of the three remaining active prismatic

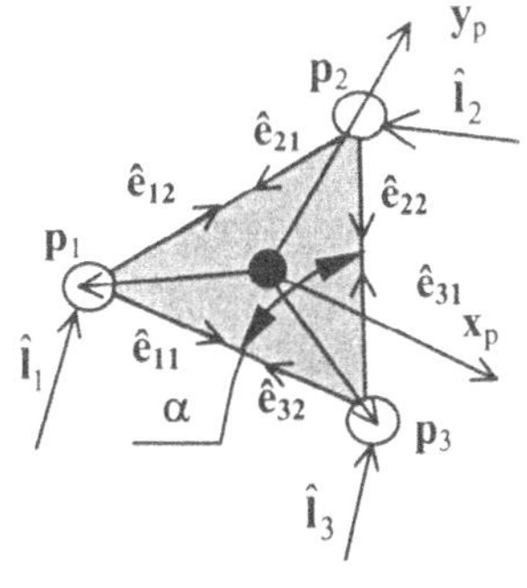

Figure 2. Geometric notations for the planar robot

actuators. These three active actuators are assumed identical having stiffness coefficient k_d. The rows of the Jacobian matrix of the robot of Fig. 1, with its sliders locked on the circular base, are the Plücker line coordinates of the three axes of the prismatic actuators, Eq. (5).

$$\mathbf{J}=\begin{bmatrix} a_1+\frac{b_1}{2} & \frac{b_1\sqrt{3}}{2} & 0 & 0 & 0 & \frac{r}{2}\left(a_1+\frac{b_1}{2}\right)-\frac{3}{4}rb_1 \\ -\frac{a_2}{2}+\frac{b_2}{2} & -\frac{a_2\sqrt{3}}{2}-\frac{b_2\sqrt{3}}{2} & 0 & 0 & 0 & -r\left(-\frac{a_2}{2}+\frac{b_2}{2}\right) \\ -b_3-\frac{a_3}{2} & \frac{a_3\sqrt{3}}{2} & 0 & 0 & 0 & \frac{r}{2}\left(-\frac{a_3}{2}-b_3\right)+\frac{3}{4}ra_3 \end{bmatrix} \tag{5}$$

The 6×6 Stiffness matrix of the planar robot with three identical prismatic actuators is given by $\mathbf{K}=k_d\mathbf{J}^T\mathbf{J}$, Tsai (1999), and the reduced planar 3×3 stiffness matrix is then, obtained. Notations of the elements of this symmetric stiffness matrix are given in Eq. (6).

$$\mathbf{K}=\begin{bmatrix} k_{xx} & k_{xy} & k_{x\theta} \\ k_{xy} & k_{yy} & k_{y\theta} \\ k_{x\theta} & k_{y\theta} & k_{\theta\theta} \end{bmatrix} \tag{6}$$

4. Stiffness synthesis with limited number of variable geometry parameters

Given a desired triplet of stiffness elements from the upper triangular part of **K** in Eq. (6), the problem of stiffness synthesis with limited number of variable geometry parameters deals with finding the geometry of the base platform (i.e., finding a_i, b_i, i=1,2,3) of the robot in Fig. 1.

To fully depict the 3×3 stiffness matrix, all the six equations in Eq. (7) given below must be fulfilled together with the three equations in Eq (4). Since the planar robot of Fig. 1 has a kinematic redundancy of order three, only three stiffness equations from Eq. (7) can be simultaneously fulfilled.

$$k_{n_i}-k_{n_{i_{desired}}}=0,\ i=1,2,3 \quad | \quad n_1\neq n_2\neq n_3\in\{xx,xy,x\theta,yy,y\theta,\theta\theta\} \tag{7}$$

Note that this results in $\binom{6}{3}$=20 systems of six polynomials with each having a total degree of 2 in a_i, b_i (i=1,2,3).

Generally, we ask whether it is possible to solve this problem for any such triplet and, if so, is changing the directions of the lines in Fig. 2 enough to allow controlling all such triplets?

To solve the polynomial systems derived from Eq. (4) and Eq. (7) ,the method of multiplication tables' eigenvalues (see Stetter (1993)) is used. The following sub-section briefly describes this method. Further details can be found in Möller and Stetter (1995) and Cox et al. (1998).

5. The method of eigenvalues for solving polynomial systems

Let $\mathscr{C}[x_1..x_m]$ represent the *ring* of polynomials with variables $x_1..x_m$, and coefficients over the complex field, $\mathscr{C}$. Let also $S=\{p_1,p_2,..p_n \mid p_1,p_2..p_n \in \mathscr{C}[x_1..x_m]\}$ be a system of n polynomials with a corresponding zero-dimensional Ideal $\mathscr{I} = \langle p_1,p_2,..p_n \rangle$, $\mathscr{I} \subset \mathscr{C}[x_1..x_m]$. The *variety* $V(\mathscr{I})$ of solution is defined by all the m-tuples of $x_1..x_m$ such that $p_1=p_2=..p_n=0$, i.e., $V(\mathscr{I})=\{[x_1..x_m] \in \mathscr{C}^m \mid p_1=p_2=..p_n=0\}$. We seek all the solutions of S.

The original system of polynomial equations, S, can be replaced by another minimal set of polynomials, $G=\{g_1...g_t\}$, called *standard basis* (or *Gröbner basis*) of the ideal $\mathscr{I}$ using Buchberger's algorithm, Buchberger (1965), which is not reviewed here for lack of space. Questions regarding ideal-membership of a given polynomial to $\mathscr{I}$, solubility of S, and finiteness of the dimension of $V(\mathscr{I})$ are readily answered when using this basis, Heck (1997). Also, for lexographic ordering G is a system of polynomials with successively eliminated number of variables as in the result of Gauss-Jordan elimination method for linear equations. However, this elimination method is unfavorable for large systems due to the computation effort associated with this ordering, Cox et al. (1998).

Two polynomials $f, g \in \mathscr{C}[x_1..x_m]$, are said to be *congruent* modulo $\mathscr{I}$, $f \equiv g \bmod \mathscr{I}$, if $f-g \in \mathscr{I}$. Consequently, f and g have the same *normal form* with respect to G and equal cosets [g]=[f]. A coset [f] of $f \in \mathscr{C}[x_1..x_m]$ is the sub-group of $\mathscr{C}[x_1..x_m]$ in which all its elements have the same normal form with respect to G, $[f]=f+\mathscr{I}=\{f+h \mid h \in \mathscr{I}\}$. The totality of cosets of the polynomials in $\mathscr{C}[x_1..x_m]$ is the quotient ring of $\mathscr{C}[x_1..x_m]$ modulo $\mathscr{I}$ indicated by $\mathscr{C}[x_1..x_m]/\mathscr{I}$, i.e., $\mathscr{C}[x_1..x_m]/\mathscr{I}=\{f+\mathscr{I} \mid f \in \mathscr{C}[x_1..x_m]\}$.

Given two polynomials $f, g \in \mathscr{C}[x_1..x_m]$ then a normal form arithmetic similar to number arithmetic exists for addition and multiplication: $n_f(f+g)=n_f(f)+n_f(g)$ and $n_f(f\,g)=n_f(n_f(f)\,n_f(g))$. Since every normal form is associated with a coset and vise-versa, this arithmetic is also translated to an associated coset arithmetic in the ring $\mathscr{C}[x_1..x_m]/\mathscr{I}$ resulting in the fact that $\mathscr{C}[x_1..x_m]/\mathscr{I}$ is a vector space in $\mathscr{C}^n$. Let B be a basis of monomials for this space $B=\{b_i, i=1..n\}$. This means that the remainder (or normal form) of any $f \in \mathscr{C}[x_1..x_m]$ is given by:

$$n_f(f)=n_f\left(\sum_{i=1..n} c_i b_i\right) \mid c_i \in \mathscr{C}, b_i \in B \tag{8}$$

or in congruence terms:

$$f \equiv \sum_{i=1..n} c_i b_i \bmod \mathscr{I} \tag{9}$$

Define the monomial basis vector $\mathbf{b}=[b_1, b_2, ...b_n]^t$, $b_i \in B$, i=1..n, then each polynomial $f \in \mathscr{C}[x_1..x_m]$ has a multiplication table $\mathbf{M}_f$ such that:

$$f\,\mathbf{b} \equiv \mathbf{M}_f \mathbf{b} \bmod \mathscr{I} \tag{10}$$

Since the congruence relation in Eq. (10) indicates that $f\mathbf{b}-\mathbf{M}_f\mathbf{b}\in\mathscr{I}$ then $f\mathbf{b}-\mathbf{M}_f\mathbf{b}=\mathbf{0}$ for all the points of $V(\mathscr{I})$. Consequently, the eigenvalues method is based on Eq. (11):

$$(\mathbf{M}_f - f\mathbf{I})\mathbf{b} = \mathbf{0} \tag{11}$$

Equation (11) shows that the eigenvalues of $\mathbf{M}_f$ are the values of f for all the points of $V(\mathscr{I})$. If f is taken as $f=x_1$ then Eq. (11) gives all the coordinates x_1 of the points of $V(\mathscr{I})$. Thus, by constructing the multiplication tables $\mathbf{M}_{x1}...\mathbf{M}_{xm}$ and solving for their eigenvalues, all the values of the coordinates $x_1...x_m$ for the points of $V(\mathscr{I})$ are obtained. The minimal polynomials of $\mathbf{M}_{xi}$, i=1..m, when written with x_i as its variable, give the monic generators of the elimination ideals $\mathscr{I}\cap\mathscr{C}[x_i]$.

There are both symbolic and numerical advantages of this method compared to standard sequential elimination of variables by resultants, Raghavan and Roth (1995), Neilsen and Roth (1999). Since this method is based on Gröbner basis construction, solvability of the system of polynomial equations is readily determined. Moreover, this method is unaffected by the term order used for the computation of G, which reduces the computation time when using more efficient term orders such as total degree order, Cox et al. (1998). Compared to sequential elimination methods, in this method the numerical computation is kept to a minimum since numerical values are used only in the computation procedure of eigenvalues and the solution of each coordinate x_i is independent of the numerical solutions of the other variables and, thus, it is unaffected by computation errors in the other variables x_j.

6. Application to the parallel planar robot

To answer the questions of section 4, the reduced Gröbner bases associated with all the 20 possible systems of equations in the form of Eq. (7) were computed. A total-degree ordering with $a_1>b_1>a_2>b_2>a_3>b_3$ was used for reducing the computation effort of these bases.

The non-solubility of a polynomial system is determined by checking whether its reduced Gröbner basis is {1}, Adams and Loustanau (1994). Performing this task using Maple® shows that all the polynomial systems including equations for simultaneously fulfilling the desired values of k_{xx} and k_{yy} are unsolvable. Consequently, changing the directions of the prismatic actuators relative to the moving platform is not sufficient for simultaneously achieving these stiffness elements.

Next, the problem of Eq. (7) for k_{xx}, k_{xy}, and $k_{x\theta}$ is solved, i.e., all the stiffness elements in the x direction of PCS are prescribed based on task requirements. The reduced Gröbner basis for this problem, hereafter called G, with total degree ordering $a_1>b_1>a_2>b_2>a_3>b_3$ has 28 generators of degrees ranging from 1 to 5 in the variables. The ith column in table 1 presents the degrees of the ith basis polynomial in the variables corresponding to a_1,b_1,a_2,b_2,a_3,b_3. It can be seen that the total degree ranges from 4 to 8. This basis is not presented here due to lack of space.

Table 1. Degrees of the 28 polynomials of G in the variables

a_1	0	0	0	0	1	2	1	1	0	1	1	0	0	1	1	0	1	1	1	0	1	0	0	0	0	1	1	1
b_1	0	0	0	2	1	0	1	1	0	1	1	0	0	1	1	0	1	1	1	0	1	0	0	0	0	1	1	1
a_2	0	1	2	0	0	0	1	1	1	0	0	1	1	0	0	0	1	1	1	0	1	0	0	1	1	0	0	1
b_2	0	2	2	2	2	2	1	1	3	2	2	1	1	0	0	2	1	1	1	2	1	2	2	1	1	0	0	1
a_3	2	1	1	0	1	1	1	1	1	1	1	1	1	1	1	1	0	0	0	1	1	1	1	1	1	1	1	1
b_3	2	2	2	2	2	1	1	1	2	2	2	3	3	3	3	4	2	2	2	4	2	5	4	4	4	4	4	3

The leading terms of G are given by:

$$\begin{aligned} LT = [&a_3^2, a_2 b_2, a_2^2, b_1^2, a_1 b_1, a_1^2, a_3 a_2 b_1, a_3 a_1 a_2, b_2^3, b_2^2 b_1, b_2^2 a_1, \\ &b_3^2 a_3 b_2, b_3^2 a_3 a_2, b_3^2 b_1 a_3, b_3^2 a_3 a_1, b_2^2 b_3^2, b_3^2 b_2 a_1, b_3^2 a_2 b_1, b_3^2 a_1 a_2, \\ &b_2^2 a_3 b_3, a_3 b_2 b_3 a_1, b_3^5, b_3^4 a_3, b_3^4 b_2, b_3^4 a_2, b_3^4 b_1, a_1 b_3^4, b_3^3 b_2 b_1] \end{aligned} \tag{12}$$

Based on the finiteness theorem, Adams and Loustaunau (1994), the system of polynomials corresponding to G is solvable and has a zero-dimensional variety. This is established by examining the group of leading terms in Eq. (12) which shows that each variable among $\{a_1,b_1,a_2,b_2,a_3,b_3\}$ appears alone as a leading term in G with the corresponding degrees of {2, 2, 2, 3, 2, 5}. Consequently, the group of all the remainders in $\mathscr{C}[a_1,b_1,a_2,b_2,a_3,b_3]/\mathscr{I}$, denoted by D, has terms with maximal degrees of {1, 1, 1, 2, 1, 4} in $\{a_1, b_1, a_2, b_2, a_3, b_3\}$, respectively.

The monomial basis of $\mathscr{C}[a_1,b_1,a_2,b_2,a_3,b_3]/\mathscr{I}$, denoted by B, is found from D by extracting all the monomials in D that are equal to their own normal forms, Cox et al. (1998). This leads to the 48-dimensional monomial basis in Eq. (13):

$$\begin{aligned} B = [&1, b_3, a_3, b_2, a_2, b_1, a1, b_3^2, a_3 b_3, b_2 b_3, a_2 b_3, b_1 b_3, a1\, b_3, a_3 b_2, a_3 a_2, a_3 b_1, a_3 a1, \\ &b_2^2, b_2 b_1, b_2 a1, a_2 b_1, a1\, a_2, b_3^3, b_3^2 a_3, b_2 b_3^2, b_3^2 a_2, b_3^2 b_1, b_3^2 a1, b_3 a_3 b_2, \\ &a_3 a_2 b_3, b_3 a_3 b_1, a_3 a1\, b_3, b_3 b_2^2, b_3 b_2 b_1, b_3 b_2 a1, b_1 a_2 b_3, a1\, a_2 b_3, a_3 b_2^2, a_3 b_2 b_1, \\ &a_3 b_2 a1, b_3^4, a_3 b_3^3, b_3^3 b_2, a_2 b_3^3, b_3^3 b_1, a1\, b_3^3, b_3^2 b_2 b_1, b_1 b_2 a_3 b_3] \end{aligned} \tag{13}$$

Next, three 48×48 multiplication tables, $\mathbf{M}_{f1}$, $\mathbf{M}_{f2}$ and $\mathbf{M}_{f3}$ for $f_1 = a_1 + b_1$, $f_2 = a_2 + b_2$, $f_3 = a_3 + b_3$ are computed together with their corresponding minimal polynomials mp_{f1}, mp_{f2}, and mp_{f3}. These minimal polynomials have only even degrees, so there are at most 24 pairs of complex solutions and their conjugate solutions (48 in total). These solutions give the values of $f_1=a_1+b_1$, $f_2=a_2+b_2$, and $f_3=a_3+b_3$. The next step is solving for the values of a_1, b_1, a_2, b_2, a_3, b_3. These values establish the locations of the sliders on the circular base. The solution algorithms for obtaining the values of (a_1, b_1), (a_2, b_2), and (a_3, b_3) are identical; hence, only the algorithm for obtaining (a_1, b_1) is presented herein.

Let ±C be one of the 24 solution pairs of mp_{f1}. The matching solutions for (a_1, b_1) are the intersections of the line and the ellipse of Eq. (14).

$$a_1 + b_1 = \pm C \quad \cap \quad a_1^2 + b_1^2 + a_1 b_1 - 1 = 0 \tag{14}$$

The solutions for +C and for −C are:

$$\begin{aligned} &\text{for } +C: (a1, b1) = \left(\frac{1}{2}C \pm \frac{1}{2}\Delta, \frac{1}{2}C \mp \frac{1}{2}\Delta\right) \\ &\text{for } -C: (a1, b1) = -\left(\frac{1}{2}C \mp \frac{1}{2}\Delta, \frac{1}{2}C \pm \frac{1}{2}\Delta\right) \end{aligned} \qquad \Delta \equiv \sqrt{4 - 3C^2} \tag{15}$$

Since only real solutions for (a_1, b_1) are of interest, only the *admissible* real solution pairs of mp_{f1} satisfying $|C| \leq 2/\sqrt{3}$ are used in Eq. (15). Note that the two solutions for +C (and −C) represent a mirror image about the bisector of the angle α in Fig. 2 and that the two solutions for +C are mirror images of the two solutions for −C about the normal to the bisector of the angle α in Fig. 2.

Once this procedure is repeated for the roots of mp_{f2} and mp_{f3}, sets of solutions for (a_1, b_1), (a_2, b_2) and (a_3, b_3) are obtained and all the sextuplets (a1, b_1, a_2, b_2, a_3, b_3) satisfying Eqs. (4, 7) are found.

7. Numerical Example

To verify the solution procedure, a predefined geometry of the planar robot was selected with $[a_1,b_1,a_2,b_2,a_3,b_3]=[\sqrt{3}/3,\sqrt{3}/3,1,0,1,0]$. This corresponds to $[\theta_1, \theta_2, \theta_3]=[30°, 240°, 120°]$, where θ_1, θ_2, and θ_3 are the angles of $\hat{\mathbf{i}}_1$, $\hat{\mathbf{i}}_2$, and $\hat{\mathbf{i}}_3$ relative to $\mathbf{x}_p$ in Fig. 2. The corresponding stiffness matrix, using a platform radius of 0.1[m] and k_d=1e+5 [N/m], is:

$$K = \begin{bmatrix} 125000.00000 & 43301.270189 & -5000.00000 \\ 43301.2701 & 174999.99999 & 0. \\ -5000.00000 & 0. & 500.00000 \end{bmatrix} \tag{16}$$

The aim of the following example is finding all the solutions for (a_i, b_i), i=1,2,3, for obtaining the stiffness elements k_{xx}, k_{xy}, and $k_{x\theta}$ of Eq. (16) at a given manipulation point of the path. The solution method is validated if one of the solutions gives the values of the predefined geometry.

Three minimal polynomials, mp_{f1}, mp_{f2}, and mp_{f3} are obtained with their solutions. Table 2 lists only the admissible distinct solution pairs C_1, C_2, and C_3, of mp_{f1}, mp_{f2}, and mp_{f3}, respectively. These solutions are distinct up to 1e-3 resolution from other close solutions. Table 3 lists the distinct 16 solutions for a_i, b_i, i=1,2,3. Note that, as expected, Table 3 contains a solution corresponding to the exact values of $[\sqrt{3}/3,\sqrt{3}/3,1,0,1,0]$ of the pre-defined example. These 16 solutions are presented in Fig. 3. Note also that only the last eight solutions, (i) through (p), are non-singular.

Table 2. Admissible real distinct solutions of mp_{f1}, mp_{f2}, and mp_{f3}

C_1	±0.577350	±(1-4.4e-15)*	±1.145112	±1.154700
C_2	±0.38207e-13	±0.376135	±0.967869	±0.999999
C_3	±0.514087	±0.968432	±(1+12e-30)*	±1.154700

* All numerical computations were made with 32 digits, but results are truncated to 6 decimal digits for presesntation purposes

Table 3. 16 real solutions to the problem in the numerical example

a1	b1	a2	b2	a3	b3
0.8869987582^{-14}	-1	-1	$0.33502100\ 10^{-9}$	±0.57735027	±0.57735027
0.8869987582^{-14}	-1	1	$-0.33502100\ 10^{-9}$	±0.57735027	±0.57735027
-0.8869987582^{-14}	1	-1	$0.33502100\ 10^{-9}$	±0.57735027	±0.57735027
-0.8869987582^{-14}	1	1	$-0.33502100\ 10^{-9}$	±0.57735027	±0.57735027
0.57735027	0.57735027	-1	$0.33502100\ 10^{-9}$	±1	$\pm 0.2240\ 10^{-28}$
0.57735027	0.57735027	1	$-0.33502100\ 10^{-9}$	±1	$\pm 0.2240\ 10^{-28}$
-0.57735027	-0.57735027	-1	$0.33502100\ 10^{-9}$	±1	$\pm 0.2240\ 10^{-28}$
-0.57735027	-0.57735027	1	$-0.33502100\ 10^{-9}$	±1	$\pm 0.2240\ 10^{-28}$

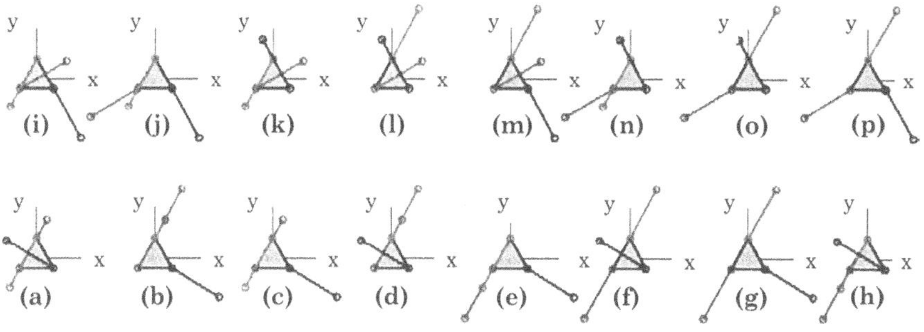

Figure 3. 16 solutions with only eight non-singular solutions (i, through p)

8. Conclusions

This investigation addresses the stiffness synthesis problem of a planar kinematically redundant 3-DOF robot by controlling a limited number of its free geometric parameters. It is shown that it is impossible to control both Cartesian stiffness elements k_{xx} and k_{yy}, by only changing the locations of the sliders on its circular base. In an example of controlling k_{xx}, k_{xy}, and $k_{x\theta}$, it is shown that, at most, there are 48 solutions in the complex plane. The numerical example solved shows only eight real non-singular solutions.

9. Acknowledgements

This research has been partially supported by the Israeli Ministry of Science under the French-Israel program.

References

Adams W., Loustanau, Ph., (1994), *An Introduction to Grobner Bases*, Graduate Studies in Mathematics, Vol. 3, American Mathematical Society.

Buchberger, B., (1965), An Algorithm for Finding a Basis for Residue Class Ring of a Zero-Dimensional Polynomial Ideals (German). PhD Thesis, University of Innsbruck, Institute for Mathematics.

Cox, D., Little, J., O'Shea, D., (1998), *Using Algebraic Geometry,* Graduate Texts in Mathematics, no. 185, Springer-Verlag.

Heck, A., (1997), "Bird's-eye View of Grobner Bases," Nuclear Instruments and Methods in Physics Research, vol. A 389, pp. 16-21.

Huang, S., Schimmels, J., (1998), "Achieving an Arbitrary Spatial Stiffness with Springs Connected in Parallel," J. of *Mechanical Design,* Vol. 120, pp.520-526.

Kim, W. K., Yi, B. J., Cho, W., (2000) "RCC Characteristics of Planar/Spherical Three Degrees of Freedom Parallel Mechanisms with Joint Compliances," ASME *Mechanical Design*, Vol. 122, pp. 10-16.

Möller H. M., Stetter, H. J., (1995), "Multivariate Polynomial Equations With Multiple Zeros Solved by Matrix Eigenproblems," *Numer. Math.,* no. 70, pp. 311-329.

Stetter, H. J., (1993), "Multivariate Polynomial Equations as Matrix Eigenproblems," In: Contributions in Numerical Mathematics, *World Scientific Series in Applicable Analyis (WSSIAA)*, vol. 2, pp. 355-371.

Merlet, J-P., (1997), "Designing a Parallel Manipulator for a Specific Workspace," *Intl. J. of Robotics Research*, Vol. 16, No. 4, September, pp. 545-556.

Merlet, J-P., (2000), *Parallel Robots,* Kluwer Academic publishers.

Merlet, J-P., Perng, M-W., Daney D., (2000), "Optimal Trajectory Planning of a 5-Axis Machine-Tool Based on a 6-Axis Parallel Manipulator," In *Advances in Robot Kinematics,* Lenarcic, J. and Stanisic M. M., Eds., Kluwer Academic Publishers, pp. 315-322.

Neilsen, J., Roth, B., (1999), "On the Kinematic Analysis of Robotic Mechanisms," *Intl. J. of Robotics Research,* vol. 18, no. 12, pp. 1147-1160.

Raghavan, M., Roth, B., (1995), "Solving Polynomial Systems for the Kinematic Analysis and Synthesis of Mechanisms and Robot Manipulators," Transactions of ASME, Special 50th Anneversary Design Issue, Vol. 117, pp. 71-79.

Roberts, G. R., (1999), "Minimal Realization of a Spatial Stiffness Matrix with Simple Springs Connected in Parallel," IEEE Transactions on *Robotics and Automation*, Vol. 15, No.,5.

Tsai, L-W., (1999), *Robot Analysis – The Mechanics of Serial and Parallel Manipulators,* John Wiley & Sins, Inc.

Yi, B. Ji. and Freeman, R. A., (1993) "Geometric Characteristics of Antagonistic Stiffness In Redundantly Actuated Mechanisms," IEEE International Conference on *Robotics and Automation*, pp. 654-661.

Zhiming, J., Song Ph., (1998), "Design of a Reconfigurable Platform Manipulators," In J. of *Robotic Systems*, Vol. 15. No. 6, pp. 341-346.

Zhiming, J., Zhenqun, L., (1999), "Identification of Placement Parameters for Modular Platform Manipulators," In J. of *Robotic Systems*, Vol. 16. No. 4, pp. 227-236.

SYNTHESIS OF ARCHITECTURALLY MOBILE DOUBLE-PLANAR PLATFORMS

K. Wohlhart
Graz University of Technology
A–8010 Graz, Austria
wohlhart@mechanik.tu–graz.ac.at

Abstract In this paper it will be shown that a double-planar platform mechanism is archititecturally mobile if it is shaky (singular) in five, finitely different positions. Knowing this mobile double-planar platform mechanisms can easily be synthesized and linear line complexes can be designed. A.Karger and M.Husty (Karger &Husty,1997) have presented four special conditions which the system parameters determining the relative positions of the spherical anchor joints on the two platforms have to comply with if a double-planar platform mechanism is to be architecturally singular. Recently a counterexample of a special double-planar platform mechanism has been found (Wohlhart, 2000), which proves that the Karger-Husty conditions are neceessary but not sufficient conditions for architectural singularity. In contrast to the four Karger-Husty conditions, the five neceessary and sufficient conditions which are proposed in the present paper need not be formulated once and for all; they can always easily be obtained ad hoc by simple computations.

1. Introduction

Already in an introductory course on statics the student of mechanics will come across the question as to how a rigid body can be safely supported by bars. In the planar case, the answer is easy to give and to understand: three bar are necessary and their axes may not intersect in one point. In the spatial case the answer is not simple and cannot be understood by only looking at the configuration: six bars are necessary to support the body and their axes may not belong to a linear line complex. A semester later our student meets with a similar question in kinematics: how many linear actuators are necessary to control the position of a rigid body and which positions have to be avoided unless controllability gets lost. In both cases, planar and spatial, the answer is the same as in statics. Apart from trivial cases a linear line complex cannot simply be visualized. The theorem of Chasles-Bricard (Chasles,1856 and Bricard,1906) provides a geometric picture of a group of linear line complexes. This theorem states: given any two conic sections in space between which a projective correspondence exists, straight lines connecting homologic points in this

J. Lenarčič and F. Thomas (eds.), Advances in Robot Kinematics, 473–482.

correspondence belong to a linear complex. Chasles formulated this theorem in the context of a static problem and R. Bricard reformulated it without proof in the context of kinematics (Fig.1).

A.Karger and M.Husty generalized the theorem of Chasles-Bricard by

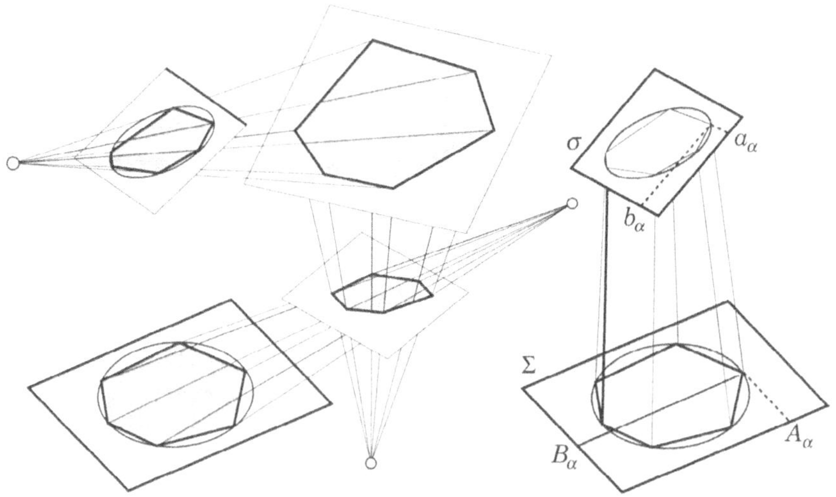

Figure 1. Chasles–Bricard Theorem

stating (Karger&Husty,1997): given two planes (σ,Σ) in space and six points in each of them $(A_\alpha, B_\alpha, 0)$, $(a_\alpha, b_\alpha, 0)$ ....$\alpha = 1 \div 6$ whose coordinates comply with the four conditions quoted below, then the straight lines which connect homologic points belong to a linear complex. In other words, a double-planar platform mechanism with such a configuration of anchor points (spherical joint) of the six actuators is architecturally singular. The four Karger-Husty conditions read with $A_1 = 0, B_1 = 0, a_1 = 0, b_1 = 0$ and $B_2 = 0, b_2 = 0$:

$$K_1 = 0, \quad K_2 = 0, \quad K_3 = 0, \quad K_4 = 0,$$ where

$$\begin{aligned}
K_1 = {} & A_3 b_3 (b_4 b_5 B_6 (B_5 - B_4) + b_4 b_6 B_5 (B_4 - B_6) + b_5 b_6 B_4 (B_6 - B_5)) + \\
& A_4 b_4 (b_3 b_5 B_6 (B_3 - B_5) + b_3 b_6 B_5 (B_6 - B_3) + b_5 b_6 B_3 (B_5 - B_6)) + \\
& A_5 b_5 (b_3 b_4 B_6 (B_4 - B_3) + b_3 b_6 B_4 (B_3 - B_6) + b_4 b_6 B_3 (B_6 - B_4)) + \\
& A_6 b_6 (b_3 b_4 B_5 (B_3 - B_4) + b_3 b_5 B_4 (B_5 - B_3) + b_5 b_4 B_3 (B_4 - B_5)) \\
K_2 = {} & a_3 B_3 (B_4 B_5 b_6 (b_5 - b_4) + B_4 B_6 b_5 (b_4 - b_6) + B_5 B_6 b_4 (b_6 - b_5)) + \\
& a_4 B_4 (B_3 B_5 b_6 (b_3 - b_5) + B_3 B_6 b_5 (b_6 - b_3) + B_5 B_6 b_3 (b_5 - b_6)) + \\
& a_5 B_5 (B_3 B_4 b_6 (b_4 - b_3) + B_3 B_6 b_4 (b_3 - b_6) + B_4 B_6 b_3 (b_6 - b_4)) + \\
& a_6 B_6 (B_3 B_4 b_5 (b_3 - b_4) + B_3 B_5 b_4 (b_5 - b_3) + B_5 B_4 b_3 (b_4 - b_5))
\end{aligned}$$

If the translation vector f, the unit vector n and the rotation angle φ are given as functions of time t, we find for the velocity of a point on σ by time derivation of the vector x:

$$v = x^{\cdot} = f^{\cdot} + R^{\cdot}.r_0 = f^{\cdot} + R^{\cdot}.R^{\mathsf{T}}.r = f^{\cdot} + \omega \times r$$
$$v = x^{\cdot} = f^{\cdot} + \omega \times (x - f),$$

or with the velocity of that point in the plane σ which coincides at the moment with the point of the origin O of $\{O, E_1, E_2, E_3\}$: $v_O = f^{\cdot} - w \times f$ we obtain for the velocity of point x on plane σ:

$$v = v_O + \omega \times x \tag{6}$$

If the distance between two homologic points P on σ and Q on Σ is held constant by a bar of given length (locked actuator), the velocity of point P can only be orthogonal to the axis of the bar, i.e.:

$$s.v = s.(v_O + \omega \times x) = 0$$

or with $x = s + y$
$$s.v_O + (y \times s).\omega = \hat{s} \circ \hat{\omega} = 0. \tag{7}$$

$\hat{s} \circ \hat{\omega}$ is the inner product of the velocity motor (ω, v_O) and the Plücker motor of the bar line

$$(s, y \times s) = (s, s_O).$$

Condition (7) is de facto independent of the length of the bar, as equation (7) contains $|s|$ as a factor.

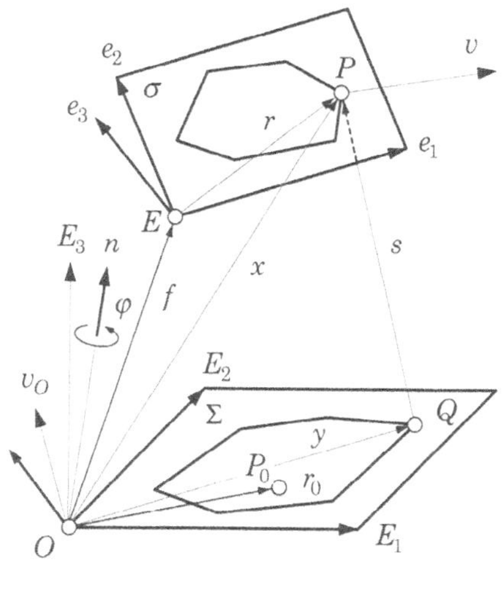

Figure 2. General Displacement σ/Σ

If there are six bars of prefixed lengths which support the platform σ against the basis Σ, then we have six linear equations

$$s_\alpha . v_O + (x_\alpha \times s_\alpha).\omega = 0 \tag{8}$$

for the six unknown components of the angular velocity ω and the linear velocity v_O of the point on σ which coincides with the origin of the coordinate system $\{O; E_1, E_2, E_3\}$. The set of linear and homogeneous equations has a non-zero solution only if the determinant of the (6×6) matrix of coefficients vanishes.

3. Synthesis of Architecturally Mobile Platforms

With the coefficient matrix M and the six-vectors $\hat{v}$ and 0 given by

$$M=\begin{bmatrix} s_{11} & s_{12} & s_{13} & (x_1\times s_1)_1 & (x_1\times s_1)_2 & (x_1\times s_1)_3 \\ s_{21} & s_{22} & s_{23} & (x_2\times s_2)_1 & (x_2\times s_2)_2 & (x_2\times s_2)_3 \\ s_{31} & s_{32} & s_{33} & (x_3\times s_3)_1 & (x_3\times s_3)_2 & (x_3\times s_3)_3 \\ s_{41} & s_{42} & s_{43} & (x_4\times s_4)_1 & (x_4\times s_4)_2 & (x_4\times s_4)_3 \\ s_{51} & s_{52} & s_{53} & (x_5\times s_5)_1 & (x_5\times s_5)_2 & (x_5\times s_5)_3 \\ s_{61} & s_{62} & s_{63} & (x_6\times s_6)_1 & (x_6\times s_6)_2 & (x_6\times s_6)_3 \end{bmatrix},\quad 0=\begin{bmatrix}0\\0\\0\\0\\0\\0\end{bmatrix},\quad \hat{v}=\begin{bmatrix}v_{O1}\\v_{O2}\\v_{O3}\\\omega_1\\\omega_2\\\omega_3\end{bmatrix},$$

equations (8) reads

$$M.\hat{v}=0. \tag{9}$$

For $f=0$ and $\varphi=0$ the two coordinate systems $\{O;E_1,E_2,E_3\}$ and$\{E;e_1,e_2,e_3\}$ coincide. Following Karger-Husty we set for thesystem parameter of the platform:

$$\begin{aligned} y_1^{T}&=(0,\ 0,\ 0),\ y_2^{T}=(A_2,0,\ 0),\ y_3^{T}=(A_3,B_3,0),\\ y_4^{T}&=(A_4,B_4,0),\ y_5^{T}=(A_5,B_5,0),\ y_6^{T}=(A_6,B_6,0),\\ r_{01}^{T}&=(0,\ 0,\ 0),\ r_{02}^{T}=(a_2,\ 0,\ 0),\ r_{03}^{T}=(a_3,b_3,0),\\ r_{04}^{T}&=(a_4,b_4,0),\ r_{05}^{T}=(a_5,b_5,0),\ r_{06}^{T}=(a_6,b_6,0). \end{aligned} \tag{10}$$

The translation vector f and Rodriques` rotation vector w determine the position of the platform, their components f_1,f_2,f_3 or w_1,w_2,w_3 determine the rotation matrix R, and with equation (5) and $s_\alpha=x_\alpha-y_\alpha$ they enter the matrix M together with the system parameters (10). As the position parameter f_1,f_2 and f_3 enter the matrix M linearly and the orientation parameter w_1,w_2 and w_3 quadratically, the order of the determinant of M as an algebraic equation will be of order 18. In a Taylor expansion of $\det M$ the coefficients of all monomials or, equivalently, all the partial derivatives up to order 18 must vanish in the case of architetural mobility. Theoretically this would lead to a the tremendous number of conditions with which the system parameter (10) would have to comply:

$$N=\sum_{k=0}^{k=18}\binom{6-k-1}{k}=131670. \tag{11}$$

But we shall see that only five of them are independent.

Now our proposition is the following:

If the determinant of matrix M is equal to zero in five different positions of the platform, then the mechanism will be architecturally mobile. Starting from any position $\{f_1,f_2,f_3,w_1,w_3,w_3\}$ the platform with blocked actuators will be movable with one degree of freedom.

Let us take the system parameters of a double-planar platform as given with the exception of four of them:

$y_1^T = (0,0,0)$, $y_2^T = (4,0,0)$, $y_3^T = (9,1,0)$, $y_4^T = (8,5,0)$, $y_5^T = (7,6,0)$, $y_6^T = (3,3,0)$, $r_{01}^T=(0,0,0)$, $r_{02}^T=(7,0,0)$, $r_{03}^T=(3,1,0)$, $r_{04}^T=(1,6,0)$, $r_{05}^T=(a_5,b_5,0)$, $r_{06}^T=(a_6,b_6,0)$.

For this mechanism we shall choose the positions $p=\{ f_1,f_2,f_3,w_1,w_2,w_3\}^T$:

$$p_1 = \{ 0,0,1,\ 1,0,0 \}^T,\ p_2 = \{ 0,0,1,\ 0,1,0\}^T,\ p_3 = \{ 0,0,1,\ 0,0,1\}^T,$$
$$p_4 = \{ 0,1,0,\ 1,0,0\}^T, p_5 = \{ 1,0,0,\ 0,1,0\}^T. \tag{12}$$

Therewith we get the following five condition $F_\alpha = (\det M)_\alpha = 0$ for the unknown four system parameters a_5,b_5 and a_6,b_6:

$$F_1 = -2835 + 56106\ a_5 - 57411\ a_6 + 180\ a_5a_6 - 4545\ b_5 + 9057\ a_6b_5 + 3750\ b_6 - 8682\ a_5b_6 + 180\ b_5b_6 = 0,$$

$$F_2 = -2583 + 3501\ a_5 - 4581\ a_6 + 207\ a_5a_6 + 246\ b_5 + 834\ a_6b_5 + 4176\ b_6 - 72\ a_5b_6 - 768\ b_5b_6 = 0,$$

$$F_3 = 1386 - 2277\ a_5 + 531\ a_6 + 144\ a_5a_6 - 726\ b_5 - 45\ a_6\ b_5 - 2071\ b_6 + 805\ a_5b_6 + 805\ a_5b_6 + 489\ b_5b_6 = 0,$$

$$F_4 = 1134\ b_5 - 702\ a_6b_5 - 19404\ b_6 + 8712\ a_5b_6 + 1710\ b_5b_6 = 0,$$

$$F_5 = 16362\ a_5 - 17712\ a_6 - 900\ a_5a_6 - 1890\ b_5 + 2676\ a_6b_5 + 1260\ b_6 - 5220\ a_5\ b_6 + 420\ b_5b_6 = 0. \tag{13}$$

The solution of this set of quadratic polynomial equations can be found e.g., with “Mathematica” by the command:

$$\text{Solve}[\{F_1 == 0, F_2 == 0, F_3 == 0, F_4 == 0, F_5 == 0\}, \{a_5, a_6, b_5, b_6\}]\ ,$$

and it reads $$a_5 = -\frac{113}{71}, a_6 = -\frac{207}{43}, b_5 = \frac{330}{71}, b_6 = \frac{135}{136}. \tag{14}$$

Thus the system parameters of the platform are complete and *independant of the position parameters* $p = \{ t_1,t_2,t_3,\ w_1,w_2,w_3\}^T$ we find:

$$\det M \equiv 0. \tag{15}$$

This proves architectual mobility of the mechanism. For any set of position parameters $p = \{ t_1,t_2,t_3,\ w_1,w_2,w_3\}^T$ a six-vector $\hat{v}$ (the null space of M) can be found from equation (9) which contains the velocity $v_O = \{ v_{O1},v_{O2},v_{O3}\}^T$ and the angular velocity $\omega = \{\omega_1,\omega_2,\omega_3\}^T$ as vector parts. This six-vector determines the infinitesimal change of position vector p. To find the time derivative $\dot{p} = \{ \dot{t}_1,\dot{t}_2,\dot{t}_3,\ \dot{w}_1,\dot{w}_2,\dot{w}_3\}^T$ we have to know the relation between the time derivative $\dot{w}$ of Rodrigues` vector w and the angular velocity vector ω. This relation can be found in the following way:

with $$R^{\cdot} = \Sigma \frac{dR}{dw_\kappa} w_\kappa^{\cdot}$$

and $$\omega\times = R^{\cdot}R^{\mathsf{T}} = \Sigma \frac{dR}{dw_\kappa} R^{\mathsf{T}} w_\kappa^{\cdot}$$

we obtain $$\omega = \frac{2}{1+w.w}(w^{\cdot} + w\times w^{\cdot})$$

or $$w^{\cdot} = G.\omega$$

where $$G = \frac{1}{2}(1+w.w)(I+w\times)^{-1}. \tag{16}$$

Together with $$f^{\cdot} = v_O + \omega\times f \tag{17}$$

the linear relation between the time derivative $p^{\cdot} = \{t_1^{\cdot}, t_2^{\cdot}, t_3^{\cdot}, w_1^{\cdot}, w_2^{\cdot}, w_3^{\cdot}\}^{\mathsf{T}}$ and the solution vector $\hat{v} = \{v_{O1}, v_{O2}, v_{O3}, \omega_1, \omega_2, \omega_3\}^{\mathsf{T}}$ is thus established and can be written in matrix form

$$\begin{bmatrix} f^{\cdot} \\ w^{\cdot} \end{bmatrix} = \begin{bmatrix} I & -f\times \\ G & 0 \end{bmatrix} . \begin{bmatrix} v_O \\ \omega \end{bmatrix}$$

or $$p^{\cdot} = Q.\hat{v} = Q.nullspace(M) = q(p). \tag{18}$$

This matrix differential equation of first order could be numerically integrated for any starting position $p(0)$ yielding with the a dimensionless time parameter $\tau = t\Omega$:

$$p = p(\tau, p(0)).$$

If we enter the same incomplete system parameter (as above)

$$y_1^{\mathsf{T}} = (0,0,0),\ y_2^{\mathsf{T}} = (4,0,0),\ y_3^{\mathsf{T}} = (9,1,0),\ y_4^{\mathsf{T}} = (8,5,0),\ y_5^{\mathsf{T}} = (7,6,0),\ y_6^{\mathsf{T}} = (3,3,0),$$
$$r_{01}^{\mathsf{T}} = (0,0,0),\ r_{02}^{\mathsf{T}} = (7,0,0),\ r_{03}^{\mathsf{T}} = (3,1,0),\ r_{04}^{\mathsf{T}} = (1,6,0),\ r_{05}^{\mathsf{T}} = (a_5,b_5,0),\ r_{06}^{\mathsf{T}} = (a_6,b_6,0)$$

into the four Karger-Husty conditions, we get a set of four equations which yield the following two solutions:

$$\{a_5 = \frac{120}{253} + \frac{21\,a_6}{253},\ b_5 = \frac{144}{23},\ b_6 = \frac{36}{7}\},$$
$$\{a_5 = -\frac{113}{71},\ a_6 = -\frac{207}{43},\ b_5 = \frac{330}{71},\ b_6 = \frac{135}{136}\}.$$

In the first solution a_6 remains indeterminate, i.e. we get with the complete set of system parameters for the determinant of the matrix M an algebraic equation of order two in a_6 which depends on the chosen position set $p = \{f_1, f_2, f_3, w_1, w_2, w_3\}^{\mathsf{T}}$. By solving this equation for a_6 we can make $\det M$ vanish for this position only. The second solution is identical with (14) and yields the same mobile mechanism.

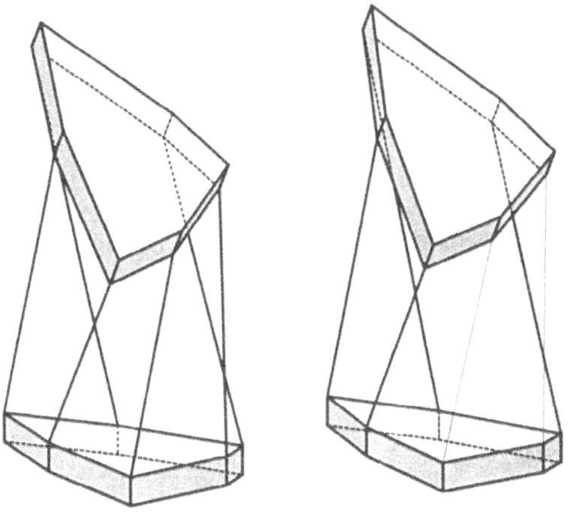

Figure 3. Stereographic Picture of the Synthezised Mobile Double-Planar Platform

Fig.3 shows the synthesized mobile platform mechanism in which the platform is translated by the translation vector

$$f = \{0.9, 1.5, 9\}$$

and rotated through the angle

$$\varphi = -27^{\circ}$$

about the rotation axis given with the unit vector

$$n = \{0.424264, -0.9, 0.1\}.$$

4. Platforms with Affine Point Configurations

If the anchor point coordinates in the two planes are related according to equation (2), then the Karger-Husty conditions would indicate architectural singularity independant of m_1 and m_2. But an arbitrary choice of the anchor point coordinates in the basis Σ results in an architecturally rigid mechanism.

However for a (sufficiently) general position $p = \{f_1, f_2, f_3, w_1, w_2, w_3\}^T$ we obtain from $\det M = 0$ in this special case one and only one condition which must be satisfied by the system parameters for architectural mobility. With $m_1 = 10/13, m_2 = 6/13$ and $A_2 = 5, A_3 = 7, B_3 = 3, A_4 = 4, B_4 = 6, A_5 = -1, B_5 = 7, A_6 = 3, B_6 = B_6$ we get for:

$$a_2 = \frac{50}{13}, a_3 = \frac{70}{13}, b_3 = \frac{18}{13}, a_4 = \frac{40}{13}, b_4 = \frac{36}{13}, a_5 = -\frac{10}{13}, b_5 = \frac{42}{13}, a_6 = -\frac{30}{13}, b_6 = \frac{6}{13} B_6.$$

Entering this set of system parameters into the equation $\det M = 0$ together with any set of position parameters which can be chosen arbitrarily $p = \{f_1 = 0, f_2 = 0, f_3 = 1, w_1 = 1, w_2 = 1, w_3 = 1\}$, we obtain for B_6:

$$36288 - 31344\, B_6 + 3936\, B_6^2 = 0,$$

from which $$B_6 = \{\frac{653 - \sqrt{178441}}{164}, \frac{653 + \sqrt{178441}}{164}\} = \{1.40596, 6.55746\}$$

follows. Both values of B_6 complete the set of system parameters and, *independent of all positions* $p = \{f_1, f_2, f_3, w_1, w_2, w_3\}^T$ we find for each of them that the determinant of the matrix M vanishes identically : $\det M \equiv 0$, which indicates architectural mobility for two mechanisms which only differ in one bar. Merging these two mechanisms yields a platform on seven bars (actuators) as shown in Fig.4. It is evident that any number of additional bars can be constructed by choosing five of the bars in existence (which already determine the path of the platform), fixing the anchor point A_α and determining two values of B_α as shown above.

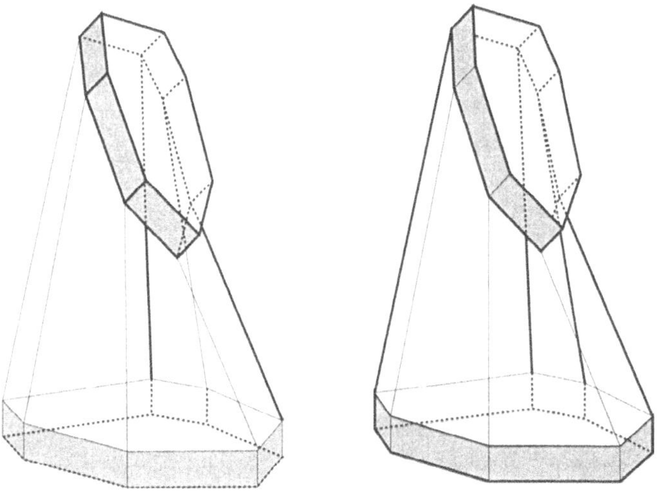

Figure 4 Stereographic Picture of a Double-Planar Platform Mechanism with Orthogonally Affine Anchor Point Configurations on Seven Actuators

References

Karger,A., Husty, M., (1997), Singularities and Self-Motions of Stewart–Gough Platforms, *Proceedings of the NATO–ASI Workshop on Comp. Meth. in Mach. Design,* vol. 2, pp. 270-279.

Ma, O., Angeles, J. (1992), Architecture Singularities of Parallel Manipulators, *International Journal of Robotic Automation,* vol.7, No. 1, pp. 23-29.

Wohlhart, K. (2000), A Spinning Platform, *Proceedings of the 8th Conference on the Theory of Machines and Mechanisms* , pp.829-838

Wohlhart, K. (2000), Architectural Shakiness or Architectural Mobility of Platforms , *Advances in Robot Kinematics* , ARK 2000 (eds : J..Lenarcic, M.M.Stanisic), Kluwer Akademik Press, pp.365-374

Chasles, M. (1851), Sur les six droites qui peuvent être les directions de six forces en équilibre, *Comptes rendus,* **t. 52**, p. 1093.

Bricard,.R., (1906), Mémoire sur les déplacements à trajectoires sphériques, *Journ. École Polyt.,* 11 , p.3, théorème V.

APPROXIMATE MOTION SYNTHESIS USING AN SVD BASED DISTANCE METRIC

Pierre M. Larochelle
Mechanical and Aerospace Engineering Department
Florida Institute of Technology
pierrel@fit.edu

Stacy L. Dees
Autonomous Systems Laboratory
University of Hawaii at Manoa
sdees@wiliki.eng.hawaii.edu

Abstract This paper presents a novel technique for approximate spatial motion generation. Finite spatial locations are represented using homogeneous transforms, which can then be approximated by rotations using products derived from the singular value decomposition. A bi-invariant metric on rotations calculates the *distance* between two locations. A multidimensional nested optimization procedure determines the optimum assembly and associated design variables of a spatial mechanism to minimize the *distances* from the moving body to a finite number of desired locations. The result is a methodology for performing the dimensional synthesis of spatial mechanisms for approximate motion generation. A ten location spatial mechanism design problem is presented to illustrate the method.

Keywords: Kinematic synthesis, spatial mechanisms, motion generation, distance metrics

1. Introduction

In this paper we present a novel dimensional synthesis technique for approximate motion synthesis of spatial linkages. The methodology utilizes an analytic representation of the linkage's workspace that is parameterized by its joint variables. Nonlinear optimization techniques are then employed to minimize the distance from the workspace to a finite number of desired locations. The result is an approximate motion dimensional synthesis technique that is applicable to open or closed spatial kinematic chains. Here, we specifically address the design of spatial 4C mechanisms, see Fig. 1.

J. Lenarčič and F. Thomas (eds.), Advances in Robot Kinematics, 483–490.

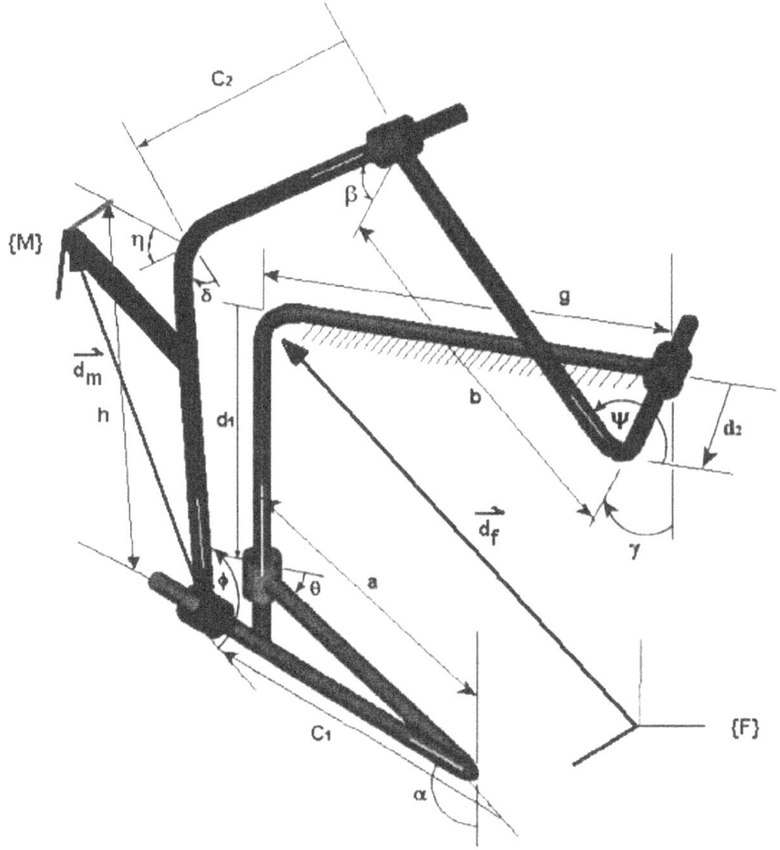

Figure 1. Spatial 4C mechanism

The dimensional synthesis goal is to optimize the design variables such that all of the prescribed locations are either in the workspace or the workspace comes as close as possible to all of the desired locations. Previous works discussing approximate motion synthesis for an arbitrary number of locations include Ravani and Roth, 1983, Bodduluri, 1990, and Larochelle, 1994. In Larochelle, 2000 a methodology for approximate planar motion synthesis that utilizes a parameterized representation of the workspace was reported. That method required a projection of the workspace into the *image space* (see McCarthy, 1990) of planar displacements. Here, we build upon that work and present a general synthesis method that does not require a mapping of the workspace and that is applicable to spatial, spherical, and planar approximate motion

Table 1. Link parameters of the spatial 4C mechanism

Link	*Dual Angle*	*Twist*	*Length*
Driving	$\hat{\alpha}$	α	a
Coupler	$\hat{\eta}$	η	h
Driven	$\hat{\beta}$	β	b
Fixed	$\hat{\gamma}$	γ	g

synthesis. Moreover, we also introduce a novel spatial distance metric which is based upon the use of the singular value decomposition(SVD) of spatial displacements.

2. Spatial 4C Mechanism Workspace

A spatial 4C mechanism has four cylindrical joints, each joint permitting relative rotation and translation along a line, see Fig. 1. The link parameters that define the mechanism are listed in Tbl. 1 and the joint variables are defined Tbl. 2.

A spatial 4C mechanism may be viewed as a combination of two CC dyads. The driving CC dyad has four independent joint variables, referred to as θ, d_1, ϕ and c_1. The driven dyad also has four independent joint variables, ψ, d_2, δ and c_2. When adjoined by the coupler link, the two dyads form a closed chain spatial 4C mechanism with two degrees of freedom. We chose θ and d_1 to be the independent joint variables. Note that ϕ and c_1 as well as the driven dyad's joint variables are now explicit functions of θ and d_1; these functions are found in Larochelle, 1998.

Table 2. Joint variables of the spatial 4C mechanism

Joint	*Dual Angle*	*Rotation*	*Translation*
Driving fixed	$\hat{\theta}$	θ	d_1
Driving moving	$\hat{\phi}$	ϕ	c_1
Driven moving	$\hat{\delta}$	δ	c_2
Driven fixed	$\hat{\psi}$	ψ	d_2

We generate a representation of the workspace that is parameterized by the driving joint variables by performing a forward kinematic analysis of the driving dyad. Fig. 2 illustrates the sequence of transformations in Eq. 1

$$ {}^F_M T \;=\; F\; Z(\theta, d_1)\; X(\alpha, a)\; Z(\phi, c_1)\; M \tag{1} $$

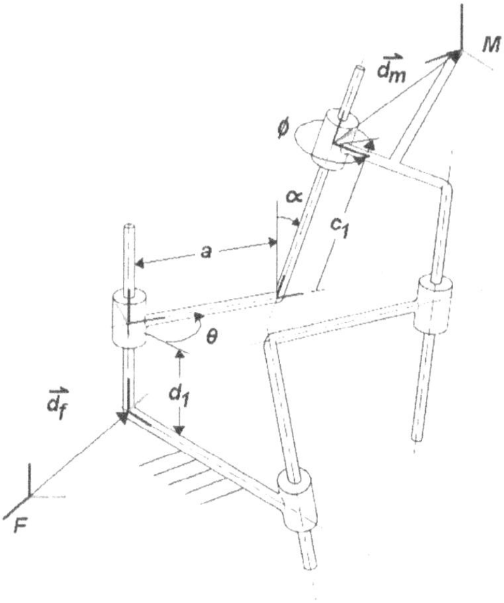

Figure 2. Link coordinate frames

where F and M are the homogeneous transforms from the fixed frame to the first frame of the driving dyad, and from the last frame of the dyad to the moving frame, respectively. Eq. 1 yields the workspace of the spatial 4C mechanisms parameterized by the two driving joint variables θ and d_1 since $\phi(\theta, d_1)$ and $c_1(\theta, d_1)$ are known.

3. SVD Based Distance Metric

We wish to find the optimum design parameters for a rigid body guidance through a finite number of desired locations. The optimization objective function is the sum of the minimum distances between the n desired locations and the workspace of the mechanism; both represented as homogeneous transforms. Hence, we require a metric on spatial displacements. It is well known that no bi-invariant spatial displacement metric exists. However, recently several researchers have made significant contributions towards finding approximately bi-invariant metrics: Chirikjian, 1998, Larochelle, 2000, Gupta, 1997, Martinez and Duffy, 1995, Etzel and McCarthy, 1996, and Kazerounian and Rastegar, 1992. Our approach uses hyper-dimensional rotations to approximate three dimensional displacements. A novel method of approximating spatial motion with rotations in E^4 that uses products derived from the singular value decomposition(SVD) of the homogeneous transform is used,

Dees, 2001. We then use biquaternions to represent the rotations in E^4. A bi-invariant metric on biquaternions is then used. The result is an approximately bi-invariant metric on spatial displacements.

A matrix $[P]$ can be factored by using a singular value decomposition:

$$[P] = [U] \begin{bmatrix} s_1 & 0 & \cdots & 0 \\ 0 & s_2 & \cdots & 0 \\ \cdots & \cdots & \cdots & \cdots \\ 0 & 0 & 0 & s_n \end{bmatrix} [V^T] \tag{2}$$

where the diagonal matrix contains the singular values of $[P]$ and $[U]$ & $[V]$ are either rotations or reflections. The following algorithm (based upon related work by Hanson and Norris, 1981) uses the SVD to construct a 4×4 rotation matrix given a 4×4 homogeneous transform.

Step 1: Find the SVD of the homogeneous transform [T]:

$$[T] = [U]\ [diag(s_1, s_2, ..., s_n)]\ [V]^T,$$

where $n = 4$.

Step 2: Find the determinant of $[X]$:

$$\sigma = det(x).$$

where

$$[X] = [U]^T\ [V].$$

Step 3: Construct $[A]$, the 4×4 rotation matrix:

$$[A] = [U]\ [diag(1, 1, 1, \sigma)]\ [V]^T,$$

where $[A]$ is a rotation in E^4.

3.1 Distance between 4×4 Rotations

Ge's (1994) construction is used to obtain the biquaternions representing the two 4×4 rotation matrices, see Ge, 1994 and Dees, 2001. A bi-invariant metric on biquaternions is:

$$d(\hat{Q}, \hat{R}) = \sqrt{(Q - R)^T(Q - R) + (S - T)^T(S - T)} \tag{3}$$

where $\hat{Q} = (Q, S)$ and $\hat{R} = (R, T)$ are biquaternions, and Q, S, R, and T are the associated quaternions. Etzel and McCarthy, 1996 prove that this metric is bi-invariant.

4. Approximate Motion Synthesis

The goal here is dimensional synthesis of spatial 4C mechanisms for approximate motion synthesis. The $(20+2n)$ dimensional design vector is

$$\vec{r} = \left[\alpha\ a\ \gamma\ g\ \beta\ b\ \eta\ h\ \vec{d}_m\ lng_m\ lat_m\ rol_m\ \vec{d}_f\ lng_f\ lat_f\ rol_f\ \vec{\theta}^T\ \vec{d_1}^T\right]^T \tag{4}$$

where the first eight elements are the link parameters(see Fig. 1 and Tbl. 1). Next are the terms which define F and M; the location of the mechanism with respect to the fixed frame and the location of the moving frame with respect to the coupler link frame. The 3×1 column vectors $\vec{d}_*$ and the angles lng_*, lat_* and rol_* define the position and orientations of these two transformations, respectively. Last are the n dimensional vectors $\vec{\theta}$ and $\vec{d_1}$. These joint pairs are optimized for each of the n desired locations.

The nested optimization strategy is as follows:

1 A candidate design (i.e. the first 20 components of $\vec{r}$) is determined.

2 Utilizing the techniques of Murray and Larochelle, 1998 verify that the candidate solution has a crank-rocker spherical image-Larochelle, 2000 showed that such mechanisms will not suffer from rotational or translational singular configuration defects.

3 The steps below are performed for each of the two circuits of the candidate solution:

   (a) For each desired location the values of θ and d_1 are optimized by utilizing the svd based distance metric. Note that an optimized θ and d_1 pair determine the location in the mechanism's workspace nearest a desired location. This is the inner optimization.

   (b) The sum of the distances to each of the n desired locations is determined.

4 The distance sums of the two circuits of the mechanism are compared; and the lesser is utilized. If the distance sum is acceptable then the design is complete. Otherwise, the distance sum and the first 20 components of $\vec{r}$ are sent to the outer optimization to determine a better candidate design and the above steps are repeated.

5. Design Case Study

We address a 10 location rigid body guidance problem taken from Larochelle, 1994. The desired locations (#) and the locations in the workspace of the solution 4C mechanism nearest them (#') are listed in Tbl. 3. The joint variables that guide the moving body to these locations as well as the distance between these locations to each of the desired locations are also listed. The sum of these distances (3.53E-1) is the minimum value of the objective function found and corresponds to the total distance from the workspace of this solution mechanism to the ten desired locations. The dimensional synthesis parameters of the

Table 3. Spatial 4C synthesis results for 10 locations

Loc.	*X*	*Y*	*Z*	*Lng.*	*Lat.*	*Rol*	θ	d_1	*Dist.*
1	1.0	0.0	5.0	100.0	0.0	0.0			
1'	-0.85	0.22	1.71	54.6	11.8	-0.20	-24.8	7.0	8.26E-2
2	2.0	0.0	4.0	90.0	0.0	10.0			
2'	-0.56	0.42	2.61	55.8	17.2	-0.5	-12.1	6.6	4.18E-2
3	3.0	0.0	3.0	80.0	0.0	20.0			
3'	0.12	0.76	3.41	58.0	25.1	0.3	13.2	5.7	4.49E-2
4	4.0	0.0	2.0	70.0	0.0	30.0			
4'	1.04	0.81	3.28	60.5	33.2	2.8	7.2	3.5	3.50E-2
5	5.0	0.0	1.0	60.0	0.0	40.0			
5'	1.40	0.05	2.12	61.3	48.3	18.7	-11.1	-2.6	6.63E-2
6	6.0	0.0	-1.0	50.0	0.0	50.0			
6'	4.97	-0.40	-0.56	45.1	-7.4	42.7	-1.5	1.5	2.48E-2
7	7.0	0.0	-2.0	40.0	0.0	60.0			
7'	5.18	-0.09	-1.45	37.4	-3.6	56.8	-14.5	5.4	1.85E-2
8	8.0	0.0	-3.0	30.0	0.0	70.0			
8'	6.43	0.03	-2.44	25.0	-0.7	69.9	22.8	9.6	1.68E-2
9	9.0	0.0	-4.0	20.0	0.0	80.0			
9'	8.94	-0.02	-3.94	19.9	0.4	80.3	10.0	15.5	2.30E-2
10	10.0	0.0	-5.0	10.0	0.0	90.0			
10'	12.89	0.11	-6.75	14.0	0.5	87.1	9.8	25.0	2.01E-2

solution 4C mechanism, i.e. the first 20 components of $\vec{r}$, are: 149.9, -2.31, 88.2, 0.81, 72.6, -0.25, 49.2, -2.89, -2.40, -0.17, 1.28, 118.2, 226.6, -15.8, 1.07, 0.70, 1.28, 26.4, 133.8, 276.3.

6. Conclusions

We have presented our development of a new method for approximate motion synthesis for rigid body guidance through n desired spatial locations. This method utilizes a parameterized representation of the mechanism's workspace that is exploited to yield mechanisms that will not suffer from rotational or translational singular configuration defects. Moreover, this dimensional synthesis procedure employs a novel approximately bi-invariant svd-based distance metric. The result is a nested multidimensional direct search optimization strategy that yields the optimal design parameters such that the mechanism comes as close as possible to all of the n desired locations in a single circuit or assembly.

7. Acknowledgements

This work was made possible by NSF Grant No. #9816611. The contributions of Prof. A. Murray of the University of Dayton to the svd-based spatial metric are gratefully acknowledged.

References

Bodduluri, R.M.C., (1990), *Design and planned movement of multi-degree of freedom spatial mechanisms*, PhD Dissertation, University of California, Irvine.

Chirikjian, G.S. (1998), Convolution metrics for rigid body motion, *Proc. of the ASME Design Engineering Technical Conferences*, Atlanta, USA.

Dees, S.L. (2001), *Spatial mechanism design using an svd-based distance metric*, Master's Thesis, Florida Institute of Technology.

Etzel, K., and McCarthy, J.M. (1996), A metric for spatial displacements using biquaternions on SO(4), *Proc. of the IEEE International Conference on Robotics and Automation*, Minneapolis, USA.

Ge, Q.J. (1994), On the matrix algebra realization of of the theory of biquaternions, *Proc. of the ASME Design Engineering Technical Conferences*, Minneapolis, USA.

Gupta, K.C. (1997), Measures of positional error for a rigid body, *ASME Journal of Mechanical Design*, vol. 119, pp. 346-349.

Hanson and Norris, (1981), Analysis of measurements based upon the singular value decomposition, *SIAM Journal of Scientific and Computations*, no. 3, vol. 2, pp. 308-313.

Kazerounian, K., and Rastegar, J., (1992), Object Norms: A class of coordinate and metric independent norms for displacements, *Proc. of the ASME Design Engineering Technical Conferences*, Scotsdale, USA.

Larochelle, P. (2000), On the geometry of approximate bi-invariant projective displacement metrics, *Proc. of the World Congress on the Theory of Machines and Mechanisms*, Oulu, Finland.

Larochelle, P. (2000), Circuit and branch rectification of the spatial 4c mechanism, *Proc. of the ASME Design Engineering Technical Conferences*, Baltimore, USA.

Larochelle, P. (2000), Approximate motion synthesis via parametric constraint manifold fitting, *Proc. of Advances in Robot Kinematics*, Piran, Slovenia.

Larochelle, P. (1998), Spades: software for synthesizing spatial 4c mechanisms, *Proc. of the ASME Design Engineering Technical Conferences*, Atlanta, USA.

Larochelle, P., (1994), *Design of cooperating robots and spatial mechanisms*, PhD Dissertation, University of California, Irvine.

Martinez, J.M.R., and Duffy, J. (1955), On the metrics of rigid body displacements for infinite and finite bodies, *ASME Journal of Mechanical Design*, vol. 117, pp. 41-47.

McCarthy, J.M., (1990), *An Introduction to Theoretical Kinematics*, MIT Press.

Murray, A.P., and Larochelle, P. (1998), A classification scheme for planar 4r, spherical 4r, and spatial rccc linkages to facilitate computer animation, *Proc. of the ASME Design Engineering Technical Conferences*, Atlanta, USA.

Ravani, B., and Roth, B., (1983), Motion synthesis using kinematic mappings, ASME *Journal of Mechanisms, Transmissions, and Automation in Design*, vol. 105, pp. 460-467.

NAVIGATION AND GAIT PLANNING FOR PLANAR WALKER

I-Ming Chen, Song Huat Yeo and R. S. Senanayake
School of Mechanical and Production Engineering
Nanyang Technological University
{ michen, myeosh, mrssenan } @ntu.edu.sg

Abstract Locomotion and navigation of a surface walking/climbing robot – Planar Walker, based on a novel planar 8-bar mechanism are investigated. The robot moves on a surface through decoupled *transverse gaits* and *turning gaits* with finite lengths and finite rotation angles. Motions of the gaits are modeled using planar rigid motion group *SE(2)*. Three point-to-point navigation methods are developed for various situations: *Simple Line of Sight* (SLS) algorithm, *Simulated Annealing Accurate Planning* (SAAP) algorithm, and *Localized Hybrid Accurate Planning* (LHAP) algorithm. Computer simulation shows that SAAP produces accurate gait sequences and LHAP saves computation time and resources for long-range targets. However, experimental results show that the inherent positional errors in individual gait movements can be substantial when accumulated and can render a particular algorithm less effective.

Keywords: robot locomotion, gait kinematics, navigation and path planning

1. Introduction

In the past decade, there has been a strong interest in developing climbing/walking maintenance robotic systems for construction, shipbuilding, chemical, petroleum, and nuclear industries. In this article, we describe a newly developed surface walking/climbing system for large surface inspection and maintenance tasks on ship hulls and oil tanks. This system is able to traverse along the surface in a swift way with minimal turning radius so as to cover a very large working area in a short amount of time. A prototype system based on pneumatic power, called *Planar Walker* is constructed (Fig.1) (Wong, 1999, Ting, 2000). The Planar Walker can traverse on a plane or a smooth surface with a large radius of curvature. This system features a closed-loop planar 8-bar locomotion mechanism formed by 4 linear cylinders and 4 revolute joints. (Fig.2) The cylinders are not only used as actuating elements but also as chassis of the system. When the four cylinders are actuated independently, the shape of the mechanism may change to a square, a rectangle, or an irregular quadrilateral. One gripper module (suction pad) is mounted below each revolute joint to hold the robot to the

J. Lenarčič and F. Thomas (eds.), Advances in Robot Kinematics, 491–500.

working surface. The cylinders and grippers are designed to allow only two-way motions, i.e., on/off (gripper) or fully retracted/extended states (cylinder). In this way, the robot may traverse (forward, backward, and sideway) a fixed distance or turn at a fixed angle after completing a series of cylinder and gripper actions termed *gaits*. Due to symmetric arrangement, translation and rotation of the robot are decoupled, and the robot can swiftly change its direction of travel.

Basic characteristics (stride lengths, turning angles) of gaits are experimentally verified and reported by Chen and Yeo (2002). This article studies gait sequence planning for navigating the Planar Walker. Motions of the gaits are formulated using planar rigid motion group - *SE(2)*. Based on gait motions, 3 point-to-point navigation methods are developed: 1) *Simple Line of Sight* (SLS) algorithm – a heuristic method based on the line of sight; 2) *Simulated Annealing Accurate Planning* (SAAP) algorithm – a method based on combinatorial optimization to minimize the final positioning error; 3) *Localized Hybrid Accuate Planning* (LHAP) algorithm – a hybrid method combining SLS and SAAP to reduce computational load in long range travel. These algorithms are validated through computer simulation and experiment using the developed Planar Walker prototype.

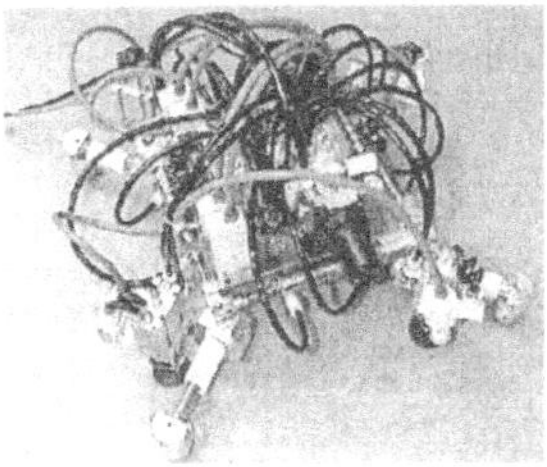

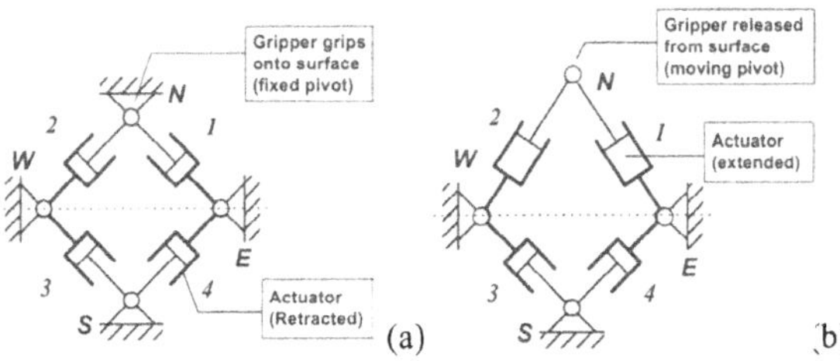

Figure 1. Planar walker. *Figure 2.* Kinematic diagram of locomotion mechanism.

2. Basic locomotion gaits

Locomotion of the Planar Walker is based on the shape-changing mechanism driven by the four identical cylinders and properly activated grippers. The cylinders produce linear motions only. The four gripper units can be activated perpendicular to the plane independently providing the friction needed for locomotion. As the gripper unit is built under the pivot joint, when activated, the associated joint becomes a fixed pivot joint. Otherwise, the pivot joint can move freely. Due to symmetry in the mechanical structure, two types of gaits are identified for locomotion purpose: the *transverse gait* and the *turning gait*. A transverse gait enables the mechanism to travel forward (backward, left, or right) without changing the orientation. A turning gait rotates the

machine with a finite angle about its center of geometry in the clockwise or counter clockwise direction.

- **Transverse gaits:** The complete sequence of a transverse gait moving forward is shown in Fig.3. Similar sequences can be generated for the other 3 directions. When the sequence is completed, the state of actuators goes back to the initial one and the robot moves forward with a fixed distance. The orientation of the robot remains unchanged.

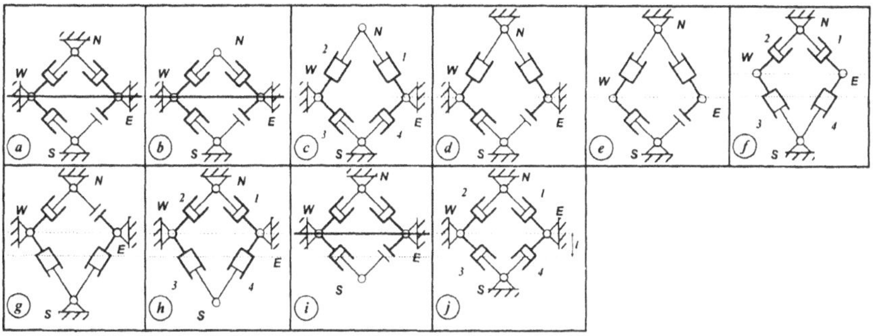

Figure 3. Transverse gait (forward, initially retracted).

The *stride* of a transverse gait is the displacement of the center of the robot after completing *one* gait sequence. From Fig. 3, we can see that the stride length of a transverse gait with initially retracted cylinders is

$$l = \sqrt{l_{max}^2 - \frac{1}{2} l_{min}^2} - \frac{1}{\sqrt{2}} l_{min} \tag{1}$$

where the lengths of a fully retracted and extended cylinder are l_{min} and l_{max} respectively. The stride lengths of a transverse gait (initially retracted) moving backward or sideways are all the same.

- **Turning gaits:** Figure 4 shows the sequences of a CW turning gait. A CCW turning gait can be deduced with similar analogy. The states of the diagonal grippers or opposite cylinders change alternatively to complete the turning motion. When the sequence is completed, the robot would rotate clockwise with angle α about its center relative to the original configuration. The center of the robot remains unchanged after the turns. The angle of rotation of a turning gait (init. retracted) is

$$\alpha = 2\sin^{-1} \frac{l_{max}^2 - l_{min}^2}{l_{max} l_{min}} \tag{2}$$

The turning gait re-orients the robot about its center of geometry and is *decoupled from* the displacement of the transverse gaits. Therefore, the robot can do zero-radius turn at any position. When navigating the Planar Walker according to a path or a goal point, the orientation and position of the robot can be planned individually. Because a gait is a cyclic sequence of actuator states that goes back to its initial state, it is initial-state dependent. Gaits with identical initial states can be used alternatively right away because the final state of the prior gait becomes the initial state of the subsequent gait.

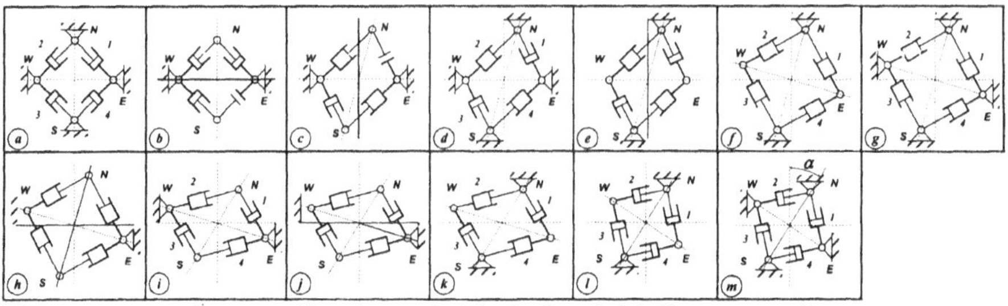

Figure 4. Turning gait (clockwise, initially retracted).

3. Kinematics of basic gaits

In this section we study the kinematics of the robot based on the execution of a series of different basic gaits. Let A be the inertia reference frame and B the body frame attached to the robot as shown in Fig.5. The +y-axis of frame B along the body is forward direction and the +x-axis along the body is right direction. The net position and orientation (termed a pose) change caused by a gait can be described as a rigid motion $g_{ab} \in SE(2)$ on a 2D plane. The rigid motion g_{ab} has a 3x3 homogeneous matrix form:

$$g_{ab} = \begin{bmatrix} R_{ab} & p_{ab} \\ 0 & 1 \end{bmatrix} \tag{3}$$

where $R_{ab} \in SO(2)$ is the 2x2 orientation matrix of frame B relative to A and $p_{ab} \in \mathbb{R}^2$ is the position vector of the origin of frame B from the origin of A. For example, executing a forward transverse gait (Fig.3), a right transverse gait, or a CW turning gait (Fig.4) have the following pose changes respectively:

$$g_{ab}=T_Y \triangleq \begin{bmatrix} 1 & 0 & 0 \\ 0 & 1 & l \\ 0 & 0 & 1 \end{bmatrix};\ g_{ab}=T_X \triangleq \begin{bmatrix} 1 & 0 & l \\ 0 & 1 & 0 \\ 0 & 0 & 1 \end{bmatrix};\ g_{ab}=R_Z^{-1},\ \text{where } R_Z \triangleq \begin{bmatrix} \cos\alpha & -\sin\alpha & 0 \\ \sin\alpha & \cos\alpha & 0 \\ 0 & 0 & 1 \end{bmatrix}$$

The net pose changes of other basic gaits (backward and left transverse gaits; CCW turning gait) can be found in this manner. In fact, these pose changes are all based on *three basic rigid motions*: T_X, T_Y, and R_Z. The poses of forward and backward transverse gaits (or right and left gaits) with identical initial states would have the property: $T_Y^{-1} = -T_Y$ and $T_X^{-1} = -T_X$. From now on, we use the term gait and its corresponding pose change g interchangeably. Let $\{g_1, g_2, \cdots, g_k\}$ be a sequence of gaits the robot executes. The final pose change of the robot can be described by the *product of all rigid motions*: $g_{Final} = g_1 g_2 \cdots g_k \in SE(2)$.

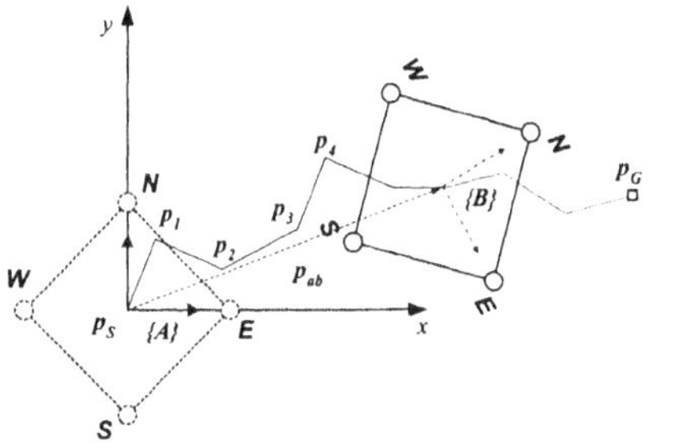

Figure 5. Pose change of the robot.

4. Navigation and gait planning

Because the robot structure decouples its rotation and translation, here we focus on a point-to-point navigation problem, i.e., determining a sequence of gaits to reach a given goal position (point). After the robot reaches the goal point, the robot then uses turning gaits to rotate to the desired orientation. Let $\overline{p}_a$ be the homogeneous coordinate of the robot center relative to frame *A*. The location of the robot after executing a series of gaits $\{g_1, g_2, \cdots, g_k\}$ would be $\overline{p}_a = g_{Final}\overline{p}_b$, where $\overline{p}_b = (0,0,1)^T$. As the robot moves or rotates with a finite distance or a finite angle for each gait, the robot's center basically moves along a zigzag path (Fig.5). At each waypoint p_i, the robot may execute several turns and then proceed to the next transverse gait. The distance between any two neighboring waypoints is the stride length produced by a transverse gait.

DEFINITION : *Point-to-Point Navigation Problem.*

Given: a starting point $\overline{p}_S$ and a goal point $\overline{p}_G$ (in frame *A*)

Find: a set of triplet (l_i, m_i, n_i), $i = 1$ to N such that

$$\overline{p}_G = \prod_{i=1}^{N}\left(R_Z^{l_i} T_X^{m_i} T_Y^{n_i}\right)\overline{p}_S \tag{4}$$

where $l_i \in \mathbb{Z}$ and $m_i, n_i \in \{-1, 0, 1\}$, $|m_i + n_i| = 1$. Note that N denotes the total number of transverse gaits (or the number of path segments $\overline{P_iP_{i+1}}$ in Fig.5); $|l_i|$ is the total number of turning gaits at waypoint p_{i-1}. A triplet (l_i, m_i, n_i) represents types of consecutive turning gaits and a transverse gait to be executed at waypoint p_{i-1}. The series of (l_i, m_i, n_i) then represents the gait sequence to be executed to reach the goal point.

As the robot can only move or rotate with finite lengths or angles, it may not reach the exact goal point. Thus, we propose three navigation algorithms here to move the robot to the vicinity of the goal position.

4.1 Simple line-of-sight

The Simple Line-of-Sight (SLS) algorithm is a heuristic navigation method. The basic idea of the algorithm is to follow the line of sight formed by the goal point and the start point within a fixed error range defined by the stride length l and the angle of rotation α (Fig.6).

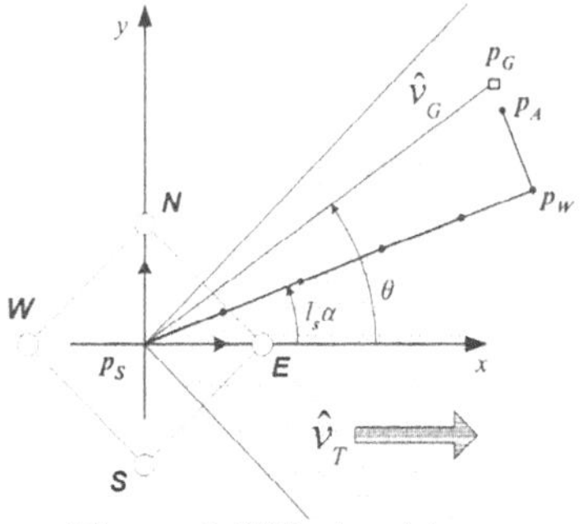

Figure 6. SLS algorithm.

Simple Line-of-Sight Algorithm

Step 1: Let $\hat{v}_G$ be the unit vector of $\overline{p}_G - \overline{p}_S$. (Line-of-sight)

Step 2: Let $\hat{v}_T$ be the unit vector in the direction of one the 4 transverse gaits: (1,0), (0,1), (-1,0), or (0,-1), where $-45^\circ \le \theta = \sin^{-1}(\hat{v}_T \times \hat{v}_G) \le 45^\circ$. (Find primary direction)

Step 3: Rotate the robot l_s times using the turning gait such that $|\theta - l_s\alpha| \le \frac{1}{2}\alpha$, and $l_s \in \mathbb{Z}$. If $l_s > 0$, rotate in CCW direction; $l_s < 0$, rotate in CW direction.

Step 4: Move the robot m_s steps using the transverse gait corresponding to $\hat{v}_T$ such that $\left| \|\overline{p}_G - \overline{p}_S\| - m_s l \right| \le \frac{1}{2} l$ and $m_s \in \mathbb{Z}$.

Step 5: Let $\overline{p}_W$ be the waypoint after the robot executes m_s transverse gaits.

Step 6: Find n_s such that $|m_s l \tan(\theta - l_s\alpha) - n_s l| \le \frac{1}{2} l$.

Step 7: If $n_s > 0$, move the robot n_s steps using the transverse gait to the *left* of $\hat{v}_T$. If $n_s < 0$, move the robot n_s steps using the transverse gait to the *right* of $\hat{v}_T$. If $n_s = 0$, go to Step 8.

Step 8: Stop.

The algorithm uses the turning gait l_s times to orient the robot to the line-of-sight with angular error less than 0.5α. Next, the algorithm determines the number (m_s) of transverse gaits to be executed. Due to this angular error, using a neighboring transverse gait (with $|n_s|$ steps) for lateral compensation may be necessary (Step 5 to 7). In total, the robot would execute $|l_s|$ turning gaits and (m_s + $|n_s|$) transverse gaits to reach the goal. The SLS algorithm only assures that a minimal number of turning gaits is executed by choosing the closest transverse gait as the primary direction of movement. It does not guarantee to take the shortest path to the goal.

Example 1: Figure 7(a) illustrates the use of SLS algorithm. Based on the Planar Walker prototype, stride length l = 32mm and turning angle α

= 25°. The goal point is (200mm, 200mm). The cylinders are initially retracted. The SLS algorithm would take 2 CW turning gaits and 10 transverse gaits to reach the final position of (200.05mm, 209.64mm) with a positioning error of 9.640mm.

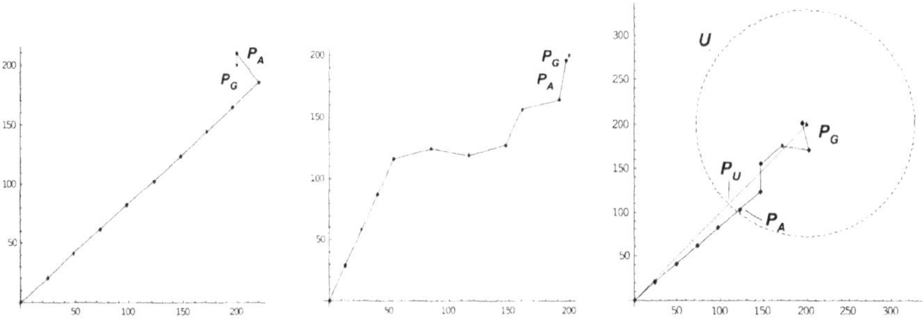

(a) SLS (b) SAAP (c) LHAP

Figure 7. Point-to-point navigation and gait planning results.

4.2 Simulated annealing accurate planning

The positioning error occurred in the SLS algorithm can be minimized if more rotations or more transverse gaits are executed. Thus, the point-to-point navigation problem can be posed as an optimization problem:

Given: Starting point $\overline{p}_S$, Goal point $\overline{p}_G$ (in frame *A*), and *N*

Minimize:

$$\left\| \overline{p}_G - \prod_{i=1}^{N} \left(R_Z^{l_i} T_X^{m_i} T_Y^{n_i} \right) \overline{p}_S \right\| \tag{5}$$

Subject to: $l_i \in \mathbb{Z}$ and $m_i, n_i \in \{-1, 0, 1\}$, $|m_i + n_i| = 1$, for $i = 1$ to N.

The configuration space, the set of gait sequences, is discrete. As *N* grows, the size of the configuration space grows exponentially. This becomes a typical combinatorial optimization problem. We propose a Simulated Annealing Accurate Planning algorithm (SAAP) after comparing the computing time and stability of the solution using Simulated annealing (Kirkpatrick et al., 1983) and Simple genetic algorithms (Goldberg, 1989). The SAAP algorithm uses Metropolis algorithm (Press et al., 1992) with the following elements:

- *Description of the configuration:* A sequence of gaits represented as a sequence of (l_i, m_i, n_i), $i = 1$ to *N*. We further restrict the number of turning gaits by setting $|l_i| \le \lceil 45^\circ / \alpha \rceil$.
- *Configuration generator:* To generate a new solution in the neighborhood of the current solution, each time we randomly pick one triplet (l_i, m_i, n_i) in the current sequence and randomly alter their values. *To speed up the convergence rate, we use the result from the SLS algorithm as the initial configuration.*
- *Objective function:* The energy *E* is naturally Eq. (5).

- *Annealing schedule:* Through experimentation, the starting value of the temperature T_0 is chosen to be *multiples of the stride length*, which is sufficiently large than ΔE. The temperature decreases according to the rule: $T_{k+1} = 0.9T_k$.

Example 2: Following Example 1, the result of using SAAP algorithm for gait planning is shown in Fig. 7(b). The initial temperature is $T_0 = 10*l$. The no. of temperature steps is 50. The SAAP algorithm takes 10 turning gaits and 10 transverse gait to reach the final position of (197.42mm, 195.81mm) with a positioning error of 4.921mm.

4.3 Localized hybrid accurate planning

The SAAP algorithm optimizes gait path globally. When the distance to the goal point is very long compared to the robot's stride length, the optimization would take long time and resources to find the solution. This poses a constraint for stand-alone implementation on the robot. On the other hand, the SLS algorithm is a heuristic approach that uses very few computational resources. Therefore, the Localized Hybrid Accurate Planning (LHAP) algorithm is proposed to combine the advantages of both SLS and SAAP algorithms for gait planning with long distance. The idea is to localize the accurate gait planning in the vicinity of the goal point. We call this region a *localized planning region U*. The size of the region can be predefined by the user. The robot uses the SLS algorithm outside of *U* and switches to the SAAP algorithm once it gets into *U*.

Localized Hybrid Accurate Planning

Step 1: Let *r* be a user input value. Define $U = \{\bar{x} : \|\bar{x} - \bar{p}_G\| \le r\}$.

Step 2: Let $\bar{p}_U$ be the intersecting point of the line of sight and the boundary of U.

Step 3: Let $\bar{p}_U$ be the new goal point and use SLS algorithm to compute the necessary gait sequence.

Step 4: Let $\bar{p}_A$ be the actual robot position after completing SLS gaits and make $\bar{p}_A$ the new starting point. Start SAAP algorithm using $\bar{p}_G$ as the final goal point.

Step 5: Stop.

Similar to the SLS algorithm, the LHAP algorithm does not guarantee to be the most accurate planning algorithm. However, the time spent on the gait planning would be reduced significantly.

Example 3: Following Example1, let the radius of *U* be 4*l*. Figure 7(c) illustrates the use of LHAP algorithm. The robot executes 9 turning gaits and 10 transverse gaits to reach the final position of (194.83mm, 201.34mm) with a positioning error of 9.134mm. Computing times of the SAAP and LHAP algorithms are compared based on *Mathematica* codes running on a Pentium III 850Hz PC. We use 50 temperature steps in both algorithms. The SAAP algorithm takes 55.58sec and the LHAP algorithm takes only 4.72sec on average. The computing time of LHAP algorithm depends on the size of U. The smaller U is, the less trial solutions are explored in the simulated annealing algorithm, hence, the less time is used. In the SAAP algorithm, the trial solutions are explored globally and this slows the computation.

6. Experiments on navigation algorithms

Table 1 and 2 summarize actual characteristics of basic gaits (Chen and Yeo 2002) based on repetitive testing of individual gaits. The errors in stride lengths and center position offset in a turning gait are mainly due to fabrication/assembly errors and open-loop control of the cylinders. The SLS and SAAP algorithms were also implemented on the Planar Walker prototype. The point-to-point navigation experiments are based on gaits in Fig.7(a) and 7(b). For each gait sequence produced from the SLS and SAAP algorithms, the experiment is repeated three times. And in total, three different target points are tested. From Table 3 we can see that the final position errors from the computation can be contained in less than 1/3 l. Whereas the errors found from the experiments could be as large as 2/3 l relative to the computed final position and the target position. We can also observe the effect of accumulated errors of individual gaits in algorithms with differn number of gaits.

Table 1. Stride lengths of transverse gaits (Time: 30 sec/cycle)

	Transverse gait stride l (unit: mm)			
	Fwd	**Bck**	**Right**	**Left**
Designed	32.00	32.00	32.00	32.00
Experiment	33.33±0.43	32.92±0.67	31.08±0.37	32.00±0.33

Table 2. Center position offset and angles of turning gaits (Time: 60 sec/cycle)

	CCW Turning gait			**CW Turning gait**		
	X (mm)	Y (mm)	α	X (mm)	Y (mm)	α
Designed	0.	0.	25°	0.	0.	-25°
Exprmt	-2.53±1.02	-2.99±0.83	25°	-3.45±2.09	-0.15±0.52	-25°

7. Conclusions

In this paper, kinematics and locomotion of the Planar Walker are studied using *SE(2)*. The point-to-point navigation problem is posed as a gait planning problem. Three methods, the SLS, SAAP, and LHAP algorithms, are proposed to solve the navigation problem for different circumstances. Simulation results show that the SAAP algorithm has significant improvement in positioning accuracy over the SLS algorithm. When operating the robot over a long-range target compared to its stride length, the LHAP algorithm outperforms the SAAP algorithm in terms of computation time. Both SLS and SAAP algorithms were implemented and tested on the Planar Walker. Experimental results show that the SAAP algorithm is no superior than the SLS algorithm due to uncertainties in the actual gait movements. This is because the accuracy of gait planning algorithms can be significantly affected by the inherent positional errors in the individual gait movements. Nevertheless, the final positioning accuracy of all implemented algorithms were contained within one gait stride length. Under this circumstance, the total number of gaits becomes the major evaluation criteria. Thus, the SLS algorithm stands out because it uses the least number of gaits.

Table 3. Point-to-point navigation – Target point (200mm, 200mm)

	SLS		SAAP		LHAP	
Gaits	(10 trans, 2 turn)		(10 trans, 10 turns)		(10 trans, 9 turns)	
FinPos	X	Y	X	Y	X	Y
Comp	*200.05*	*209.64*	*197.42*	*195.81*	*194.83*	*201.34*
Exp 1	205	198	181	199	NA	NA
Exp 2	203	207	194	193	NA	NA
Exp 3	208	201	192	176	NA	NA
PosErr	ToComp	ToTarget	ToComp	ToTarget	ToComp	ToTarget
Comp	*NA*	*9.640*	*NA*	*4.921*	*NA*	*9.134*
Exp 1	12.65	5.385	16.73	19.03	NA	NA
Exp 2	3.959	7.616	4.426	9.220	NA	NA
Exp 3	11.74	8.062	20.54	25.30	NA	NA

References

Chen, I-.M. and Yeo, S.H. (2002). *Gait Generation and Mechatronic Design of Planar Walker*. Tecnical Report, School of MPE, Nanyang Technological University, Singapore.

Goldberg, D.E. (1989). *Genetic Algorithms in Search, Optimization & Machine Learning*. New York, Addison-Wesley.

Kirkpatrick, S., Gelatt, C.D., Vecchi, M.P. (1983). Optimization by simulated annealing. *Science*, vol. 220, no. 4598, pp.671-680.

Press, W.H., et al. (1992). *Numerical Recipes in C, 2ed*, Cambridge Univ. Press.

Ting, T.B. (2000). *Locomotion Gait Planning for a Planar Inchworm Robot*, Final Year Project Report, School of Mechanical and Production Engineering, Nanyang Technological University.

Wong, P.S. (1999). *Design and Fabrication of A Planner Inchworm Robot*, Final Year Project Report, School of Mechanical and Production Engineering, Nanyang Technological University.

NEAR OPTIMAL COORDINATION OF LEGGED VEHICLES OVER LARGE OBSTACLES

Kenneth J. Waldron
Department of Mechanical Engineering
Stanford University, California, USA
Waldron@cdr.stanford.edu

Abstract The theoretical basis of a new approach to force allocation in legged vehicles in which the locomotion elements fall approximately into two support planes is presented. The Adaptive Suspension Vehicle was an example of a legged locomotion system with this type of architecture. It is demonstrated that the approach presented here overcomes deficiencies in the zero interaction force allocation principle used in the Adaptive Suspension Vehicle, particularly with respect to trafficability over large obstacles.

The approach presented here has been implemented in simulation, and is being reduced to practive on a prototype vehicle. Space limitations and proprietary considerations preclude presentation of that part of the work here. It will be presented elsewhere.

Keywords: robotic vehicle, legged locomotion, optimal traction

1. Introduction

In this paper we present a new approach to force allocation for active locomotion systems with certain geometric and architectural characteristics. The method is applicable both to legged systems, and to wheeled systems with actively controllable suspensions. Geometrically, it is necessary that the legs, or wheels approximate two planar motion systems. That is the legs on each side of the vehicle should normally move in, or approximately in, a plane. Likewise, the wheels on each side of the vehicle should be coplanar. The presentation here is specialized for the legged vehicle case.

The fundamental principle upon which the coordination of such locomotion systems rests is that of minimizing the maximum ratio of tangential to normal force at any contact between a locomotion element and the terrain. Absent any knowledge of the effective friction properties, this principle minimizes the tendency for slip to occur at any foot. With the current state of sensor technology, soil properties in the field must be regarded as unknowable in advance of contact with a foot.

J. Lenarčič and F. Thomas (eds.), Advances in Robot Kinematics, 501–515.

Figure 1. The Adaptive Suspension Vehicle. The force allocation approach presented here applies to vehicle architectures in which the legs can be regarded as working in two planes, such as this one.

Any coordination algorithm based on achieving, or approaching satisfaction of this principle must satisfy the additional requirement of being capable of computationally efficient implementation. This is because the force allocation computation must be performed during every coordination update cycle of the system. While it is true that on-board computers are becoming much more powerful, real-time requirements still make efficient algorithms extremely desirable.

Several different types of force allocation algorithm have been studied, or applied on past systems. It is possible, with relatively minor linearization, to formulate the force allocation problem as a linear programming problem (Klein and Chung, 1987; Kumar and Waldron, 1990). The resulting solutions are very near optimal, and this approach has been used as a baseline against which to measure the performances of alternative algorithms. It has not, to the author's knowledge been implemented on a vehicle. The reason for this has as much to do with the nature of the solution as the open, iterative character of the algorithm. For linear programming problems of this dimension, the appropriate solution algorithm is the Simplex algorithm. It is the nature of this algorithm that the solution jumps from vertex to vertex of a polyhedral solution space. The resulting force commands exhibit discontinuities in gradient and "chatter" which makes them undesirable as servo system commands.

The Adaptive Suspension Vehicle, shown in Figure 1, (Pugh *et al.*, 1990) used an algorithm based on requiring that the interaction force between any pair of feet be zero. This algorithm was shown to be isomorphic with, but much more computationally efficient than, Moore-Penrose pseudo-inversion of the system Jacobian (Kumar and Waldron,

1988). That is, it has the effect of minimizing the Euclidean Norm of the contact force magnitudes.

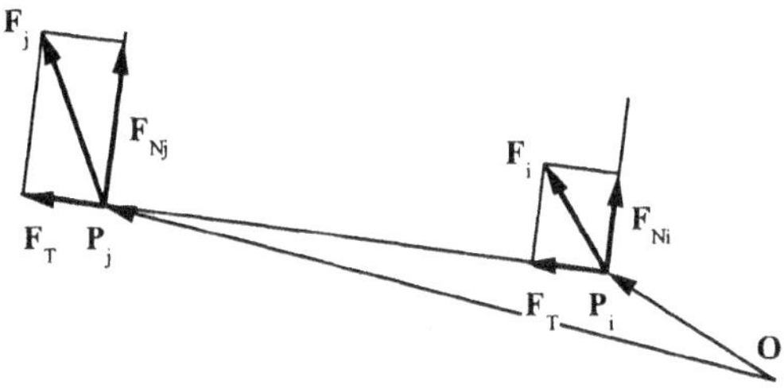

Figure 2. Geometric form of the zero interaction force condition. The components, $\mathbf{F}_T$ of the contact forces at the points P_i and P_j along the line P_iP_j must be equal for every pair of contact points in the system.

Geometrically, the zero interaction force condition is as shown in Figure 2. It means that the components of the contact forces at any two feet along the line joining the contact points are equal. Imposing this condition removes any tendency of pairs of feet or wheels to work against each other. As has been demonstrated, it leads, *via* a straightforward geometric argument, to an elegantly simple set of equations that can be computed with great efficiency in real time (Kumar and Waldron, 1988).

The zero interaction force algorithm is, however, sub-optimal (Klein and Chung, 1987). It approaches optimality in terrain that is reasonably close to planar. In more extreme terrain it may perform poorly. The reason this happens is illustrated in Figure 3, which shows the mechanics of two locomotion elements in contact with a step obstacle: admittedly an extreme example. Equalizing the components of the contact forces along the line joining the contact points in this case produces a very large ratio of tangential to normal force at one contact, with the traction forces at the two contacts actually opposed.

An algorithm based on the theory formulated below, while restricted in the vehicle architectures to which it is applicable, much more nearly approaches optimality for locomotion over large obstacles. The solution will be developed geometrically. However, the resulting equations, while quite different to those produced by the zero interaction force approach, are similarly simple and conducive to efficient, real-time solution. The algorithm can also be adapted relatively easily for application to non-holonomic, wheeled systems, as well as to legged systems.

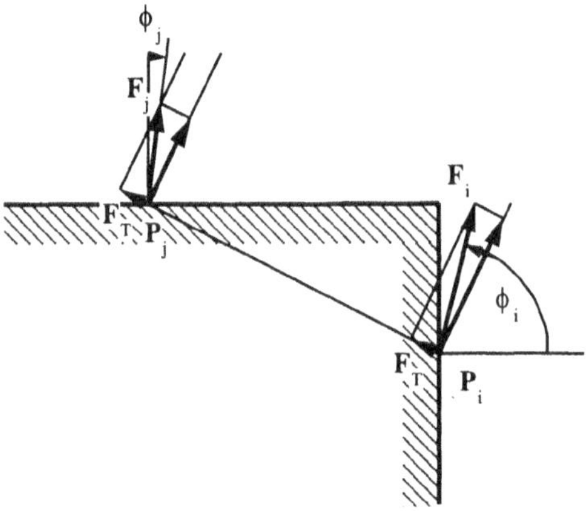

Figure 3. The zero interaction force condition applied to a pair of feet in contact with a step obstacle. The resulting contact forces are far from optimal. The friction angle at point P_i is very large, and is actually opposed by that at point P_j.

2. The Commanded Force System

It is appropriate to command vehicle motion in the position/rate domain, whether the commands come from a navigation system, or from a human operator. However, it is also appropriate to control the interactions between the locomotion elements and the terrain in the force domain. The reason for this is that there are usually more than the minimum number of locomotion elements needed to provide static stability acting in parallel to one another. The system is then kinematically overconstrained. Commanding position for such a system results in the same effect as setting a four-legged table on an uneven floor. In general no more than three legs can contact the floor at any time and there is no way of controlling the force distribution between the legs that are in contact. Commanding contact forces results in an underconstrianed problem that allows optimization of the contact conditions.

Given these, apparently conflicting, system requirements it is necessary to translate the position/ rate commands to the vehicle reference frame into force commands at the contacting locomotion elements. In the Adaptive Suspension Vehicle this was accomplished by means of a control architecture that featured two nested loops. The outer loop was a rate loop that accepted a six-axis commanded velocity system from the operator and was closed by a simple, inertial measurement unit. The inner loop was a force loop closed by the leg actuator force sensors. The two loops were connected by a simple, dynamic model of the vehicle, which converted accelerations from the outer loop into inertia forces.

These were combined with the vehicle weight to produce a force system to be equilibrated by the contact force system. For the purposes of the algorithm, this equilibrating force system, which is the negative of the force system obtained by combining the inertia force system with the weight, were regarded as being commanded on the vehicle as a whole. The force allocation algorithm described above was then applied to decompose this force system into contact forces to be commanded at the feet that were in contact with the ground.

It is assumed, in the following, that the contact force at each foot that is in contact is sensed with respect to magnitude, direction and position of line of action. When we refer to the position of the "contact point" we are, strictly speaking, referring to the intersection of the line of action of the contact force with the contacting surfaces.

3. Lateral Forces

The contact forces are optimized in the planes of motion of the legs. Therefore, lateral forces are not optimized. For the vehicle architectures for which this approach is suited, lateral forces will, very often be small. When crossing large obstacles speeds are expected to be low, so even lateral forces due to turns will be small. Thus, in most instances little is given up by not including lateral forces in the optimization process, and much is gained in terms of efficiency.

The procedures proposed for handling the lateral force components ensures that the lateral components are not incompatible with the in-plane components of contact force, even if they are of significant magnitude. In particular, since the lateral forces do contribute to roll and yaw moments, and since the planar force systems of the support planes do not have roll or yaw moment components, it is important that the line of action on which each lateral force component be chosen carefully and known *a priori* so that those moments can be fully accounted for. We hare adopt the practice of considering the line of action of the lateral force on each side of the vehicle to pass through the centroid of the constellation of contact points for that side. This ensures that the lateral force can be equally distributed among the contats.

4. Decomposition into Two Planar Force Systems

The approach that is the subject of this paper applies to the decomposition of the force system commanded on the vehicle as a whole into forces to be commanded at the feet. The first stage of this process is decomposition of the three-dimensional commanded system into two planar systems in the motion planes of the locomotion elements on either side of the vehicle. The geometry and notation used in this decomposition are shown in Figure 4. F_x, F_y, F_z, M_x, M_y, M_z are the components of the force system to be commanded on the vehicle as a whole. The L subscripts denote components of the force system in the left support plane, and the R components denote those of the force system in the right support plane. $\mathbf{p}_\mathrm{L} = x_\mathrm{L}\mathbf{i} + z_\mathrm{L}\mathbf{k}$ is the calculated position of the centroid of the constellation of contact points in the left support plane, and $\mathbf{p}_\mathrm{R}$ is the corresponding point in the right support plane. The vehicle is assumed to be bilaterally symmetric, with the left and right support planes each distant b from the origin of the vehicle coordinate system.

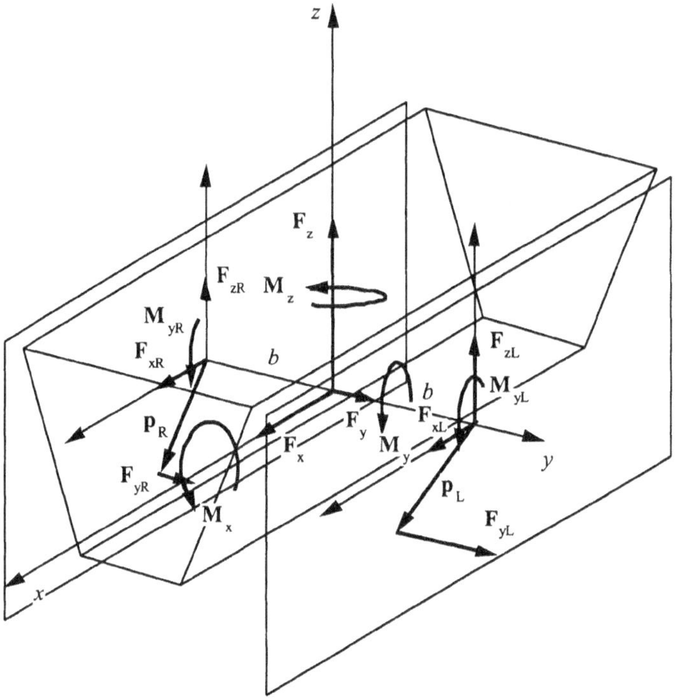

Figure 4. Vehicle reference frame, and left and right support planes.

The allocations of F_y and M_y between the support planes are indeterminate. We assume that

$$F_{yL} = F_{yR} = \frac{F_y}{2}, \ M_{yL} = M_{yR} = \frac{M_y}{2} \tag{1}$$

The resulting expressions for the other components of the support plane force systems are:

$$F_{xL} = \frac{1}{4b}(2bF_x - 2M_z + (x_L + x_R)F_y)$$
$$F_{xR} = \frac{1}{4b}(2bF_x + 2M_z - (x_L + x_R)F_y) \quad (2)$$

and

$$F_{zL} = \frac{1}{4b}(2bF_z + 2M_x + (z_L + z_R)F_y)$$
$$F_{zR} = \frac{1}{4b}(2bF_z - 2M_x - (z_L + z_R)F_y) \quad (3)$$

5. Force Allocation Within the Support Planes

Optimal force allocation within the support planes requires that the ratio of traction force to radial force at each contact point be minimized. This ratio determines the tendency of the foot or wheel to slip. Minimizing the maximum value of this ratio: the tangent of the "quasi-friction angle" is viewed as minimizing the tendency to slip. As will be seen, the optimal solution of this problem occurs when the ratio is the same at all contact points.

There is a second, overriding constraint on the force allocation. No contact force may have a negative compnent in the direction of the contact normal. Since a contact cannot generate a force that pulls the vehicle toward the ground, a negative normal contact force implies instability.

Optimal selection of the quasi-friction angle is discussed in detail below. In most circumstances it amounts to identifying the two contact points, out of the planar set in one support plane, that are best placed to provide the commanded force, and placing the load on them with equal quasi-friction angles. Physically, what is happening is that weight load is needed to generate traction. For this reason, the best traction is provided by placing the load on the contacts that are best placed to provide traction, and not "wasting" gravity load on contactsthat are not in position to provide traction. This is a simplification since the contactsdo interact to produce support and traction forces.

There are a number of exceptional conditions requiring special handling. They include zero or negative contact forces, indicating that the corresponding foot or wheel is off the ground, and contact force directions close to parallel, indicating planar terrain, which is a singular case.

In the former case, it is necessary to zero the contact force at the relevant contact, and re-calculate. In the latter case, weight loads can be distributed according to vehicle attitude, and traction forces made proportional to radial force in all cases. In order to increase speed the traction forces are proportionately increased, and conversely to decrease speed.

6. Optimal Selection of Quasi-Friction Angle

As indicated above, a strategy that can be used to resolve the issue of optimally setting the quasi-friction angle (called simply "friction angle" below for convenience) is to identify the two best placed contacts from each planar set of three, and commit as much of the load as feasible to them. It can be proved that if only two wheels are in contact, the optimal solution is that in which the friction angles are equal. Given that, techniques for identifying the best placed contacts in any given situation are discussed below.

(a) Two-Point Contact.

When only two points are in contact the basic geometry is as shown in Figure 5. Since the forces $\mathbf{R}$, $\mathbf{F}_i$ and $\mathbf{F}_j$ form a three force set they must be concurrent. That is, the lines of action of $\mathbf{F}_i$ and $\mathbf{F}_j$ must intersect on the line of action of $\mathbf{R}$.

If we draw the circle that passes through the two contact points and the intersection of the contact normals, as shown on Figure 5, the angle subtended at all points on the circumference of that circle by the line P_1P_2 is the same. This ensures that, if the lines of action of the contact forces intersect at the intersection of the line of action of the commanded force, $\mathbf{R}$, with this circle, the quasi-friction angles, ϕ, are equal.

Further, if the concurrency point is moved outward along the line of action of $\mathbf{R}$, beyond the circle, the friction angle at one contact decreases, while that at the other increases. The same happens if the concurrency point is moved inward to the interior of the circle. Thus, the solution shown in Figure 5 is, in fact, the minimax solution: that solution that minimizes the maximum of the two friction angles.

There are, in fact, two intersections of the line of action of $\mathbf{R}$. The second solution is shown on Figure 6. Since the angle subtended by the line P_iP_j on the shorter of the two arcs P_iP_j is actually the supplement of that subtended on the longer arc, $\phi > \pi/2$, in this case. Further, as may be visualized from Figure 6, it is necessary that the radial component of $\mathbf{F}_j$ be negative. Thus, this is not a valid solution.

For computation it is most convenient to work from the equilibrium equations. In view of the above result we make the quasi-coefficient of

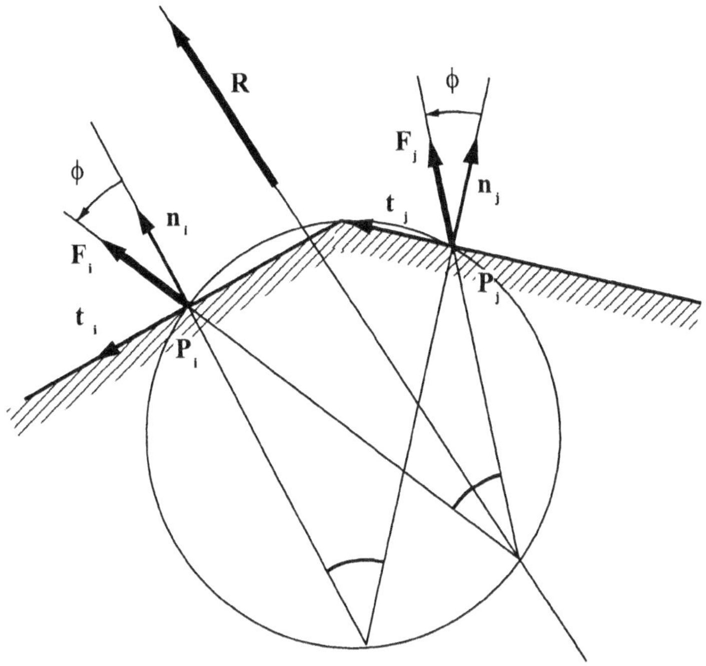

Figure 5. Solution of the two contact problem for which the quasi-friction angles, ϕ, are equal.

friction, $\mu = tan\phi$ the same at both contacts, resulting in

$$F_i\mathbf{n}_i + \mu F_i\mathbf{t}_i + F_j\mathbf{n}_j + \mu F_j\mathbf{t}_j = \mathbf{R}$$
$$\mathbf{p}_i \times (F_i\mathbf{n}_i + \mu F_i\mathbf{t}_i) + \mathbf{p}_j \times (F_j\mathbf{n}_j + \mu F_j\mathbf{t}_j) = T\mathbf{j} \quad (4)$$

where T is the moment of $\mathbf{R}$ about the origin of the vehicle reference frame. $\mathbf{p}_i$ and $\mathbf{p}_j$ are the positions of the respective contact points relative to that frame.

Note that, although the equations 4 have the form of static equilibrium equations, they are really dynamic since $\mathbf{R}$ and T include inertia forces and torques.

Equations 4 constitute three scalar equations in the variables $\mathbf{F}_{\mathrm{i}}$, $\mathbf{F}_{\mathrm{j}}$ and μ. Elimination of $\mathbf{F}_i$, and $\mathbf{F}_{\mathrm{j}}$ gives the quadratic equation

$$A\mu^2 + B\mu + C = 0 \quad (5)$$

where

$$\begin{aligned} A = {} & R_x((x_i - x_j)n_{xi}n_{xj} + z_i n_{zi}n_{xj} - z_j n_{xi}n_{zj}) \\ & + R_z(x_i n_{xi}n_{zj} - x_j n_{zi}n_{xj} + (z_i - z_j)n_{zi}n_{zj}) \\ & - T(n_{zi}n_{xj} - n_{xi}n_{zj}) \end{aligned} \quad (6)$$

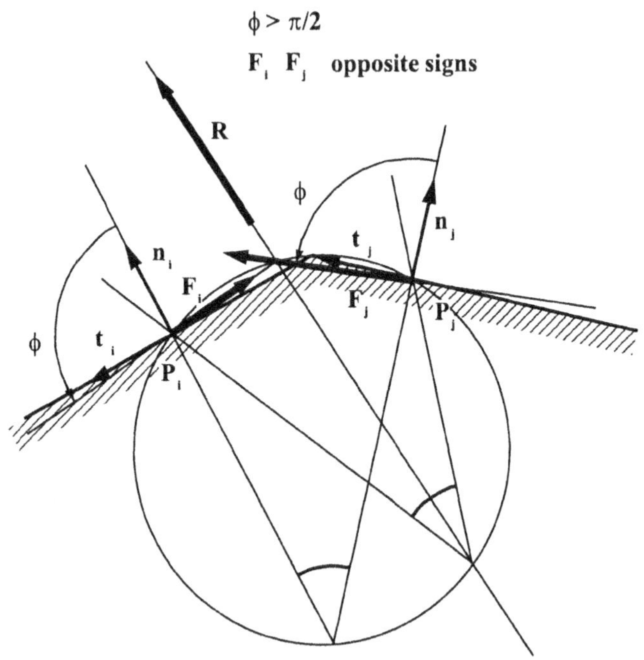

Figure 6. The second solution to the two-point contact configuration. Note that this is an invalid solution because the radial component of $\mathbf{F}_j$ must be negative.

$$\begin{aligned} B = & R_x(-(x_i - x_j)(n_{zi}n_{xj} + n_{xi}n_{zj}) + (z_i - z_j)(n_{xi}n_{xj} - n_{zi}n_{zj})) \\ & + R_z((x_i - x_j)(n_{xi}n_{xj} - n_{zi}n_{zj}) + (z_i - z_j)(n_{zi}n_{xj} + n_{xi}n_{xj})) \end{aligned} \tag{7}$$

$$\begin{aligned} C = & R_x((x_i - x_j)n_{zi}n_{zj} - z_i n_{xi}n_{zj} - z_j n_{zi}n_{xj}) \\ & + R_z(-x_i n_{zi}n_{xj} + x_j n_{xi}n_{zj} + (z_i - z_j)n_{xi}n_{xj}) \\ & - T(n_{zi}n_{xj} - n_{xi}n_{zj}) \end{aligned} \tag{8}$$

Solution of Equation 5 gives two values of μ. These values correspond to the two solutions found geometrically. A negative value of μ means that ϕ is clockwise from the normal, rather than anticlockwise as shown in Figure 5. This means that the traction component of the contact force is in the reverse direction. That is, the system is braking rather than driving forward.

F_i and F_j can now be found from the equations

$$F_i = \frac{R_z(n_{xj} + \mu n_{zj}) - R_x(n_{zj} - \mu n_{xj})}{(1 + \mu^2)(n_{zi}n_{xj} - n_{xi}n_{zj})} \tag{9}$$

and

$$F_j = \frac{R_x(n_{zi} - \mu n_{xi}) - R_z(n_{xi} + \mu n_{zi})}{(1 + \mu^2)(n_{zi}n_{xj} - n_{xi}n_{zj})} \tag{10}$$

If the line of action of **R** passes between $\mathrm{P_i}$ and $\mathrm{P_j}$ it must intersect the circle, so the two solutions of Equation 5 are necessarily real. It can be shown geometrically that only one of the two resulting solutions are acceptable in the sense that the normal components of both contact forces are positive. It is necessary to substitute into Equations 9 and 10 to identify the valid solution.

If the line of action of **R** does not pass between $\mathrm{P_i}$ and $\mathrm{P_j}$ then it can be shown geometrically that no solution for which the radial components are both positive exists. Therefore, there is no valid solution regardless of whether the line of action intersects the circle or not. It is useful to test if the line of action passes between $\mathrm{P_i}$ and $\mathrm{P_j}$. This can be done by calculating h_i and h_j from Equations 11 and 12 below. If h_i and h_j have opposite signs, then the line of action of **R** passes between $\mathrm{P_i}$ and $\mathrm{P_j}$ and a stable solution exists. If they have the same sign, the system is unstable.

$$h_i = \mathbf{n} \bullet \mathbf{p}_\mathrm{i} + \frac{T}{R} \tag{11}$$

$$h_j = \mathbf{n} \bullet \mathbf{p}_\mathrm{j} + \frac{T}{R} \tag{12}$$

This check will be particularly useful when considering three contact points.

(b) Three-Point Contact

In this case, in principle, we can choose any value of the friction angle and impose that value at all three contact points, and solve for the resulting magnitudes of the contact forces. The equilibrium equations, in this case, become determinate if the directions of the contact forces are specified. However, if this is done, there is no guarantee that the solution will include only positive values of the radial components of the contact forces. The treatment below ensures that the solutions derived include only positive radial components, and allows the solution with minimal friction angles subject to that constraint to be identified.

It is instructive to consider the system as three sets of two-point contacts. This geometry is shown on Figure 7.

Here, since the forces $\mathbf{R}$, $\mathbf{F}_i$, $\mathbf{F}_j$ and $\mathbf{F}_k$ constitute a four force system, there is no requirement that they be concurrent. Nevertheless, the three circles drawn through pairs of contact points and the intersections of the corresponding contact normals have a special significance. Let the lines of action of the contact forces at two contact points, say P_i and P_j, be chosen to intersect on the circle through those contact points and the intersection of their contact normals, thereby ensuring that the friction angles, ϕ, at those points are equal Then choosing the line of action of the contact force at P_k as the line passing through the intersection, of P_iL with the circle through P_i and P_k and the intersection of their common normals ensures that the friction angle at P_k is also ϕ.

Assuming that such a solution has been identified for which all the radial components are positive, let us explore the variation in ϕ as we move the intersection point of the contact force lines of action along the circle P_iP_j. At some point the direction of $\mathbf{F}_k$ will reverse. Thus, beyond that point the solution is no longer valid. Since $\mathbf{F}_i$, $\mathbf{F}_j$ and $\mathbf{F}_k$ vary continuously, $\mathbf{F}_k$ can only reverse sign by passing through a point at which its magnitude is zero. At that point $\mathbf{F}_i$ and $\mathbf{F}_j$ sum to $\mathbf{R}$. In other words, we have the two point contact solution for P_i and P_j. Since we know that that solution is given by selecting the lines of action of the contact forces at P_i and P_j to pass through L: the intersection of the line of action of $\mathbf{R}$ with the ij circle, that point on the ij circle marks a limit of the range of points on the circle that give valid solutions.

Moving along the ij circle in the opposite direction, we eventually come to point M. This is the point at which the intersection of the lines of action of the lines of action of the forces at P_i and P_k on the ik circle coincides with the intersection of that circle with the line of action of $\mathbf{R}$. At this point we have the two-point contact solution for P_i and P_k. It follows that $\mathbf{F}_j$ has become zero, and will reverse sign if we choose an intersection point on the circle beyond M. It follows that the arc LM, shown as a heavy line, is that on which the lines of action of the contact forces at points P_i and P_j must intersect for valid solutions. Further, since ϕ varies monotonically along this segment, its maximum and minimum values occur at the ends of the segment. They are identified in Figure A5. That is, the optimal traction solution in the sense of minimizing the maximum friction angle is the two-point ik solution in this case. Likewise, the valid solution for which the friction angle is largest is the two-point ij solution.

Since a given line can only cut at most two sides of a triangle between the vertices, it follows that the condition that the line of action of $\mathbf{R}$ pass between the contact points can be satisfied by at most two of the three pairs of contacts that can be selected from three contacts. In the case of the third pair, jk in the example of Figure 7, the line of action of **R** does not pass between the contact points. Thus, this pair of contacts, taken alone, does not give a stable solution. It follows that $\mathbf{F}_i$ cannot be zero in any stable solution of this three-point problem.

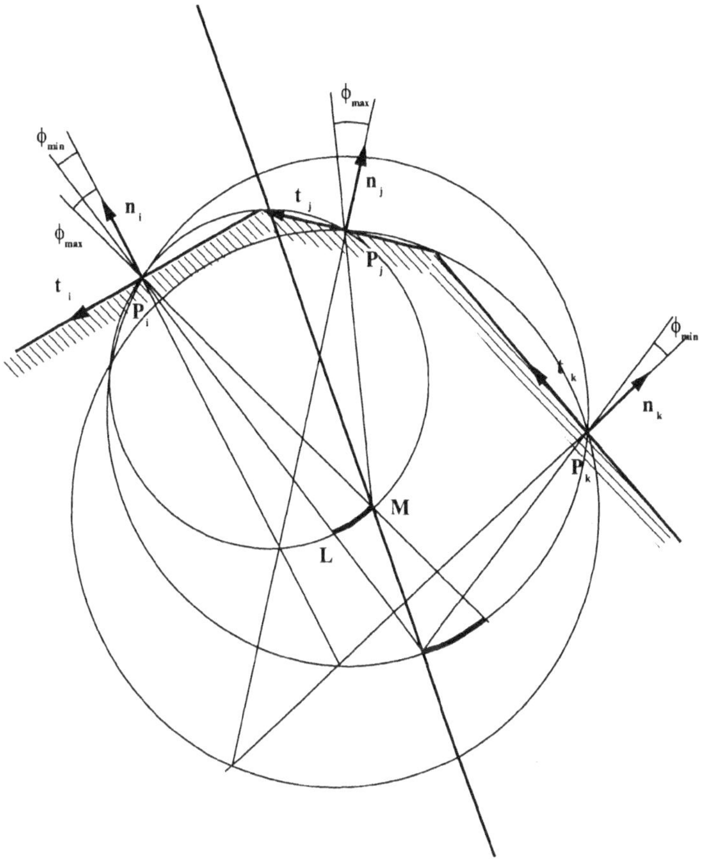

Figure 7. Geometry associated with three points of contact.

Hence, the solution procedure is as follows:
(a) Apply equations 11 and 12 to determine which pairs of contacts the line of action of **R** passes between.
(b) For those two pairs of points apply the two-point contact solution to find the valid solution in each case.
(c) Select the solution for which the magnitude of the friction angle is least.

The situation discussed above applies whenever the signs of the bounding values of ϕ are the same, either positive or negative. However, it is also possible for the intersection of the contact normals to lie within the range LM. Since that point corresponds to $\phi = 0$, the bounding values of ϕ will have opposing signs, and the solution that minimizes the friction angle will be that for which $\phi = 0$ at all three contacts. Such a case is shown on Figure 8.

In this instance, the solution is best obtained directly by use of the equilibrium equations with the assumption that each contact force acts along the contact normal. The equilibrium equations are then

$$F_i \mathbf{n}_\mathrm{i} + F_j \mathbf{n}_\mathrm{j} + F_k \mathbf{n}_\mathrm{k} = \mathbf{R} \tag{13}$$

$$\mathbf{p}_\mathrm{i} \times F_i \mathbf{n}_\mathrm{i} + \mathbf{p}_\mathrm{j} \times F_j \mathbf{n}_\mathrm{j} + \mathbf{p}_\mathrm{k} \times F_k \mathbf{n}_\mathrm{k} = T\mathbf{j} \tag{14}$$

This gives the three scalar linear equations in F_i, F_j, F_k:

$$F_i n_{xi} + F_j n_{xj} + F_k n_{xk} = R_x \tag{15}$$

$$F_i n_{zi} + F_j n_{zj} + F_k n_{zk} = R_z \tag{16}$$

$$\begin{aligned} &F_i(z_i n_{xi} - x_i n_{zi}) + F_j(z_j n_{xj} - x_j n_{zj}) \\ &+F_k(z_k n_{xk} - x_k n_{zk}) = T \end{aligned} \tag{17}$$

These equations may be written in matrix form:

$$\mathbf{R} = \mathbf{JF} \tag{18}$$

where

$$\mathbf{R} = \begin{bmatrix} R_x \\ R_z \\ T \end{bmatrix}$$

$$\mathbf{F} = \begin{bmatrix} F_i \\ F_j \\ F_k \end{bmatrix} \tag{19}$$

$$\mathbf{J} = \begin{bmatrix} n_{xi} & n_{xj} & n_{xk} \\ n_{zi} & n_{zj} & n_{zk} \\ z_i n_{xi} - x_i n_{zi} & z_j n_{xj} - x_j n_{zj} & z_k n_{xk} - x_k n_{zk} \end{bmatrix}$$

which is solved by inversion of the matrix, $\mathbf{J}$, to give

$$\mathbf{F} = \mathbf{J}^{-1}\mathbf{R} \tag{20}$$

Unfortunately, at the present time we have no *a priori* method for determining whether the operative solution will be obtained as the two-point solution giving the minimum friction angle, or the zero friction angle solution of Equation A34. Thus, there is an issue of which one to try first to minimize computation. The zero friction angle solution will probably be computationally more efficient, and in the absence of other clues, should probably be tried first.

However, at most times it will be appropriate to assume that the type of solution will be similar to that of the preceding time step. It might also be useful to apply a heuristic based on the direction of the line of action of the commanded force relative to the normal to the mean ground plane based on the contact point locations. If the commanded force has a substantial forward component, indicating climbing or acceleration, or a substantial rearward component, then the best two-point solution is likely to be the relevant one. There will be a mid-range in which the direction of the commanded force is close to the normal, for which the zero friction angle solution is likely to be available.

7. Summary

We have presented the theoretical basis for a new approach to force allocation among feet in contact with the ground for legged vehicles with architectures generally similar to that of the Adaptive Suspension Vehicle. It has been shown that the solution that minimizes the maximum ratio of tangential to normal force among all the contacts is that for which those ratios are everywhere the same.

It has been demonstrated that, in many cases, optimal traction is achieved by placing all load on the two feet out of a set of three that are best placed to provide traction. However, it has also been demonstrated that this is not always the optimal solution, and that, in some cases, a solution with zero traction components is possible, and optimal.

References

Klein, C. A. and Chung, T. S., (1987), Force interaction and allocation for the legs of a walking vehicle, *IEEE Transactions on Robotics and Automation*, vol. RA-3, no. 6, pp. 546–555.

Kumar, V. and Waldron, K.J. (1990), Force distribution in walking vehicles, *ASME Trans. Journal of Mechanical Design*, vol. 112, pp. 90–99.

Pugh, D.R., Ribble E.A., Vohnout V.J., Bihari, T.E., Walliser, T.M., Patterson, M.R., and Waldron, K.J. (1990), Technical description of the adaptive suspension vehicle, *International Journal of Robotics Research*, vol. 9, vo. 2, pp. 24–42.

Kumar, V. and Waldron, K. J. (1988), Force distribution in closed kinematic chains, *IEEE Transactions on Robotics and Automation*, vol. 4, no. 6, pp. 657–664.

Author Index